Basic Mathematical Skills

Books by Streeter

Streeter and Alexander: Basic Mathematical Skills
Streeter, Hutchison, and Alexander: Beginning Algebra
Streeter and Hutchison: Intermediate Algebra

Also Available From McGraw-Hill

Schaum's Outline Series in Mathematics & Statistics

Each outline includes basic theory, definitions, and hundreds of solved problems and supplementary problems with answers.

Current List Includes:

Analytic Geometry
College Algebra
Elementary Algebra
Review of Elementary Mathematics
First Year College Mathematics
Mathematical Handbook
Modern Elementary Algebra
Technical Mathematics
Trigonometry

AVAILABLE AT YOUR COLLEGE BOOKSTORE

Second Edition

Basic Mathematical Skills

James Streeter
Clackamas Community College

Gerald Alexander
Clackamas Community College

McGraw-Hill Publishing Company

New York St. Louis San Francisco Auckland Bogotá
Caracas Hamburg Lisbon London Madrid Mexico Milan
Montreal New Delhi Oklahoma City Paris San Juan
São Paulo Singapore Sydney Tokyo Toronto

Basic Mathematical Skills

Copyright © 1989, 1984 by McGraw-Hill, Inc. All rights reserved. Printed in the United States of America. Except as permitted under the United States Copyright Act of 1976, no part of this publication may be reproduced or distributed in any form or by any means, or stored in a data base or retrieval system, without the prior written permission of the publisher.

5 6 7 8 9 0   VNH VNH   9 4 3 2 1 0

ISBN 0-07-062435-6

This book was set in Aster by York Graphic Services, Inc.
The editors were Robert Weinstein, David A. Damstra, and Jack Maisel; the production supervisor was Friederich W. Schulte.
The design was done by Caliber Design Planning, Inc.
Von Hoffmann Press, Inc., was printer and binder.

Library of Congress Cataloging-in-Publication Data

Streeter, James (James A.)
Basic mathematical skills.
Includes index.
1. Arithmetic (date).   I. Alexander, Gerald.
II. Title.
QA107.S78 1989b   513   88-37736
ISBN 0-07-062435-6
ISBN 0-07-062436-4 (instructor's manual)

Contents

Preface xi
To the Student xv

PART 1 Whole Numbers   1

PRETEST Chapter 1   3

Chapter 1 Addition and Subtraction of Whole Numbers   5

- **1.1** The Decimal Place-Value System   5
- **1.2** The Language of Addition   10
- **1.3** The Properties of Addition   13
- **1.4** Adding Whole Numbers   17
- **1.5** Carrying   22
- Using Your Calculator   26
- **1.6** Rounding, Estimation, and Order   28
- **1.7** Subtraction of Whole Numbers   35
- **1.8** Borrowing   39
- Using Your Calculator   46
- **1.9** Solving Word Problems with Addition and Subtraction   48

SELF-TEST Chapter 1   55

PRETEST Chapter 2   57

Chapter 2 Multiplication of Whole Numbers   59

- **2.1** The Language of Multiplication   59
- **2.2** The Properties of Multiplication   64
- **2.3** Multiplying by One-Digit Numbers   68
- **2.4** Multiplying by Numbers with More Than One Digit   72
- **2.5** Multiplying by Numbers Ending in Zero   77
- **2.6** The Order of Operations   82
- Using Your Calculator   85
- **2.7** Solving Word Problems Involving Multiplication   88
- **2.8** Powers of Whole Numbers   95
- Using Your Calculator   100

SELF-TEST Chapter 2   103

PRETEST Chapter 3   105

Chapter 3 Division of Whole Numbers   107

- **3.1** The Language of Division   107
- **3.2** Division with Zero and One   112

3.3	Long Division with One-Digit Divisors	115
3.4	Divisors with More than One Digit	120
3.5	Short Division	127
3.6	The Order of Operations	131
	Using Your Calculator	134
3.7	Solving Word Problems with Division	136
3.8	Finding the Average	140
SELF-TEST	Chapter 3	145

PRETEST   Chapter 4   147

Chapter 4   Factors and Multiples   149

4.1	Prime and Composite Numbers	149
4.2	Writing Composite Numbers as a Product of Prime Factors	154
4.3	Finding the Greatest Common Factor (GCF)	159
4.4	Finding the Least Common Multiple (LCM)	164
SELF-TEST	Chapter 4	171

Summary   Part 1   Whole Numbers   173
Summary Exercises   Part 1   179
Cumulative Test   Part 1   183

PART 2   Fractions   185

PRETEST   Chapter 5   187

Chapter 5   An Introduction to Fractions   189

5.1	The Language of Fractions	189
5.2	Proper Fractions, Improper Fractions, and Mixed Numbers	195
5.3	Converting Improper Fractions and Mixed Numbers	198
5.4	Equivalent Fractions	202
5.5	Simplifying Fractions	205
5.6	Building Fractions	212
SELF-TEST	Chapter 5	221

PRETEST   Chapter 6   223

Chapter 6   Multiplication and Division of Fractions   225

6.1	Multiplying Fractions	225
6.2	Multiplying Mixed Numbers	229
6.3	Simplifying in Multiplication	223
6.4	Dividing Fractions	244
SELF-TEST	Chapter 6	255

Contents

PRETEST Chapter 7   257

Chapter 7 **Addition and Subtraction of Fractions**   259

- **7.1** Adding Fractions with a Common Denominator   259
- **7.2** Finding the Least Common Denominator   264
- **7.3** Adding Fractions with Different Denominators   267
- **7.4** Subtracting Fractions   273
- **7.5** Adding and Subtracting Mixed Numbers   280
- **7.6** Applying Fractions   289

SELF-TEST Chapter 7   297

Summary Part 2 Fractions   299
Summary Exercises Part 2   303
Cumulative Test Part 2   307

PART 3   Decimals   311

PRETEST Chapter 8   313

Chapter 8 **Addition, Subtraction, and Multiplication of Decimals**   315

- **8.1** Understanding Place Value in Decimal Fractions   315
- **8.2** Adding Decimals   321
- **8.3** Subtracting Decimals   327
-      Using Your Calculator   333
- **8.4** Multiplying Decimals   335
-      Using Your Calculator   340
- **8.5** Multiplying Decimals by Powers of 10   341
- **8.6** Rounding Decimals   345

SELF-TEST Chapter 8   351

PRETEST Chapter 9   353

Chapter 9 **Division of Decimals**   355

- **9.1** Dividing Decimals by Whole Numbers   355
- **9.2** Dividing by Decimals   361
- **9.3** Dividing Decimals by Powers of 10   368
-      Using Your Calculator   371
- **9.4** Converting Common Fractions to Decimals   374
-      Using Your Calculator   381
- **9.5** Converting Decimals to Common Fractions   382

SELF-TEST Chapter 9   385

Summary Part 3 Decimals   387
Summary Exercises Part 3   391
Cumulative Test Part 3   395

PART 4 Ratios, Proportions, and Percents 397

PRETEST Chapter 10 399

Chapter 10 Ratio and Proportion 400

- 10.1 Using Ratios 400
- 10.2 The Language of Proportions 406
- 10.3 Solving Proportions 412
- 10.4 Using Proportions to Solve Word Problems 421
- Using Your Calculator 429

SELF-TEST Chapter 10 435

PRETEST Chapter 11 437

Chapter 11 Percent 439

- 11.1 The Meaning of Percent 439
- 11.2 Changing a Percent to a Fraction or a Decimal 442
- 11.3 Changing a Decimal or a Fraction to a Percent 447
- 11.4 Identifying the Rate, Base, and Amount 452
- 11.5 The Three Types of Percent Problems 457
- 11.6 Applications of Percent 468
- Using Your Calculator 481

SELF-TEST Chapter 11 487

Summary Part 4 Ratios, Proportions, and Percents 489
Summary Exercises Part 4 493
Cumulative Test Part 4 497

PART 5 Measurement 501

PRETEST Chapter 12 503

Chapter 12 The English System of Measurement 505

- 12.1 The Units of the English System 505
- 12.2 Denominate Numbers 508
- 12.3 Finding Perimeter and Circumference 515
- 12.4 Finding Area 524
- 12.5 Finding Volume 535

SELF-TEST Chapter 12 543

PRETEST Chapter 13   545

Chapter 13 The Metric System of Measurement   547

- **13.1** Metric Units of Length   547
- **13.2** Metric Units of Weight   559
- **13.3** Metric Units of Volume   564
- **13.4** Measuring Temperature in the Metric System   570

SELF-TEST   Chapter 13   575

Summary   Part 5   Measurement   577
Summary Exercises   Part 5   581
Cumulative Test   Part 5   585

PART 6 An Introduction To Algebra   589

PRETEST   Chapter 14   591

Chapter 14 The Integers   593

- **14.1** Signed Numbers   593
- **14.2** Adding Signed Numbers   600
- **14.3** Subtracting Signed Numbers   608
- **14.4** Multiplying Signed Numbers   613
- **14.5** Dividing Signed Numbers   620

SELF-TEST   Chapter 14   625

PRETEST   Chapter 15   627

Chapter 15 Algebraic Expressions and Equations   629

- **15.1** Evaluating Algebraic Expressions   629
- **15.2** Solving Equations   636
- **15.3** More on Solving Equations   648
- **15.4** Applying Algebra   659

SELF-TEST   Chapter 15   671

Summary   Part 6   An Introduction to Algebra   673
Summary Exercises   Part 6   677
Cumulative Test   Part 6   681

Answers   683
Index   697

Preface

Basic Mathematical Skills is the first in a series of three developmental level college texts which includes *Beginning Algebra* and *Intermediate Algebra*. The design of these texts was based on our experiences with students at Clackamas Community College. This second edition of *Basic Mathematical Skills* retains the format and style of the first edition.

It is clear that students entering an introductory program at the college level bring with them a considerable variety of needs and preparation. Some, upon entering college classes, need only a careful review of the basic arithmetic skills and their applications to continue on to a beginning algebra program or to meet the requirements for entry into a technical program. Others find that they will require a much more extensive program of diagnosis, instruction, and study of these same skills.

A primary objective for both editions was to accommodate this wide range of student needs. Four years of class-testing the original manuscript and our students' use of the first edition have elicited many helpful comments from students of all ages and from a wide variety of backgrounds. These comments, along with suggestions from our teaching colleagues, have guided us in the process of revision for this second edition.

To encourage the involvement of all levels of students, we have directed our attention to readability. We have tried to present topics in a straightforward manner and format with numerous examples to clarify the ideas being presented. Margin notes are used extensively to guide students through the development of examples and to reinforce earlier concepts that are being applied. Examples are generally followed by parallel *Check Yourself* exercises to provide the student with immediate feedback as to his or her success in mastering each new concept.

New to this second edition is an increased emphasis on the skills of estimation. This concept is developed in the first chapter and then reinforced as the student works with fractions, decimals, and percents. Similarly, the ideas of order on the number line and the symbols for inequality are introduced in the first chapter and then applied in later work with fractions and decimals.

An expansion of Chapter 15 for this second edition provides experience in translating word phrases into algebraic expressions and then considers some basic applications of algebra. Our emphasis on applied or word problems, begun in the first edition, is continued here with increased attention to a step-by-step process that encourages a thoughtful and organized approach to problem solving. It is our hope that these additions will ease the student's transition into a subsequent beginning algebra course.

Organization

This text covers all of the topics traditionally included in a college level course in arithmetic. The text material is divided into fifteen chapters, organized into six parts:

Whole Numbers

Fractions

Decimals

Ratios, Proportions, and Percents

Measurement

An Introduction to Algebra

To help students as they work their way through the text, the following diagnostic, review, and evaluation features have been incorporated throughout.

- CHAPTER PRETESTS—Each chapter begins with a pretest designed to diagnose student difficulties. Answers and section references are provided on the following page. If desired, students may use this diagnostic test to determine which sections of the chapter need particular emphasis and study.

- CHECK-YOURSELF EXERCISES—These exercises accompany most of the text examples and are designed to involve the student actively in the learning process. Answers are provided at the end of each chapter section for immediate feedback.

- SECTION EXERCISES—Each chapter is divided into short sections followed by two exercise sets of the same level of difficulty. All answers for the first exercise set appear in the text. (The odd answers appear with the exercises and the even answers are in the answer section of the text.)

  A second (supplementary) set of exercises is provided for students who want further review of the topics. Answers are provided in the instructor's manual. Some instructors may prefer to use this second set for homework exercises to be handed in as assignments.

  Included within many section exercises are two features new to this second edition—the *Skillscan* and *Extension* exercises. *Skillscan* draws problems from previous sections of the text, designed to aid the student in the process of reviewing concepts that will be applied in the following section. The *Extension* exercises, identified by a shaded bar at the left, are intended to further challenge the student. In all cases the necessary concepts have been covered in the text section.

- CHAPTER SELF-TESTS—Each chapter concludes with a self-test to give students guidance in preparing for a parallel in-class test. Answers are provided with section references to aid in the process of review. Three parallel test forms are provided in the instructor's manual.

- PART SUMMARIES—Each of the six main parts is followed by a careful and concise summary of the important terms, techniques, and rules that have been developed in the preceding chapters. A set of comprehensive exercises on the material follows each of those summaries. Again section references are provided with the answers to encourage the review process.

- CUMULATIVE TESTS—Also new to the second edition are *Cumulative Tests* after Chapters 4, 7, 9, 11, 13, and 15. These are designed to give the student further opportunity for building skills in the process of a cumulative review. These should be especially useful aids to students preparing for midterm and final examinations.

Supplements

Available for this text is an *Instructor's Manual* containing the answers to all of the supplementary exercises. Additional sample tests for each chapter are also included, as are sample midterm and final examinations.

A *Teacher's Edition* includes answers to all exercises and tests. These answers are printed in a second color for easier use. Other supplements aimed especially at the instructor include both print and computerized testing. The computerized testing provides over 1800 test questions from throughout the text. Several types of test questions are used, including multiple-choice, open-ended, matching, true-false, and vocabulary. The testing system enables the teacher to find the questions by section, topic, question type, difficulty level, and other criteria. In addition, instructors may add their own criteria and edit their own questions. As noted above, sample tests are provided in the *Instructor's Manual*. The Print Test Bank is a hard-copy listing of the questions found in the computerized version.

Several supplements are available for the student. First is a *Student Solutions Manual*, which includes answers and, when appropriate, solutions to all odd-numbered exercises. Students may purchase this manual through their local bookstores.

Videotapes are also available to adopters for student use in a learning laboratory. For the entire series, twenty-four hours of video lessons have been developed, each broken into smaller sections that follow the three texts. These videos provide students with additional instructional and visual support of the lessons.

Finally, a *Computerized Study Guide* is also available to adopters. This tutorial provides additional coverage and support for all sections of each text in the series. Students can work additional problems of many different types and receive constructive feedback based on their answers. Virtually no computer training is needed for the student to work with this supplement.

We trust that this range of supplements will support both teachers and students in a variety of instructional settings. At the same time, both the authors and McGraw-Hill welcome any recommendations from students and teachers for the continual development of the package as the teaching environment and technology change.

Acknowledgments

Many of the features discussed have been developed and refined in response to the comments of our colleagues at Clackamas Community College and we would like to especially acknowledge their helpful support.

We are grateful to the following reviewers for their many valuable suggestions which have been incorporated into this second edition. Michael Brozinsky, Queensborough Community College; Deborah Dalton, Santa Fe Community College; Sandra Evans, Cumberland County College; John Garlow, Tarrant County Junior College; Nancy Hyde, Broward Community College; Nancy Johnson, Broward Community College; Frank Kelly, University of New Mexico; Virginia Lee, Brookdale Community College; Shirley Markus, University of Louisville; Yvonne Martinez, Southwest Texas Junior College; Elise Price, Tarrant County Junior College; Roberta L. Simmons, Lansing Community College; and Paul Wozniak, El Camino College.

Our work in this revision has been greatly eased by the helpful staff at McGraw-Hill. In particular, our special appreciation goes to Robert Weinstein, David Damstra, and Jack Maisel.

Also to our wives, Sharon and Ann, our thanks for their patience during the processes of writing and revision. Finally to our students, our acknowledgment of their many useful ideas, comments, and support during the processes of writing and revision.

James Streeter
Gerald Alexander

To the Student

You are about to begin a course in arithmetic. We have made every attempt to provide a text that will help you build your skills in arithmetic and effectively apply those skills. We have made no assumptions about your previous experience in mathematics. Your rate of progress through the course will depend both upon the amount of time and effort that you give to the course and to your previous background in mathematics.

There are some specific features in this textbook that will aid you in your studies. Here are some suggestions about how to use those features. Keep in mind that a review of *all* of the chapter material will further enhance your ability to grasp later topics and to move more effectively through the following chapters.

1. Whatever your instructional setting, a *Pretest* starts each chapter. Answers and section references are provided on the following page. This will help you determine which sections of the chapter need particular emphasis and study.

2. If you are in a lecture class, make sure that you take the time to read the appropriate text section *before* your instructor's lecture on the subject. Then take careful notes on the examples that your instructor presents during class.

3. Your next step is to work through the examples presented in the text, making sure that you understand each of the steps. Examples are generally followed by *Check Yourself* exercises. Algebra is best learned by being involved in the process and that is the purpose of these exercises. Always have a pencil and paper in hand, work out the problems that are presented, and check your results immediately. If you have difficulty, go back and carefully review the previous examples. Make sure that you understand what you are doing and why. One good test of whether you do understand a concept lies in your ability to explain that concept to one of your fellow students. Try working together.

4. At the end of each chapter section you will find a set of exercises. Work these carefully in order to check your progress on the section you have just finished. You will find the answers for the odd-numbered exercises following the problem set and the even-numbered answers in the answer section of the text. If you have had difficulties with any of the exercises, review the appropriate

parts of the chapter section. If your questions are not completely cleared up, by all means do not become discouraged. Ask your instructor or an available tutor for further assistance. A word of caution: Work the exercises on a regular (preferably daily) basis. Again, learning mathematics requires becoming involved. As is the case with learning any skill, the main ingredient is practice.

5. When you have completed a chapter, try the *Self-Test* that appears at the end of each chapter. This will give you an actual practice test to work as you review for in-class testing. Again, answers with section references are provided.

6. At the end of each of the six parts of the text, you will find a *Part Summary*. The summary includes all of the important terms and definitions of the preceding chapters, along with examples illustrating the techniques that have been developed. Following each summary are *Summary Exercises* for further practice. These exercises are keyed to chapter sections, so you will know where to turn if you are still having problems.

7. Finally, an important element of success in studying mathematics is the process of regular review. We have provided a series of *Cumulative Tests* (they are located after Chapters 4, 7, 9, 11, 13 and 15). These will help you review not only the concepts of the chapter that you have just completed, but those of previous chapters. Use these tests in preparation for any mid-term or final examinations. If it appears that you have forgotten some concepts, don't worry. Go back and carefully review the sections where the idea was initially explained or the appropriate summary.

We hope that you will find our suggestions helpful as you work through this material, and we wish you the best of luck in the course.

J. S.
G. A.

Basic Mathematical Skills

Part 1

Whole Numbers

PRETEST for Chapter One

Addition and Subtraction of Whole Numbers

This pretest will point out any difficulties you may be having in adding and subtracting whole numbers. Do all the problems. Then check your answers on the following page.

1. What is the place value of 5 in 25,943? *Thousand*

2. Write 35,743 in words.

3. Write two hundred seven thousand, five hundred thirty-eight as a numeral.

4. The statement 5 + 9 = 9 + 5 is an illustration of which property of addition?

5. 213 + 131 + 435 =

6. 2369 + 589 + 57,643 + 75 =

7. Round 24,871 to the nearest thousand.

8. 352 − 137 =

9. 9703 − 2685 =

10. Suppose that you must have a total of 540 points on five tests during a semester to receive an A for the course. Your scores on the first four tests have been 92, 87, 82, and 89. What is the lowest you can score on the 200-point final and still receive the A?

ANSWERS TO PRETEST

Be sure to read the "To the Student" for instructions on the use of your test results. For help with similar problems, turn to the section indicated.

1. Thousands (Section 1.1)
2. Thirty-five thousand, seven hundred forty-three (Section 1.1)
3. 207,538 (Section 1.1)
4. Commutative property
5. 779 (Sections 1.4, 1.5)
6. 60,676 (Sections 1.4, 1.5)
7. 25,000 (Section 1.6)
8. 215 (Sections 1.7, 1.8)
9. 7018 (Sections 1.7, 1.8)
10. 190 points (Section 1.9)

Chapter One

Addition and Subtraction of Whole Numbers

1.1 The Decimal Place-Value System

OBJECTIVES
1. To determine the place value of a digit
2. To write a numeral in expanded form
3. To write a numeral in words
4. To write a numeral, given its word name

Number systems have been developed throughout human history. Starting with simple tally systems used to count and keep track of possessions, more and more complex systems developed. The Egyptians used a set of picturelike symbols called *hieroglyphics* to represent numbers. The Romans and Greeks had their own systems of numeration. We see the Roman system today in the form of roman numerals. Some examples of these systems are shown in Figure 1.

NUMERALS	EGYPTIAN	GREEK	ROMAN
1	l	I	I
10	∩	Δ	X
100	૧	H	C

Figure 1

The prefix *deci* means ten.
Our word digit comes from the Latin word *digitus*, which means finger.

Any number system provides a way of naming numbers. The system we use is described as a *decimal place-value system*. This system is based on the number 10 and uses symbols called *digits*. (Other numbers have also been used as bases. The Mayans used 20, and the Babylonians used 60.)

Example 1

Any number, no matter how large, can be represented as a numeral using the 10 digits of our system.

The basic symbols of our system are the digits:

0, 1, 2, 3, 4, 5, 6, 7, 8, and 9

These basic symbols, or digits, were first used in India and then adopted by the Arabians. For this reason, our system is called the Hindu-Arabic numeration system.

Numbers are represented by symbols called *numerals*. Numerals may consist of one or more *digits*.

Example 2

3, 45, 567, and 2359 are numerals, or symbols that *name* numbers. We say that 45 is a two-digit numeral, 567 is a three-digit numeral, and so on.

As we said, our decimal system uses a *place-value* concept based on the number 10. Understanding how this system works will help you see the reasons for the rules and methods of arithmetic that we will be introducing. Let's look first at the basic idea behind place value.

PLACE VALUE

The position (*place*) of an individual digit in a numeral tells us the *value* that the digit represents.

Example 3

Look at the numeral 438.

Each digit in a numeral has its own place value.

8 represents 8 individual items. We call 8 the *ones digit*.

Moving to the left, the digit 3 represents 3 groups of 10 objects. 3 is the *tens digit*.

Again moving to the left, 4 represents 4 groups of one hundred. 4 is the *hundreds digit*.

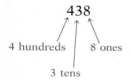

In symbols, we can write 438 as

(4 × 100) + (3 × 10) + 8

This is called the *expanded form* of the numeral.

(4 × 100) means 4 is multiplied by 100.

(3 × 10) means 3 is multiplied by 10.

The parentheses ( ) tell you to multiply as the first step.

CHECK YOURSELF 1

Write 593 in expanded form.

Section 1.1: The Decimal Place-Value System

Look at the following diagram, which shows the place value of digits as we write numerals that represent larger numbers. For the numeral 3,156,024,798, we have

Of course, the naming of place values continues for larger and larger numbers beyond the chart.

For the numeral 3,156,024,798, the place value of 4 is thousands. As we move to the left, each place value is 10 times the value of the previous place. For instance, the place value of 2 is ten thousands, the place value of 0 is hundred thousands, and so on.

CHECK YOURSELF 2

Use a place-value diagram to answer the following questions for the numeral 6,831,425,097.

(1) What is the place value of 2?
(2) What is the place value of 4?
(3) What is the place value of 3?

Understanding place value will help you read or write numerals in word form. Look at the numeral

7 2, 3 5 8, 6 9 4
Millions Thousands Ones

A four-digit numeral, such as 3456, can be written with or without a comma. We have chosen to omit the comma in these materials.

Commas are used to set off groups of three digits in the numeral. The name of each group—millions, thousands, and so on—is then used as we write the numeral in words. To write a word name for a numeral we work from left to right, writing the numerals in each group, followed by the group name. The following example illustrates.

Example 4

27,345 is written in words as twenty-seven *thousand*, three hundred forty-five.

2,305,273 is two *million*, three hundred five *thousand*, two hundred seventy-three.

Note We do *not* write the name of the ones group. Also, the word "and" is not used. It will have a special meaning later.

CHECK YOURSELF 3

Write the word name for each of the following numerals.

(1) 658,942
(2) 2305

We reverse the process to write numerals for numbers given in word form. Consider the following.

Example 5

Forty-eight thousand, five hundred seventy-nine in numeral form is

48,579

Five hundred three thousand, two hundred thirty-eight in numeral form is

503,238 — Note the use of zero as a placeholder in writing the numeral.

CHECK YOURSELF ANSWERS

1. $(5 \times 100) + (9 \times 10) + 3$.
2. (1) Ten thousands; (2) hundred thousands; (3) ten millions.
3. (1) Six hundred fifty-eight thousand, nine hundred forty-two; (2) two thousand, three hundred five.

1.1 Exercises

Write each numeral in expanded form.

1. 352 $(3 \times 100) + (5 \times 10) + 2$
2. 549 $(5 \times 100) + (4 \times 10) + 9$
3. 5073 $(5 \times 1000) + (7 \times 10) + 3$
4. 20,721

Give the place values for the indicated digits.

5. 3 in the numeral 327
6. 5 in the numeral 5469
7. 6 in the numeral 56,489
8. 0 in the numeral 203,456
9. 7 in the numeral 27,243,012
10. 5 in the numeral 3,527,213
11. 2 in the numeral 523,010,000
12. 3 in the numeral 317,008,000

Give word names for the following numerals.

13. 3456

14. 23,567

15. 200,304

16. 502,200

Write each of the following as numerals.

17. Two hundred fifty-three thousand, four hundred eighty-three

18. Three hundred seven thousand, three hundred fifty-nine

19. Five hundred two million, seventy-eight thousand

20. One billion, two hundred thirty million

Answers We will provide the answers (with some worked out in detail) for the odd-numbered exercises at the end of each exercise set. The answers for the even-numbered exercises are provided in the back of the book.
1. $(3 \times 100) + (5 \times 10) + 2$ **3.** $(5 \times 1000) + (7 \times 10) + 3$ **5.** Hundreds   **7.** Thousands   **9.** Millions
11. Ten millions   **13.** Three thousand, four hundred fifty-six
15. Two hundred thousand, three hundred four   **17.** 253,483   **19.** 502,078,000

1.1 Supplementary Exercises

Write each numeral in expanded form.

1. 638

2. 3725

Give the place values for the indicated digits.

3. 2 in the numeral 32,785

4. 8 in the numeral 584,123

5. 3 in the numeral 23,000,000

6. 0 in the numeral 1,205,567,293

Give word names for the following numerals.

7. 25,489

8. 300,257

Write each of the following as numerals.

9. Fifty-two thousand, three hundred eighty-four

10. Five hundred three million, six hundred eighty-seven thousand

1.2 The Language of Addition

OBJECTIVES
1. To use the language of addition
2. To add single-digit numbers

For convenience we will not worry about the distinction between number and numeral in the remainder of this text.

The three dots (. . .) are called ellipses *and mean that the set continues without end.*

A *natural or counting number* is used to count or to indicate how many objects there are in a set or a group of objects.

The natural numbers are 1, 2, 3, . . .

Example 1

3, 7, 23, and 256 are natural numbers.

When we include the number 0, we then have the set of *whole numbers*.

The whole numbers are 0, 1, 2, 3, . . .

Example 2

0, 5, 189, and 34,508 are whole numbers.

Do you see that every natural number is also a whole number?

We are now ready to look at the operation of *addition* on the whole numbers.

> Think of addition as combining two or more groups of the same kinds of objects.

This is extremely important, as we will see in our later work with fractions. We can only combine or add numbers that represent the same kinds of objects.

The first printed use of the symbol + dates back to 1500.

Each operation of arithmetic has its own special terms and symbols. The addition symbol + is read *plus*. When we write 3 + 4, 3 and 4 are called the *addends*.

We can use a number line to illustrate the addition process. To construct a number line, we pick a point on the line and label it 0. We then mark off evenly spaced units to the right, naming each point marked off with a successively larger whole number.

The point labeled 0 is called the origin *of the number line.*

A Number Line

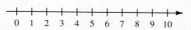

We use an arrowhead to show the direction of increase.

Example 3

To represent an addition, such as 3 + 4, on the number line, start by moving 3 spaces to the right of the origin. Then move 4 more spaces to the right to arrive at 7. The number 7 is called the *sum* of the addends.

Again, addition corresponds to combining groups of the same kind of objects

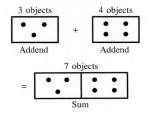

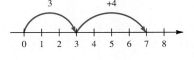

We can write 3 + 4 = 7
                Addend Addend Sum

CHECK YOURSELF 1

Represent 5 + 6 on the number line.

A statement such as 3 + 4 = 7 is one of the *basic addition facts*. These facts include adding all combinations of the numbers 0 through 9. Before you can add larger numbers correctly and quickly, you must memorize these basic facts.

Use the table of Basic Addition Facts if you feel that you need to review.

To find the sum 5 + 8, start with the row labeled 5. Move along that row to the column headed 8 to find the sum, 13.

BASIC ADDITION FACTS									
+	0	1	2	3	4	5	6	7	8	9
0	0	1	2	3	4	5	6	7	8	9
1	1	2	3	4	5	6	7	8	9	10
2	2	3	4	5	6	7	8	9	10	11
3	3	4	5	6	7	8	9	10	11	12
4	4	5	6	7	8	9	10	11	12	13
5	5	6	7	8	9	10	11	12	13	14
6	6	7	8	9	10	11	12	13	14	15
7	7	8	9	10	11	12	13	14	15	16
8	8	9	10	11	12	13	14	15	16	17
9	9	10	11	12	13	14	15	16	17	18

CHECK YOURSELF ANSWERS

1.   5 + 6 = 11.

1.2 Exercises

1. In the statement $5 + 4 = 9$
   5 is called the _____
   4 is called the _____
   9 is called the _____

2. In the statement $7 + 8 = 15$
   7 is called the _____
   8 is called the _____
   15 is called the _____

Add.

3. 4
   $+3$

4. 3
   $+4$

5. 9
   $+5$

6. 5
   $+9$

7. 7
   $+3$

8. 1
   $+6$

9. 4
   $+4$

10. 9
    $+7$

11. 8
    $+6$

12. 6
    $+8$

13. 7
    $+4$

14. 2
    $+6$

15. 6
    $+6$

16. 0
    $+9$

17. 9
    $+1$

18. 5
    $+8$

19. 8
    $+7$

20. 6
    $+7$

21. 5
    $+4$

22. 5
    $+7$

23. 3
    $+5$

24. 4
    $+8$

25. 5
    $+5$

26. 6
    $+6$

27. 7
    $+0$

28. 8
    $+8$

29. 6
    $+9$

30. 4
    $+7$

31. 3
    $+8$

32. 9
    $+9$

Answers

1. 5 is the addend, 4 is the addend, 9 is the sum   3. 7   5. 14   7. 10   9. 8   11. 14   13. 11
15. 12   17. 10   19. 15   21. 9   23. 8   25. 10   27. 7   29. 15   31. 11

Note To be successful in the next section, you should have been able to work quickly through these exercises. If you made errors or had to refer very often to the table, stop and review the Basic Addition Facts now!

1.2 Supplementary Exercises

1. In the statement 9 + 4 = 13
   9 is called the
   4 is called the
   13 is called the

Add.

2. 5
   +8

3. 9
   +8

4. 8
   +9

5. 2
   +9

6. 6
   +5

7. 8
   +0

8. 6
   +3

9. 9
   +4

10. 4
    +9

11. 7
    +7

12. 6
    +1

13. 3
    +9

14. 8
    +2

15. 6
    +8

16. 0
    +9

1.3 The Properties of Addition

OBJECTIVE
To identify the properties of addition

Examining the table of Basic Addition Facts in Section 1.2 leads us to several important properties of addition on the whole numbers. For instance, we know that the sum 3 + 4 is 7. What about the sum 4 + 3? It is also 7. This is an illustration of the fact that addition is a *commutative* operation.

THE COMMUTATIVE PROPERTY FOR ADDITION

The order in which you add two whole numbers does not affect the sum.

Example 1

The *order* does not affect the sum.

$8 + 5 = 5 + 8 = 13$

$6 + 9 = 9 + 6 = 15$

CHECK YOURSELF 1

Show that

$7 + 8 = 8 + 7$

If we wish to add *more* than two numbers, we can group them and then add. In mathematics this grouping is indicated by a set of parentheses ( ). This symbol tells us to perform the operation in the parentheses first.

Example 2

We add 3 and 4 as the first step and then add 5.

$(3 + 4) + 5 = 7 + 5 = 12$

We also have

Here we add 4 and 5 as the first step and then add 3. Again the final sum is 12.

$3 + (4 + 5) = 3 + 9 = 12$

As Example 2 suggests,

THE ASSOCIATIVE PROPERTY FOR ADDITION

The way in which you group whole numbers in addition does not affect the final sum.

Example 3

$(6 + 7) + 5 = 13 + 5 = 18$

We can also write

$6 + (7 + 5) = 6 + 12 = 18$

Thus

The *grouping* does not affect the sum.

$(6 + 7) + 5 = 6 + (7 + 5)$

Section 1.3: The Properties of Addition

CHECK YOURSELF 2

Find

(4 + 8) + 3   and   4 + (8 + 3)

The number 0 has a special property in addition. Looking at the table of Basic Addition Facts, we see that:

> The sum of zero and any whole number is just that whole number.

Example 4

3 + 0 = 3
0 + 8 = 8

Because of this property, we call 0 the *identity* for the addition operation.

USING VARIABLES

We can use letters in stating the properties of addition. The letters used, here *a*, *b*, and *c*, may represent any whole numbers and are called *variables*.

THE COMMUTATIVE PROPERTY OF ADDITION   If *a* and *b* represent any whole numbers, then

$$a + b = b + a$$

The order in which we add does not affect the sum.

THE ASSOCIATIVE PROPERTY OF ADDITION   If *a*, *b*, and *c* represent any whole numbers, then

$$(a + b) + c = a + (b + c)$$

The way in which we group the numbers does not affect the sum.

THE ADDITIVE IDENTITY   If *a* is any whole number, then

$$a + 0 = a$$

The sum of any whole number and 0 is just that whole number.

CHECK YOURSELF ANSWERS

1. 7 + 8 = 15 and 8 + 7 = 15.
2. (4 + 8) + 3 = 12 + 3 = 15; 4 + (8 + 3) = 4 + 11 = 15.

1.3 Exercises

Do the indicated addition.

1. $5 + 7$
2. $7 + 8$
3. $7 + 5$
4. $8 + 7$

5. $(2 + 4) + 6$
6. $(3 + 7) + 5$
7. $2 + (4 + 6)$
8. $3 + (7 + 5)$

9. $3 + 0$
10. $0 + 7$
11. $0 + 8$
12. $5 + 0$

13. $2 + 7 + 9$
14. $3 + 4 + 8$
15. $2 + 3 + 4 + 9$
16. $3 + 6 + 9 + 5$

Name the property of addition that is illustrated.

17. $5 + 8 = 8 + 5$
18. $2 + (7 + 9) = (2 + 7) + 9$
19. $(4 + 5) + 8 = 4 + (5 + 8)$

20. $9 + 7 = 7 + 9$
21. $3 + (7 + 5) = (3 + 7) + 5$
22. $5 + 0 = 5$

23. $3 + (4 + 0) = 3 + 4$
24. $(3 + 6) + 4 = 3 + (6 + 4)$

Answers

1. 12   3. 12. The answers to 1 and 3 must be the same, since addition is commutative.   5. 12
7. 12. The answers to 5 and 7 must be the same, since addition is associative.   9. 3   11. 8   13. 18
15. 18   17. Commutative property   19. Associative property   21. Associative property
23. Additive identity

1.3 Supplementary Exercises

Do the indicated addition.

1. $8 + 9$
2. $9 + 8$
3. $(3 + 8) + 5$
4. $3 + (8 + 5)$

5. $6 + 0$
6. $0 + 7$
7. $3 + 8 + 5$
8. $2 + 5 + 6 + 8$

Name the property of addition that is illustrated.

9. $6 + 9 = 9 + 6$
10. $8 + 0 = 8$

11. $(4 + 8) + 3 = 4 + (8 + 3)$
12. $5 + (7 + 6) = (5 + 7) + 6$

1.4 Adding Whole Numbers

OBJECTIVE
To add groups of whole numbers without carrying

Let's turn now to the process of adding larger numbers. We will apply the following rule.

> We can add the digits of the same place value, since they represent the same quantities.

Example 1

Remember that 25 means 2 tens and 5 ones; 34 means 3 tens and 4 ones.

$$
\begin{aligned}
25 &= 2 \text{ tens} + 5 \text{ ones} \\
+34 &= 3 \text{ tens} + 4 \text{ ones} \\
&= 5 \text{ tens} + 9 \text{ ones} \\
&= 59
\end{aligned}
$$
↓ Add down

In actual practice we use a more convenient short form to perform the addition.

Example 2

Step 1 Add first in the ones column.

In using the short form, be very careful to line up the numbers correctly so that each column contains digits of the same place value.

```
  25
 +34
 ———
   9
```

Step 2 Now add in the tens column.

```
  25
 +34
 ———
  59
```

CHECK YOURSELF 1

Add:

```
  46
 +32
```

The process is easily extended to even larger numbers. Again we begin in the ones column.

Example 3

Step 1 Add in the ones column.

```
  352
 +546
    8
```

Step 2 Add in the tens column.

```
  352
 +546
   98
```

Step 3 Add in the hundreds column.

```
  352
 +546
  898
```

CHECK YOURSELF 2

Add.

```
  245
 +632
```

If you want to add more than two numbers, use the idea illustrated in the following example.

Example 4

This uses the associative property. Group the first two numbers and then add the third.

Step 1
$$\left.\begin{array}{r} 531 \\ 142 \\ +\ 25 \\ \hline 8 \end{array}\right\} \begin{array}{c} 1 + 2 = 3 \\ \\ \end{array} \Big\} \; 3 + 5 = 8$$

Think: "In the ones column, $1 + 2 = 3$. Then add the 5 for the sum, 8."

You can complete the addition by using the same idea in the tens and hundreds columns.

Step 2
```
   531
   142
 +  25
    98
```

Step 3
```
   531
   142
 +  25
   698
```

Section 1.4: Adding Whole Numbers

CHECK YOURSELF 3

Add.

```
  423
   42
+ 332
```

Many problems will require you to "set up" the addition. Let's work through an example.

Example 5

Add: 21, 362, 1403, and 3.

Start by lining up like place values under each other.

```
   21
  362
 1403
+   3
```

You must be very careful to line up the numbers correctly so that each column contains digits of the same place value.

Now you can add in each column as before.

```
   21
  362
 1403
+   3
 1789
```

CHECK YOURSELF ANSWERS

1.
```
  46          46
 +32         +32
   8          78
```
Add the ones. Then add the tens.

2. 877

3.
```
  423
   42
 +332
  797
```

1.4 Exercises

Add.

1. 24
   + 3

2. 13
   + 5

3. 23
   +56

4. 75
   +20

5. 332
   + 54

6. 620
   + 67

7. 307
   +232

8. 349
   +420

9. 2792 + 205	10. 5463 + 435	11. 2345 +6053	12. 3271 +4715
13. 2531 +5354	14. 5003 +4205	15. 21,314 +43,042	16. 12,325 +35,403
17. 13 21 +35	18. 24 31 +32	19. 3462 213 + 24	20. 2430 356 + 12

21. 35 + 432

22. 527 + 62

23. 4 + 12 + 340 + 1213

24. 534 + 2 + 31 + 2322

You have already seen that the word *sum* indicates addition. There are other words that also tell you to use the addition operation.

The *total* of 12 and 5 is written as

12 + 5 or 17

8 *more than* 10 is written as

10 + 8 or 18

12 *increased by* 3 is written as

12 + 3 or 15

With this information, find each of the following.

25. The total of 23 and 31

26. 7 more than 22

27. The sum of 562 and 231

28. 123 increased by 45

29. 34 more than 125

30. The total of 124 and 2351

31. The sum of 23, 122, and 451

32. The total of 112, 24, and 532

Do each of the following problems.

33. A golfer shot a score of 42 on the first nine holes and a score of 46 on the second nine holes. What was her total score for the round?

34. A bowler scored 201, 153, and 215 pins in three games. What was the total score for those games?

35. Janet had a score of 73 on her first mathematics test. On the second test, she increased her score by 23 points. What was her score on the second test?

36. Jesse drove 244 miles from St. Louis to Indianapolis on the first day of a business trip. He then drove 113 more miles to Louisville on the second day of the trip. How far did he travel?

Answers
1. 27   3. 79   5. 386   7. 539   9. 2997   11. 8398   13. 7885   15. 64,356   17. 69   19. 3699
21.   35   23.   4   25. 54   27. 793   29. 159   31. 596   33. 88   35. 96
    +432   12
     467   340
                +1213
                 1569

1.4 Supplementary Exercises

Add.

1.   34
       + 4

2.   40
        +38

3.   450
        + 27

4.   328
        +431

5.   240
       +3157

6.   5315
        +2463

7.   1253
        +3644

8.   31,304
        +52,583

9.   73
         12
        +13

10.   2450
           134
         + 15

11. 13 + 21 + 34

12. 545 + 31 + 3 + 2020

1.5 Carrying

OBJECTIVE
To add any group of whole numbers

Carrying in addition is also called *regrouping* or *renaming*. Of course, the name makes no difference as long as you understand the process.

In the examples and exercises of the last section, the digits in each column added to 9 or less. Let's look at the situation where a column has a two-digit sum. This will involve the process of *carrying*. Let's look at the process in expanded form.

Example 1

$$\begin{aligned} 67 &= 6 \text{ tens} + 7 \text{ ones} \\ +28 &= 2 \text{ tens} + 8 \text{ ones} \\ \hline &8 \text{ tens} + 15 \text{ ones} \end{aligned}$$

or   8 tens + 1 ten + 5 ones   We have written 15 ones as 1 ten and 5 ones. The 1 ten is then combined with the 8 tens.

or   9 tens   + 5 ones

or   95

The more convenient short form carries the excess units from one column to the next column left. Recall that the place value of the next column left is 10 times the value of the original column. (Of course this is true for any size number. The place value thousands is 10 times the place value hundreds, and so on.) It is this property of our decimal place-value system that makes carrying work. Let's look at the problem of example 1 again, this time done in the short or "carrying" form.

Example 2

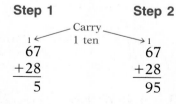

Step 1   Step 2

             Carry
  1   1 ten   1
 67   67
+28   +28
   5   95

The sum of the digits in the ones column is 15, so write 5 and carry 1 to the tens column. Now add in the tens column, being sure to include the carried 1.

CHECK YOURSELF 1

Add.

$$\begin{aligned} 58 \\ +36 \\ \hline \end{aligned}$$

Section 1.5: Carrying

The addition process often requires more than one carrying step as is shown in our next example.

Example 3

$$\overset{1}{}\leftarrow \text{Carry 1 ten}$$
$$285$$
$$+378$$
$$\overline{3}$$

The sum of the digits in the ones column is 13, so write 3 and carry 1 to the tens column.

Carry ⟶ 1 1
1 hundred   285
           +378
           ─────
             63

Now add in the tens column, being sure to include the carry. We have 16 tens, so write 6 in the tens place and carry 1 to the hundreds column.

1 1
285
+378
─────
663

Finally add in the hundreds column.

CHECK YOURSELF 2

Add.

479
+287
─────

The carrying process is the same if we want to add more than two numbers.

Example 4

Add 53, 2678, 587, and 9.

1 2 2 ⟵ Carries
   53
 2678
  587
+   9
─────
 3327

Add in the ones column, 3 + 8 + 7 + 9 = 27. Write 7 in the sum and carry 2 to the tens column.

Now add in the tens column, being sure to include the carry. The sum is 22. Write 2 tens and carry 2 to the hundreds column. Complete the addition by adding in the hundreds column and the thousands column.

CHECK YOURSELF ANSWERS

1. $\overset{1}{}$
    58
    +36
    ───
    94

    The sum in the ones column is 14, so write 4 and carry 1 to the tens column. Then add in the tens column.

2. $\overset{1}{}$ ⟵ Carry 1 ten
    479
    +287
    ────
      6

    1 1 ⟵ Carry 1 hundred
    479
    +287
    ────
    766

1.5 Exercises

Add.

1. 47
   + 9

2. 64
   + 8

3. 23
   +48

4. 96
   +57

5. 31
   27
   +35

6. 69
   27
   +58

7. 213
   + 78

8. 392
   + 58

9. 703
   +287

10. 898
    +457

11. 589
    306
    + 42

12. 257
    18
    +504

13. 590
    345
    +758

14. 358
    271
    +595

15. 2578
    +3455

16. 8295
    +4927

17. 3490
    548
    + 25

18. 678
    4533
    + 70

19. 2289
    38
    578
    +3489

20. 3678
    259
    27
    +2356

21. 23,458
    +32,623

22. 52,591
    +59,739

23. 26,735
    259
    3,056
    +35,489

24. 35,607
    2,345
    456
    +81,247

25. Find the sum of 79 and 735.

26. Add 28 and 386.

27. What is the total of 38, 354, and 8?

28. Find the sum of 23, 57, and 236.

29. Add 23, 2845, 5, and 589.

30. Find the sum of 3295, 9, 427, and 56.

31. What is the total of 2195, 348, 640, 59, and 23,785?

32. Add 5637, 78, 690, 28, and 35,589.

Section 1.5: Carrying 25

The sequences below are called *arithmetic sequences*. Determine the pattern and write the next four numbers in each sequence.

33. 5, 12, 19, 26, _____ , _____ , _____ , _____

34. 8, 14, 20, 26, _____ , _____ , _____ , _____

35. 7, 13, 19, 25, _____ , _____ , _____ , _____

36. 9, 17, 25, 33, _____ , _____ , _____ , _____

Answers

1. 56   **3.** 71   **5.** 93   **7.** 291   **9.** 990   **11.** 937   **13.** 1693   **15.** 6033

17. 4063   **19.** $\overset{1\,2\,3}{2289}$ **21.** 56,081   **23.** 65,539   **25.** 814   **27.** 400   **29.** 3462   **31.** $\overset{2\,\,3\,2}{2,195}$

                     38   348
                    578   23,785
                 +3489   640
                   6394   +   59
                                                                                      27,027

33. 33, 40, 47, 54   **35.** 31, 37, 43, 49

1.5 Supplementary Exercises

Add.

1.   58
     + 7

2.   48
     +76

3.   35
       29
     +32

4.   987
     + 84

5.   783
     +529

6.   604
       489
     + 54

7.   705
       583
     +435

8.   4538
     +2759

9.   5832
       539
     +2470

10.   684
      5372
      2358
    +   56

11.   24,078
     +58,254

12.   48,032
      2,509
        358
    +53,645

13. Find the sum of 527 and 43.

14. Add 29, 273, and 155.

15. What is the total of 3245, 300, 2891, and 78?

16. Find the sum of 5848, 39, 583, 157, and 29,875.

Using Your Calculator

Calculators use a variety of logic systems. Since *algebraic logic* is used on the majority of calculators being sold today, the examples of this text will be illustrated with that type of calculator.

Even though the reason for this book is to help you review the basic skills of arithmetic, many of you will want to be able to use a hand-held calculator for some of the problems that are presented. Ideally you should learn to do the basic operations *by hand*. So in each section of this book, start by learning to do the work *without* your calculator. We will then provide these special calculator sections to show how you can use the calculator.

First, to enter a number in your calculator, simply press the digits in order *from left to right*.

Example 1

To enter 23,456; press (one at a time)

| 2 | 3 | 4 | 5 | 6 |

We start with the digit of the largest place value, in this case 2, meaning 2 ten thousands.

Display (23456)

From this point on, we will simply say "enter the number" as a single step.

Example 2

To add whole numbers, say 23 + 45 + 67, follow the indicated steps.

This simply clears the calculator for what follows.

1. Press the clear key.   C
2. Enter the first number.   2 3
3. Press the plus key.   +
4. Enter the second number.   4 5

The sum of the first two numbers, 68, should now be in the display.

5. Press the plus key.   +
6. Continue until the last   6 7
   number is entered.
7. Press the equals key.   =

The desired sum should now be in the display.

Display (135)

Let's look at another example.

Example 3

To add 23 + 3456 + 7 + 985, enter:

23 [+] 3456 [+] 7 [+]
985 [=]   (4471)

Enter each of the first three numbers followed by the plus key. Then enter the final number and press the equals key. The sum will be in the display.

As we mentioned, calculators use a variety of patterns in performing operations in arithmetic. If you have a calculator, try the addition of Example 3 now. If you do not get 4471 as the answer, check the operating manual or ask your instructor for assistance.

Exercises Using Your Calculator

Add.

1. 23
   +78

2. 589
   +257

3. 2173
   +5899

4. 458
   273
   +568

5. 2743
   258
   35
   +5823

6. 29,753
   249
   53
   5,821
   + 4,258

7. 3,295,153
   573,128
   21,257
   2,586,241
   +   5,291

8. 507 + 359 + 259

9. 23 + 5638 + 385 + 7 + 27,345

10. 563 + 9487 + 5 + 35,600 + 39

11. The following table shows the number of customers using three branches of a bank during 1 week. Complete the calculations.

BRANCH	M	T	W	Th	F	WEEKLY TOTALS
Downtown	487	356	429	278	834	____
Suburban	236	255	254	198	423	____
Westside	345	278	323	257	563	____
DAILY TOTALS	____	____	____	____	____	GRAND TOTAL

12. The following table lists the number of possible types of poker hands. What is the total number possible?

Royal flush	4
Straight flush	36
Four of a kind	624
Full house	3,744
Flush	5,108
Straight	10,200
Three of a kind	54,912
Two pair	123,552
One pair	1,098,240
Nothing	1,302,540

Answers

1. 101 3. 8072 5. 8859 7. 6,481,070 9. 33,398 11.

						WEEKLY TOTALS
						2384
						1366
						1766
DAILY TOTALS	1068	889	1006	733	1820	5516

1.6 Rounding, Estimation, and Order

OBJECTIVES
1. To round a whole number to any place value
2. To estimate sums by rounding
3. To use the symbols < and >

It is a common practice to express numbers to the nearest hundred, thousand, and so on. For instance, the distance from Los Angeles to New York along one route is 2833 mi. We might say that the distance is 2800 mi. We have rounded the distance to the nearest hundred miles.

One way to picture this rounding process is with the use of a number line.

Example 1

To round 2833 to the nearest hundred:

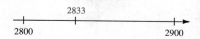

Since 2833 is closer to 2800, we round *down* to 2800.

Example 2

To round 28,734 to the nearest thousand:

Since 28,734 is closer to 29,000, we round *up* to 29,000.

CHECK YOURSELF 1

Locate 375 and round to the nearest hundred.

Instead of using a number line, we can apply the following rule.

> **ROUNDING WHOLE NUMBERS**
>
> STEP 1 To round off a whole number to a certain decimal place, look at the digit to the right of that place.

By a certain *place,* we mean tens, hundreds, thousands, and so on.

Example 3

Round 587 to the nearest ten.

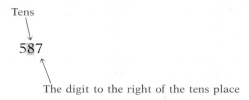

The digit to the right of the tens place

> **ROUNDING WHOLE NUMBERS**
>
> STEP 2
>
> (*a*) If that digit is 5 or more, that digit and all digits to the right become 0. The digit in the place you are rounding to is increased by 1.
>
> (*b*) If that digit is less than 5, that digit and all digits to the right become 0. The digit in the place you are rounding to remains the same.

This is called *rounding up.*

This is called *rounding down.*

Example 4

Round 587 to the nearest ten.

5 **8** 7   is rounded to 590

We shade the tens digit. The digit to the right of the tens place, 7, is 5 or more. So round up.

587 is between 580 and 590. It is closer to 590, so it makes sense to round up.

Example 5

Round 2638 to the nearest hundred:

2 6 38 is rounded to 2600   We shade the hundreds digit. The digit to the right, 3, is less than 5. So round down.

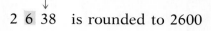

2638 is closer to 2600 than to 2700. So it makes sense to round down.

Let's look at some further examples of using the rounding rule.

Example 6

(a) Round 2378 to the nearest hundred:

2 3 78 is rounded to 2400   We have shaded the hundreds digit. The digit to the right is 7. Since this is 5 or more, the 7 and all digits to the right become 0. The hundreds digit is increased by 1.

(b) Round 53,258 to the nearest thousand:

5 3 ,258 is rounded to 53,000   We have shaded the thousands digit. Since the digit to the right is less than 5, it and all digits to the right become 0, and the thousands digit remains the same.

(c) Round 685 to the nearest ten:

6 8 5 is rounded to 690   The digit to the right of the tens place is 5 or more. Round up by our rule.

CHECK YOURSELF 2

(1) Round 568 to the nearest ten.
(2) Round 5446 to the nearest hundred.

A Special Case

Suppose we want to round 397 to the nearest ten. We shade the tens digit and look at the next digit to the right.

3 9 7   The digit to the right is 5 or more.

If this digit is 9, and it must be increased by 1, replace the 9 with 0 and increase the next digit to the *left* by 1.

So 397 is rounded to 400.

Section 1.6: Rounding, Estimation, and Order

CHECK YOURSELF 3

Round 4961 to the nearest hundred.

An estimate is basically a good guess. If your answer is close to your estimate, then your answer is reasonable.

Whether you are doing an addition problem by hand or using a calculator, rounding numbers gives you a handy way of deciding if the answer seems reasonable. The process is called *estimating*. Let's illustrate with an example.

Example 7

```
   456        500     Rounded to the nearest hundred
   235        200
   976       1000
+  344     +  300
  ────       ────
  2011       2000  ← Estimate
```

By rounding to the nearest hundred and adding quickly, we get an estimate or guess of 2000. Since this is close to the sum calculated, 2011, our answer seems reasonable.

Earlier in this section, we used the number line to illustrate the idea of rounding numbers. The number line also gives us an excellent way to picture the concept of *order* for whole numbers. Recall that the numbers become larger as we move from left to right on the line.

For instance, we know that 3 is less than 5. On the number line

3 is less than or smaller than 5.

and we see that 3 lies *to the left* of 5.

We also know that 4 is greater than 2. On the number line

4 is greater than or larger than 2.

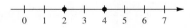

and we see that 4 lies *to the right* of 2.

Two symbols are used to indicate these relationships.

For whole numbers *a* and *b*, we can write

1. *a* < *b* (read "*a* is less than *b*") when *a* is *to the left* of *b* on the number line.
2. *a* > *b* (read "*a* is greater than *b*") when *a* is *to the right* of *b* on the number line.

The following example illustrates the use of this notation.

Example 8

Use the symbols $<$ or $>$ to complete each statement.

(a) 7 ____ 10
(b) 25 ____ 20
(c) 200 ____ 300
(d) 8 ____ 0

Solutions

(a) $7 < 10$ 7 lies to the left of 10 on the number line.
(b) $25 > 20$ 25 lies to the right of 20 on the number line.
(c) $200 < 300$
(d) $8 > 0$

CHECK YOURSELF 4

Use the symbols $<$ or $>$ to complete the following statements.

(1) 35 ____ 25   (2) 0 ____ 4   (3) 12 ____ 18   (4) 1000 ____ 100

CHECK YOURSELF ANSWERS

1. Round 375 *up* to 400.
2. (1) 570;   (2) 5400.   3. 5000.
4. (1) $35 > 25$;   (2) $0 < 4$;   (3) $12 < 18$;   (4) $1000 > 100$.

1.6 Exercises

Round the following numbers to the indicated place.

1. 38, the nearest ten

2. 72, the nearest ten

3. 253, the nearest ten

4. 578, the nearest ten

5. 696, the nearest ten

6. 683, the nearest hundred

7. 3482, the nearest hundred

8. 6741, the nearest hundred

9. 5962, the nearest hundred

10. 4352, the nearest thousand

Section 1.6: Rounding, Estimation, and Order

11. 4927, the nearest thousand

12. 39,621, the nearest thousand

13. 23,429, the nearest thousand

14. 38,589, the nearest thousand

15. 787,000, the nearest ten thousand

16. 582,000, the nearest hundred thousand

17. 21,800,000, the nearest million

18. 931,000, the nearest ten thousand

Estimate the following sums by rounding to the indicated place. Then do the addition and use your estimate to see if your actual sum seems reasonable.

Round to the nearest ten.

19. 58
     27
    +33

20. 92
     37
     85
    +64

21. 87
     53
     41
     93
    +62

22. 78
     67
     53
     42
    +86

Round to the nearest hundred.

23.   379
     1215
    + 528

24.   967
     2365
      544
    + 738

25. 1378
      519
      792
    +2041

26. 3145
      889
      259
      692
    +2518

Round to the nearest thousand.

27. 2238
     3925
    +5217

28. 3678
     4215
    +2032

29. 9137
     2315
     7643
    +3092

30. 11,548
      3,874
     14,435
    + 5,398

Use the symbols < or > to complete each statement.

31. 4 _____ 8

32. 0 _____ 5

33. 500 _____ 400

34. 20 _____ 15

35. 100 _____ 1000

36. 3000 _____ 2000

Answers

1. 40   3. 250   5. 700   7. 3500   9. 6000   11. 5000   13. 23,000   15. 790,000   17. 22,000,000
19. Estimate:
```
    60
    30
  + 30
   120
```
Actual sum: 118
21. Estimate: 330, Actual sum: 336   23. Estimate:
```
   400
  1200
  + 500
  2100
```
Actual sum: 2122
25. Estimate: 4700, Actual sum: 4730   27. Estimate: 11,000, Actual sum: 11,380
29. Estimate: 22,000, Actual sum: 22,187   31. 4 < 8   33. 500 > 400   35. 100 < 1000

1.6 Supplementary Exercises

Round the following numbers to the indicated place.

1. 59, the nearest ten
2. 73, the nearest ten
3. 583, the nearest ten
4. 768, the nearest ten
5. 896, the nearest hundred
6. 2981, the nearest hundred
7. 43,587, the nearest thousand
8. 81,243, the nearest thousand
9. 791,000, the nearest hundred thousand

Estimate the following sums by rounding to the indicated place. Then do the actual addition and use your estimate to see if your actual sum seems reasonable.

Round to the nearest ten.

10.
```
   57
   73
  +62
```

11.
```
   93
   68
   75
  +62
```

Round to the nearest hundred.

12.
```
    687
   2320
    573
  + 435
```

13.
```
    480
    515
    693
   2560
  + 727
```

Round to the nearest thousand.

14. 2788
     7215
    +11089

15. 8319
    12835
     5902
    + 3019

Use the symbols < or > to complete each statement.

16. 18 _____ 12   17. 10 _____ 0   18. 500 _____ 600

1.7 Subtraction of Whole Numbers

OBJECTIVES
1. To use the language of subtraction
2. To subtract whole numbers without borrowing

By *opposite* we mean that subtracting a number "undoes" an addition of that same number. Start with 1. Add 5 and then subtract 5. Where are you?

We are now ready to consider a second operation of arithmetic, subtraction. In Section 1.2, we described addition as the process of combining two or more groups of objects. Subtraction can be thought of as the *opposite operation* to addition. Every arithmetic operation has its own notation. The symbol for subtraction, $-$, is called a *minus sign*.

When we write $8 - 5$, we wish to subtract 5 from 8. We call 5 the *subtrahend*. This is the number being subtracted. 8 is the *minuend*. This is the number we are subtracting from. The *difference* is the result of the subtraction.

To find the *difference* of two numbers, we look for a number which, when added to the number being subtracted, will give the number that we started with.

Example 1

$8 - 5 = 3$ since $3 + 5 = 8$

The special relationship between addition and subtraction provides a method of checking subtraction.

> The sum of the difference and the subtrahend must be equal to the minuend.

Example 2

12 − 5 = 7

Check:

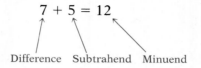

Difference Subtrahend Minuend

Our check works because 12 − 5 asks for the number that must be added to 5 to get 12.

CHECK YOURSELF 1

Subtract and check your work.

13 − 9 =

The procedure for subtracting larger whole numbers is similar to the procedure for addition. We subtract digits of the same place value.

Let's look at an example in the expanded form.

Example 3

$$\begin{array}{r} 87 = 8 \text{ tens} + 7 \text{ ones} \\ \underline{-53} = \underline{5 \text{ tens} + 3 \text{ ones}} \\ 3 \text{ tens} + 4 \text{ ones} \\ \text{or} \quad 34 \end{array}$$

We subtract the digits of the same place value, first ones and then tens.

In the short form we write:

Example 4

$$\begin{array}{r} 87 \\ \underline{-53} \\ 34 \end{array}$$

Again we subtract in the ones column and then the tens column. In practice we will always use this short form.

To check the subtraction:

34 + 53 = 87

CHECK YOURSELF 2

Subtract and check your work.

$$\begin{array}{r} 79 \\ \underline{-36} \end{array}$$

Section 1.7: Subtraction of Whole Numbers

The subtraction process is easily extended to larger numbers.

Example 5

Step 1	Step 2	Step 3
789	789	789
−246	−246	−246
3	43	543

We subtract in the ones column, then in the tens column, and finally in the hundreds column.

To check: 789
          −246 } Add
           543 } 543 + 246 = 789

The sum of the difference and the subtrahend must be the minuend.

CHECK YOURSELF 3

Subtract and check your work.

 3468
−2248

CHECK YOURSELF ANSWERS

1. 13 − 9 = 4
   Check: 4 + 9 = 13

2. 79
   −36
   43
   Check: 43 + 36 = 79

3. 1220

1.7 Exercises

1. In the statement 9 − 6 = 3
   9 is called the
   6 is called the
   3 is called the
   Write the related addition statement.

2. In the statement 7 − 5 = 2
   5 is called the
   2 is called the
   7 is called the

Do the indicated subtraction and check your results by addition.

3. 86
   −23

4. 97
   −54

5. 97 − 45

6. 86 − 32

7. Subtract 57 from 98.

8. Subtract 35 from 87.

9. 347
   −201

10. 575
    −302

11. 689
    −245

12. 598
    −278

13.   3446
    −2326
	14. 5896
	    −3862
		15. 8540
		    −2320
			16. 5830
			    −3220

17.   23,689
    − 2,523
	18. 59,786
	    − 3,214
		19. 47,235
		    −23,025
			20. 59,342
			    −27,140

You know that the word *difference* indicates subtraction. There are other words that also tell you to use the subtraction operation.

5 *less than* 12 is written as

12 − 5 or 7

20 *decreased* by 8 is written as

20 − 8 or 12

With this information, find each of the following.

21. The difference of 97 and 43

22. 58 decreased by 23

23. 25 less than 76

24. The difference of 167 and 57

25. 298 decreased by 47

26. 125 less than 265

Solve each of the following problems.

27. Danielle's score on a math test was 87 while Tony's score was 23 points less than Danielle's. What was Tony's score on the test?

28. Monica's monthly pay of $879 was decreased by $175 for withholding. What amount of pay did she receive?

29. The difference between two numbers is 134. If the larger number is 655, what is the smaller number?

30. In Jason's monthly budget, he set aside $375 for housing and $165 less than that for food. How much did he budget for food?

Answers
1. 9 is the minuend, 6 is the subtrahend, and 3 is the difference. 3 + 6 = 9   **3.** 63   **5.** 52
7. 98   Check: 41 + 57 = 98.   **9.** 146   **11.** 444   **13.** 1120   **15.** 6220   **17.** 21,166   **19.** 24,210
$$\begin{array}{r} 98 \\ -57 \\ \hline 41 \end{array}$$
21. 54   **23.** 51   **25.** 251   **27.** 64   **29.** 521

1.7 Supplementary Exercises

1. In the statement 8 − 5 = 3
   5 is called the
   3 is called the
   8 is called the
   Write the related addition statement.

Do the indicated subtraction and check your results by addition.

2. 97
     −43

3. 88 − 53

4. Subtract 36 from 78.

5. 779
     −238

6. 835
     −205

7. 2587
     −1342

8. 3489
     −2209

9. 52,486
     − 2,356

10. 43,495
      −20,462

1.8 Borrowing

OBJECTIVES
1. To use borrowing in subtracting whole numbers
2. To estimate differences by rounding

Difficulties can arise in subtraction if one or more of the digits of the subtrahend are larger than the corresponding digits in the minuend. We will solve this problem by *borrowing*.

Let's look at an example in the expanded form:

Example 1

$$\begin{array}{r} 52 = 5\text{ tens} + 2\text{ ones} \\ -27 = 2\text{ tens} + 7\text{ ones} \\ \hline \end{array}$$

Do you see that we cannot subtract in the ones column?

Regrouping, we borrow 1 ten in the minuend and write that ten as 10 ones:

$$5 \text{ tens} + 2 \text{ ones}$$

becomes $4 \text{ tens} + 1 \text{ ten} + 2 \text{ ones}$

or $4 \text{ tens} + 10 \text{ ones} + 2 \text{ ones}$

or $4 \text{ tens} + 12 \text{ ones}$

We now have

$52 = 4 \text{ tens} + 12 \text{ ones}$
$-27 = 2 \text{ tens} + 7 \text{ ones}$
$2 \text{ tens} + 5 \text{ ones}$

or 25

We can now subtract as before.

In practice we will use a more convenient short form for the subtraction.

Example 2

$$\begin{array}{r} 52 \\ -27 \\ \hline \end{array} \qquad \begin{array}{r} {}^{4\,1}\!\!\!\not{5}2 \\ -27 \\ \hline 25 \end{array}$$

We indicate the fact that we have borrowed 1 ten by putting a slash through the 5. Write 4 tens. Add 10 ones to the original 2 ones to get 12 ones.
We can then subtract.

Check: $25 + 27 = 52$

CHECK YOURSELF 1

Subtract, and check your work.

$$\begin{array}{r} 64 \\ -38 \\ \hline \end{array}$$

The subtraction process may involve more than one borrowing step.

Example 3

Step 1 $\begin{array}{r} {}^{1\,1}\!\!\!62\!\!\not{4} \\ -346 \\ \hline 8 \end{array}$

We borrow 1 ten to write 14 ones. Subtracting, we have 8 ones. Now we cannot subtract in the tens column. We must borrow again.

Section 1.8: Borrowing

Step 2 $\overset{5\ \ ^{1}1}{6\not{2}4}$
$\underline{-346}$
278

Borrow 1 hundred. This is written as 10 tens and combined with the existing 1 ten. We then have 11 tens and can continue.

CHECK YOURSELF 2

Subtract.

536
$\underline{-258}$

Let's work through another subtraction example that will require a number of borrowing steps. Here zero appears as a digit in the minuend.

Example 4

Step 1 $\overset{\ \ \ 4\ 1}{40\not{5}3}$
$\underline{-2365}$
8

In this first step we borrow 1 ten. This is written as 10 ones and combined with the original 3 ones. We can then subtract in the ones column.

Here we borrow 1 thousand; this is written as 10 hundreds.

Step 2 $\overset{3\ 14\ 1}{\not{4}0\not{5}3}$
$\underline{-2365}$
8

We must borrow again to subtract in the tens column. There are no hundreds, and so we move to the thousands column.

We now borrow 1 hundred; this is written as 10 tens and combined with the remaining 4 tens.

Step 3 $\overset{3\ ^{9}14\ 1}{\not{4}\not{0}\not{5}3}$
$\underline{-2365}$
8

The minuend is now renamed as 3 thousands, 9 hundreds, 14 tens, and 13 ones.

Step 4 $\overset{3\ ^{9}14\ 1}{\not{4}\not{0}\not{5}3}$
$\underline{-2365}$
1688

The subtraction can now be completed.

To check our subtraction: 1688 + 2365 = 4053

CHECK YOURSELF 3

Subtract, and check your work.

5024
$\underline{-1656}$

Estimation is also a useful skill when used with subtraction. Our final example illustrates.

Example 5

Subtract: 48,378
−34,429

Round the minuend to 48,000 (the nearest thousand) and the subtrahend to 34,000. Our estimate of the difference is then 14,000.
We will leave it to you to perform the subtraction and find the actual difference. Does your answer seem reasonable given our estimate?

CHECK YOURSELF 4

Estimate the difference below by rounding to the nearest thousand. Then perform the actual subtraction to see if your answer seems reasonable, given your estimate.

53,928
−38,199

CHECK YOURSELF ANSWERS

1. $\overset{5\,1}{\cancel{6}4}$ To check: 2. $\overset{4\,\overset{1}{\cancel{3}}\,1}{\cancel{5}\cancel{3}6}$ 3. 3368.
   −38   26 + 38 = 64   −258
   ―――   ―――
   26   278

4. Estimate: 16,000; difference: 15,729.

1.8 Exercises

Subtract and check your results by addition.

1. 64   2. 73   3. 50   4. 40
   −27   −36   −36   −23

5. 372   6. 500   7. 534   8. 867
   − 58   − 65   −263   −483

9. 627   10. 642   11. 280   12. 370
   −358   −367   −185   −193

Section 1.8: Borrowing

13. 603
    −259

14. 705
    −368

15. 2358
    − 562

16. 3547
    − 673

17. 3537
    −2675

18. 4693
    −2736

19. 6423
    −3678

20. 5352
    −2577

21. 6034
    −2569

22. 5206
    −1748

23. 4000
    −2345

24. 6000
    −4349

25. 33,486
    −14,047

26. 53,487
    −25,649

27. 29,400
    −17,900

28. 53,500
    −28,700

29. 59,000
    −23,458

30. 41,000
    −27,645

31. To keep track of a checking account, you must subtract the amount of each check from the current balance. Complete the following statement.

    Beginning balance $351
      Check #1   29
        Balance
      Check #2   139
        Balance
      Check #3   75
    Ending balance

32. Complete the following record of a monthly expense account.

    Monthly income $1620
      House payment   343
        Balance
      Car payment   183
        Balance
      Food   312
        Balance
      Clothing   89
    Amount remaining

Estimate each of the following differences by rounding. Use your estimate to determine which of the problems are correct and which are incorrect.

33. 5846
    −1938
    ─────
    3908

34. 7983
    −3579
    ─────
    3404

35. 29,857
    − 2,098
    ──────
    26,759

36. 53,174
    −30,098
    ──────
    23,076

Determine the pattern for the following arithmetic sequences and write the next four numbers in each sequence.

37. 53, 47, 41, 35, _____ , _____ , _____ , _____

38. 158, 141, 124, 107, _____ , _____ , _____ , _____

The figures below are called *magic squares*. Let's see why.

39.

4	9	2
3	5	7
8	1	6

Add each row.
Add each column.
Add each diagonal.
Compare the sums.

40.

1	15	14	4
12	6	7	9
8	10	11	5
13	3	2	16

Add each row.
Add each column.
Add each diagonal.
Compare the sums.

Complete the magic square.

41.

	7	2
	5	9
8		

42.

4	3	
	5	
		6

43.

16	3		13
	10	11	
9	6	7	
4			1

44.

7			14
2	13	8	11
16			
		6	15

Answers

1. 37   **3.** 14   **5.** 314   **7.** $\overset{4\,1}{\cancel{5}34}$ $\underline{-263}$ 271   **9.** 269   **11.** 95   **13.** 344   **15.** $\overset{1\,{}^{1}21}{2\cancel{3}58}$ $\underline{-562}$ 1796   **17.** 862

Section 1.8: Borrowing   45

19. $\overset{5\,13\,11}{\cancel{6}\cancel{4}\cancel{2}3}$
    -3678
    $\overline{2745}$

21. $\overset{5\,9\,21}{\cancel{6}\cancel{0}\cancel{3}\cancel{4}}$
    -2569
    $\overline{3465}$

23. 1655   25. 19,439   27. 11,500   29. 35,542   31. End balance: $108

33. Correct   35. Incorrect   37. 29, 23, 17, 11   39. 15

41.
6	7	2
1	5	9
8	3	4

43.
16	3	2	13
5	10	11	8
9	6	7	12
4	15	14	1

1.8 Supplementary Exercises

Subtract and check your results by addition.

1. 75
    −38

2. 90
    −76

3. 700
    − 54

4. 835
    −678

5. 540
    −268

6. 607
    −454

7. 900
    −246

8. 3592
    − 795

9. 6585
    −3492

10. 5372
    −2588

11. 2058
    −1379

12. 5000
    −3482

13. 42,024
    −25,509

14. 65,482
    −37,594

15. 32,000
    −17,985

16. Complete the following checking account statement.

    Beginning balance $1268
      Check #1   349
        Balance
      Check #2   57
        Balance
      Check #3   459
    Ending balance

Estimate each of the following differences by rounding. Use your estimate to determine which of the problems are correct and which are incorrect.

17. 6732
     −3843
     $\overline{2889}$

18. 73,524
     −42,678
     $\overline{20,846}$

Using Your Calculator

Particularly after working with borrowing in subtraction, you may be tempted to use your calculator for problems besides the ones in these special calculator sections.

Remember, the point is to brush up on your arithmetic skills by hand, *along with* learning to use the calculator in a variety of situations.

Now that you have reviewed the process of subtracting by hand, let's look at the use of the calculator in performing that operation. To compute 56 − 29, follow the indicated calculator steps.

1. Press the clear key. $\boxed{C}$
2. Enter the first number.   56
3. Press the minus key. $\boxed{-}$
4. Enter the second number.   29
5. Press the equals key. $\boxed{=}$ The difference, 27, will now be in the display.

Display $\left(27\right)$

The calculator can be very helpful in a problem that involves both addition and subtraction operations.

Example 1

Find:

23 − 13 + 56 − 29

Enter the numbers and the operation signs exactly as they appear in the expression.

23 $\boxed{-}$ 13 $\boxed{+}$ 56 $\boxed{-}$ 29 $\boxed{=}$ $\left(37\right)$

Display

An alternative approach would be to add 23 and 56 first and then subtract 13 and 29. The result is the same in either case.

Exercises Using Your Calculator

Do the indicated operations.

1.   89
 −48

2.   576
 −389

3.   2793
 − 345

4.   9923
 −7759

5.   5830
 −3987

6.   15,280
 − 7,595

7.   243,800
 −   5,965

8.   598,700
 − 2,934

Section 1.8: Borrowing

9. 27,983 − 1993

10. 31,234 − 2849

11. 193,243 − 49,285

12. 257,500 − 78,750

13. 125 − 75 + 40

14. 235 − 90 + 25

15. 53 − 26 + 35 − 16

16. 69 − 36 − 22 + 18

17. Subtract 235 from the sum of 534 and 678.

18. Subtract 476 from the sum of 306 and 572.

19. Readings from the Fast Service Station's storage tanks were taken at the beginning and the end of a month. How much of each type of gas was sold? What was the total sold?

	REGULAR	UNLEADED	SUPER UNLEADED	TOTAL
Beginning reading	73,255	82,349	81,258	
End reading	28,387	19,653	8,654	
Gallons used	_____	_____	_____	_____

20. Connie Jackson made three business trips in her car. She recorded the reading on the odometer (the mileage indicator on the car speedometer) at the start and the end of each of the trips. How far did she drive on each trip? What was the total number of miles driven?

	TRIP 1	TRIP 2	TRIP 3	TOTAL
Start of trip	27,345	30,204	34,821	
End of trip	28,377	31,192	35,539	
Miles driven	_____	_____	_____	_____

The land areas of three Pacific Coast states are shown below:

California   158,693 sq mi
Oregon   96,981 sq mi
Washington   68,192 sq mi

21. How much larger is California than Oregon?

22. How much larger is California than Washington?

23. How much larger than Washington is Oregon?

24. How much larger than California are Washington and Oregon combined?

Answers
1. 41   **3.** 2448   **5.** 1843   **7.** 237,835   **9.** 25,990   **11.** 143,958   **13.** 90   **15.** 46   **17.** 977
19. Regular, 44,868 gal; unleaded, 62,696 gal; super unleaded, 72,604 gal; total, 180,168 gal
21. 61,712 sq mi   **23.** 28,789 sq mi

1.9 Solving Word Problems Involving Addition and Subtraction

OBJECTIVE
To solve word problems involving addition and subtraction

You may very well be able to do some of these problems in your head. Get into the habit of writing down *all* your work, rather than just an answer.

In this section we will consider applications, or word problems, which will use the operations of addition and subtraction. An organized approach is the key to successful problem solving, and we would suggest the following strategy. First, make sure you understand the problem. Then decide upon the operation or operations that should be used for the solution. At that point you can do the necessary calculations. Always finish your work by making sure that you have answered the question asked in the problem, and that your answer seems reasonable. We can summarize this strategy with the following four basic steps.

> **SOLVING WORD PROBLEMS**
> 1. Read the problem carefully to determine the given information and what you are asked to find.
> 2. Decide upon the operation or operations to be used.
> 3. Write down the complete statement necessary to solve the problem and do the calculations.
> 4. Check to make sure you have answered the question of the problem and that your answer seems reasonable.

Let's work through some examples using these steps.

Example 1

A housing development has 38 finished homes. A builder has already started construction on 27 new homes and plans to start another 24. How many homes will there be in the development when the construction is complete?

Step 1 The given information is the number of existing homes, 38, the number already started, 27, and the number that will be started, 24. We want the total number of homes.

Section 1.9: Solving Word Problems Involving Addition and Subtraction

Step 2 Since we want a *sum*, we must use addition for the solution.

Step 3 Write

```
 38 homes
 27 homes
+24 homes
 89 homes
```

Step 4 The answer to the question of the original problem is 89 homes. It is generally a good idea to use estimation to check that your answer is reasonable. For instance, to form an estimate you might think

Round each number to the nearest 10.

40 + 30 + 20, or 90 homes

and our actual result, 89 homes, certainly seems reasonable.

Example 2

Five sections of algebra were offered in the fall quarter, with enrollments of 33, 24, 28, 41, and 22 students. What was the total number of students taking algebra?

Step 1 The given information is the number of students in each section. We want the total number.

Step 2 Since we wish a total, addition is called for.

Be sure to attach the proper unit, here *students*, to your answer.

Step 3 Write: 33 + 24 + 28 + 41 + 22 = 148 students

Step 4 Our answer is 148 students. We will leave it to you to check, by estimating, that this result is reasonable.

CHECK YOURSELF 1

Janet Jeffries won an election for city council with 2485 votes. Her two opponents had 1873 and 985 votes. How many votes were cast for that office?

Let's look at an example in which subtraction must be used for the solution.

Example 3

Note: While we will not separate the steps of our final examples, you should see that the four-step strategy remains the same.

Jack McIver bought a car with a list price of $9400. The dealer marked down the car to $7964. How much was Jack's discount? (The discount is the amount that the price was reduced.)

Solution Here we want to know the difference between the list price and the actual selling price. This will be the amount of the discount. Since we want to find a difference, subtraction must be used for the solution. Write

$9400 − $7964 = $1436

The discount was $1436.

CHECK YOURSELF 2

An evening performance for a play had 1208 people in attendance while the afternoon matinee drew 959 people. How many fewer people attended the afternoon performance?

You will need to use both addition and subtraction to solve some problems. Our final example illustrates.

Example 4

Bernard wants to buy a new piece of stereo equipment. He has $142 and can trade in his old amplifier for $135. How much more does he need if the new equipment costs $449?

Solution First we must add to find out how much money Bernard has available. Then we subtract to find out how much more money he needs.

$142 + $135 = $277   The money available
$449 − $277 = $172   Bernard still needs

CHECK YOURSELF 3

Martina spent $239 in airfare, $174 for lodging, and $108 for food on a business trip. Her company allowed her $375 for the expenses. How much of these expenses will she have to pay herself?

CHECK YOURSELF ANSWERS

1. We want to know the total number of votes cast, and so we must use addition for the solution.

   ```
     2485
     1873
   + 985
   ─────
     5343
   ```

   5343 votes were cast in the election.

2. We must subtract for the solution.

   ```
     1208
   −  959
   ─────
      249
   ```

   249 fewer people attended in the afternoon.

3. $239 $521 ← Total expenses
    174   − 375 ← Amount allowed
  + 108 $146
    $521 ← Total expenses

1.9 Exercises

Solve each of the following applications.

1. The Burton family drove 385 and 273 mi on 2 days of a vacation trip. How far did they drive in these 2 days?

2. Susan Compton buys a car with a list price of $8250. She also orders an air conditioner for $475. What will the total cost be?

3. Marilyn rolled games of 189, 212, and 208 in a three-game bowling series. What was her total score for the series?

4. A basketball player scored 35, 29, 28, and 30 points in a four-game tournament. How many points did he score in all?

5. For a band concert, an auditorium has 245 $9 seats, 350 $7 seats, and 475 $5 seats. How many tickets can be sold?

6. Four performances of a play had attendance figures of 235, 312, 244, and 285. How many people saw the play during this period?

7. An airline had 137, 179, 154, 201, and 168 passengers on their five shuttle flights between Los Angeles and San Francisco during 1 day. What was the total number of passengers?

8. A company spent $1825 on rent, $285 for heat, $178 for electricity, $428 for phone service, and $619 for cleaning during one month. What was the company's total expense for the month?

9. Alan owes $543 after buying a stereo. If he makes a payment of $175, how much does he still owe?

10. Brenda has $278 in cash and wants to buy a television set that costs $449. How much more money does she need?

11. A bond election for schools had the following results: yes, 3457 votes; no, 3189 votes. By how much of a margin did the bond pass?

12. At the beginning of a trip your odometer read 21,342 mi, and at the end it read 22,078 mi. How far did you drive?

13. The Sears Tower in Chicago is 1454 ft tall. The Empire State Building is 1250 ft tall. How much taller is the Sears Tower than the Empire State Building?

14. A college's enrollment was 2479 students in the fall of 1987 and 2653 students in the fall of 1988. What was the increase in enrollment?

15. In 1 week Margaret earned $278 in regular pay and $53 for overtime work. $49 was deducted from her paycheck for income taxes and $18 for social security. What was her take-home pay?

16. Mr. Haller opened a checking account and made deposits of $85 and $272. He wrote checks during the month for $35, $27, $89, and $178. What was his balance at the end of the month?

17. Sandra is trying to limit her calories to 1500 per day. Her breakfast was 270 calories, her lunch was 450 calories, and her dinner was 820 calories. By how much was she *under* or *over* her diet?

18. A professional basketball team scored 98, 136, and 113 points in three games. If its opponents scored 102, 109, and 93 points, by how much did the team outscore its opponents?

19. A course outline states that you must have 540 points on five tests during the term to receive an A for the course. Your scores on the first four tests have been 95, 84, 82, and 89. How many points must you score on the 200-point final to receive an A?

20. Michelle's frequent-flyer program requires 30,000 mi for a free flight. During 1987 she accumulated 13,850 mi. In 1988 she took three more flights of 2800, 1475, and 4280 mi. How much further must she fly for her free trip?

Use your calculator for the following problems.

21. The three largest cities of Texas are Houston, with a population of 1,501,000; Dallas, with 869,500; and San Antonio, with 798,500. What is the combined population of these three cities?

22. The area of Canada is 3,851,809 sq mi and the area of the United States is 3,615,122 sq mi. How much larger in area is Canada than the United States?

Answers

1. 385 + 273 = 658 mi   **3.** 609   **5.** 1070 tickets   **7.** 839 passengers   **9.** $543 − $175 = $368
11. 3457 − 3189 = 268 votes   **13.** 204 ft
15.

Total pay	Deductions	Take-home pay
$278	$49	$331
+ 53	+18	− 67
$331	$67	$264

17. Number of calories:
270
450
+ 820
1540
40 cal over.

19. 190 points   **21.** 3,169,000

1.9 Supplementary Exercises

Solve each of the following applications.

1. Pat spent $145 for tuition and $93 for books and supplies during one term. What were her expenses for tuition, books, and supplies?

2. Matt buys a television set with a list price of $495. With his payment plan, finance charges are $98. What will be the total cost of the purchase?

3. A salesman drove 68 mi on Tuesday, 114 mi on Thursday, and 79 mi on Friday. What was his mileage for those 3 days?

4. Marsha had grades of 85, 79, 93, and 89 on four tests during a course. What was her total number of points?

5. The down payment on a new car is $1560. Curt has $1385 in his savings account now. How much more does he need for the down payment?

6. Janice bought a new Toyota with a list price of $8958 and was given a discount of $789. What did she pay for the car?

7. The 1970 census showed a population of 18,387 for Byrnsville. In 1980 the population was 27,585. What was the increase in population over that period?

8. Teresa wrote checks of $54, $37, and $143. Her balance before that was $503. What is her new balance?

9. A family used 810 kilowatt-hours (kWh) of electricity in January 1988 and 875 kWh in February. In January 1987 they used 905 kWh, and in February, 935 kWh. How much less electricity did they use in 1988 than during the comparable period in 1987?

10. The Brackens had the following expenses in one month: Housing, $350; food, $185; clothing, $60; utilities, $95; and recreation, $45. If they had budgeted for expenses of $750, how much were they *over* or *under* that budget?

SELF-TEST for Chapter One

The purpose of the Self-Test is to help you check your progress and review for a chapter test in class. Allow yourself about an hour to take the test. When you are done, check your answers in the back of the book. If you missed any problems, be sure to go back and review the appropriate sections in the chapter and do the supplementary exercises provided there.

[1.1] **1.** What is the place value of 3 in 238,157?

2. Write 23,543 in words.

3. Write four hundred eight million, five hundred twenty thousand as a numeral.

[1.3] In Problems 4 to 6, name the property of addition that is illustrated.

4. $6 + 7 = 7 + 6$

5. $8 + (3 + 7) = (8 + 3) + 7$

6. $5 + 0 = 5$

[1.4] In Problems 7 and 8, add.

7. $23 + 542 =$

8. $14 + 223 + 4321 =$

[1.5] In Problems 9 to 14, perform the indicated operations.

9. $1369 + 5804 =$

10. $489 + 562 + 613 + 254 =$

11. $357 + 28 + 2346 =$

12. $13 + 2543 + 10{,}547 =$

13. Find the sum of 2459, 53, and 23,467.

14. What is the total of 392, 95, 9237, and 11,972?

[1.6] In Problems 15 and 16, complete the following table by rounding the numbers to the indicated place.

		NEAREST 10	NEAREST 100	NEAREST 1000
(Example)	6743	6740	6700	7000
	8546	15. _____	_____	_____
	2973	16. _____	_____	_____

[1.6] In Problems 17 and 18, use estimation to determine which of the problems is correct.

17. 293
    521
    +789
    ————
    1503

18. 23,875
    +14,230
    ————
    38,105

[1.6] In Problems 19 and 20, use the symbols < or > to complete the statement.

19. 12 _____ 14

20. 500 _____ 400

[1.7] In Problems 21 and 22, subtract.

21. 289 − 54

22. 53,294 − 41,074

[1.8] In Problems 23 to 27, perform the indicated operations.

23. 503 − 74 =

24. 5731 − 2492 =

25. 32,345 − 1575 =

26. 55,342 − 14,787 =

27. Find the difference of 5000 and 3428.

[1.9] In Problems 28 to 30, solve each application.

28. The attendance for the games of a playoff series in basketball was 12,438, 14,325, 14,581, 12,587, and 14,634. What was the total attendance for the series?

29. The area of Arizona is 113,417 sq mi, while the area of Nevada is 109,889 sq mi. How much larger is Arizona than Nevada?

30. The maximum load for a light plane with full gas tanks is 500 lb. Mr. Whitney weighs 215 lb, his wife 135 lb, and their daughter 78 lb. How much luggage can they take on a trip without exceeding the load limit?

PRETEST for Chapter Two

Multiplication of Whole Numbers

This pretest will point out any difficulties you may be having in multiplying whole numbers. Do all the problems. Then check your answers on the following page.

1. If $7 \times 9 = 63$, we call 7 and 9 the _____ of 63.
   63 is the _____ of 7 and 9.

2. List all the factors of 42.

3. The statement $2 \times (3 \times 5) = (2 \times 3) \times 5$ is an illustration of which property of multiplication.

4. $2538 \times 7 =$

5. $567 \times 48 =$

6. $348 \times 200 =$

7. $4 + 7 \times 7 =$

8. $5 \times (4 + 2) =$

9. A refrigerator is advertised as follows: "Pay $50 down and $30 a month for 24 months." If the cash price of the refrigerator is $619, how much extra will you pay if you buy on the installment plan?

10. $3 \times 4^3 =$

ANSWERS TO PRETEST

For help with similar problems turn to the section indicated.

1. Factors, product (Section 2.1)
2. 1, 2, 3, 6, 7, 14, 21, 42 (Section 2.1)
3. Associative property (Section 2.2)
4. 17,766 (Section 2.3)
5. 27,216 (Section 2.4)
6. 69,600 (Section 2.5)
7. 53 (Section 2.6)
8. 30 (Section 2.6)
9. $151 (Section 2.7)
10. 192 (Section 2.8)

Chapter Two

Multiplication of Whole Numbers

2.1 The Language of Multiplication

OBJECTIVES
1. To use the language of multiplication
2. To multiply single-digit whole numbers

The use of the symbol × dates back to the 1600's.

Our work in this chapter deals with multiplication, another of the basic operations of arithmetic. Multiplication is closely related to addition. In fact, we can think of multiplication as a shorthand method for repeated addition. The symbol × is used to indicate multiplication.

> There are other ways of indicating multiplication. Later we will use a raised dot or parentheses. For example, 3 · 5 is the same as 3 × 5. The dot was introduced so that the multiplication symbol × would not be confused with the letter *x*. Parentheses can also be used to indicate multiplication. For example, (3)(5) is the same as 3 × 5.

Example 1

3 × 5 means 5 multiplied by 3. It is read 3 *times* 5. To find 3 × 5 we can add 5 three times.

3 × 5 = 5 + 5 + 5 = 15

In a multiplication problem such as 3 × 5 = 15, we call 3 and 5 the *factors*. The answer, 15, is the *product* of the factors, 3 and 5.

3 × 5 = 15
Factor Factor Product

Example 2

A whole number greater than 1 will always have itself and 1 as factors.

Since 4 × 6 = 6 + 6 + 6 + 6 = 24, 4 and 6 are factors of 24. 24 is the product of 4 and 6. Also, because 1 × 24, 2 × 12, and 3 × 8 are all 24, we say that 1, 2, 3, 8, 12, and 24 are also factors of 24.

Example 3

Since 1 × 18, 2 × 9, and 3 × 6 are all 18, a list of the factors of 18 is 1, 2, 3, 6, 9, and 18. Be sure that you include the number itself, and 1, when you list the factors of a whole number.

CHECK YOURSELF 1

List all the factors of 50.

The product of two numbers is also called a *multiple* of each of the numbers.

Example 4

We know that 5 × 6 = 30. We can say that 30 is a multiple of both 5 and 6.

Example 5

We can write:

1 × 5 = 5   4 × 5 = 20
2 × 5 = 10   5 × 5 = 25
3 × 5 = 15   6 × 5 = 30

Recall that the three dots mean that the list goes on indefinitely.

The numbers 5, 10, 15, 20, 25, 30, . . . are the multiples of 5. One easy way to list these multiples is to think of "counting by fives."

Example 6

The numbers 6, 12, 18, 24, 30, . . . are the multiples of 6.

CHECK YOURSELF 2

List some multiples of 8.

Statements such as 3 × 5 = 15 and 4 × 6 = 24 are called the *basic multiplication facts*. If you have difficulty with multiplication, it may be that you do not know some of these facts. The table of Basic Multiplication Facts will help you review before going on.

Section 2.1: The Language of Multiplication   61

BASIC MULTIPLICATION FACTS

×	0	1	2	3	4	5	6	7	8	9
0	0	0	0	0	0	0	0	0	0	0
1	0	1	2	3	4	5	6	7	8	9
2	0	2	4	6	8	10	12	14	16	18
3	0	3	6	9	12	15	18	21	24	27
4	0	4	8	12	16	20	24	28	32	36
5	0	5	10	15	20	25	30	35	40	45
6	0	6	12	18	24	30	36	42	48	54
7	0	7	14	21	28	35	42	49	56	63
8	0	8	16	24	32	40	48	56	64	72
9	0	9	18	27	36	45	54	63	72	81

To use the table to find the product of 7 × 6: Move to the right in the row labeled 7 until you are in the column labeled 6 at the top. We see that 7 × 6 is 42.

CHECK YOURSELF ANSWERS

1. 1, 2, 5, 10, 25, 50.   **2.** 8, 16, 24, 32, ...

2.1 Exercises

1. Find 3 × 7 and 7 × 3 by repeated addition.

2. Find 4 × 5 and 5 × 4 by repeated addition.

3. If 6 × 7 = 42, we call 6 and 7 _____ of 42.   42 is the _____ of 6 and 7.

4. If 5 × 8 = 40, we call 5 and 8 _____ of 40.   40 is the _____ of 5 and 8.

5. List all the factors of 30.

6. List all the factors of 36.

7. The numbers 3, 6, 9, 12, 15, . . . are the _____ of 3.

8. The numbers 4, 8, 12, 16, 20, . . . are the _____ of 4.

Multiply.

9. 5
   ×3

10. 7
    ×4

11. 8
    ×1

12. 9
    ×5

13. 6
    ×0

14. 6
    ×6

15. 2
    ×9

16. 1
    ×7

17. 5
    ×6

18. 0
    ×5

19. 4
    ×9

20. 8
    ×4

21. 3
    ×8

22. 8
    ×7

23. 5
    ×7

24. 7
    ×5

25. 6
    ×9

26. 7
    ×9

27. 8
    ×8

28. 6
    ×7

29. 9
    ×8

30. 8
    ×6

31. 5
    ×8

32. 9
    ×9

33. 4
    ×6

34. 7
    ×7

35. 3
    ×9

36. 5
    ×5

Exercises 9 to 36 refer to the table of Basic Multiplication Facts. You should have been able to work quickly through these exercises. If you made errors or had to refer to the table, you should go back and review the table now!

The following set of exercises has been designed to help you review the skills that you will need as you progress through this text. Where possible, a reference to the section where the problems were introduced has been included. Go back to that section for further review if you would like.

Skillscan (Section 1.3)

Name the property of addition that is illustrated.

a. $3 + 5 = 5 + 3$

b. $(2 + 7) + 4 = 2 + (7 + 4)$

c. $5 + 0 = 5$

d. $3 + (8 + 9) = (3 + 8) + 9$

e. $8 + 9 = 9 + 8$

f. $0 + 9 = 9$

Section 2.1: The Language of Multiplication   63

Answers
Solutions for the even-numbered exercises are provided in the back of the book.
1. 21   **3.** Factors, product   **5.** 1, 2, 3, 5, 6, 10, 15, and 30. The order in which you list these factors makes no difference. Did you remember 1 and 30?   **7.** Multiples   **9.** 15   **11.** 8   **13.** 0   **15.** 18   **17.** 30   **19.** 36   **21.** 24   **23.** 35   **25.** 54   **27.** 64   **29.** 72   **31.** 40   **33.** 24   **35.** 27
a. Commutative property   **b.** Associative property   **c.** Additive identity   **d.** Associative property
e. Commutative property   **f.** Additive identity

2.1 Supplementary Exercises

1. Find 3×6 and 6×3 by repeated addition.

2. If $6 \times 8 = 48$, we call 6 and 8 _____ of 48.   48 is the _____ of 6 and 8.

3. List all the factors of 42.

4. The numbers 6, 12, 18, 24, 30, . . . are the _____ of 6.

Multiply.

5. 4
   ×5

6. 7
   ×3

7. 9
   ×1

8. 9
   ×6

9. 0
   ×6

10. 7
   ×7

11. 3
   ×8

12. 1
   ×5

13. 4
   ×9

14. 6
   ×8

15. 8
   ×0

16. 8
   ×9

17. 9
   ×5

18. 5
   ×5

2.2 The Properties of Multiplication

OBJECTIVE
To identify the properties of multiplication

From the table of Basic Multiplication Facts, we can discover some important properties of multiplication. For instance, did you see that the order in which we multiply two numbers does not affect the product?

Example 1

Looking at the table, we see

$4 \times 6 = 6 \times 4 = 24$

This illustrates the *commutative property* of multiplication.

THE COMMUTATIVE PROPERTY OF MULTIPLICATION

Multiplication, like addition, is a *commutative* operation. The order in which you multiply two whole numbers does not affect the product.

CHECK YOURSELF 1

Show that $5 \times 8 = 8 \times 5$.

The next example will lead us to another property of multiplication.

Example 2

$(2 \times 3) \times 4 = 6 \times 4 = 24$ We do the multiplication in the parentheses first, $2 \times 3 = 6$. Then multiply 6×4.

Also,

$2 \times (3 \times 4) = 2 \times 12 = 24$ Here we multiply 3×4 as the first step. Then multiply 2×12.

We see that

$(2 \times 3) \times 4 = 2 \times (3 \times 4)$

The product is the same no matter which way we *group* the factors. This is called the *associative property* of multiplication.

Section 2.2: The Properties of Multiplication

> **THE ASSOCIATIVE PROPERTY OF MULTIPLICATION**
>
> Multiplication is an *associative* operation. The way in which you group numbers in multiplication does not affect the final product.

CHECK YOURSELF 2

Find:

(1) $(5 \times 3) \times 6$
(2) $5 \times (3 \times 6)$

Examining the Multiplication Facts Table, we can also see that the two numbers 1 and 0 have special properties in multiplication.

> **THE MULTIPLICATIVE IDENTITY**
>
> Multiplying any whole number by 1 simply gives that number as a product.

We call 1 the *multiplicative identity* because of this property.

Example 3

$5 \times 1 = 5$
$1 \times 8 = 8$

> **THE MULTIPLICATION PROPERTY OF ZERO**
>
> Multiplying any whole number by 0 gives the product 0.

Example 4

$3 \times 0 = 0$
$0 \times 5 = 0$

Another property involves *both* multiplication and addition. Let's look at an example that will illustrate this law.

Example 5

$2 \times (3 + 4) = 2 \times 7 = 14$ We have added $3 + 4$ and then multiplied.

Also,

$2 \times (3 + 4) = (2 \times 3) + (2 \times 4)$ We have multiplied 2×3 and 2×4 as the first step.
$ = 6 + 8$
$ = 14$ The result is the same.

We see that 2 × (3 + 4) = (2 × 3) + (2 × 4). This is an example of the *distributive property of multiplication over addition,* because we distributed the multiplication (in this case by 2) over the sum.

> **THE DISTRIBUTIVE PROPERTY OF MULTIPLICATION OVER ADDITION**
>
> To multiply a factor by a sum of numbers, multiply the factor by each number inside the parentheses. Then add the products.

Example 6

Evaluate 5 × (6 + 3) in two ways.

(*a*) 5 × (6 + 3) = 5 × 9 = 45   We add 6 + 3. Then multiply.

(*b*) 5 × (6 + 3) = (5 × 6) + (5 × 3)   Multiply by 5 and then add the products.
           =   30   +   15   = 45

CHECK YOURSELF 3

Evaluate 3 × (4 + 6) in two ways.

The distributive law will be very important later in this chapter when we discuss multiplication involving larger numbers.

> **THE PROPERTIES OF MULTIPLICATION**
>
> We can use variables to state the properties of multiplication just as we did for addition.
>
> THE COMMUTATIVE PROPERTY OF MULTIPLICATION
> If *a* and *b* represent any whole numbers, then
>
> $a \times b = b \times a$ The order in which we multiply does not affect the product.
>
> THE ASSOCIATIVE PROPERTY OF MULTIPLICATION
> If *a*, *b*, and *c* represent any whole numbers, then
>
> $(a \times b) \times c = a \times (b \times c)$ The way in which we group the numbers does not affect the product.
>
> THE MULTIPLICATIVE IDENTITY
> If *a* is any whole number, then
>
> $a \times 1 = a$ The product of any whole number and 1 is just the whole number.

Section 2.2: The Properties of Multiplication

> **THE MULTIPLICATION PROPERTY OF ZERO**
> If a is any whole number, then
>
> $$a \times 0 = 0$$
>
> The product of any whole number and 0 is 0.
>
> **THE DISTRIBUTIVE PROPERTY OF MULTIPLICATION OVER ADDITION**
> If a, b, and c represent any whole numbers, then
>
> $$a \times (b + c) = (a \times b) + (a \times c)$$

CHECK YOURSELF ANSWERS

1. $5 \times 8 = 40$ and $8 \times 5 = 40$.
2. (1) $15 \times 6 = 90$; (2) $5 \times 18 = 90$.
3. $3 \times (4 + 6) = 3 \times 10 = 30$ or $3 \times (4 + 6) = (3 \times 4) + (3 \times 6)$
   $\phantom{3 \times (4 + 6) = 3 \times 10 = 30 \text{ or } 3 \times (4 + 6)} = 12 + 18$
   $\phantom{3 \times (4 + 6) = 3 \times 10 = 30 \text{ or } 3 \times (4 + 6)} = 30$

2.2 Exercises

Name the property of addition and/or multiplication that is illustrated.

1. $5 \times 8 = 8 \times 5$
2. $8 \times 1 = 8$
3. $5 \times 0 = 0$
4. $7 \times 6 = 6 \times 7$
5. $2 \times (3 \times 5) = (2 \times 3) \times 5$
6. $0 \times 8 = 0$
7. $9 \times 3 = 3 \times 9$
8. $2 \times (5 + 7) = (2 \times 5) + (2 \times 7)$
9. $1 \times 5 = 5$
10. $4 \times (3 \times 5) = (4 \times 3) \times 5$
11. $0 \times 9 = 0$
12. $1 \times 7 = 7$
13. $5 \times (2 \times 3) = (5 \times 2) \times 3$
14. $9 \times 8 = 8 \times 9$
15. $3 \times (2 + 8) = (3 \times 2) + (3 \times 8)$
16. $4 \times 0 = 0$

Skillscan (Section 1.1)
Give the place value for the indicated digits.

a. 5 in 45

b. 4 in 541

c. 9 in 928

d. 4 in 4392

e. 7 in 27,800

f. 3 in 35,500

Answers
1. Commutative property 3. Multiplication property of zero 5. Associative property
7. Commutative property 9. Multiplicative identity 11. Multiplication property of zero
13. Associative property 15. Distributive property of multiplication over addition **a.** Ones **b.** Tens
c. Hundreds **d.** Thousands **e.** Thousands **f.** Ten thousands

2.2 Supplementary Exercises

Name the property of addition and/or multiplication that is illustrated.

1. $9 \times 6 = 6 \times 9$

2. $7 \times 1 = 7$

3. $5 \times (2 \times 7) = (5 \times 2) \times 7$

4. $(2 \times 8) \times 3 = 2 \times (8 \times 3)$

5. $8 \times 0 = 0$

6. $4 \times (3 + 8) = (4 \times 3) + (4 \times 8)$

7. $0 \times 3 = 0$

8. $3 \times (5 + 6) = (3 \times 5) + (3 \times 6)$

2.3 Multiplying by One-Digit Numbers

OBJECTIVE
To multiply by single-digit whole numbers

Let's look at the multiplication of larger numbers. To multiply 2×34, recall that 34 means 3 tens and 4 ones. So in the expanded form,

$2 \times 34 = 2 \times (3 \text{ tens} + 4 \text{ ones})$
$ = 2 \times 3 \text{ tens} + 2 \times 4 \text{ ones}$
$ = 6 \text{ tens} + 8 \text{ ones}$
$ = 68$

Here's where we use the distributive property to multiply over the addition.

We multiply each digit of 34 by 2, keeping track of the place value. A more convenient short form will allow us to do the same thing.

Section 2.3: Multiplying by One-Digit Numbers

Example 1

To multiply 2 × 34, write

Step 1
```
  34
×  2
```
Line up the ones digits.

Step 2
```
  34
×  2
   8
```
Multiply 2 × 4. Write the product, 8, in the ones place of the product.

Step 3
```
  34
×  2
  68
```
Now multiply 2 × 3. Write the product in the tens place.

CHECK YOURSELF 1

Multiply.

```
  21
×  3
```

Often carrying must be used to multiply larger numbers. Let's see how carrying works in multiplication by looking at an example in the expanded form.

Example 2

$3 \times 25 = 3 \times (2 \text{ tens} + 5 \text{ ones})$

$\qquad = 3 \times 2 \text{ tens} + 3 \times 5 \text{ ones}$ We use the distributive property again.

$\qquad = 6 \text{ tens} \quad + 15 \text{ ones}$ Write the 15 ones as 1 ten and 5 ones.

$\qquad = 6 \text{ tens} + \overbrace{1 \text{ ten} + 5 \text{ ones}}$

$\qquad = 7 \text{ tens} + 5 \text{ ones}$ Carry 10 ones or 1 ten to the tens place.

$\qquad = 75$

Here is the same multiplication problem done in the short form.

Example 3

Step 1
```
   1  ← Carry
  25
×  3
   5
```
Multiplying 3 × 5 gives us 15 ones. Write 5 ones and carry 1 ten.

Step 2
```
   1
  25
×  3
  75
```
Now multiply 3 × 2 tens and add the carry to get 7, the tens digit of the product.

CHECK YOURSELF 2

Multiply.

```
  34
×  6
─────
```

When a multiplication problem involves larger whole numbers, you may have to carry more than once while doing the multiplication.

Example 4

Step 1
$$\begin{array}{r} \overset{3}{}\text{← Carry} \\ 438 \\ \times4 \\ \hline 2 \end{array}$$

We multiply 4 × 8 to get 32. Write 2 ones and carry 3 tens.

Step 2
$$\begin{array}{r} \overset{1\,3}{}\text{← Carry} \\ 438 \\ \times4 \\ \hline 52 \end{array}$$

Now multiply 4 × 3 tens. The product is 12 tens, and we add the carry of 3 tens. The result is 15 tens. Write 5 as the tens digit of the product and carry 1 to the hundreds place.

Step 3
$$\begin{array}{r} \overset{1\,3}{}\text{← Carry} \\ 438 \\ \times4 \\ \hline 1752 \end{array}$$

Multiply again and add the carry. We have 17 hundreds, and the multiplication is complete.

CHECK YOURSELF 3

Multiply.

```
  527
×   5
─────
```

CHECK YOURSELF ANSWERS

1. 63.   **2.** $\overset{2}{}$ ← Carry
         34
       × 6
       ────
        204

3. 2635.

2.3 Exercises

Multiply.

1. 23
    × 2

2. 32
    × 3

3. 48
    × 4

4. 53
    × 5

5. 508
   × 6

6. 903
   × 9

7. 523
   × 8

8. 635
   × 7

9. 2035
   × 9

10. 5018
    × 7

11. 5478
    × 7

12. 6893
    × 9

13. 26,555
    × 7

14. 30,524
    × 6

15. 20,108
    × 7

16. 31,015
    × 5

17. 245 × 8

18. 9 × 306

19. 3249 × 5

20. 7 × 1258

21. Find the product of 304 and 7.

22. Find the product of 409 and 4.

23. Find the product of 8 and 5679.

24. Find the product of 23,452 and 5.

Skillscan (Section 1.5)
Add.

a. 56
   +420

b. 72
   +840

c. 28
   840
   +9100

d. 34
   850
   +10200

e. 521
   2450
   +31700

f. 782
   3150
   +72800

Answers

1. 46

3. 192

5. 3048

7. 4184

9. 18,315

11. $\overset{355}{5478}$
    × 7
    38,346

13. 185,885

15. 140,756

17. 1960

19. $\overset{124}{3249}$
    × 5
    16,245

21. 2128

23. 45,432

a. 476   b. 912   c. 9968   d. 11,084   e. 34,671   f. 76,732

2.3 Supplementary Exercises

Multiply.

1. 56
   × 7

2. 528
   × 5

3. 409
   × 7

4. 2048
   × 5

5. 4823
   × 8

6. 6587
   × 9

7. 23,271
   × 6

8. 35,478
   × 7

9. 243 × 9

10. 7 × 2573

11. Find the product of 3582 and 8.

12. Find the product of 5 and 13,487.

2.4 Multiplying by Numbers with More Than One Digit

OBJECTIVE
To multiply any two whole numbers

To multiply by numbers with more than one digit, we must multiply each digit of the first factor by each digit of the second. To do this, we form a series of partial products and then add them to arrive at the final product.

Example 1

Step 1
```
   32
 × 13
   96
```
Always start the multiplication with the ones digit. The first partial product is 3 × 32, or 96.

Step 2
```
   32
 × 13
   96
  320
```
The second partial product is 1 ten × 32, or 320.

Step 3
```
   32
  ×13
   96
  320
  416
```
Add

As the final step of the process, we add the partial products to arrive at 416.

Section 2.4: Multiplying by Numbers with More Than One Digit   73

CHECK YOURSELF 1
Multiply.

$$\begin{array}{r} 31 \\ \times 23 \\ \hline \end{array}$$

It may be necessary to carry in forming one or both of the partial products.

Example 2

Step 1
$$\begin{array}{r} \overset{4}{}56 \\ \times 47 \\ \hline 392 \end{array}$$
The first partial product is 7 × 56 or 392. Note that we had to carry 4 to the tens column.

Step 2
$$\begin{array}{r} \overset{2}{}56 \\ \times 47 \\ \hline 392 \\ 224 \end{array}$$
The second partial product is 4 tens × 56, or 224 tens. We must carry 2 during the process.

Note that we shift the product *one place to the left* to indicate the multiplication by 4 *tens*. This means that the digit 4 is placed below the tens digit of 47. This is more efficient than writing the unnecessary zero.

Step 3
$$\begin{array}{r} 56 \\ \times 47 \\ \hline 392 \\ 224 \\ \hline 2632 \end{array}$$
We add the partial products for our final result.

CHECK YOURSELF 2
Multiply.

$$\begin{array}{r} 38 \\ \times 76 \\ \hline \end{array}$$

Let's work through another multiplication example.

Example 3

Multiply 56 × 673.

Step 1
$$\begin{array}{r} 673 \\ \times56 \\ \hline \end{array}$$
Write the product as shown.

Step 2
$$\begin{array}{r}\overset{4\ 1}{673}\\ \times\ 56\\ \hline 4038\end{array}$$

Multiply 6 × 673 for the first partial product. We must carry to the tens and hundreds columns.

Note: This places 5 below 3, the *tens* digit of the first partial product.

Step 3
$$\begin{array}{r}\overset{3\ 1}{673}\\ \times\ 56\\ \hline 4038\\ 3365\end{array}$$

Multiply 5 tens × 673 for the second partial product.

Again we shift the product one place left to show the multiplication by 5 tens.

Step 4
$$\begin{array}{r}673\\ \times\ 56\\ \hline 4038\\ 3365\\ \hline 37688\end{array}$$

We now add the partial products for our result.

In practice these steps are combined, and your work should look like that shown in step 4.

CHECK YOURSELF 3

Multiply.

74 × 538

If multiplication involves two three-digit numbers, another step is necessary. In this case we form three partial products. This will ensure that each digit of the first factor is multiplied by each digit of the second.

The three partial products are formed when we multiply by the ones, tens, and then the hundreds digits.

Example 4

$$\begin{array}{r}278\\ \times 343\\ \hline 834\\ 1112\\ 834\\ \hline 95354\end{array}$$

In forming the third partial product, we must multiply by 3 hundreds. To indicate this, we shift that product *two* places left. Again the unnecessary zeros are omitted.

CHECK YOURSELF 4

Multiply.

$$\begin{array}{r}352\\ \times 249\end{array}$$

Let's look at an example of multiplying by a number involving 0 as a digit. There are several ways to arrange the work as our example shows.

Example 5

Method 1
```
   573
  ×205
  ────
  2865
   000  ← We can write the second partial product as 000 to
  1146    indicate the multiplication by 0 in the tens place.
  ─────
  117465
```

Let's look at a second approach to the problem.

Method 2
```
   573
  ×205
  ────
  2865
  11460  ← We can write a single 0 as our second step. If we
  ─────    place the third partial product on the same line, that
  117465   product will be shifted *two* places left, indicating
           that we are multiplying by 2 hundreds.
```

Since this second method is more compact, it is usually used.

CHECK YOURSELF 5

Multiply.

```
  489
 ×304
```

CHECK YOURSELF ANSWERS

1. 713.
2.
```
   5  ← Carry (tens)
   4  ← Carry (ones)
   38
  ×76
  ───
  228
  266
  ────
  2888
```
3. 39,812.   4. 87,648.   5. 148,656.

2.4 Exercises

Multiply.

1. 47
   ×38

2. 58
   ×49

3. 98
   ×57

4. 75
   ×68

5. 235
   × 49

6. 327
   × 59

7. 2364
   × 67

8. 4075
   × 84

9. 315
   ×243

10. 124
   ×225

11. 345
   ×267

12. 639
   ×358

13. 547
　　×203

14. 668
　　×305

15. 2458
　　× 135

16. 3219
　　× 207

17. 1208
　　× 305

18. 2407
　　× 521

19. 2534
　　×3106

20. 3158
　　×2043

21. Find the product of 203 and 57.

22. What is the product of 588 and 32?

23. What is the product of 135 and 507?

24. Find the product of 2409 and 68.

Skillscan (Section 1.6)
Round each of the numbers as indicated.

a. 78 to the nearest ten

b. 533 to the nearest hundred

c. 92 to the nearest ten

d. 689 to the nearest hundred

e. 2541 to the nearest hundred

f. 23,745 to the nearest thousand

Answers
1. 1786　3. 5586　5. 11,515　7. 2364
　　　　　　　　　　　　　　　　× 67
　　　　　　　　　　　　　　　　16548
　　　　　　　　　　　　　　　　14184
　　　　　　　　　　　　　　　　158388
9. 76,545　11. 92,115　13. 547
　　　　　　　　　　　　　　×203
　　　　　　　　　　　　　　1641
　　　　　　　　　　　　　　10940
　　　　　　　　　　　　　　111041
15. 331,830
17. 368,440　19. 7,870,604　21. 11,571　23. 68,445　a. 80　b. 500　c. 90　d. 700　e. 2500
f. 24,000

2.4 Supplementary Exercises

Multiply.

1. 45
　×36

2. 93
　×48

3. 587
　× 47

4. 2345
　× 85

5. 457
　×348

6. 397
　×503

7. 933
　×566

8. 1357
　× 236

9. 2493
    × 307

10. 2536
    ×2048

11. Find the product of 708 and 29.

12. What is the product of 456 and 205?

2.5 Multiplying by Numbers Ending in Zero

OBJECTIVES
1. To multiply by whole numbers ending in 0
2. To use rounding to estimate a product

There are some shortcuts that will let you simplify your work when you are multiplying by a number that ends in 0. Let's see what we can discover by looking at some examples.

Example 1

```
   67
  ×10
  670
```
$10 \times 67 = 670$

Example 2

```
    537
   ×100
  53700
```
$100 \times 537 = 53{,}700$

Example 3

```
     489
   ×1000
  489000
```
$1000 \times 489 = 489{,}000$

Do you see a pattern? Rather than writing out the multiplication, there is an easier way!

We call the numbers 10, 100, 1000, and so on, *powers of 10*.

We'll talk about powers of 10 in more detail later in Section 2.8.

> When a whole number is multiplied by a power of 10, the product is just that number followed by as many zeros as there are in the power of 10.

Example 4

There are two zeros in 100. $45 \times 100 = 4500$ — Two zeros

There are three zeros in 1000. $1000 \times 67 = 67{,}000$ — Three zeros

10,000 has four zeros. $543 \times 10{,}000 = 5{,}430{,}000$ — Four zeros

This method works because multiplying by a power of 10 is the same as multiplying by 10 a certain number of times, each time adding a 0 to the product.

CHECK YOURSELF 1

Multiply.

(1) 100×58
(2) 395×1000

Let's see how we can use this to simplify any multiplication problem in which one of the factors ends in a zero or zeros.

Example 5

Multiply 30×56.

Solution We could write

$$\begin{array}{r} 56 \\ \times 30 \\ \hline \end{array}$$ We have lined up the ones digits as before.

This is better:

$$\begin{array}{r} 56 \\ \times 30 \\ \hline \end{array}$$ Shift 30 to the right so that the zero is *to the right* of the digits above.

$$\begin{array}{r} 56 \\ \times 30 \\ \hline 1680 \end{array}$$ Now bring down the zero and multiply 3×56 as before to write the product.

Section 2.5: Multiplying by Numbers Ending in Zero

CHECK YOURSELF 2

Multiply.

50 × 49

Example 6

Multiply 400 × 678.

Solution Write

$$\begin{array}{r} 678 \\ \times 400 \end{array}$$

Shift 400 so that the two zeros are *to the right* of the digits above.

$$\begin{array}{r} 678 \\ \times 400 \\ \hline 271200 \end{array}$$

Bring down the two zeros, then multiply 4 × 678 to find the product.

There is no mystery about why this works. We know that 400 is 4 × 100. In this method, we are multiplying 678 by 4 and then by 100, adding two zeros to the product by our earlier rule.

CHECK YOURSELF 3

Multiply.

300 × 574

Your work in this section, together with our earlier rounding techniques, provides a convenient means of using estimation to check the reasonableness of our results in multiplication. Our final example illustrates.

Example 7

Estimate the product below by rounding each factor to the nearest hundred.

Rounded

$$\begin{array}{r} 512 \rightarrow 500 \\ \times 289 \rightarrow \times 300 \\ \hline 150{,}000 \end{array}$$

You might want to now find the *actual* product and use our estimate to see if your result seems reasonable.

CHECK YOURSELF 4

Estimate the product by rounding each factor to the nearest hundred.

$$\begin{array}{r} 689 \\ \times 425 \\ \hline \end{array}$$

CHECK YOURSELF ANSWERS

1. (1) 5800 (Two zeros); (2) 395,000 (Three zeros).   2. 2450.   3. 172,200.
4. 700 × 400, or 280,000.

2.5 Exercises

Multiply.

1. 53 × 10
2. 10 × 67
3. 89 × 100

4. 1000 × 73
5. 1000 × 567
6. 3456 × 100

7. 236 × 10,000
8. 10,000 × 42
9. $\begin{array}{r} 43 \\ \times 70 \\ \hline \end{array}$

10. $\begin{array}{r} 58 \\ \times 40 \\ \hline \end{array}$
11. $\begin{array}{r} 562 \\ \times 400 \\ \hline \end{array}$
12. $\begin{array}{r} 907 \\ \times 900 \\ \hline \end{array}$

13. $\begin{array}{r} 345 \\ \times 230 \\ \hline \end{array}$
14. $\begin{array}{r} 362 \\ \times 310 \\ \hline \end{array}$
15. $\begin{array}{r} 157 \\ \times 3200 \\ \hline \end{array}$

16. $\begin{array}{r} 253 \\ \times 5300 \\ \hline \end{array}$
17. 367 × 20
18. 30 × 563

19. 249 × 300
20. 700 × 612
21. 238 × 4000

22. 134 × 2500
23. 5000 × 408
24. 3200 × 265

Estimate the following products by rounding each factor to the nearest ten.

25. 36
× 23

26. 27
× 34

27. 93
× 48

28. 74
× 57

Estimate the following products by rounding each factor to the nearest hundred.

29. 212
× 278

30. 179
× 431

31. 391
× 531

32. 729
× 481

Skillscan (Section 1.3)
Evaluate by adding inside the parentheses and then multiplying.

a. 2 × (3 + 5)

b. 3 × (5 + 7)

c. 4 × (2 + 8)

Evaluate by distributing the multiplication over the sum.

d. 2 × (3 + 5)

e. 3 × (5 + 7)

f. 4 × (2 + 8)

Answers
1. 530 **3.** 8900; 100 has *two* zeros **5.** 567,000 **7.** 2,360,000 **9.** 3010 **11.** 562
× 400
224800

13. 79,350 **15.** 502,400 **17.** 7340 **19.** 74,700 **21.** 238
× 4000
952000
23. 2,040,000 **25.** 800

27. 4500 **29.** 60,000 **31.** 200,000 **a.** 16 **b.** 36 **c.** 40 **d.** 16 **e.** 36 **f.** 40

2.5 Supplementary Exercises

Multiply.

1. 48 × 10

2. 100 × 57

3. 1000 × 63

4. 275 × 10,000

5. 93
× 70

6. 587
× 800

7. 478
× 320

8. 354
× 2700

9. 30 × 42

10. 57 × 600

11. 400 × 538

12. 2400 × 108

Estimate by rounding to the nearest ten.

13. 37
    ×22

14. 86
    ×42

Estimate by rounding to the nearest hundred.

15. 195
    ×321

16. 529
    ×278

2.6 The Order of Operations

OBJECTIVE
To evaluate an expression using the rules for the order of operations.

If multiplication is combined with addition or subtraction, you must know which operation to do first in finding the expression's value. Our first example illustrates this problem.

Example 1

$3 \times 4 + 5 = ?$

Here multiplication and addition are involved in the same expression. Which should you do first to find the answer?

(a) Multiply first   or   (b) Add first
    $12 + 5 = 17$ $3 \times 9 = 27$

The answers differ depending on which operation is done first!

Only one of these results can be correct, and we need to agree on a rule to tell us the order in which the operations should be performed. The rules are as follows.

THE ORDER OF OPERATIONS

If multiplication, addition, and subtraction are involved in the same expression, do the operations in the following order:

1. Do all multiplication in order from left to right.
2. Do all addition or subtraction in order from left to right.

Section 2.6: The Order of Operations

Example 2

By this rule, we see that strategy (a) in Example 1 was correct.

(a) $5 + 3 \times 6 = 5 + 18 = 23$ Multiply *first*, then add or subtract.
(b) $16 - 2 \times 3 = 16 - 6 = 10$
(c) $7 \times 8 - 20 = 56 - 20 = 36$
(d) $5 \times 6 + 4 \times 3 = 30 + 12 = 42$

CHECK YOURSELF 1

Evaluate.

(1) $8 + 3 \times 5$ (2) $15 \times 5 - 3$ (3) $4 \times 3 + 2 \times 6$

We now want to extend our rule for the order of operations. Let's see what happens when parentheses are involved in an expression.

Example 3

Evaluate $3 \times (4 + 5)$ We saw this expression earlier. In this case, add $4 + 5$ as the first step.

$3 \times (4 + 5) = 3 \times 9 = 27$

The following explains the rules concerning parentheses.

Do you see the importance of understanding parentheses?

$3 \times (4 + 5) \neq 3 \times 4 + 5$

The notation $\neq$ means not equal to. Check the statement above for yourself.

> **THE ORDER OF OPERATIONS**
>
> If an expression contains an operation in parentheses, do that operation first. Then follow the order of our previous rule.

Example 4

(a) Evaluate: $(12 - 5) \times 6$ Perform the subtraction in the parentheses as the first step.

$(12 - 5) \times 6 = 7 \times 6 = 42$

(b) Evaluate: $(8 + 3) \times 4 + 3$ Add $8 + 3$ to get 11 as the first step. Then multiply 11×4. As the final step, we add.

$(8 + 3) \times 4 + 3 = \underbrace{11 \times 4} + 3$
$= 44 + 3 = 47$

CHECK YOURSELF 2

Evaluate.

$5 \times (8 - 3)$

CHECK YOURSELF ANSWERS

1. (1) $8 + 3 \times 5 = 8 + 15 = 23$; (2) $15 \times 5 - 3 = 75 - 3 = 72$;
   (3) $4 \times 3 + 2 \times 6 = 12 + 12 = 24$.
2. 25.

2.6 Exercises

Evaluate.

1. $4 \times 5 + 7$
2. $5 \times 2 + 6$
3. $3 + 6 \times 4$

4. $7 - 3 \times 2$
5. $8 \times 5 - 20$
6. $7 + 4 \times 8$

7. $48 - 8 \times 5$
8. $3 \times 5 + 8$
9. $9 + 6 \times 8$

10. $8 \times 6 - 48$
11. $20 \times 6 - 5$
12. $8 \times 7 + 2$

13. $20 \times (6 - 5)$
14. $8 \times (7 + 2)$
15. $9 \times (4 + 3)$

16. $7 \times (9 - 5)$
17. $4 \times 5 + 7 \times 3$
18. $8 \times 5 - 7 \times 4$

19. $9 \times 8 - 12 \times 6$
20. $4 \times 5 + 2 \times 10$
21. $5 \times (3 + 4)$

22. $7 \times (6 + 8)$
23. $5 \times 3 + 5 \times 4$
24. $7 \times 6 + 7 \times 8$

Skillscan (Sections 1.3 and 1.7)
Find each of the following.

a. 12 more than 17

b. 9 less than 32

c. 23 increased by 15

d. The difference of 49 and 23

e. The sum of 53 and 29

f. 58 decreased by 39

Answers
1. 27   3. $3 + 6 \times 4 = 3 + 24 = 27$ 5. 20   7. 8   9. 57   11. $20 \times 6 - 5 = 120 - 5 = 115$
13. $20 \times (6 - 5) = 20 \times 1 = 20$ 15. 63   17. $4 \times 5 + 7 \times 3 = 20 + 21 = 41$ 19. 0   21. 35
23. 35   a. 29   b. 23   c. 38   d. 26   e. 82   f. 19

2.6 Supplementary Exercises

Evaluate.

1. $4 \times 6 + 5$
2. $7 + 3 \times 2$
3. $6 + 8 \times 3$
4. $36 - 6 \times 3$
5. $5 \times 4 - 2$
6. $6 \times 5 - 2$
7. $5 \times (4 - 2)$
8. $6 \times (5 - 2)$
9. $3 \times 2 + 4 \times 5$
10. $8 \times 3 - 4 \times 6$
11. $6 \times (4 + 5)$
12. $6 \times 4 + 6 \times 5$

Using Your Calculator

As we pointed out earlier, there are many differences among calculators. If any of the examples we consider in this section doesn't come out the same on the model you are using, check your operating manual for details.

Multiplication, like addition and subtraction, is easy on your calculator. (We hope you haven't been using one in this chapter so far!) Let's outline the steps of using a calculator to do a multiplication problem.

To multiply 23×37 on the calculator:

1. Press the clear key. $\boxed{C}$
2. Enter the first factor. 23
3. Press the times key. $\boxed{\times}$
4. Enter the second factor. 37
5. Press the equals key. $\boxed{=}$ Your display will now show the desired product, 851.

Display $ 851$

It is also possible to use the calculator to multiply more than two numbers in a chain of calculations.

Example 1

To multiply $3 \times 5 \times 7$, use this sequence:

$3 \;\boxed{\times}\; 5 \;\boxed{\times}\; 7 \;\boxed{=}$ When you press the times key the second time, note that the product of 3 and 5 is in the display.

Display $ 105$

The calculator can also be used to evaluate expressions involving different operation signs.

Example 2

To evaluate $3 \times 4 + 5$, use this sequence:

3 ⨯ 4 + 5 =

Display   17

Note Our examples show the results of mixed calculations using a "scientific" calculator.

The calculator has done the multiplication and then the addition. So it operates according to our rule of the previous section for the order of operations.

Let's change the order of the addition and multiplication in an expression and see what the calculator does. Consider our next example.

Example 3

To evaluate $6 + 3 \times 5$, use this sequence:

6 + 3 ⨯ 5 =

Display   21

Again the calculator does the multiplication and then the addition. Try this with yours to see if it works this way. You'll have to change the order of operations if it works strictly from left to right.

Many calculators have parentheses keys that will allow you to evaluate more complicated expressions easily. The parentheses tell the calculator to do any operations in the parentheses first according to the rule in Section 2.6.

Example 4

To evaluate $3 \times (4 + 5)$ use this sequence:

3 ⨯ (4 + 5) =

Display   27

Now the calculator does the addition in the parentheses as the first step. Then it does the multiplication.

Section 2.6: The Order of Operations

Exercises Using Your Calculator

Multiply.

1. 57
   ×89

2. 98
   ×25

3. 256
   ×508

4. 285
   ×820

5. 23,456
   × 2358

6. 18,569
   × 3286

7. $12 \times 15 \times 8 =$

8. $32 \times 5 \times 18 =$

9. $78 \times 145 \times 36 =$

10. $358 \times 39 \times 928 =$

11. $24 \times 35 \times 48 \times 36 =$

12. $37 \times 15 \times 42 \times 29 =$

Evaluate these expressions by hand, then use your calculator to verify your results.

13. $4 \times 5 - 7 =$

14. $3 \times 7 + 8 =$

15. $9 + 3 \times 7 =$

16. $6 \times 0 + 3 =$

17. $4 + 5 \times 0 =$

18. $23 - 4 \times 5 =$
    23 - 20 = 3.

19. $5 \times (4 + 7) =$

20. $8 \times (6 + 5) =$

21. $5 \times 4 + 5 \times 7 =$

22. $8 \times 6 + 8 \times 5 =$

23. A car dealer kept the following record of a month's sales. Complete the table.

MODEL	NUMBER SOLD	PROFIT PER SALE	MONTHLY PROFIT
Subcompact	38	$528	____
Compact	33	647	____
Standard	19	912	____
		MONTHLY TOTAL PROFIT	____

24. You take a job paying $1 the first day. On each following day your pay doubles. That is, on day 2 your pay is $2, on day 3 the pay is $4, and so on. Complete the table.

DAY	DAILY PAY	TOTAL PAY
1	$1	$1
2	2	3
3	4	7
4	_____	_____
5	_____	_____
6	_____	_____
7	_____	_____
8	_____	_____
9	_____	_____
10	_____	_____

Answers

1. 5073   3. 130,048   5. 55,309,248   7. 1440   9. 407,160   11. 1,451,520   13. 13   15. 30
17. 4   19. 55   21. 55
23. Monthly Profit
    $20,064
     21,351
     17,328
    $58,743

2.7 Solving Word Problems Involving Multiplication

OBJECTIVE
To solve word problems involving multiplication

Remember that it is best to write down the complete statement necessary for the solution of any word problem.

Let's review our discussion of word problems from the last chapter. Read the problem carefully. Reread it if necessary. Pick out the important facts and write them down. Know what you are being asked to find, then decide which operation or operations must be used for the solution.

As you will see, the process of solving applications is the same no matter which operation is required for the solution. In fact the four-step procedure we suggested in Section 1.9 can be applied effectively here and we restate it for your reference.

Section 2.7: Solving Word Problems Involving Multiplication

> **SOLVING WORD PROBLEMS**
> 1. Read the problem carefully to determine the given information and what you are asked to find.
> 2. Decide upon the operation or operations to be used.
> 3. Write down the complete statement necessary to solve the problem and do the calculations.
> 4. Check to make sure you have answered the question of the problem and that your answer seems reasonable.

Let's apply the procedure to our first example.

Example 1

A car rental agency orders a fleet of 37 new subcompact cars at a cost of $7258 per automobile. What will the company pay for the entire order?

Step 1 We know the number of cars and the price per car. We want to find the total cost.

Step 2 Multiplication is the best approach to the solution.

Step 3 Write

37 × $7258 = $268,546   We could, of course, *add* $7258, the cost, 37 times, but multiplication is certainly more efficient.

Step 4 The total cost of the order is $268,546. To see if this answer seems "reasonable," estimate the product by rounding:

Round
37 to 40
and
$7258 to $7000

40 × 7000 = 280,000

Our actual product does seem reasonable.

CHECK YOURSELF 1

If Richard's income from an apartment building is $1175 per month, what will his yearly income be?

Applications of multiplication come up naturally in finding the area of rectangles. Figure 1 shows a rectangle. The length of the rectangle is 4 inches (in) and the width is 3 in. The area of the rectangle is measured in terms of square inches. We can simply count to find the area, 12 square inches (sq in). However, since each of the four vertical strips contains 3 sq in, we can multiply:

Area = 4 in × 3 in = 12 sq in

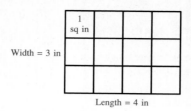

Figure 1

Note that area is always measured in *square units,* and the length and width must be in terms of the same unit of measure.

A *formula* is a mathematical rule or statement written with algebraic symbols.

In general, we can write the formula for *the area of a rectangle:* If the length of a rectangle is L units and the width is W units then the formula for the area, A, of the rectangle can be written as

$A = L \times W$ (square units)

Example 2

A tennis court has dimensions 78 ft by 36 ft. Find its area.

Solution
Area = 78 ft × 36 ft = 2808 sq ft

CHECK YOURSELF 2

A classroom has dimensions 28 ft by 22 ft. Find its area.

Some word problems may require more than one multiplication step for their solution. Consider our next example.

Example 3

Tom owns three rental units. They rent for $385 per month, and all three are occupied for an entire year. What is Tom's income for the year from the three units?

Solution We multiply $385 by 12 to find the *yearly income* for *each* unit. Then multiply by 3 to find the combined income for the three units.

Note: You could have written

$385 × 12 × 3 = $13,860

for the solution.

$385 × 12 = $4620   Tom will receive 12 monthly payments during the year for each unit.

$4620 × 3 = $13,860   There were three units.

To solve a word problem, you may also have to combine multiplication with other operations. Our final examples illustrate.

Section 2.7: Solving Word Problems Involving Multiplication

Example 4

Suzanne buys a used car on the following terms. She pays $1500 down and agrees to make payments of $80 per month for 24 months. What will be her total cost for the car?

Solution There are three pieces of information which you should write down: the amount of the down payment, the monthly payment, and the number of months that the payment must be made. Write

Do you see that two operations are necessary? We must multiply to find the total of the monthly payments. Then we add to find the total amount Suzanne paid for the car.

Cost = $1500 + 24 × $80   Note that we must multiply first and then add.
     = $1500 + $1920
     = $3420   Total cost

Example 5

A farmer harvested 58 bushels of wheat per acre from 27 acres in July. He then harvested 69 bushels per acre from 39 acres in August. What was the total harvest?

Solution We must multiply to find the harvest for each month, then add to find the total.

July: 27 × 58 bushels = 1566 bushels
August: 39 × 69 bushels = 2691 bushels
Total harvest: 4257 bushels

CHECK YOURSELF 3

Larry buys a $699 stereo system but decides to finance the system by paying $115 down and making $30 monthly payments for 24 months. How much would he have saved by paying cash?

CHECK YOURSELF ANSWERS

1. $14,100.   2. Area = 616 sq ft.   3. $136.

2.7 Exercises

1. A painter can finish 850 sq ft/h. What can he complete in an 8-hour shift?

2. A tennis player burns an average of 275 cal/h. How many calories will be burned in playing a 3-hour match?

3. Jan saves $125 per month in a payroll savings plan. What amount will she save in 1 year?

4. The gas tank of a Honda holds 12 gal. If the car gets 42 mi/gal for highway driving, how far will it travel on one tankful of gas?

5. Sharon shot 18 rolls of 24-exposure slide film on a trip to England. How many pictures did she take?

6. A convoy company can transport eight new cars on one of its trucks. If 34 truck shipments were made in one week, how many cars were shipped?

7. A computer printer can print 40 mailing labels per minute. How many labels can be printed in one hour?

8. A rectangular parking lot has 14 rows of parking spaces and each row contains 24 spaces. How many cars can be parked in the lot?

9. A college purchased 28 computers for a new laboratory at a cost of $879 per computer. What was the total cost of the order?

10. A ream of paper is 500 sheets. If 29 reams of paper were used in a copy machine during one week, how many copies were made?

11. If sound waves travel at a rate of 1088 ft/s and you hear thunder 23 seconds after seeing a lightning flash, how far away did the lightning flash?

12. Margaret has a job which pays $356 per week. What will she earn in one year? (Use 52 weeks per year.)

13. A rectangular house has dimensions 35 ft by 45 ft. What is the floor area of the house?

14. A sheet of paper has dimensions 28 cm by 21 cm. Find the area of one side of the sheet.

15. A classroom has dimensions 24 ft by 28 ft. What is its area?

16. Tile for a kitchen counter will cost $7 per square foot to install. If the counter is 12 ft by 3 ft, what will the tile cost?

17. You wish to cover a floor 4 yd by 5 yd with a carpet costing $13 per square yard. What will the carpeting cost?

Section 2.7: Solving Word Problems Involving Multiplication

18. What will be the cost of glass to replace a window which measures 6 ft by 5 ft if the glass costs $3 per square foot?

19. You buy a stereo for $125 down and agree to pay $25 per month for 12 months on the balance that is owed. What is your total cost?

20. Mr. Ramsey owns two rental units. During one year the first unit was rented for 9 months at $245 per month. The other was occupied for 11 months at $285 per month. What was the income from the two units for the year?

21. A shopping center has two rectangular parking lots. One is 250 ft by 225 ft, and the other is 300 ft by 275 ft. What is the total parking area?

22. A television set is advertised for $399 cash or $39 for 12 months on a credit plan. How much will you save by paying cash?

23. Cassie worked for the first 22 weeks of the year for a salary of $412 per week. At that point she received a raise of $70 per week and continued at that new salary for the remainder of the year. What amount did she earn for the year?

24. A restaurant has 16 booths that will seat four people each, 10 tables that will seat six people, and 6 tables that will seat eight people. How many people can be seated in the restaurant if every space is full?

Use your calculator for the following problems.

25. The number of different ways the teams in a 16-team basketball tournament can finish 1st, 2nd, 3rd, and 4th is given by 16 × 15 × 14 × 13. Calculate the number of ways the top four places can be decided.

26. Suppose the tournament has 32 teams. In how many ways can the top four places be decided?

27. There are 60 min in 1 hour, 24 hours in a day, and 365 days in a year. How many minutes are there in a year?

28. There are 60 seconds in 1 min. How many seconds are there in a year?

Skillscan

Find the following products.

a. $2 \times 2 \times 2$

b. $3 \times 3 \times 3 \times 3$

c. $5 \times 5 \times 5$

d. $9 \times 9 \times 9$

e. $4 \times 4 \times 4 \times 4$

f. $2 \times 2 \times 2 \times 2 \times 2$

Answers

1. 6800 sq ft **3.** $1500 **5.** 432 pictures **7.** 2400 labels **9.** $24,612 **11.** 25,024 ft
13. 1575 sq ft **15.** 672 sq ft **17.** $260 **19.** $425 **21.** 138,750 sq ft **23.** $23,524
25. 43,680 ways **27.** 525,600 min **a.** 8 **b.** 81 **c.** 125 **d.** 729 **e.** 256 **f.** 32

2.7 Supplementary Exercises

1. A student averages $485 per month for living expenses. What will she spend during the 9-month school year?

2. The average person in this country eats 270 eggs in 1 year. How many eggs will a family of four eat in 1 year?

3. An auditorium has 32 rows of seats. If there are 24 seats in each row, how many seats are there in the auditorium?

4. Paul shot 21 rolls of 36-exposure film on a trip to Europe. How many pictures did he take?

5. A school purchases 18 television sets for $369 each. What is the cost of the order?

6. How many hours are there in a 365-day year? (Use 24 hours per day.)

7. A television screen is 18 in by 14 in. What is its area?

8. A football field has dimensions 55 yd by 120 yd. What is the area of the field?

9. Jane wants to cover the floor of a rectangular room with linoleum costing $7 per square yard. If the dimensions of the room are 3 yd by 5 yd, what will be the cost of the linoleum?

10. The overall dimensions of a plot of land are 225 ft by 150 ft. A garden measuring 45 ft by 70 ft is planted. How much land remains?

11. In 1 year, David works at one job for 4 months, earning $1350 per month. He then works 7 months on a new job, this time earning $1525 per month. What are his total earnings for the year?

12. A VW Rabbit is advertised for sale at $8850. You agree to purchase the car by making monthly payments of $194 for 5 years. What would you save by paying cash for the car rather than making monthly payments?

2.8 Powers of Whole Numbers

OBJECTIVES
1. To use exponent notation
2. To evaluate expressions involving powers of whole numbers

Earlier we described multiplication as a shorthand for repeated addition. There is also a shorthand for repeated multiplication. It uses the idea of *powers of a whole number*.

Example 1

$3 + 3 + 3 + 3 = 4 \times 3$ Here repeated addition is written as multiplication.

Example 2

$3 \times 3 \times 3 \times 3$ can be written as 3^4. ⟵ This is read as "three to the fourth power."

In this case, repeated multiplication is written as the power of a number.

In Example 2, 3 is the *base* of the expression and the raised number, 4, is the *exponent* or *power*.

$$3^4 = \underbrace{3 \times 3 \times 3 \times 3}_{4 \text{ factors}}$$

Base ↗   Exponent or power ↖

René Descartes, a French philosopher and mathematician, is generally credited with first introducing our modern exponent notation (about 1637).

> The *exponent* tells us the number of times the base is to be used as a factor.

Example 3

2^5 is read "two to the fifth power."

$2^5 = 2 \times 2 \times 2 \times 2 \times 2 = 32$

Here 2 is the base and 5 is the exponent.

5 times

2^5 tells us to use 2 as a factor 5 times. The result is 32.

Example 4

$5^3 = 5 \times 5 \times 5 = 125$ Use three factors of 5

$5^3 = 125$ while $5 \times 3 = 15$

Be Careful! 5^3 is *entirely different* from 5×3.

5^3 is read "five to the third power" or "five cubed."

Example 5

$8^2 = 8 \times 8 = 64$ Use two factors of 8

8^2 is read "eight to the second power" or "eight squared."

CHECK YOURSELF 1

Evaluate.

(1) 6^2 (2) 2^4

We will need two special definitions.

> A whole number raised to the first power is just that number.
>
> For any whole number a, $a^1 = a$.

Example 6

$5^1 = 5$
$3^1 = 3$

This definition may look a bit strange. Don't concern yourself at this point. Later, in algebra, you will see that we must define the 0 exponent this way to be consistent with the laws for exponents.

> A whole number, other than 0, raised to the zero power is 1.
>
> For any whole number a, where $a \neq 0$, $a^0 = 1$.

Example 7

$8^0 = 1$

$4^0 = 1$

CHECK YOURSELF 2

Evaluate.

(1) 7^0 (2) 7^1

We talked about *powers of 10* earlier when we multiplied by numbers that end in zero. Since the powers of 10 have a special importance, let's list some of them.

Note that 10^3 is just a 1 followed by *three zeros*.

10^5 is a 1 followed by *five zeros*.

$10^0 = 1$
$10^1 = 10$
$10^2 = 10 \times 10 = 100$
$10^3 = 10 \times 10 \times 10 = 1000$
$10^4 = 10 \times 10 \times 10 \times 10 = 10,000$
$10^5 = 10 \times 10 \times 10 \times 10 \times 10 = 100,000$

Do you see why the powers of 10 are so important?

Archimedes (about 250 B.C.) reportedly estimated the number of grains of sand in the universe to be 10^{63}. This would be a 1 followed by 63 zeros!

> The powers of 10 correspond to the place values of our number system, ones, tens, hundreds, thousands, and so on.

This is what we meant earlier when we said that our number system was based on the number 10.

To deal with powers of numbers in an expression, we must again extend our rule from Section 2.6 for the order of operations.

> **THE ORDER OF OPERATIONS**
>
> Mixed operations in an expression should be done in the following order:
>
> 1. Do any operations inside parentheses.
> 2. Evaluate any powers.
> 3. Do all multiplication in order from left to right.
> 4. Do all addition or subtraction in order from left to right.

Example 8

$4^3 + 5 = 64 + 5 = 69$

CHECK YOURSELF 3

Evaluate $2 + 3^3$

Example 9

$4 \times 2^3 = 4 \times 8 = 32$ Find 2^3 as the first step and then multiply by 4.

CHECK YOURSELF 4

Evaluate 3×3^2

Example 10

$(2 + 3)^2 + 4 \times 3 = 5^2 + 4 \times 3$ Perform the operation in parentheses first, $2 + 3 = 5$. Then evaluate the power, $5^2 = 25$.

$ = 25 + 4 \times 3$ Multiply.

$ = 25 + 12$ Add.

$ = 37$

CHECK YOURSELF 5

Evaluate $4 + (8 - 5)^2$

CHECK YOURSELF ANSWERS

1. (1) $6^2 = 6 \times 6 = 36$;   (2) $2^4 = 2 \times 2 \times 2 \times 2 = 16$.
2. (1) 1;   (2) 7.
3. $2 + 3^3 = 2 + 27 = 29$.   4. $3 \times 3^2 = 3 \times 9 = 27$.
5. $4 + (8 - 5)^2 = 4 + 3^2 = 4 + 9 = 13$.

2.8 Exercises

Evaluate.

1. 3^2
2. 2^3
3. 2^4
4. 5^2

5. 8^3
6. 3^5
7. 1^5
8. 4^4

9. 5^1
10. 6^0
11. 9^0
12. 7^1

Section 2.8: Powers of Whole Numbers   99

13. 10^3 **14.** 10^2 **15.** 10^6 **16.** 10^7

17. 2×4^3 **18.** $(2 \times 4)^3$ **19.** 2×3^2 **20.** $(2 \times 3)^2$

21. $5 + 2^2$ **22.** $(5 + 2)^2$ **23.** $(3 \times 2)^4$ **24.** 3×2^4

25. 2×6^2 **26.** $(2 \times 6)^2$ **27.** $14 - 3^2$ **28.** $12 + 4^2$

29. $(3 + 2)^3 - 20$ **30.** $5 + (9 - 5)^2$ **31.** $(7 - 4)^4 - 30$ **32.** $(5 + 2)^2 + 20$

Numbers such as 3, 4, and 5 are called *Pythagorean triples*, after the Greek mathematician Pythagoras (sixth century B.C.), because

$3^2 + 4^2 = 5^2$

Which of the following sets of numbers are Pythagorean triples?

33. 6, 8, 10   **34.** 6, 11, 12   **35.** 5, 12, 13

36. 7, 24, 25   **37.** 8, 16, 18   **38.** 8, 15, 17

Answers
1. 9   3. 16   5. $8^3 = 8 \times 8 \times 8 = 512$ 7. 1   9. 5   11. 1   13. 1000   15. 1,000,000
17. $2 \times 4^3 = 2 \times 64 = 128$ 19. 18   21. 9   23. 1296   25. 72   27. 5   29. 105   31. 51
33. PT   35. PT   37. Not a PT

2.8 Supplementary Exercises

Evaluate.

1. 4^2 **2.** 5^3 **3.** 3^4 **4.** 9^2

5. 4^3 **6.** 2^5 **7.** 6^1 **8.** 7^0

9. 10^4 **10.** 10^8 **11.** 3×2^3 **12.** $(3 \times 2)^3$

13. $(2 + 3)^2$ **14.** $2 + 3^2$ **15.** $(7 - 3)^2 + 4$ **16.** $(2 + 4)^2 - 30$

Using Your Calculator

y^x is a common labeling for the exponent or power key.

Many calculators have exponent or power keys that will allow you to raise a number to a power. Let's illustrate the use of this feature with an example.

Example 1

To evaluate 5^4, use the following sequence:

Remember:

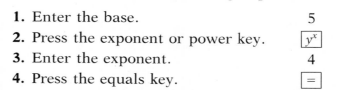

1. Enter the base.   5
2. Press the exponent or power key. $\boxed{y^x}$
3. Enter the exponent.   4
4. Press the equals key. $\boxed{=}$

The display will now show 5^4, or 625.

Display $\boxed{625}$

Again your calculator will follow our rules for the order of operations in evaluating an expression.

Example 2

To evaluate 2×5^3, use the following sequence:

2 $\boxed{\times}$ 5 $\boxed{y^x}$ 3 $\boxed{=}$

Display $\boxed{250}$

The calculator evaluates the power as the first step again using our rules for the order of operations.

The key labeled x^2 will "square a number" in a single step. This multiplies the number by itself.

Many calculators also have a special key which allows you to evaluate the square of a number in a single step.

Example 3

To evaluate 7^2, you can use the following sequence:

7 $\boxed{x^2}$

Display $\boxed{49}$

Exercises Using Your Calculator

Evaluate.

1. 9^2
2. 8^2
3. 6^3
4. 7^3

5. 6^4
6. 4^5
7. 9^0
8. 9^1

9. 6^1
10. 4^0
11. 10^4
12. 10^6

13. 3^6
14. 9^3
15. 5^8
16. 25^4

17. 3×2^5
18. 3×2^4
19. $3^4 - 20$
20. $25 + 4^3$

Answers
1. 81   3. 216   5. 1296   7. 1   9. 6   11. 10,000   13. 729   15. 390,625   17. 96   19. 61

SELF-TEST for Chapter Two

The purpose of the Self-Test is to help you check your progress and review for a chapter test in class. Allow yourself about an hour to take the test. When you are done, check your answers in the back of the book. If you missed any problems, be sure to go back and review the appropriate sections in the chapter and do the supplementary exercises provided there.

[2.1] **1.** List all the factors of 48.

2. List the first five multiples of 6.

[2.2] In Problems 3 to 7, name the property of addition and/or multiplication that is illustrated.

3. $7 \times 9 = 9 \times 7$

4. $5 \times 1 = 5$

5. $8 \times 0 = 0$

6. $3 \times (2 \times 7) = (3 \times 2) \times 7$

7. $4 \times (3 + 6) = (4 \times 3) + (4 \times 6)$

[2.3] In Problems 8 to 15, do the indicated operations.

8. 58
  $\times\ 3$

9. Find the product of 273 and 7.

[2.4] **10.** 89
  $\times 56$

11. 538
  $\times 103$

12. Find the product of 4568 and 537.

[2.5] **13.** 53
  $\times 1000$

14. 567
  $\times 400$

15. 894
  $\times 360$

In Problems 16 and 17, estimate the products by rounding each factor to the nearest hundred.

[2.5] **16.** 325
            ×468

17. 2345
       × 389

In Problems 18 to 22, evaluate the expressions.

[2.6] **18.** $5 + 6 \times 2$ **19.** $(5 + 6) \times 2$ **20.** $2 \times 8 - 3 \times 5$

21. $5 \times (7 + 8)$ **22.** $5 \times 7 + 5 \times 8$

[2.7] In Problems 23 to 26, solve each application.

23. A college lecture hall has 14 rows of seats with 18 seats per row. How many people will the room seat?

24. A truck rental firm has ordered 25 new vans at a cost of $12,350 per van. What will be the total cost of the order?

25. A rectangular field has dimensions 72 meters by 45 meters. What is its area?

26. Mr. Jackson has invested in the stocks of two companies. His dividend is $59 per month on the stock of the first company and $128 per month on the stock of the second. What is his yearly income from the two stocks?

[2.8] In Problems 27 to 30, evaluate the expressions.

27. 5^4 **28.** 8^0 **29.** 3×6^2 **30.** $(4 + 3)^2 - 40$

PRETEST for Chapter Three

Division of Whole Numbers

This pretest will point out any difficulties you may be having in dividing whole numbers.
Do all the problems. Then check your answers on the following page.

1. In the division problem shown, identify the divisor, the dividend, the quotient, and the remainder.

    $$\begin{array}{r} 8 \\ 9\overline{)77} \\ \underline{72} \\ 5 \end{array}$$

2. $5 \div 1 =$

3. $9 \div 0 =$

Divide by long division.

4. $6\overline{)6358}$

5. $23\overline{)645}$

6. $415\overline{)23,517}$

Divide by short division.

7. $7\overline{)9823}$

Evaluate.

8. $6 + 4^2 \div 2 =$

9. Marcia purchases furniture costing $614. She agrees to pay $50 down and the balance in 12 equal monthly payments. Find the amount of each monthly payment.

10. Five company-owned cars are driven 243, 181, 357, 193, and 411 mi in 1 week. What is the average number of miles driven during the week per car?

ANSWERS TO PRETEST

For help with similar problems, turn to the section indicated.

1. Divisor, 9; dividend, 77; quotient, 8; remainder, 5; (Section 3.1)
2. 5 (Section 3.2)
3. Undefined (Section 3.2)
4. 1059, remainder 4 (Section 3.3)
5. 28, remainder 1 (Section 3.4)
6. 56, remainder 277 (Section 3.4)
7. 1403, remainder 2 (Section 3.5)
8. 14 (Section 3.6)
9. $47 (Section 3.7)
10. 277 mi (Section 3.8)

Chapter Three

Division of Whole Numbers

3.1 The Language of Division

OBJECTIVES
1. To use the language of division
2. To divide using repeated subtraction

Let's look at the fourth of the basic arithmetic operations, division. Division asks *how many times* one number is contained in another. Just as multiplication was thought of as repeated addition, division can be carried out by repeated subtraction steps.

Example 1

How many times is 7 contained in 35?

To answer the question, subtract 7 repeatedly until the difference is smaller than 7.

$$\begin{array}{r}35\\-7\\\hline 28\end{array} \quad \begin{array}{r}28\\-7\\\hline 21\end{array} \quad \begin{array}{r}21\\-7\\\hline 14\end{array} \quad \begin{array}{r}14\\-7\\\hline 7\end{array} \quad \begin{array}{r}7\\-7\\\hline 0\end{array}$$

We can see that there are five 7s in 35. This means that 35 *divided* by 7 is 5.

The division sign ÷ and the symbol ⟌ are two common ways to indicate division.

The symbol ⟌ was used first in the 1500s. The division sign ÷ was introduced about a hundred years later.

Example 2

A third way to indicate division will be considered later in the text. It is the fraction bar, giving us:

$$\frac{35}{7} = 5$$

35 divided by 7 can be written as

$$35 \div 7 = 5 \quad \text{or} \quad 7\overline{)35}^{\,5}$$

The number being divided (35 in our example) is called the *dividend*. The number we are dividing by (7 in the example) is the *divisor*. The result of the division, here 5, is the *quotient*.

107

Example 3

By repeated subtraction we can easily show that 8 is contained four times in 32.

```
  32      24      16      8
－ 8    － 8    － 8    －8
  24      16       8      0
```

As a division statement we can write

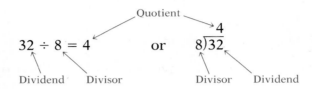

$32 \div 8 = 4$ or $8\overline{)32}$

Dividend Divisor   Divisor Dividend

In either notation we see that the divisor is 8. The dividend is 32, and the answer, 4, is the quotient.

CHECK YOURSELF 1

In $40 \div 5 = 8$, identify the dividend, divisor, and quotient.

Because of the special relationship between multiplication and division we say that they are *inverse* operations.

Division is closely related to multiplication. In fact, for every division statement there is a related multiplication operation.

Example 4

$32 \div 8 = 4$ because $32 = 8 \times 4$.
$36 \div 4 = 9$ because $36 = 4 \times 9$.

$7\overline{)56}^{\,8}$ because $56 = 7 \times 8$.

Remember that we can check subtraction using its inverse operation, addition. Similarly, division can be checked by multiplication.

Example 5

For a division problem to check, the *product* of the divisor and the quotient *must equal the dividend*.

(a) $7\overline{)21}^{\,3}$ Check: $7 \times 3 = 21$
(b) $48 \div 6 = 8$ Check: $6 \times 8 = 48$

CHECK YOURSELF 2

Complete the division statements and check your results.

(1) $9\overline{)45}$ (2) $28 \div 7 =$

Section 3.1: The Language of Division

Since 36 ÷ 9 = 4, we say that 36 is *exactly divisible* by 9.

In our examples so far, the product of the divisor and the quotient has been equal to the dividend. This means that the dividend is *exactly divisible* by the divisor. That is not always the case. Let's look at another example using the repeated subtraction process.

Example 6

How many times is 5 contained in 23?

Note that the remainder must be smaller than the divisor or we could subtract again.

$$\begin{array}{cccc} 23 & 18 & 13 & 8 \\ -5 & -5 & -5 & -5 \\ \hline 18 & 13 & 8 & 3 \end{array}$$

We see that 5 is contained four times in 23, but 3 is "left over."

23 is not exactly divisible by 5. The "left over" 3 is called the *remainder* in the division.

To check the division operation when a remainder is involved, we have the following rule:

Dividend = divisor × quotient + remainder

Example 7

Using the work of the previous example, we can write

$5\overline{)23}$ quotient 4 with remainder 3

An easier way to write the result is

$5\overline{)23}$ 4 r3   The "r" stands for remainder.

To apply our previous rule, we have

Note that the rule for the order of operations must be used.

Dividend ⟶ 23 = 5 × 4 + 3 ⟵ Remainder
             (Divisor) (Quotient)
23 = 20 + 3
23 = 23   The division checks.

CHECK YOURSELF ANSWERS

1. Dividend, 40; divisor, 5; quotient, 8.
2. (1) $9\overline{)45}$ with quotient 5; check: 9 × 5 = 45; (2) 28 ÷ 7 = 4; check: 7 × 4 = 28.

3.1 Exercises

1. If 48 ÷ 8 = 6, 8 is the _____, 48 is the _____, and 6 is the _____.

2. In the statement $5\overline{)45}^{\,9}$, 9 is the _____, 5 is the _____, and 45 is the _____.

3. Find 36 ÷ 9 by repeated subtraction.

4. Find 40 ÷ 8 by repeated subtraction.

5. If 30 ÷ 6 = 5, write the related multiplication statement.

   30 = 6 · 5 + 0
   dividend = divisor · quotient + Remainder

6. If 35 ÷ 5 = 7, write the related multiplication statement.

   35 = 7 · 5 + 0

7. Check: $4\overline{)35}^{\,8\ r3}$

8. Check: $7\overline{)46}^{\,6\ r4}$

Do the indicated division and check your results.

9. 35 ÷ 7   5
10. $6\overline{)36}$
11. $5\overline{)40}$
12. 54 ÷ 9

13. 21 ÷ 3   7
14. $6\overline{)42}$
15. $7\overline{)63}$
16. $4\overline{)32}$

17. 56 ÷ 8   7
18. 75 ÷ 5
19. $7\overline{)84}$
20. 49 ÷ 7

21. 16 ÷ 3
22. $5\overline{)28}$
23. 27 ÷ 8
24. 40 ÷ 7

25. $7\overline{)39}$
26. $8\overline{)50}$
27. $5\overline{)43}$
28. 40 ÷ 9

29. $9\overline{)65}$
30. $6\overline{)51}$
31. 57 ÷ 8
32. 90 ÷ 8

   $8\overline{)58}$

Skillscan (Section 2.2)
Find the following products.

a. 8×1 **b.** 1×9 **c.** 0×7

d. 5×0 **e.** 10×1 **f.** 12×0

Answers
Solutions for the even-numbered exercises are provided in the back of the book.
1. Divisor, dividend, quotient   3. 4   5. $30 = 6 \times 5$ 7. $4 \times 8 + 3 = 32 + 3 = 35$ 9. 5   11. 8
13. 7   15. 9   17. 7   19. 12   21. 5 r1   23. 3 r3   25. 5 r4   27. 8 r3   29. 7 r2   31. 7 r1
a. 8   **b.** 9   **c.** 0   **d.** 0   **e.** 10   **f.** 0

3.1 Supplementary Exercises

1. If $36 \div 9 = 4$, 9 is the _____, 36 is the _____, and 4 is the _____.

2. Find $35 \div 5$ by repeated subtraction.

3. If $42 \div 7 = 6$, write the related multiplication statement.

4. Check: $8\overline{)39}$ with quotient 4 r7

Do the indicated division and check your results.

5. $45 \div 9$ **6.** $8\overline{)48}$ **7.** $4\overline{)28}$ **8.** $72 \div 8$

9. $7\overline{)56}$ **10.** $54 \div 6$ **11.** $5\overline{)39}$ **12.** $51 \div 8$

13. $43 \div 7$ **14.** $37 \div 9$ **15.** $75 \div 9$ **16.** $7\overline{)65}$

3.2 Division with Zero and One

OBJECTIVE
To perform division involving 0 and 1

The relationship between division and multiplication allows us to illustrate some special division results. First:

For any whole number a, $a \neq 0$,

$a \div a = 1$

In words: Any whole number (except 0) divided by itself is 1.

Since the related multiplication statement is

$a = a \times 1$

Example 1

$8 \div 8 = 1$ because $8 = 8 \times 1$.

Here is our second result.

For any whole number a,

$a \div 1 = a$

In words: Any whole number divided by 1 is just that number.

Since the related multiplication statement is

$a = 1 \times a$

Example 2

$6 \div 1 = 6$ because $6 = 1 \times 6$.

CHECK YOURSELF 1

(1) $9 \div 9 =$ \hspace{2cm} (2) $7 \div 1 =$

We must be careful when zero is involved in a division problem. Again there are two special cases. First:

For any whole number a, $a \neq 0$,

$0 \div a = 0$

In words: Zero divided by any whole number (except 0) is 0.

Example 3

Since $0 = a \times 0$

$0 \div 5 = 0$ because $0 = 5 \times 0$.

CHECK YOURSELF 2

(1) $0 \div 7 =$ (2) $0 \div 12 =$

Now our second case is when zero is the *divisor*. Here we have a special problem. Our next example illustrates.

Example 4

$8 \div 0 = ?$ This means that $8 = 0 \times ?$

Can 0 times some number ever be 8? From our multiplication facts, the answer is *no!* There is no answer to this problem.

Let's look at another example of division involving zero.

Example 5

$0 \div 0 = ?$

Suppose that we write

In this case, both $0 = 0 \times 5$ and $0 = 0 \times 9$ are *true* statements.

$0 \div 0 = 5$ Because $0 = 0 \times 5$

or

$0 \div 0 = 9$ Because $0 = 0 \times 9$

Do you see the problem? We now have two answers to the same question. In fact, any number would work as a quotient. There is *no* specific answer.

From Examples 4 and 5, we see that 0 presents a special problem in division. To solve this problem we agree to the following rule: *Division by zero is not allowed.* We say that:

> For any whole number a,
>
> $a \div 0$
>
> is undefined.
>
> In words: Division by zero is undefined.

Example 6

$4 \div 0$ is undefined.
$0 \div 0$ is undefined.

CHECK YOURSELF 3

(1) $4 \div 0 =$ (2) $15 \div 0 =$

CHECK YOURSELF ANSWERS

1. (1) 1; (2) 7. **2.** (1) 0; (2) 0. **3.** (1) Undefined; (2) undefined.

3.2 Exercises

Divide if possible.

1. $5 \overline{)5}$
2. $9 \div 1$
3. $0 \div 5$
4. $7 \overline{)7}$

5. $5 \overline{)5}$
6. $1 \overline{)10}$
7. $1 \overline{)6}$
8. $5 \div 0 =$

9. $9 \div 9$
10. $1 \overline{)8}$
11. $4 \div 0 =$ undefined
12. $12 \div 0 =$ Unde.

13. $8 \overline{)8}$
14. $0 \div 7 = 0$
15. $8 \div 1$
16. $3 \div 0 =$ unde

17. $0 \div 6 = 0$
18. $6 \div 6$
19. $10 \div 1$
20. $0 \div 8 = 0$

Skillscan (Section 1.8)
Subtract.

a. 45
   −40

b. 53
   −48

c. 73
   −64

d. 61
   −56

e. 41
   −35

f. 62
   −54

Answers
1. 1 **3.** 0 **5.** 1 **7.** 6 **9.** 1 **11.** Undefined **13.** 1 **15.** 8 **17.** 0 **19.** 10 **a.** 5 **b.** 5
c. 9 **d.** 5 **e.** 6 **f.** 8

3.2 Supplementary Exercises

Divide if possible.

1. $1\overline{)5}$

2. $8 \div 1$

3. $7 \div 0 =$ undefi

4. $6\overline{)6}$

5. $0 \div 4 = 0$

6. $9 \div 0 =$ undef

7. $8 \div 8$

8. $7 \div 7$

9. $7 \div 1$

10. $0 \div 5 = 0$

3.3 Long Division with One-Digit Divisors

OBJECTIVE
To divide by single-digit numbers using long division

With larger numbers, repeated subtraction is just too time-consuming to be practical.

It is easy to divide when small whole numbers are involved, since much of the work can be done mentally. In working with larger numbers, we turn to a process called *long division*. This is a shorthand method for performing the steps of repeated subtraction.

To start, let's look at an example in which we subtract multiples of the divisor.

Example 1

Divide 176 by 8.

Solution Since 20 eights are 160, we know that there are at least 20 eights in 176.

Step 1 Write

```
          20
       8)176
20 eights→160
           16
```

Subtracting 160 is just a shortcut for subtracting eight 20 times.

After subtracting the 20 eights or 160, we are left with 16. There are 2 eights in 16, and so we continue.

Step 2

$$\left.\begin{array}{r} 2 \\ 20 \end{array}\right\} 22$$

Adding 20 and 2 gives us the quotient 22.

$$\begin{array}{r} 8\overline{)176} \\ \underline{160} \\ 16 \end{array}$$

2 eights ⟶ $\underline{16}$
0

Subtracting the 2 eights, we have a 0 remainder. So

$176 \div 8 = 22$

The next step is to simplify this repeated subtraction process one step further. The result will be the long division method.

Example 2

Divide 358 by 6.

Solution The dividend is 358. We look at the first digit, 3. We cannot divide 6 into 3, and so we look at the *first two digits*, 35. There are 5 sixes in 35, and so we write 5 above the tens digit of the dividend.

$$\begin{array}{r} 5 \\ 6\overline{)358} \end{array}$$

When we place 5 as the tens digit, we really mean 5 tens or 50.

Now multiply 5 × 6, place the product below 35, and subtract.

$$\begin{array}{r} 5 \\ 6\overline{)358} \\ \underline{30} \\ 5 \end{array}$$

We have actually subtracted 50 sixes (300) from 358.

To continue the division, we bring down 8, the ones digit of the dividend.

$$\begin{array}{r} 5 \\ 6\overline{)358} \\ \underline{30}\downarrow \\ 58 \end{array}$$

Now divide 6 into 58. There are 9 sixes in 58, and so 9 is the ones digit of the quotient. Multiply 9 × 6 and subtract to complete the process.

$$\begin{array}{r} 59 \\ 6\overline{)358} \\ \underline{30} \\ 58 \\ \underline{54} \\ 4 \end{array}$$

We now have:
$358 \div 6 = 59$ r4

Section 3.3: Long Division with One-Digit Divisors

To check:

$358 = 6 \times 59 + 4$ Verify that this is true and that the division checks.

CHECK YOURSELF 1
Divide.

$7 \overline{)453}$

Let's work through an example with a slight difference.

Example 3

Divide 863 by 7.

Solution We look at the first digit of the dividend, 8. In this case we can divide 7 into 8. Place 1 in the quotient above the 8, multiply, and subtract.

$$\begin{array}{r} 1 \\ 7{\overline{\smash{\big)}\,863}} \\ \underline{7} \\ 1 \end{array}$$

By placing 1 in the hundreds place, we are saying that there are at least 100 sevens in 863.

Now bring down 6, the next digit of the dividend. There are 2 sevens in 16, and so 2 is the tens digit of the quotient. Multiply and subtract as before.

$$\begin{array}{r} 12 \\ 7{\overline{\smash{\big)}\,863}} \\ \underline{7}\downarrow \\ 16 \\ \underline{14} \\ 2 \end{array}$$

By placing 2 in the tens place, we remove 20 more sevens.

Bring down 3, the last digit of the dividend, and complete the division.

$$\begin{array}{r} 123 \\ 7{\overline{\smash{\big)}\,863}} \\ \underline{7} \\ 16 \\ \underline{14}\downarrow \\ 23 \\ \underline{21} \\ 2 \end{array}$$

The quotient is 123 r2.
The check:
$863 = 7 \times 123 + 2$

CHECK YOURSELF 2

Divide.

6)952

Here is an example in which a special problem comes up during the division.

Example 4

Divide 3278 by 8.

Solution We start by dividing 8 into the first two digits of the dividend. Multiply 4×8 and subtract the product.

```
    4
8)3278
   32
    0
```

We continue by bringing down 7, the tens digit of the dividend. Do you see the problem? We cannot divide 8 into 7, so we place a 0 in the quotient.

```
   40
8)3278
   32 ↓
    7
```

To continue, we bring down 8, the last digit of the dividend, and complete the process

```
   409
8)3278
   32 ↓
    78
    72
     6
```

The quotient is 409 r6.

Check: $3278 = 8 \times 409 + 6$

CHECK YOURSELF 3

Divide 5641 by 7.

Section 3.3: Long Division with One-Digit Divisors

CHECK YOURSELF ANSWERS

1. $\begin{array}{r} 64 \\ 7\overline{)453} \\ \underline{42} \\ 33 \\ \underline{28} \\ 5 \end{array}$ We have: $453 \div 7 = 64\ r5$

2. $158\ r4$

3. $\begin{array}{r} 805 \\ 7\overline{)5641} \\ \underline{56} \\ 41 \\ \underline{35} \\ 6 \end{array}$ We have: $5641 \div 7 = 805\ r6$

3.3 Exercises

Divide, and check your work.

1. $5\overline{)83}$
2. $9\overline{)78}$
3. $7\overline{)93}$
4. $6\overline{)88}$

5. $3\overline{)162}$
6. $4\overline{)232}$
7. $8\overline{)293}$
8. $7\overline{)346}$

9. $5\overline{)738}$
10. $4\overline{)953}$
11. $8\overline{)975}$
12. $7\overline{)884}$

13. $6\overline{)649}$
14. $7\overline{)758}$
15. $8\overline{)3136}$
16. $5\overline{)4938}$

17. $8\overline{)5438}$
18. $9\overline{)3527}$
19. $4\overline{)5804}$
20. $6\overline{)6703}$

21. $5\overline{)8643}$
22. $4\overline{)6490}$
23. $7\overline{)2153}$
24. $8\overline{)3268}$

25. $8\overline{)22{,}153}$
26. $5\overline{)43{,}287}$
27. $7\overline{)82{,}013}$
28. $6\overline{)73{,}108}$

Skillscan (Section 1.6)
Round each number to the indicated place.

a. 87 to the nearest ten

b. 134 to the nearest ten

c. 758 to the nearest hundred

d. 1225 to the nearest hundred

Answers

1. 16 r3   **3.** 13 r2   **5.** 54   **7.** 36 r5   **9.**
$$5\overline{)738} \begin{array}{l}147\end{array}$$
```
      147
   5)738
     5
     ‾
     23
     20
     ‾
      38
      35
      ‾
       3
```
738 ÷ 5 = 147 r3
Check:
738 = 5 × 147 + 3
11. 121 r7   **13.** 108 r1   **15.** 392

17. 679 r6   **19.** 1451   **21.** 1728 r3   **23.**
```
      307
   7)2153
     21
     ‾
      53
      49
      ‾
       4
```
25. 2769 r1   **27.** 11,716 r1   **a.** 90   **b.** 130

c. 800   **d.** 1200

3.3 Supplementary Exercises

Divide and check your work.

1. $4\overline{)93}$ 2. $5\overline{)83}$ 3. $5\overline{)485}$ 4. $3\overline{)654}$

5. $8\overline{)953}$ 6. $7\overline{)894}$ 7. $4\overline{)831}$ 8. $8\overline{)2704}$

9. $9\overline{)5327}$ 10. $6\overline{)7354}$ 11. $9\overline{)2783}$ 12. $7\overline{)3545}$

13. $8\overline{)23,410}$ 14. $6\overline{)19,059}$

3.4 Divisors with More than One Digit

OBJECTIVE
To divide any two whole numbers using long division

Long division becomes a bit more complicated when we have a two-digit divisor. It is now a matter of trial and error. We round the divisor and dividend to form a *trial divisor and a trial dividend*. We then estimate the proper quotient and must determine whether or not our estimate was correct.

Example 1

Divide

$38\overline{)293}$

Think: $4\overline{)29}^{\,7}$

Round the divisor and dividend to the nearest ten. 38 is rounded to 40, and 293 is rounded to 290. The trial divisor is then 40, and the trial dividend is 290.

Now look at the nonzero digits in the trial divisor and dividend. They are 4 and 29. We know that there are 7 fours in 29, and so 7 is our first estimate of the quotient. Now let's see if 7 works.

$$\begin{array}{r} 7 \\ 38\overline{)293} \\ \underline{266} \\ 27 \end{array}$$ ←— Your estimate

Multiply 7 × 38. The product, 266, is less than 293, and so we can subtract.

The remainder, 27, is less than the divisor, 38, and so the process is complete.

$293 \div 38 = 7\ r27$

Check: $293 = 38 \times 7 + 27$ You should verify that this statement is true.

CHECK YOURSELF 1

$57\overline{)482}$ 8 r 26

Since this process is based on estimation, we can't expect our first guess to always be right.

Example 2

Divide

$46\overline{)342}$

Round to the nearest ten to form the trial divisor, 50, and the trial dividend, 340.

Using the nonzero digits of our trial divisor and dividend, think, "How many fives are in 34?" There are 6, and this is our first estimate.

$$\begin{array}{r} 6 \\ 46\overline{)342} \end{array}$$

Now multiply 6 × 46 and subtract.

```
      6
46)342
   276      Do you see the problem? The remainder, 66,
    66      is larger than the divisor, 46.
```

Since the remainder was larger than the divisor after our first guess, we must use a *larger quotient*. We will now try 7 as our quotient.

```
      7
46)342
   322
    20      We can complete the problem.
```

342 ÷ 46 = 7 r20

Check: 342 = 46 × 7 + 20

CHECK YOURSELF 2

Divide.

57)463

Let's look at a second example in which our estimate of the quotient must be changed.

Example 3

Divide

54)428

Think: 5)43 with 8 above

Rounding to the nearest ten, we have a trial divisor of 50 and a trial dividend of 430.

Looking at the nonzero digits, how many fives are in 43? There are 8. This is our first estimate.

```
      8
54)428
   432  ←—— Too large
```

We multiply 8 × 54. Do you see what's wrong? The product 432 is too large. We can't subtract. Our estimate of the quotient must be adjusted *downward*.

We adjust the quotient downward to 7. We can now complete the division.

```
      7
54)428
   378
    50
```

We have

428 ÷ 54 = 7 r50

Check: 428 = 54 × 7 + 50

CHECK YOURSELF 3

Divide.

63)̄557

Example 4

Divide

43)̄948

> There is a difference between this and our previous examples. Look at the first two digits of the dividend. Our divisor, 43, will divide into 94.

Rounding to the nearest ten, the trial divisor is 40 and the trial dividend is 90. (We use only the first two digits of the dividend.)
  Our first estimate of the quotient is 2.

$$\begin{array}{r} 2 \\ 43\overline{)948} \\ \underline{86} \\ 8 \end{array}$$

We bring down 8, the next digit of the dividend, and repeat the process of estimating the quotient.

$$\begin{array}{r} 22 \\ 43\overline{)948} \\ \underline{86} \\ 88 \\ \underline{86} \\ 2 \end{array}$$

948 ÷ 43 = 22 r2

Check: 948 = 43 × 22 + 2

CHECK YOURSELF 4

Divide.

37)̄859

  We saw earlier that we have to be careful when a zero appears as a digit in the quotient. Let's look at an example in which this happens with a two-digit divisor.

Example 5

Again our divisor, 32, will divide into 98, the first two digits of the dividend.

Divide

$$32 \overline{)9871}$$

Rounding to the nearest ten, the trial divisor is 30, and the trial dividend is 100. Think, "How many threes are in ten?" There are 3, and this is our first estimate of the quotient.

$$\begin{array}{r} 3 \\ 32{\overline{\smash{\big)}\,9871}} \\ \underline{96} \\ 2 \end{array}$$

Everything seems fine so far!

Bring down 7, the next digit of the quotient.

$$\begin{array}{r} 30 \\ 32{\overline{\smash{\big)}\,9871}} \\ \underline{96}\downarrow \\ 27 \end{array}$$

Now do you see the difficulty? We cannot divide 32 into 27, and so we place zero in the tens place of the quotient to indicate this fact.

We continue by bringing down 1, the last digit of the dividend.

$$\begin{array}{r} 30 \\ 32{\overline{\smash{\big)}\,9871}} \\ \underline{96}\downarrow \\ 271 \end{array}$$

Another problem develops here. We round 32 to 30 for our trial divisor, and we round 271 to 270, which is the trial dividend at this point. Our estimate of the last digit of the quotient must be 9.

$$\begin{array}{r} 309 \\ 32{\overline{\smash{\big)}\,9871}} \\ \underline{96} \\ 271 \\ \underline{288} \end{array}$$

← Too large

We can't subtract. The trial quotient must be adjusted downward to 8. We can now complete the division.

$$\begin{array}{r} 308 \\ 32{\overline{\smash{\big)}\,9871}} \\ \underline{96} \\ 271 \\ \underline{256} \\ 15 \end{array}$$

$9871 \div 32 = 308 \text{ r}15$

Check: $9871 = 32 \times 308 + 15$

Section 3.4: Divisors with More than One Digit 125

CHECK YOURSELF 5

Divide.

$43 \overline{)8857}$

We can also simplify long division problems involving three-digit divisors by using trial divisors and dividends. In this case we round those divisors and dividends to the nearest hundred.

Example 6

Divide

$205 \overline{)6585}$

Round 205 to the nearest hundred for a trial divisor of 200. To find the trial dividend, round 658 to the nearest hundred, 700.

Think: $2 \overline{)7}$ with 3 above

Using the nonzero digits, think "How many twos are in 7?" There are 3, and this is our estimate.

```
       3
205)6585
     615
      43
```
We multiply and subtract. Our estimate, 3, was correct.

Continue the process by bringing down 5, the last digit of the dividend.

```
       3
205)6585
     615
     435
```

The trial divisor is still 200, but the trial dividend is now 400. Our estimate for the last digit of the quotient is 2.

```
      32
205)6585
     615
     435
     410
      25
```
We multiply and subtract to complete the process.

$6585 \div 205 = 32 \text{ r}25$

Check: $6585 = 205 \times 32 + 25$

CHECK YOURSELF 6

Divide.

$321\overline{)7582}$

CHECK YOURSELF ANSWERS

1. 8 r26.
2. The trial divisor is 60; the trial dividend is 460.
   Let's try 7 as a quotient.   Try a larger quotient, 8.

   $$\begin{array}{r}7\\57\overline{)463}\\399\\\hline 64\end{array}$$ Too large

   $$\begin{array}{r}8\\57\overline{)463}\\456\\\hline 7\end{array}$$

3. 8 r53.   4. $859 \div 37 = 23$ r8.   5. 205 r42.   6. 23 r199

3.4 Exercises

Divide and check your results.

1. $58\overline{)345}$ 2. $39\overline{)821}$ 3. $63\overline{)429}$ 4. $49\overline{)379}$

5. $48\overline{)892}$ 6. $54\overline{)372}$ 7. $23\overline{)534}$ 8. $67\overline{)939}$

9. $45\overline{)2367}$ 10. $53\overline{)3480}$ 11. $34\overline{)8748}$ 12. $27\overline{)9335}$

13. $42\overline{)7952}$ 14. $53\overline{)8729}$ 15. $28\overline{)8547}$ 16. $38\overline{)7892}$

17. $763\overline{)3871}$ 18. $871\overline{)4321}$ 19. $326\overline{)7564}$ 20. $229\overline{)8312}$

21. $432\overline{)8770}$ 22. $375\overline{)7610}$ 23. $454\overline{)32,751}$ 24. $527\overline{)27,563}$

25. $103\overline{)21,185}$ 26. $205\overline{)61,825}$ 27. $234\overline{)125,000}$ 28. $179\overline{)126,500}$

Skillscan (Section 3.1)
Divide.

a. $5\overline{)35}$ b. $7\overline{)63}$ c. $8\overline{)65}$

d. $9\overline{)84}$ e. $6\overline{)40}$ f. $4\overline{)39}$

Answers

1. 5 r55   3. 6 r51   5. 18 r28   7. 23 r5   9. $45\overline{)2367}$ with quotient 52:
   $$\begin{array}{r} 52 \\ 45\overline{)2367} \\ \underline{225} \\ 117 \\ \underline{90} \\ 27 \end{array}$$
   Check: $2367 = 45 \times 52 + 27$ 11. 257 r10

13. 189 r14   15. 305 r7   17. 5 r56   19. $326\overline{)7564}$ with quotient 23:
    $$\begin{array}{r} 23 \\ 326\overline{)7564} \\ \underline{652} \\ 1044 \\ \underline{978} \\ 66 \end{array}$$
    Check: $7564 = 326 \times 23 + 66$ 21. 20 r130

23. 72 r63   25. $103\overline{)21185}$ with quotient 205:
    $$\begin{array}{r} 205 \\ 103\overline{)21185} \\ \underline{206} \\ 585 \\ \underline{515} \\ 70 \end{array}$$
    Check: $21{,}185 = 103 \times 205 + 70$ 27. 534 r44   a. 7   b. 9   c. 8 r1

d. 9 r3   e. 6 r4   f. 9 r3

3.4 Supplementary Exercises

Divide, and check your results.

1. $47\overline{)392}$
2. $54\overline{)429}$
3. $36\overline{)837}$
4. $28\overline{)692}$

5. $63\overline{)3542}$
6. $45\overline{)2791}$
7. $52\overline{)7621}$
8. $31\overline{)9427}$

9. $563\overline{)2935}$
10. $218\overline{)7473}$
11. $329\overline{)8527}$
12. $434\overline{)35{,}202}$

13. $228\overline{)69{,}750}$
14. $325\overline{)275{,}000}$

3.5 Short Division

OBJECTIVE
To divide using short division

Short division is a way of simplifying the long-division process. You can use it whenever you have a single-digit divisor. Look at the following example comparing the two approaches.

Example 1

LONG DIVISION	SHORT DIVISION	
132	132	**Step 1** 3 divides into 3 one time. Write 1 in the quotient as shown.
$3\overline{)396}$	$3\overline{)396}$	**Step 2** 3 divides into 9 three times. Write 3 in the quotient.
$\underline{3}$		**Step 3** 3 divides into 6 two times. Write 2 as the final digit of the quotient.
9		
$\underline{9}$		
6		
$\underline{6}$		
0		

When you use short division, just carry out the steps mentally without writing down all the steps of the long-division process.

Example 2

Step 1 $\quad 5\overline{)375}^{\,7}$ 5 will not divide into 3, so look at the first two digits. 37 divided by 5 is 7 with a remainder of 2. Write 7 in the tens place of the quotient.

Step 2 $\quad 5\overline{)3^{2}75}^{\,75}$ Insert the remainder, 2, from the last step before the ones digit of the dividend. Then divide 5 into 25 for the last digit of the quotient.

Example 3

Divide

$6\overline{)327}$

Step 1 $\quad 6\overline{)32^{2}7}^{\,5}$ To start the short-division process, think, "32 divided by 6 is 5 with remainder 2." Insert 2 before the ones digit of the dividend.

Step 2 $\quad 6\overline{)32^{2}7}^{\,54\ r3}$ Now divide 6 into 27. The last digit of the quotient is 4, and we have a remainder of 3.

CHECK YOURSELF 1

Divide.

$8\overline{)425}$

Section 3.5: Short Division   129

Example 4

Divide

4)947

Step 1   4)9⁴7   Divide 4 into 9. Write 2 in the quotient
         (2 above, 1 before 4)   and place the remainder, 1, before
                        the next digit of the dividend.

Step 2   4)9¹4²7   Divide 4 into 14. Write 3 in the quotient
         (23 above)   with remainder 2.

Step 3   4)9¹4²7   Divide 4 into 27. The final digit of the
         (236 r3 above)   quotient is 6, and we have a
                        remainder of 3.

CHECK YOURSELF 2

Divide.

5)873

Let's look at an example in the final form we will use in practice.

Example 5

Divide

7)2298

Write: 7)2²29⁵8   Divide 7 into 22. Write 3 and place
        (328 r2 above)   the remainder, 1, before the 9. Then
                        divide 7 into 19. Write 2 and place
                        the remainder, 5, before the 8. Divide
                        7 into 58. We have 8 with remainder 2.

CHECK YOURSELF 3

Divide.

8)2583

CHECK YOURSELF ANSWERS

   1. 53 r1.   **2.** 174 r3.   **3.** 322 r7

3.5 Exercises

Divide, using short division.

1. 4)85
2. 5)78
3. 3)88
4. 6)79

5. 4)848
6. 2)486
7. 7)342
8. 3)954

9. 6)625
10. 7)918
11. 4)634
12. 6)853

13. 5)2364
14. 7)3421
15. 4)4351
16. 8)3251

17. 4)7321
18. 7)8923
19. 6)2453
20. 5)6314

21. 3)13,421
22. 4)34,293
23. 7)32,128
24. 9)28,155

Skillscan (Sections 2.6 and 2.8)

a. $2 \times 3 + 6$
b. $4 \times 5 - 10$
c. $2 + 3 \times 6$

d. $(2 + 3) \times 6$
e. $2 \times 3^2 + 5$
f. $20 - 3^2 \times 2$

Answers

1. 21 r1 3. 29 r1 5. 212 7. 7)342 with 48 r6 9. 104 r1 11. 158 r2 13. 5)2364 with 472 r4 15. 1087 r3
17. 1830 r1 19. 408 r5 21. 3)13,421 with 4473 r2 23. 4589 r5 a. 12 b. 10 c. 20 d. 30
e. 23 f. 2

3.5 Supplementary Exercises

Divide, using short division.

1. 3)42
2. 5)94
3. 3)396
4. 9)658

5. 4)438
6. 8)971
7. 5)3744
8. 8)4059

9. 3)3412
10. 4)4357
11. 7)35,263
12. 9)37,243

3.6 The Order of Operations

OBJECTIVE

To evaluate expressions involving division using the rules for the order of operations

Now that we have introduced division, we can write our rule for the order of operations in its final form.

THE ORDER OF OPERATIONS

Mixed operations in an expression should be done in the following order:

1. Do any operations inside parentheses.
2. Evaluate any powers.
3. Do all multiplication and division in order from left to right.
4. Do all addition and subtraction in order from left to right.

Example 1

Evaluate $2 \times 3 + 8 - 4$.

$$\underline{2 \times 3} + 8 - 4 \qquad \text{Do the multiplication as the first step.}$$
$$= \underline{6 + 8} - 4$$
$$= 14 - 4 \qquad \text{Then add and subtract working from left to right.}$$
$$= 10$$

So $2 \times 3 + 8 - 4 = 10$

Example 2

Evaluate $36 \div 3^2 - 4$.

$$36 \div 3^2 - 4$$
$$= \underline{36 \div 9} - 4 \qquad \text{First, } 3^2 \text{ is 9. Now divide 36 by 9.}$$
$$= 4 - 4 \qquad \text{Subtract.}$$
$$= 0$$

So $36 \div 3^2 - 4 = 0$

CHECK YOURSELF 1

Evaluate.

$24 \div 2^3 + 7$

Example 3

Evaluate $20 \div 2 \times 5$.

$\underbrace{20 \div 2}_{} \times 5$
$= 10 \times 5$ Since the multiplication and division appear next to each other, work in order from left to right. Try it the other way and see what happens!
$= 50$

So $20 \div 2 \times 5 = 50$

Example 4

Evaluate $(5 + 13) \div 6$.

$\underbrace{(5 + 13)}_{} \div 6$ Do the addition in the parentheses as the first step.
$= 18 \div 6$
$= 3$

So $(5 + 13) \div 6 = 3$

CHECK YOURSELF 2

Evaluate.

$(8 + 22) \div 5$

CHECK YOURSELF ANSWERS

1. 10   **2.** 6

3.6 Exercises

Do the indicated operations.

1. $8 \div 4 + 2$
2. $3 \times 5 + 2$
3. $24 - 6 \div 3$

4. $3 + 9 \div 3$
5. $(24 - 6) \div 3$
6. $(3 + 9) \div 3$

7. $12 + 3 \div 3$
8. $6 \times 12 \div 3$
9. $18 \div 6 \times 3$

10. $30 \div 5 \times 2$
11. $30 \div 6 - 12 \div 3$
12. $5 + 8 \div 4 - 3$

Section 3.6: The Order of Operations   133

13. $4^2 \div 2$ **14.** 2×4^3 **15.** $5^2 \times 3$

16. $6^2 \div 3$ **17.** 3×3^3 **18.** $2^5 \times 3$

19. $(3^3 + 3) \div 10$ **20.** $(2^4 + 4) \div 5$ **21.** $15 \div (5 - 3 + 1)$

22. $20 \div (3 + 4 - 2)$ **23.** $27 \div (2^2 + 5)$ **24.** $48 \div (2^3 + 4)$

Skillscan
Find each of the following.

a. The quotient of 75 and 5   **b.** The sum of 20 and 30

c. The product of 8 and 9   **d.** 48 divided by 8

e. The difference of 45 and 30   **f.** 20 more than 30

Answers
1. 4   3. $24 - 6 \div 3 = 24 - 2 = 22$ 5. 6   7. 13   9. $18 \div 6 \times 3 = 3 \times 3 = 9$ 11. 1
13. $4^2 \div 2 = 16 \div 2 = 8$ 15. 75   17. 81   19. 3   21. 5   23. $27 \div (2^2 + 5) = 27 \div (4 + 5) = 27 \div 9 = 3$
a. 15   b. 50   c. 72   d. 6   e. 15   f. 50

3.6 Supplementary Exercises

Do the indicated operations.

1. $9 + 6 \div 3$ **2.** $(9 + 6) \div 3$ **3.** $36 \div 9 \times 2$

4. $48 \div 6 \times 2$ **5.** $12 \div 3 - 15 \div 5$ **6.** $10 - 18 \div 3 + 4$

7. $6^2 \div 3$ **8.** 3×4^2 **9.** 5×2^4

10. $64 \div 4^2$ **11.** $(2^3 + 7) \div 5$ **12.** $40 \div (3^2 + 1)$

Using Your Calculator

Of course, division is easily done using your calculator. However, as we will see, some special things come up when we use a calculator to divide. First let's outline the steps of division as it is done on a calculator.

Divide $35\overline{)2380}$

1. Enter the dividend.   2380
2. Press the divide key. $\boxed{\div}$
3. Enter the divisor.   35
4. Press the equals key. $\boxed{=}$ The desired quotient is now in your display.

Display $\boxed{68}$

We mentioned some difficulties with zero earlier. Let's experiment on the calculator.

Example 1

To find $0 \div 5$, we use this sequence:

$0 \;\boxed{\div}\; 5 \;=\; \boxed{0}$

There is no problem with this. Zero divided by any whole number other than zero is just zero.

Example 2

To find $5 \div 0$, we use this sequence:

$5 \;\boxed{\div}\; 0 \;=\; \boxed{\text{ERROR}}$

You may find that you must "clear" your calculator after trying this.

If we try this sequence, the calculator gives us an error! Do you see why? Division by zero is not allowed. Try this on your calculator to see how this error is indicated.

Another special problem comes up when a remainder is involved in a division problem.

Example 3

In a previous section, we divided 293 by 38 and got 7 with remainder 27.

We will say more about this later. For now, just be aware that the calculator will not give you a remainder in the form we have been using in this chapter.

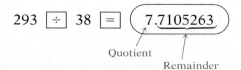

7 is the whole number part of the quotient as before.

.7105263 is the *decimal form* of the *remainder as a fraction*.

Section 3.6: The Order of Operations

The calculator can also help you combine division with other operations.

Example 4

To find 18 ÷ 2 + 3, use this sequence:

18 ÷ 2 + 3 =

Display 12   Do you see that the calculator has done the division as the first step according to our rules for the order of operations?

Example 5

To find 6 ÷ 3 × 2, use this sequence:

6 ÷ 3 × 2 =

Display 4

Again the calculator has followed our rules for the order of operations, working from *left to right* to do the division first and then the multiplication.

Exercises Using Your Calculator

Do the indicated operations.

1. 5940 ÷ 45
2. 2808 ÷ 36
3. 36,182 ÷ 79
4. 36,232 ÷ 56
5. 583,467 ÷ 129
6. 464,184 ÷ 189
7. 6 + 9 ÷ 3
8. 18 − 6 ÷ 3
9. 24 ÷ 6 × 4
10. 32 ÷ 8 × 4
11. 4368 ÷ 56 + 726 ÷ 33
12. 1176 ÷ 42 − 1572 ÷ 524
13. $3 \times 8^3 \div 12$
14. $5 \times 6^2 \div 18$
15. (18 + 87) ÷ 15
16. (89 − 14) ÷ 25

Answers
1. 132   3. 458   5. 4523   7. 9   9. 16   11. 100   13. 128   15. 7

3.7 Solving Word Problems Involving Division

OBJECTIVE
To solve word problems involving division

You have now looked at the four operations of arithmetic: addition, subtraction, multiplication, and division. So, in dealing with word problems, it becomes even more important that you read the problem carefully, and then decide which operation or operations must be used for the solution.
Let's look at some examples.

Example 1

Six people estimate that the car expenses for a trip will be $282. What will be the amount of each person's share?

Solution The needed information from the problem is the total expense, $282, and the number of people sharing that expense, here 6.
To solve the problem, think, "If the total expense is $282, we must divide that expense into 6 equal parts." Write:

$282 ÷ 6 = $47

Each person's share is $47.

Division is the necessary operation for the solution.

CHECK YOURSELF 1

There are 135 students enrolled in five algebra classes. What is the average number of students per class?

Example 2

The Barton family took a trip of 2394 mi in their new car, using 63 gal of gas. What was their gas mileage (miles per gallon)?

Solution The information given here is the total number of miles driven and the amount of gas used. We divide to find the desired gas mileage.

2394 ÷ 63 = 38 mi/gal

In reading a word problem, you can think of the word "per" as meaning division. So if we want gas mileage, miles per gallon, we divide the number of miles by the number of gallons of gas used.

CHECK YOURSELF 2

Ron flew a light plane on a trip of 1740 mi. If that trip took 12 hours, what was his average speed (miles per hour)?

Section 3.7: Solving Word Problems Involving Division

As before, we may have to combine operations to solve a word problem.

Example 3

Chuck purchases a stereo for $598. Interest charges are $98. He agrees to make one payment each month for 24 months. How much will each payment be?

Interest is the amount charged for the use of money.

Solution First, we must find the amount that Chuck owes.

$598 + $98 = $696   The interest is added to the amount of the purchase to find the total that Chuck owes.

Now, to find the amount of the monthly payments, we divide the total amount owed by 24.

$696 ÷ 24 = $29   He will make 24 equal payments

CHECK YOURSELF 3

One bag of fertilizer will cover 350 sq ft. How many bags should be purchased to cover a garden which is 70 ft by 20 ft?

CHECK YOURSELF ANSWERS

1. 27.   2. 145 mi/h.   3. Four bags.

3.7 Exercises

1. Ticket receipts for a play were $552. If the tickets were $4 each, how many tickets were purchased?

2. Construction of a fence section requires 8 boards. If you have 256 boards available, how many sections can you build?

3. A basketball player scored 476 points in 17 games. On the average, how many points did she score per game?

4. You drive 559 mi in your new Toyota, using 13 gal of gas. What is your gas mileage (miles per gallon)?

5. There were 522 students enrolled in 18 sections of English composition. What was the average enrollment per class?

6. The bill for a luncheon was $560. If 35 people attended the luncheon, what was the cost per person?

7. The homeowners along a street must share the $2030 cost of new street lighting. If there are 14 homes, what amount will each owner pay?

8. A bookstore ordered 325 copies of a text at a cost of $7800. What was the cost to the store for an individual text?

9. The records of an office show that 1702 calls were made in 1 day. If there are 37 phones in the office, on the average how many calls were placed per phone?

10. A television dealer purchased 23 sets, each the same model, for $5267. What was the cost of each set?

11. A computer printer can print 340 lines per minute. How long will it take to complete a report of 10,880 lines?

12. A train traveled 1364 mi in 22 hours. What was the average speed of the train? *Hint:* Average speed is the distance traveled divided by the time.

13. A company distributes $16,488 in year-end bonuses. If each of the 36 employees receives the same amount, what bonus will each receive?

14. A total of 17,949 cars use a toll bridge in 1 month (31 days). What is the average number of cars using the bridge in a single day?

15. You purchase a new car for $9852, make a down payment of $1500, and agree to pay off the balance in equal monthly payments for 36 months. What is the amount of each monthly payment?

16. Brad borrows $1000 and is charged $92 interest. If he arranges to pay off the loan and the interest charge in monthly payments over 1 year, what will his monthly payments be?

17. A gallon of paint should cover 450 sq ft. How many gallons must be purchased to paint a wall of a warehouse which is 135 ft long and 20 ft high?

Section 3.7: Solving Word Problems Involving Division 139

18. Sandra worked for 5 months with a salary of $1260 per month. She then received a $96 monthly raise and continued at that rate for the remainder of the year. What was her average monthly income for the year?

19. Arrange the digits 1, 3, 5, 7, and 9 in any order that you wish to form a five-digit number. Divide that number by 9. What is the remainder? Repeat the experiment two more times. What do you find?

20. Consecutive whole numbers are numbers which follow one another, like 10, 11, and 12. Choose any three consecutive whole numbers, find their product, and divide that product by 6. What is the remainder? Repeat the experiment. What do you find?

Use your calculator for the following exercises.

21. The state of Arizona has an area of 114,000 sq mi and a population of 3,078,000. What is the population per square mile?

22. The state of New York has an area of 47,800 sq mi and a population of 17,208,000. What is the population per square mile?

Skillscan (Section 1.5)
Find the following sums.

a. 43 + 27 + 34 + 40

b. 73 + 82 + 91 + 77 + 92

c. 29 + 35 + 41 + 27 + 38 + 28

d. 123 + 98 + 105 + 117 + 112

Answers
1. 138 tickets   3. 28 points   5. 29 students   7. $145   9. 46 calls   11. 32 min   13. $458
15. $232   17. 6 gal   19. You should have found the same remainder, 7, in all three experiments.
21. 27   a. 144   b. 415   c. 198   d. 555

3.7 Supplementary Exercises

1. Ticket receipts during a basketball tournament were $11,102. If the tickets were $7 each, how many were sold?

2. The Goodsons drove 2072 mi on a vacation trip using 56 gal of gas. What was their gas mileage during the trip?

3. Lee earns $23,256 per year. What is her monthly salary?

4. A typist averages 65 words per minute. How long will he need to type a manuscript of 31,200 words? Give your answer in minutes and then in hours.

5. The airline distance from New York to Paris is 3640 mi. If a plane can average 520 mi/h, how long will the flight take?

6. A bookstore ordered 225 copies of a text at a cost of $4050. What was the cost per book?

7. A machine can produce 137 items per hour. How long will it take to produce 2192 items?

8. A charity collected $5888 in donations from 256 people. What was the average donation per person?

9. Alan buys a color television set for $595 and agrees to pay for the set with monthly payments for 1 year. If the interest charge is $89, find the amount of each monthly payment.

10. Arrange the digits 2, 4, 6, and 8 in any order that you wish to form a four-digit number. Divide that number by 9. What is the remainder? Repeat the experiment two more times. What do you find?

3.8 Finding the Average

OBJECTIVE
To find the average of a group of whole numbers

A very useful concept is the *average* of a group of numbers. The average of a group of numbers is the number which is generally the most representative of all the numbers in the group.

The average that we are finding is actually called the *arithmetic mean* or just the *mean* of the numbers.

FINDING THE AVERAGE

To find the average:

STEP 1 Add all the numbers in the group.

STEP 2 Divide that sum by the number of items in the group.

Example 1

Find the average of 12, 19, 15, and 14.

Step 1 First add the four numbers.

12 + 19 + 15 + 14 = 60

Step 2 Now divide by 4, the number of items in the group.

60 ÷ 4 = 15

The average of the group of numbers is 15.

CHECK YOURSELF 1

Find the average of 33, 27, 41, 29, and 30.

Example 2

Susan had scores of 83, 91, 78, 59, and 84 on five tests. What was her average score?

Step 1 First we add the five test scores.

83 + 91 + 78 + 59 + 84 = 395

Step 2 Now divide the total by 5, the number of scores, to find the average score.

395 ÷ 5 = 79
         ↑
      The average

This means that if Susan had scored 79 on each of the five tests, she would have had the same total score, 395.

CHECK YOURSELF 2

A basketball player scored 21, 17, 29, and 25 points in the four games of a tournament. How many points did she average?

Example 3

Suppose that to earn a B in your mathematics course, you must have an average of 80 on all the tests. So far you have scores of 75, 84, and 78. How high a score must you have on the fourth and last test to get a B for the course?

Solution To have an 80 average on the four tests, the total number of points you will need is

4 × 80 = 320

The sum of the scores thus far is

75 + 84 + 78 = 237

So the score you need on the last test is

320 − 237 = 83

You can easily check this by finding the average of 75, 84, 78, and 83.

CHECK YOURSELF 3

To earn an A in your course, you must have an average of 90 on all tests. Thus far your scores are 91, 83, 95, and 89. What must you score on the fifth test to earn the A?

Your calculator can be very useful in finding the average of a group of numbers, especially if large numbers are involved or if there are many numbers in the group. However, you do have to be careful of one thing, as the following example shows.

Example 4

To find the average of 12, 15, and 21, try:

12 $\boxed{+}$ 15 $\boxed{+}$ 21 $\boxed{\div}$ 3 $\boxed{=}$

Display $\boxed{34}$ Certainly 34 does not seem like a "reasonable" average for the three numbers. Do you see what went wrong?

Many scientific calculators will do the division *first,* according to our rules for the order of operations. So the calculator added 12, 15, and 7!

To avoid this use the sequence:

12 $\boxed{+}$ 15 $\boxed{+}$ 21 $\boxed{=}$ $\boxed{\div}$ 3 $\boxed{=}$

The *first* use of the equals key adds the three numbers *before* the division.

Display $\boxed{16}$ We now have the correct average.

Section 3.8: Finding the Average   143

CHECK YOURSELF ANSWERS

1. 32.   2. 23.   3. 92.

3.8 Exercises

Find the averages of the following groups of numbers.

1. 8, 12, and 13

2. 18, 23, and 22

3. 5, 8, 11, and 12

4. 7, 19, 11, 12, and 16

5. 23, 34, 25, 19, 31, and 24

6. 7, 13, 9, 12, 13, and 12

7. High temperatures of 86, 91, 92, 103, and 98° were recorded for the first 5 days of July. What was the average high temperature?

8. A salesperson drove 238, 159, 87, 163, and 198 mi on a 5-day trip. What was the average number of miles driven per day?

9. Highway mileage ratings for seven new diesel cars were 43, 29, 51, 36, 33, 42, and 32 mi/gal. What was the average rating?

10. The enrollments in the four elementary schools of a district are 278, 153, 215, and 198 students. What is the average enrollment?

11. To get an A in history you must have an average of 90 on five tests. Your scores thus far are 83, 93, 88, and 91. How many points must you have on the last test to receive an A?

   Total points required to get an A = 5(90) = 450 points. 83+93+88+91 = 355
   450 − 355 = 95 points

12. To pass biology you must have an average of 70 on six quizzes. So far your scores have been 65, 78, 72, 66, and 71. How many points must you have on the last quiz to pass biology?
   6(70) = 420 = 65+78+72+66+71 = 352.
   420 − 352 = 68 points

13. Louis had scores of 87, 82, 93, 89, and 84 on five tests. Cheryl had scores of 92, 83, 89, 94, and 87 on the same five tests. Who had the higher average score? By how much?

14. The Sheehan family had heating bills of $85, $78, $64, and $57 in the first 4 months of 1987. The bills for the same months of 1988 were $82, $86, $68, and $56. In which year was the average monthly bill higher? By how much?

Use your calculator for the following problems.

15. Fred kept the following records of his utility bills for 12 months: $53, $51, $43, $37, $32, $29, $34, $41, $58, $55, $49, and $58. What was the average monthly bill?

16. The following scores were recorded on a 200-point final examination: 193, 185, 163, 186, 192, 135, 158, 174, 188, 172, 168, 183, 195, 165, 183. What was the average of the scores?

Answers
1. 11   3. 9   5. 26   7. 94°   9. 38 mi/gal   11. 95 points   13. Louis's average score was 87, Cheryl's was 89. Cheryl's average score was 2 points higher than Louis's.   15. $45

3.8 Supplementary Exercises

Find the averages of the following groups of numbers.

1. 17, 23, and 14
2. 7, 15, 12, and 14
3. 38, 20, 29, 32, and 26

4. A city recorded rainfall of 8, 12, 3, 9, and 13 inches over a 5-month period. What was the average monthly rainfall?

5. Doug's expenses for textbooks over six terms were $38, $61, $49, $72, $54, and $62. What was the average cost of books for one term?

6. The attendance figures at five performances of a play were 215, 228, 190, 201, and 231. What was the average attendance per performance?

7. A grade of B in chemistry requires an average of 80 on five tests. You have scores of 83, 78, 76, and 81 on the first four tests. How many points must you have on the fifth test to receive a B?

8. The test scores of two groups of students were recorded to compare different teaching methods. Group A had scores of 85, 73, 82, 88, and 77. Group B had scores of 91, 71, 84, 87, and 82. Which group had the higher average score? By how much?

SELF-TEST for Chapter Three

The purpose of the Self-Test is to help you check your progress and review for a chapter test in class. Allow yourself about an hour to take the test. When you are done, check your answers in the back of the book. If you missed any problems, be sure to go back and review the appropriate sections in the chapter and do the supplementary exercises provided there.

[3.1] 1. In the division problem:

$$\begin{array}{r} 5 \\ 8\overline{)43} \\ \underline{40} \\ 3 \end{array}$$

8 is the Divisor.

43 is the dividend.

5 is the quotient.

3 is the rest.

[3.2] In Problems 2 to 5, divide if possible.

2. $8 \div 1$ 3. $5\overline{)5}$ 4. $0 \div 6$ 5. $3 \div 0$

[3.3] In Problems 6 to 14, divide using long division.

6. $6\overline{)738}$ 7. $8\overline{)3942}$ 8. $9\overline{)27,371}$

[3.4] 9. $38\overline{)251}$ 10. $53\overline{)783}$ 11. $28\overline{)2135}$

12. $281\overline{)6935}$ 13. $571\overline{)12,583}$ 14. $293\overline{)61,382}$

[3.5] In Problems 15 to 17, divide using short division.

15. $6\overline{)786}$ 16. $8\overline{)951}$ 17. $7\overline{)4549}$

[3.6] In Problems 18 to 23, evaluate the expressions.

18. $12 \div 6 + 3 =$ 19. $4 + 12 \div 4 =$ 20. $3^3 \div 9 =$

21. $28 \div 7 \times 4 =$ 22. $5 \times 8 \div 2 =$ 23. $36 \div (3^2 + 3) =$

145

[3.7] In Problems 24 to 27, solve each application.

24. Eight people estimate that the expenses for a trip will be $1784. If each person pays an equal amount, what will be that share?

25. On a trip Ray drove 2345 mi while using 67 gal of gas. What was his mileage on the trip?

26. Sharon's income was $12,896 last year. What was her weekly pay? (Use 52 weeks per year.)

27. Tom buys a television set for $722. He pays $50 down and agrees to pay the balance in 24 monthly payments. What will be the amount of each monthly payment?

[3.8] In Problems 28 to 30, solve each application.

28. Find the average of the numbers 12, 19, 15, 20, 11, and 13.

29. A bus carried 234 passengers on the first day of a newly scheduled route. The next 4 days there were 197, 172, 203, and 214 passengers. What was the average number of riders per day?

30. To earn an A in biology you must have an average of 90 on four tests. Your scores thus far are 87, 89, and 91. How many points must you have on the last test to earn the A?

PRETEST
for
Chapter Four

Factors and Multiples

This pretest will point out any difficulties you may be having with the factors and multiples of whole numbers. Do all the problems. Then check your answers on the following page.

1. List all the factors of 50.   1, 50, 25, 10, 2,

2. For the group of numbers 2, 3, 6, 7, 9, 17, 18, 21, and 23, list the prime and the composite numbers.

    2, 3, 7, 17, 23.

Using divisibility tests, determine which, if any, of the numbers 2, 3, and 5 are factors of the following numbers.

3. 684   = 18 ÷ 3 by 2

4. 3495

Write the prime factorizations for the following numbers.

5. 60

6. 315

Find the greatest common factor (GCF) for the following groups of numbers.

7. 16 and 28

8. 24, 36, and 42

Find the least common multiple (LCM) for the following groups of numbers.

9. 25 and 30

10. 12, 16, and 18

ANSWERS TO PRETEST

For help with similar problems, turn to the section indicated.

1. 1, 2, 5, 10, 25, and 50 (Section 4.1)
2. 2, 3, 7, 17, 23 (primes); 6, 9, 18, 21 (composites) (Section 4.1)
3. 2 and 3 (Section 4.1)
4. 3 and 5 (Section 4.1)
5. $2 \times 2 \times 3 \times 5$ (Section 4.2)
6. $3 \times 3 \times 5 \times 7$ (Section 4.2)
7. 4 (Section 4.3)
8. 6 (Section 4.3)
9. 150 (Section 4.4)
10. 144 (Section 4.4)

Chapter Four

Factors and Multiples

4.1 Prime and Composite Numbers

OBJECTIVES
1. To find the factors of a number
2. To determine whether a number is prime, composite, or neither
3. To determine whether or not a number is divisible by 2, 3, or 5

2 and 5 can also be called *divisors* of 10. Again this means that they divide 10 exactly.

In Section 2.1 we said that since $2 \times 5 = 10$, we call 2 and 5 *factors* of 10. A factor of a whole number is another whole number that will *divide exactly* into that number. This means that the division will have a remainder of zero.

Example 1

Note that the factors of 18, except for 18 itself, are *smaller* than 18.

$3 \times 6 = 18$ — Since $3 \times 6 = 18$, 3 and 6 are factors (or divisors) of 18.

$2 \times 9 = 18$ — 2 and 9 are also factors of 18.

$1 \times 18 = 18$ — 1 and 18 are factors of 18.

1, 2, 3, 6, 9, and 18 are all the factors of 18. — This is a complete list of the factors. There are no other whole numbers that divide 18 exactly.

CHECK YOURSELF 1

Remember that a whole number greater than 1 will always have itself and 1 as factors. Sometimes these will be the *only* factors. For instance, 1 and 3 are the only factors of 3.

List all the factors of 24.

Listing factors leads us to an important classification of whole numbers. Any whole number larger than 1 will be either a *prime* or a *composite* number. Let's look at the following definitions.

> **PRIME NUMBER**
>
> A *prime number* is any whole number greater than 1 that has only 1 and itself as factors.

As examples, 2, 3, 5, and 7 are prime numbers. Their only factors are 1 and themselves.

How large can a prime number be? There is no largest prime number. To date, the largest *known* prime is $2^{216,091} - 1$. This is a number with 65,050 digits if you are curious.

Of course, a computer had to be used to verify that a number of this size is prime. By the time you read this, someone may very well have found an even larger prime number.

To check whether a number is prime, one approach is simply to divide the smaller primes, 2, 3, 5, 7, and so on into the given number. If no factors other than 1 and the given number are found, the number is prime.

Example 2

17 is a prime number.   1 and 17 are the only factors.

29 is a prime number.   1 and 29 are the only factors.

33 is *not* prime.   1, 3, 11, and 33 are all factors of 33.

Note For two-digit numbers, if the number is *not* a prime, it will have one or more of the numbers 2, 3, 5, or 7, as factors.

CHECK YOURSELF 2

Which of the following numbers are prime numbers?
2, 6, 9, 11, 15, 19, 23, 35, 41

We can now define a second class of whole numbers.

This definition tells us that a composite number *does* have factors other than 1 and itself.

> **COMPOSITE NUMBER**
>
> A composite number is any whole number greater than 1 that is not prime.

Example 3

18 is a composite number.   1, 2, 3, 6, 9, and 18 are all factors of 18.

25 is a composite number.   1, 5, and 25 are factors.

38 is a composite number.   1, 2, 19, and 38 are factors.

23 is not a composite number. It is a prime number.   In this case, 1 and 23 are the only factors.

CHECK YOURSELF 3

Which of the following numbers are composite numbers?
2, 6, 10, 13, 16, 17, 22, 27, 31, 35

By the definitions of prime and composite numbers:

> The whole numbers 0 and 1 are neither prime nor composite.

Section 4.1: Prime and Composite Numbers

This is simply a matter of the way in which prime and composite numbers are defined in mathematics. 0 and 1 are the *only* two whole numbers that can't be classified as one or the other.

For our work in this and the following sections, it is very useful to be able to tell whether or not a given number is divisible by 2, 3, or 5. The tests that follow will give you some tools to check divisibility without actually having to divide.

Tests for divisibility by other numbers are also available. However, we have limited this section to those for 2, 3, and 5 because they are very easy to use and occur frequently in our work.

> **DIVISIBILITY BY 2**
>
> A whole number is divisible by 2 if its last digit is 0, 2, 4, 6, or 8.

This of course means that the number is even.

Example 4

2346 is divisible by 2.

13,254 is divisible by 2.

23,573 is *not* divisible by 2.

57,085 is *not* divisible by 2.

CHECK YOURSELF 4

Which of the following are divisible by 2?

274   3587   7548   13,593

> **DIVISIBILITY BY 3**
>
> A whole number is divisible by 3 if the sum of its digits is divisible by 3.

Example 5

345 is divisible by 3.   The sum of the digits, $3 + 4 + 5$, is 12, and 12 is divisible by 3.

1243 is *not* divisible by 3.   The sum of the digits, $1 + 2 + 4 + 3$, is 10, and 10 is not divisible by 3.

25,368 is divisible by 3.   The sum of the digits, $2 + 5 + 3 + 6 + 8$, is 24, and 24 is divisible by 3. Note that 25,368 is also divisible by 2.

CHECK YOURSELF 5

(1) Is 372 divisible by 2? By 3?
(2) Is 5493 divisible by 2? By 3?

Let's look now at a third divisibility test.

> **DIVISIBILITY BY 5**
>
> A whole number is divisible by 5 if its last digit is 0 or 5.

Example 6

2435 is divisible by 5.   Its last digit is 5.
23,123 is *not* divisible by 5.   Its last digit is 3.
123,240 is divisible by 5.   Its last digit is 0. Do you see that 123,240 is also divisible by 2 and 3?

CHECK YOURSELF 6

(1) Is 12,585 divisible by 5? By 2? By 3?
(2) Is 5890 divisible by 5? By 2? By 3?

CHECK YOURSELF ANSWERS

1. 1, 2, 3, 4, 6, 8, 12, and 24.
2. 2, 11, 19, 23, and 41 are prime numbers.
3. 6, 10, 16, 22, 27, and 35 are composite numbers.
4. 274 and 7548.
5. (1) Yes in both cases; (2) only by 3.
6. (1) By 5 and by 3; (2) by 5 and by 2.

4.1 Exercises

List the factors of the following numbers.

1. 4
2. 6   1, 2, 3, 4.
3. 10   1, 2, 5, 10.
4. 12   1, 2, 3, 4, 6, 12
5. 15   1, 3, 5, 15.
6. 21   1, 3, 7, 21.
7. 24   1, 2, 4, 6, 8, 10, 12, 24.
8. 32   1, 2, 4, 8, 16, 32.
9. 64
10. 66   1, 2, 3, 6, 66, 33, 22, 11
11. 11
12. 37   1, 37

Label each of the following numbers as prime or composite.

13. 15
14. 19   1, 19.
15. 23
16. 31   1, 31.
17. 55
18. 49   49 is divisibily by 7
19. 87
20. 97 — 16   Prime.

Section 4.1: Prime and Composite Numbers   153

21. 59   **22.** 91   **23.** 103   **24.** 105

25. List all the prime numbers between 30 and 50.

26. List all the prime numbers between 55 and 75.

Which of the following numbers are divisible by 2? By 3? By 5?

27. 72   **28.** 45   **29.** 158   **30.** 260   **31.** 569

32. 378   **33.** 570   **34.** 585   **35.** 8300   **36.** 3541

37. 2544   **38.** 3345   **39.** 13,875   **40.** 53,291

Skillscan (Section 3.5)
Divide using short division.

a. 3)72   **b.** 5)90   **c.** 4)84

d. 2)384   **e.** 3)693   **f.** 5)750

Answers
Solutions for the even-numbered exercises are provided in the back of the book.
1. 1, 2, and 4   **3.** 1, 2, 5, and 10   **5.** 1, 3, 5, and 15   **7.** 1, 2, 3, 4, 6, 8, 12, and 24
9. 1, 2, 4, 8, 16, 32, and 64   **11.** 1 and 11   **13.** Composite   **15.** Prime   **17.** Composite
19. Composite   **21.** Prime   **23.** Prime   **25.** 31, 37, 41, 43, 47   **27.** 2, 3   **29.** 2   **31.** None
33. 2, 3, and 5   **35.** 2, 5   **37.** 2, 3   **39.** 3, 5   **a.** 24   **b.** 18   **c.** 21   **d.** 192   **e.** 231   **f.** 150

4.1 Supplementary Exercises

List the factors of the following numbers.

1. 15   **2.** 16   **3.** 30

4. 40   **5.** 42   **6.** 53

Label each of the following numbers as prime or composite.

7. 11   **8.** 33   **9.** 35

10. 37   **11.** 99   **12.** 73

13. List all the prime numbers between 45 and 65.

Which of the following numbers are divisible by 2? By 3? By 5?

14. 65   **15.** 78   **16.** 952   **17.** 1259

18. 5490   **19.** 22,390   **20.** 14,685

4.2 Writing Composite Numbers as a Product of Prime Factors

OBJECTIVE
To find the prime factorization of a number

To *factor* a number means to write the number as a product of its whole-number factors.

Example 1

To factor 10, write

$10 = 2 \times 5$

The order in which you write the factors does not matter, so $10 = 5 \times 2$ would also be correct.

Of course, $10 = 10 \times 1$ is also a correct statement. However, in this section we are interested in factors other than 1 and the given number.

To factor 21, write

$21 = 3 \times 7$

In writing composite numbers as a product of factors, there will be a number of different possible factorizations.

Example 2

There have to be at least two different factorizations, since a composite number has factors other than 1 and itself.

$72 = 8 \times 9$ (1)
$ = 6 \times 12$ (2)
$ = 3 \times 24$ (3)

    We want now to write composite numbers as a product of their *prime factors*. Look again at line 1 of Example 2. The process of factoring can be continued until all the factors are prime.

Section 4.2: Writing Composite Numbers as a Product of Prime Factors   155

Example 3

This is often called a *factor tree*.

$72 = 8 \times 9$
$= 2 \times 4 \times 3 \times 3$ 4 is still not prime, and so we continue by factoring 4.
$= 2 \times 2 \times 2 \times 3 \times 3$ 72 is now written as a product of prime factors.

Finding the prime factorization of a number will be important in our later work in adding fractions.

When we write 72 as $2 \times 2 \times 2 \times 3 \times 3$, no further factorization is possible. This is called the *prime factorization* of 72.

Now what if we start with line 2 from the same example, $72 = 6 \times 12$?

Example 3 (*continued*)

$72 = 6 \times 12$ Continue to factor 6 and 12.
$= 2 \times 3 \times 3 \times 4$ Continue again to factor 4. Other choices for the factors of 12 are possible. As we shall see, the end result will be the same.
$= 2 \times 3 \times 3 \times 2 \times 2$

No matter which pair of factors you start with, you will find the same prime factorization. In this case there are three factors of 2 and two factors of 3. Since multiplication is commutative, the order in which we write the factors does not matter.

This is called the fundamental theorem of arithmetic.

> There is exactly one prime factorization for any composite number.

Example 4

$108 = 9 \times 12$
$= 3 \times 3 \times 3 \times 4$ Continue to factor 9 and 12.
$= 3 \times 3 \times 3 \times 2 \times 2$ This is the prime factorization for 108.

It makes no difference which factors of 108 you start with. Try the following exercise.

CHECK YOURSELF 1

$108 = 36 \times 3$

Continue the factorization.

The method of the previous examples and exercises will always work. As numbers get larger, however, factoring with the use of a factor tree is more difficult, and so we turn to a process based on division.

Note: The prime factorization is then the product of all the prime divisors and the final quotient.

> To find the prime factorization of a number, divide the number by a series of primes until the final quotient is a prime number.

Example 5

To write 60 as a product of prime factors, divide 2 into 60 for a quotient of 30. Continue to divide by 2 again for the quotient 15. Since 2 won't divide evenly into 15, we try 3. Since the quotient 5 is prime, we are done.

$$2\overline{)60} \rightarrow 2\overline{)30} \rightarrow 3\overline{)15} \leftarrow \text{Prime}$$
$$30 15 5$$

Do you see how the divisibility tests are used here? 60 is divisible by 2, 30 is divisible by 2, and 15 is divisible by 3.

Our factors are the prime divisors and the final quotient. We have

$$60 = 2 \times 2 \times 3 \times 5$$

CHECK YOURSELF 2

Complete the process to find the prime factorization of 90.

$$2\overline{)90} \rightarrow ?\overline{)45}$$
$$45 ?$$

Remember to continue until the final quotient is prime.

Writing composite numbers in their completely factored form can be simplified if we use a format called *continued division*. This method is based on the short-division process that you studied in Chapter 3.

Example 6

Use the short-division method to divide 60 by a series of prime numbers.

$$\text{Primes} \longrightarrow \begin{array}{r} 2\overline{)60} \\ 2\overline{)30} \\ 3\overline{)15} \\ 5 \end{array}$$

Stop when the final quotient is prime.

In each short division we write the quotient *below* rather than above the dividend. This is just a convenience for the next division.

To write the factorization of 60, we list each divisor used and the final prime quotient. In our example, we have

$$60 = 2 \times 2 \times 3 \times 5$$

Section 4.2: Writing Composite Numbers as a Product of Prime Factors

CHECK YOURSELF 3

Find the prime factorization of 234.

Here is a similar example.

Example 7

To find the prime factorization of 126:

$2 \overline{)126}$
$3 \overline{)63}$
$3 \overline{)21}$
$7 \longleftarrow$ Prime

There is no particular reason for the order of the division. We simply find an easy prime factor, here 2, as our first step, and then keep dividing until the quotient is prime. Note again how useful the divisibility tests are.

We can write

$126 = 2 \times 3 \times 3 \times 7$ The original number, 126, is the product of the series of divisors and the final prime quotient.

CHECK YOURSELF 4

Find the prime factorization of 150.

Let's look at one more complete example of finding the prime factorization of a composite number.

Example 8

To find the prime factorization of 400:

$2 \overline{)400}$
$2 \overline{)200}$
$2 \overline{)100}$
$2 \overline{)50}$
$5 \overline{)25}$
5

So

$400 = 2 \times 2 \times 2 \times 2 \times 5 \times 5$

You can, of course, use the exponent notation to write

$400 = 2^4 \times 5^2$

In each of the examples presented, we have removed the *smallest* available prime factor first, since that is often the easiest way to begin. However, it makes no difference in what order the factors are removed. The result will be the same, for there is only one prime factorization for any whole number.

CHECK YOURSELF 5

Find the prime factorization of 780.

CHECK YOURSELF ANSWERS

1. $108 = 36 \times 3 = 6 \times 6 \times 3 = 2 \times 3 \times 2 \times 3 \times 3$ You should have two factors of 2 and three factors of 3. Compare with Example 4.

2. $2\overline{)90} \rightarrow 3\overline{)45} \rightarrow 3\overline{)15}$; $90 = 2 \times 3 \times 3 \times 5$

3. $2 \times 3 \times 3 \times 13$ 4. $2 \times 3 \times 5 \times 5$

5. $2 \times 2 \times 3 \times 5 \times 13$

4.2 Exercises

Find the prime factorization of each number.

1. 18
2. 22
3. 30
4. 35

5. 51
6. 42
7. 63
8. 94

9. 70
10. 90
11. 66
12. 100

13. 130
14. 88
15. 315
16. 400

17. 225
18. 132
19. 189
20. 330

21. 336
22. 500
23. 840
24. 1170

Often in later mathematics courses you will want to find factors of a number with a given sum or difference. The following problems use this technique.

25. Find two factors of 24 with a sum of 10.

26. Find two factors of 15 with a difference of 2.

27. Find two factors of 30 with a difference of 1.

28. Find two factors of 28 with a sum of 11.

Section 4.3: Finding the Greatest Common Factor (GCF)

Skillscan (Section 4.1)
List all factors of the following numbers.

a. 12 **b.** 20 **c.** 30

d. 45 **e.** 17 **f.** 29

Answers
1. $2 \times 3 \times 3$ **3.** $2 \times 3 \times 5$ **5.** 3×17 **7.** $3 \times 3 \times 7$ **9.** $2 \times 5 \times 7$ **11.** $2 \times 3 \times 11$
13. $\begin{array}{r}2\overline{)130}\\5\overline{)65}\\13\end{array}$ **15.** $3 \times 3 \times 5 \times 7$ **17.** $3 \times 3 \times 5 \times 5$ **19.** $\begin{array}{r}3\overline{)189}\\3\overline{)63}\\3\overline{)21}\\7\end{array}$
$130 = 2 \times 5 \times 13$ $189 = 3 \times 3 \times 3 \times 7$
21. $2 \times 2 \times 2 \times 2 \times 3 \times 7$ **23.** $2 \times 2 \times 2 \times 3 \times 5 \times 7$ **25.** 4, 6   **27.** 5, 6   **a.** 1, 2, 3, 4, 6, 12
b. 1, 2, 4, 5, 10, 20   **c.** 1, 2, 3, 5, 6, 10, 15, 30   **d.** 1, 3, 5, 9, 15, 45   **e.** 1, 17   **f.** 1, 29

4.2 Supplementary Exercises

Find the prime factorization of each number.

1. 12 **2.** 26 **3.** 50 **4.** 52

5. 78 **6.** 110 **7.** 200 **8.** 105

9. 154 **10.** 252 **11.** 300 **12.** 1260

4.3 Finding the Greatest Common Factor (GCF)

OBJECTIVE
To find the greatest common factor of a group of numbers

We know that a factor or a divisor of a whole number divides that number exactly.

Example 1

Again the factors of 20, other than 20 itself, are less than 20.

The factors or divisors of 20 are

1, 2, 4, 5, 10, 20

Each of these numbers divides 20 exactly, that is, with no remainder.

CHECK YOURSELF 1

List the factors of 30.

Our work in this section involves common factors or divisors. A *common factor or divisor* for two or more numbers is any number that divides *each* of the given numbers exactly.

Example 2

Look at the numbers 20 and 30. Is there a common factor for the two numbers? First we list the factors. Then we circle the ones that appear in both lists.

FACTORS

20: ①, ②, 4, ⑤, ⑩, 20
30: ①, ②, 3, ⑤, 6, ⑩, 15, 30

We see that 1, 2, 5, and 10 are common factors of 20 and 30. Each of these numbers divides both 20 and 30 exactly.

Our later work with fractions will require finding the greatest common factor of a group of numbers.

> The *greatest common factor* (GCF) of a group of numbers is the *largest* number that will divide each of the given numbers exactly.

Example 3

In the previous example the common factors of 20 and 30 were listed:

1, 2, 5, 10 were the common factors.

The greatest common factor of the two numbers is then 10, because 10 is the largest of the four common factors.

CHECK YOURSELF 2

List the factors of 30 and 36, and then find the greatest common factor.

The method of listing all the factors for each of the numbers in a group and then picking out the largest number that appears as a common factor will also work in finding the greatest common factor of a group of more than two numbers.

Section 4.3: Finding the Greatest Common Factor (GCF)   **161**

Example 4

To find the GCF of 24, 30, and 36, we list the factors of the three numbers.

FACTORS

Looking at the three lists, we see that 1, 2, 3, and 6 are common factors.

24: ①, ②, ③, 4, ⑥, 8, 12, 24
30: ①, ②, ③, 5, ⑥, 10, 15, 30
36: ①, ②, ③, 4, ⑥, 9, 12, 18, 36

6 is the greatest common factor of 24, 30, and 36.

CHECK YOURSELF 3

Find the greatest common factor (GCF) of 16, 24, and 32.

While you will always be able to use the process shown in the previous examples and problems, it is very time-consuming where larger numbers are involved. A better approach to the problem of finding the greatest common factor of a group of numbers uses the prime factorization of each of the numbers. Let's outline the process.

FINDING THE GREATEST COMMON FACTOR

STEP 1   Write the prime factorization for each of the numbers in the group.

STEP 2   Locate the prime factors that are *common* to all the numbers.

If there are no common prime factors, the GCF is 1.

STEP 3   The greatest common factor (GCF) will be the *product* of all of the common prime factors.

Example 5

Let's look at the numbers of our second example.

Step 1   Write the prime factorization of 20 and 30.

$20 = 2 \times 2 \times 5$
$30 = 2 \times 3 \times 5$

Step 2   Find the prime factors common to each number.

$20 = ② \times 2 \times ⑤$
$30 = ② \times 3 \times ⑤$ 2 and 5 are the common prime factors.

Step 3 Form the product of the common prime factors.

2 × 5 = 10

10 is the greatest common factor.

CHECK YOURSELF 4

Find the greatest common factor (GCF) of 30 and 36.

To find the greatest common factor of a group of more than two numbers, we use the same process.

Example 6

Using the numbers of Example 4:

24 = ② × 2 × 2 × ③
30 = ② × ③ × 5
36 = ② × 2 × ③ × 3

2 and 3 are the prime factors common to *all three numbers*.

2 × 3 = 6 is the GCF.

CHECK YOURSELF 5

Find the greatest common factor (GCF) of 15, 30, and 45.

Example 7

Find the greatest common factor of 36, 84, and 120.

36 = ② × ② × ③ × 3
84 = ② × ② × ③ × 7
120 = ② × ② × 2 × ③ × 5

In this case, 2 appears *twice* as a common factor of the three numbers.

12 = 2 × 2 × 3 is the GCF

CHECK YOURSELF 6

Find the greatest common factor (GCF) of 16, 24, and 32.

Example 8

Find the greatest common factor of 15 and 28.

$15 = 3 \times 5$
$28 = 2 \times 2 \times 7$

There are no common prime factors listed. But remember that 1 is a factor of every whole number.

If two numbers, such as 15 and 28, have no common factor other than 1, they are called *relatively prime*.

The greatest common factor of 15 and 28 is 1.

CHECK YOURSELF 7

Find the greatest common factor (GCF) of 30 and 49.

CHECK YOURSELF ANSWERS

1. The factors or divisors of 30 are 1, 2, 3, 5, 6, 10, 15, 30.
2. 30: 1, 2, 3, 5, ⑥, 10, 15, 30
   36: 1, 2, 3, 4, ⑥, 9, 12, 18, 36
   6 is the greatest common factor.
3. 16: ①, ②, ④, ⑧, 16
   24: ①, ②, 3, ④, 6, ⑧, 12, 24
   32: ①, ②, ④, ⑧, 16, 32
   The GCF is 8.
4. $30 = ② \times ③ \times 5$
   $36 = ② \times 2 \times ③ \times 3$
   The GCF is $2 \times 3 = 6$.
5. 15.   6. 8.   7. GCF is 1; 30 and 49 are relatively prime.

4.3 Exercises

Find the greatest common factor (GCF) for the following groups of numbers.

1. 4 and 6
2. 6 and 9
3. 10 and 15
4. 12 and 14
5. 21 and 24
6. 22 and 33
7. 20 and 21
8. 28 and 42
9. 18 and 24
10. 35 and 36
11. 18 and 54
12. 12 and 48
13. 36 and 48
14. 36 and 54
15. 84 and 105
16. 70 and 105
17. 45, 60, 75
18. 36, 54, and 180

19. 12, 36, and 60 **20.** 15, 45, and 90 **21.** 105, 140, and 175

22. 32, 80, and 112 **23.** 25, 75, and 150 **24.** 36, 72, and 144

Skillscan (Section 2.3)
Find the following products.

a. 7 × 1 **b.** 7 × 2 **c.** 7 × 3

d. 7 × 4 **e.** 7 × 5 **f.** 7 × 6

Answers
1. 2 **3.** 5 **5.** 3 **7.** 1 **9.** 6 **11.** 18 **13.** 12 **15.** 21 **17.** 15 **19.** 12 **21.** 35 **23.** 25
a. 7 **b.** 14 **c.** 21 **d.** 28 **e.** 35 **f.** 42

4.3 Supplementary Exercises

Find the greatest common factor (GCF) for the following groups of numbers.

1. 16 and 20 **2.** 15 and 20 **3.** 24 and 42

4. 15 and 16 **5.** 21 and 25 **6.** 30 and 40

7. 36 and 48 **8.** 40 and 56 **9.** 30, 45, and 60

10. 36, 72, and 90 **11.** 24, 36, and 60 **12.** 70, 105, and 140

4.4 Finding the Least Common Multiple (LCM)

OBJECTIVE
To find the least common multiple of a group of numbers

Another idea that will be important in our work with fractions is the concept of *multiples*. Every whole number has an associated group of multiples.

> The multiples of a number are the product of that number with the natural numbers 1, 2, 3, 4, 5, ...

Example 1

The multiples of 3 are

$3 \times 1, 3 \times 2, 3 \times 3, 3 \times 4, 3 \times 5, 3 \times 6, 3 \times 7, \ldots$

or

Note that the multiples, except for 3 itself, are *larger* than 3.

3, 6, 9, 12, 15, 18, 21, . . .   The three dots indicate that the list will go on without stopping.

An easy way of listing the multiples of 3 is to think of *counting by threes*.

Example 2

The multiples of 5 are

5, 10, 15, 20, 25, 30, 35, 40, 45, 50, . . .

CHECK YOURSELF 1

List the first six multiples of 4.

If a number is a multiple of each of a group of numbers, it is called a *common multiple* of the numbers; that is, it is a number which is evenly divisible by all of the numbers in the group.

Example 3

Some common multiples of 3 and 5 are

15, 30, 45, and 60 are multiples of *both* 3 and 5.

15, 30, 45, 60

These numbers will occur in the lists of both our previous examples. Of course there will be many others if you extend the lists.

CHECK YOURSELF 2

List the first six multiples of 6. Then look at your list from Check Yourself 1 and list some common multiples of 4 and 6.

For our later work we will use the *least common multiple* of a group of numbers.

The least common multiple is often abbreviated as the LCM.

> The *least common multiple* (LCM) of a group of numbers is the *smallest* number that is a multiple of each number in the group.

It is possible to simply list the multiples of each number and then find the LCM by inspection.

Example 4

To find the least common multiple of 6 and 8:

MULTIPLES

6: 6, 12, 18, ⓐ24, 30, 36, 42, 48, . . .

8: 8, 16, ⓐ24, 32, 40, 48, . . .

48 is also a common multiple of 6 and 8, but we are looking for the smallest such number.

We see that 24 is the smallest number common to both lists. So 24 is the LCM of 6 and 8.

CHECK YOURSELF 3

Find the least common multiple of 20 and 30 by listing the multiples of each number.

The technique of the last example will work for any group of numbers. However, it becomes tedious for larger numbers. Let's outline a different approach.

> **FINDING THE LEAST COMMON MULTIPLE**
>
> STEP 1 Write the prime factorization for each of the numbers in the group.
>
> STEP 2 Find all the prime factors that appear in any one of the prime factorizations.
>
> STEP 3 Form the product of those prime factors, using each factor the greatest number of times it occurs in any one factorization.

For instance, if a number appears three times in the factorization of a number, it must be included at least three times in forming the least common multiple.

Example 5

Let's try this method on the numbers of the last example. To find the LCM of 6 and 8:

Step 1 We write the prime factorizations.

This first step is exactly the same as the first step in finding the GCF.

$6 = 2 \times 3$
$8 = 2 \times 2 \times 2$

Step 2 The prime factors that appear are 2 and 3.

Section 4.4: Finding the Least Common Multiple (LCM)

Step 3 Look at the factor 2. 2 appears three times in the factorization of 8. Look at the factor 3. 3 appears only once in any factorization.

$2 \times 2 \times 2 \times 3 = 24$ 2 is included three times, and 3 is used once as a factor.

So 24 is the LCM of 6 and 8.

CHECK YOURSELF 4
Find the LCM of 12 and 18.

Some students prefer a slightly different method of lining up the factors to help in remembering the process of finding the LCM of a group of numbers.

Example 6

To find the LCM of 10 and 18, factor:

Line up the *like* factors vertically.

$10 = 2 \times 5$
$18 = \underline{2 \times 3 \times 3}$
$ 2 \times 3 \times 3 \times 5$ Bring down the factors.

2 and 5 appear, at most, one time in any one factorization. 3 appears two times in one factorization.

$2 \times 3 \times 3 \times 5 = 90$

So 90 is the LCM of 10 and 18.

CHECK YOURSELF 5
Use the prime factorization to find the LCM of 24 and 36.

The procedure is the same for a group of more than two numbers.

Example 7

To find the LCM of 12, 18, and 20, we factor:

The different factors that appear are 2, 3, and 5.

$12 = 2 \times 2 \times 3$
$18 = 2 \times 3 \times 3$
$20 = \underline{2 \times 2 \times 5}$
$ 2 \times 2 \times 3 \times 3 \times 5$

2 and 3 appear twice in one factorization. 5 appears just once.

$2 \times 2 \times 3 \times 3 \times 5 = 180$

So 180 is the LCM of 12, 18, and 20.

Let's work through a second example of finding the least common multiple of three numbers, again using the vertical format.

Example 8

To find the LCM of 15, 20, and 25, factor:

Line up the like factors vertically.

$$
\begin{array}{l}
15 = 3 \times 5 \\
20 = 2 \times 2 \times 5 \\
25 = \times 5 \times 5 \\
\hline
 2 \times 2 \times 3 \times 5 \times 5
\end{array}
$$

Bring down the factors and multiply.

$= 300 \longleftarrow$ The LCM

CHECK YOURSELF 6

Find the LCM of 6, 8, and 20.

If the numbers have no prime factors in common, the LCM of the group of numbers is just the product of the numbers.

Example 9

To find the LCM of 10 and 21, factor:

$$
\begin{array}{l}
10 = 2 \times 5 \\
21 = 3 \times 7 \\
\hline
 2 \times 3 \times 5 \times 7
\end{array}
$$

No factor appears more than once; so the least common multiple is

Note that 10 and 21 have *no* prime factors in common, and so the LCM, 210, is just the product of 10 and 21.

$2 \times 3 \times 5 \times 7 = 210$

CHECK YOURSELF 7

Find the LCM of 9 and 20.

CHECK YOURSELF ANSWERS

1. The first six multiples of 4 are: 4, 8, 12, 16, 20, 24.
2. 6, 12, 18, 24, 30, 36. Some common multiples of 4 and 6 are 12, 24, 36.
3. The multiples of 20 are 20, 40, 60, 80, 100, 120, . . . ; the multiples of 30 are 30, 60, 90, 120, 150, . . . ; the least common multiple of 20 and 30 is 60, the smallest number common to both lists.
4. $12 = 2 \times 2 \times 3$; $18 = 2 \times 3 \times 3$. The LCM is $2 \times 2 \times 3 \times 3$, or 36.
5. $2 \times 2 \times 2 \times 3 \times 3 = 72$.   6. 120.   7. 180.

4.4 Exercises

Find the least common multiple (LCM) for the following groups of numbers. Use whichever method you wish.

1. 2 and 3
2. 3 and 5
3. 4 and 6

4. 6 and 9
5. 10 and 20
6. 12 and 36

7. 9 and 12
8. 20 and 30
9. 12 and 16

10. 10 and 15
11. 12 and 15
12. 12 and 21

13. 18 and 36
14. 25 and 50
15. 25 and 40

16. 10 and 14
17. 30 and 40
18. 18 and 24

19. 8 and 15
20. 20 and 21
21. 30 and 150

22. 36 and 72
23. 8 and 48
24. 15 and 60

25. 2, 3, and 5
26. 3, 4, and 7
27. 3, 5, and 6

28. 2, 8, and 10
29. 18, 21, and 28
30. 8, 15, and 20

31. 20, 30, and 45
32. 12, 20, and 35

Answers
1. 6   3. 12   5. 20   7. 36   9. 48   11. $12 = 2 \times 2 \times 3$; $15 = 3 \times 5$; the LCM is $2 \times 2 \times 3 \times 5 = 60$.
13. 36   15. 200   17. 120   19. 120   21. 150   23. 48   25. 30   27. 30   29. $18 = 2 \times 3 \times 3$; $21 = 3 \times 7$; $28 = 2 \times 2 \times 7$; the LCM is $2 \times 2 \times 3 \times 3 \times 7 = 252$ 31. 180

4.4 Supplementary Exercises

Find the least common multiple (LCM) for the following groups of numbers. Use whichever method you wish.

1. 4 and 12
2. 8 and 16
3. 6 and 9
4. 15 and 20
5. 5 and 25
6. 14 and 21
7. 8 and 12
8. 20 and 25
9. 18 and 24
10. 12 and 18
11. 7 and 9
12. 8 and 11
13. 9, 12, and 24
14. 8, 12, and 15
15. 14, 21, and 28
16. 6, 18, and 33

SELF-TEST for Chapter Four

The purpose of the Self-Test is to help you check your progress and review for a chapter test in class. Allow yourself about an hour to take the test. When you are done, check your answers in the back of the book. If you missed any problems, be sure to go back and review the appropriate sections in the chapter and do the supplementary exercises provided there.

[4.1] 1. List all the factors of 18.

2. List all the factors of 42.

3. List all the factors of 17.

4. From the group of numbers {13, 21, 29, 37, 51, 1, and 91}, list the prime numbers.

5. List the composite numbers from the group of Problem 4.

6. List the prime numbers between 40 and 60.

[4.1] In Problems 7–10, use the divisibility tests to determine which, if any, of the numbers 2, 3, and 5 are factors of the given numbers.

7. 90   8. 341   9. 774   10. 3585

[4.2] In Problems 11–14, find the prime factorization.

11. 42 =   12. 72 =

13. 210 =   14. 792 =

[4.3] In Problems 15–22, find the greatest common factor (GCF) for the given numbers.

15. 10 and 15
16. 30 and 48
17. 8 and 24
18. 15 and 16
19. 10, 20, and 50
20. 8, 36, and 60
21. 28, 42, and 70
22. 66, 110, and 154

[4.4] In Problems 23–30, find the least common multiple (LCM) for the given numbers.

23. 5 and 7
24. 6 and 9
25. 12 and 60
26. 15 and 18
27. 4, 5, and 10
28. 6, 8, and 12
29. 8, 14, and 21
30. 36, 20, and 30

SUMMARY for Part One

Our Decimal Place Value System

Digits   Digits are the basic symbols of the system

0, 1, 2, 3, 4, 5, 6, 7, 8, and 9 are digits.

Numerals   Numerals name numbers. They may have one or more digits.

5, 27, and 5345 are numerals.

Place value   The value of a digit in a numeral depends on its position or place.

7,352,589
- Ones
- Tens
- Hundreds
- Thousands
- Ten thousands
- Hundred thousands
- Millions

The value of a numeral is the sum of each digit multiplied by its place value.

$2345 = (2 \times 1000) + (3 \times 100) + (4 \times 10) + (5 \times 1)$

Addition

Addends   The numbers that are being added.

Sum   The result of the addition.

$\begin{array}{r} 5 \\ +8 \\ \hline 13 \end{array}$ Addends / Sum

The Properties

For any whole numbers a, b, and c:

The Commutative Property

$a + b = b + a$

$5 + 4 = 4 + 5$

The Associative Property

$(a + b) + c = a + (b + c)$

$(2 + 7) + 8 = 2 + (7 + 8)$

The Additive Identity

$a + 0 = 0 + a = a$

$6 + 0 = 0 + 6 = 6$

Rounding Whole Numbers

Step 1 To round a whole number to a certain decimal place, look at the digit to the *right* of that place.

Step 2 (a) If that digit is 5 or more, that digit and all digits to the right become 0. The digit in the place you are rounding to is increased by 1.

      (b) If that digit is less than 5, that digit and all digits to the right become 0. The digit in the place you are rounding to remains the same.

Round 43,578 to the nearest hundred

43,578 is rounded to 43,600.

Round 273,212 to the nearest thousand.

273,212 is rounded to 273,000.

Order on the Whole Numbers

For the numbers a and b, we can write

1. $a < b$ (read "a is less than b") when a is *to the left* of b on the number line.

    $8 < 12$

2. $a > b$ (read "a is greater than b") when a is *to the right* of b on the number line.

    $15 > 10$

Subtraction

Minuend The number we are subtracting from.

Subtrahend The number that is being subtracted.

Difference The result of the subtraction.

15 ← Minuend
$-\ 9$ ← Subtrahend
$\ \ 6$ ← Difference

Multiplication

Factors The numbers being multiplied.

Product The result of the multiplication.

$\underbrace{7 \times 9}_{\text{Factors}} = 63$ ← Product

The Properties

For any whole numbers a, b, and c:

The Commutative Property

$a \times b = b \times a$

$7 \times 9 = 9 \times 7$

The Associative Property

$(a \times b) \times c = a \times (b \times c)$

$(3 \times 5) \times 6 = 3 \times (5 \times 6)$

The Multiplicative Identity

$a \times 1 = 1 \times a = a$ $7 \times 1 = 1 \times 7 = 7$

The Multiplication Property of Zero

$a \times 0 = 0 \times a = 0$ $9 \times 0 = 0 \times 9 = 0$

The Distributive Property

$a \times (b + c) = (a \times b) + (a \times c)$ $2 \times (3 + 7) = (2 \times 3) + (2 \times 7)$

Using Exponents

Base   The number that is raised to a power.

Exponent   The exponent is written to the right and above the base. The exponent tells the number of times the base is to be used as a factor.

$5^3 = \underbrace{5 \times 5 \times 5}_{\text{Three factors}} = 125$

Base — Exponent

This is read "five to the third power" or "five cubed."

Division

Divisor   The number we are dividing by.

Dividend   The number being divided.

Quotient   The result of the division.

Remainder   The number "left over" after the division.

Divisor   Quotient

$\,\,5$
$7\overline{)38}$ ← Dividend
35
3 ← Remainder

The Role of Zero

Zero divided by any whole number (except 0) is 0. $0 \div 7 = 0$

Division by 0 is undefined. $7 \div 0$ is undefined.

The Order of Operations

Mixed operations in an expression should be done in the following order:

Step 1   Do any operations inside parentheses. $4 \times (2 + 3)^2 - 7$

Step 2   Evaluate any powers. $= 4 \times 5^2 - 7$
$\phantom{\text{Step 2 Evaluate any powers.}\,\,\,\,\,\,\,\,\,\,\,\,}= 4 \times 25 - 7$

Step 3   Do all multiplication and division in order from left to right. $= 100 - 7$

Step 4   Do all addition and subtraction in order from left to right. $= 93$

Prime and Composite Numbers

Prime number Any whole number greater than 1 that has only 1 and itself as factors.

7, 13, 29, and 73 are prime numbers.

Composite number Any whole number greater than 1 that is not prime.

8, 15, 42, and 65 are composite numbers.

Zero and one 0 and 1 are not classified as prime or composite numbers.

Divisibility Tests

By 2 A whole number is divisible by 2 if its last digit is 0, 2, 4, 6, or 8.

932 is divisible by 2; 1347 is not.

By 3 A whole number is divisible by 3 if the sum of its digits is divisible by 3.

546 is divisible by 3; 2357 is not.

By 5 A whole number is divisible by 5 if its last digit is 0 or 5.

865 is divisible by 5; 23,456 is not.

Prime Factorization

To find the prime factorization of a number, divide the number by a series of primes until the final quotient is a prime number. The prime factors include each prime divisor and the final quotient.

$$\begin{array}{r}2\overline{)630}\\3\overline{)315}\\3\overline{)105}\\5\overline{)35}\\7\end{array}$$

So $630 = 2 \times 3 \times 3 \times 5 \times 7$

The Greatest Common Factor

Greatest common factor (GCF) The GCF is the *largest* number that is a factor of each of a group of numbers.

To Find the GCF

Step 1 Write the prime factorization for each of the numbers in the group.

To find the GCF of 24, 30, and 36:

Step 2 Locate the prime factors that are common to all the numbers.

$24 = ②\times 2 \times 2 \times ③$
$30 = ②\times ③\times 5$
$36 = ②\times 2 \times ③\times 3$

Step 3 The greatest common factor (GCF) will be the product of all of the common prime factors. If there are no common prime factors, the GCF is 1.

The GCF $= 2 \times 3 = 6$

The Least Common Multiple

Least common multiple (LCM) The LCM is the *smallest* number that is a multiple of each of a group of numbers.

To Find the LCM

Step 1 Write the prime factorization for each of the numbers in the group.

Step 2 Find all the prime factors that appear in any one of the prime factorizations.

Step 3 Form the product of those prime factors, using each factor the greatest number of times it occurs in any one factorization.

To find the LCM of 12, 15, and 18:

$$\begin{array}{l} 12 = 2 \times 2 \times 3 \\ 15 = \phantom{2 \times 2 \times{}} 3 \times \phantom{3 \times{}} 5 \\ \underline{18 = 2 \times 3 \times 3 } \\ \phantom{18 = {}} 2 \times 2 \times 3 \times 3 \times 5 \end{array}$$

The LCM = $2 \times 2 \times 3 \times 3 \times 5$, or 180.

SUMMARY EXERCISES for Part One

You should now be reviewing the material in Part 1 of the text. The following exercises will help in that process. Work all the exercises carefully. Then check your answers against the ones in the back of the book. References are provided there to the chapter and section for each problem. If you made an error, go back and review the related material and do the supplementary exercises for that section.

[1.1] Give the place values of the indicated digits.

   1. 6 in the numeral 5674

   2. 5 in the numeral 543,400

Give word names for the following numerals.

   3. 27,428

   4. 200,305

Write each of the following as a numeral.

   5. Thirty-seven thousand, five hundred eighty-three

   6. Three hundred thousand, four hundred

[1.3] In Problems 7 and 8, name the property of addition that is illustrated.

   7. $4 + 9 = 9 + 4$

   8. $(4 + 5) + 9 = 4 + (5 + 9)$

[1.5] In Problems 9 to 11, perform the indicated operations.

   9. 784
      358
     +247

   10. 2,570
         498
       21,456
     +   28

   11. Give the total of 578, 85, 1235, and 12,824.

[1.6] In Problems 12 to 14, round the numbers to the indicated place.

   12. 6975 to the nearest hundred

   13. 15,897 to the nearest thousand

   14. 548,239 to the nearest ten thousand

[1.6] In Problem 15, estimate the sum by rounding to the indicated place.

15.
```
   897
  2335
   591
  1215
+  884
```
(Hundreds)

[1.6] In Problems 16 and 17, complete the statements using the symbols < or >.

16. 60   70

17. 38   35

[1.8] In Problems 18 to 20, perform the indicated operations.

18.
```
  5325
-  847
```

19.
```
  38,400
- 19,600
```

20. Find the difference of 7342 and 5579.

[1.9] In Problems 21 and 22, solve the applications.

21. An airline had 173, 212, 185, 197, and 202 passengers on five morning flights between Washington D.C. and New York. What was the total number of passengers?

22. Chuck owes $795 on a credit card after a trip. He makes payments of $75, $125, and $90. Interest of $31 is charged. How much remains to be paid on the account?

[2.1] In Problems 23 and 24, complete the statements using the word "factor" or the word "multiple."

23. 6 is a _____ of 36.

24. 35 is a _____ of 5.

[2.2] In Problems 25 to 27, name the property of addition and/or multiplication that is illustrated.

25. $4 \times (3 \times 5) = (4 \times 3) \times 5$

26. $7 \times 8 = 8 \times 7$

27. $3 \times (4 + 7) = 3 \times 4 + 3 \times 7$

[2.4] In Problems 28 to 30, perform the indicated operations.

28.
```
   25
  ×43
```

29.
```
   378
  ×409
```

30. Find the product of 59 and 723.

Summary Exercises for Part One

[2.5] In Problem 31, perform the indicated operation.

31. 129
 ×240

[2.5] In Problem 32, estimate the product by rounding each factor to the nearest hundred.

32. 1217
 × 494

[2.6] In Problems 33 to 35, evaluate the expressions.

33. $4 + 8 \times 3$
34. $(4 + 8) \times 3$
35. $4 \times 3 + 8 \times 3$

[2.7] In Problems 36 and 37, solve the applications.

36. Mr. Parks owns five apartments that rent for $225, $255, $285, $295, and $305 per month, respectively. What is his rental income for 1 year if all are occupied?

37. You wish to carpet a room 5 yd by 7 yd. The carpet costs $18 per square yard. What will be the total cost of the materials?

[2.8] In Problems 38 and 39, evaluate the expressions.

38. 5×2^3
39. $(5 \times 2)^3$

[3.2] In Problems 40 and 41, divide if possible.

40. $0 \div 8$
41. $5 \div 0$

[3.3] In Problem 42, divide using long division.

42. $8 \overline{)2469}$

[3.4] In Problems 43 and 44, divide using long division.

43. $39 \overline{)2157}$
44. $353 \overline{)37,589}$

[3.5] In Problem 45, divide using short division.

45. $5 \overline{)2139}$

[3.6] In Problem 46, evaluate the expression.

46. $48 \div (2^3 + 4)$

[3.7] In Problems 47 and 48, solve the applications.

47. Terry's odometer read 25,235 mi at the beginning of a trip and 26,215 mi at the end. If she used 35 gal of gas for the trip, what was her mileage (miles per gallon)?

48. Nicolas bought a new Mazda which, with interest charges included, will cost $11,360. He paid $2000 down and agreed to pay the balance in 36 monthly payments. Find the amount of each monthly payment.

[3.8] In Problem 49, solve the application.

49. Sally had test scores of 82, 85, 93, 95, 79, and 94 in her biology course. What was her average test score?

[4.1] In Problems 50 and 51, list all the factors of the given numbers.

50. 52 **51.** 41

[4.1] In Problem 52, use the group of numbers 2, 5, 7, 11, 14, 17, 21, 23, 27, 39, and 43.

52. List the prime numbers.
List the composite numbers.

[4.1] In Problems 53 and 54, use the divisibility tests to determine which, if any, of the numbers 2, 3, and 5 are factors of the following numbers.

53. 2350 **54.** 33,451

[4.2] In Problems 55 and 56, find the prime factorization for the given numbers.

55. 48 **56.** 420

[4.3] In Problems 57 and 58, find the greatest common factor (GCF) for the given numbers.

57. 15 and 20 **58.** 30 and 31

[4.4] In Problems 59 and 60, find the least common multiple (LCM) for the given numbers.

59. 18 and 54 **60.** 8, 12, and 20

CUMULATIVE TEST for Part One

This test is provided to help you in the process of reviewing Chapters 1 to 4. Answers are provided in the back of the book. If you missed any problems, be sure to go back and review the appropriate chapter sections.

1. Give the place value of 7 in 3,738,500.

2. Give the word name for 302,525.

3. Write two million, four hundred thirty thousand as a numeral.

In Problems 4 to 6, name the property of addition that is illustrated.

4. $5 + 12 = 12 + 5$ 5. $9 + 0 = 9$ 6. $(7 + 3) + 8 = 7 + (3 + 8)$

In Problems 7 and 8, perform the indicated operations.

7.
```
   593
   275
 + 98
```

8. Find the sum of 58, 673, 5325, and 17,295.

In Problems 9 and 10, round the numbers to the indicated place value.

9. 5873 to the nearest hundred

10. 953,150 to the nearest ten thousand

In Problem 11, estimate the sum by rounding to the nearest hundred.

11.
```
    943
   3281
    778
   2112
 +  570
```

In Problems 12 and 13, complete the statements using the symbols $<$ or $>$.

12. 49   47   13. 80   90

In Problems 14 and 15, perform the indicated operations.

14.
```
   4834
 -  973
```

15. Find the difference of 25,000 and 7535.

In Problems 16 and 17, solve the applications.

16. Attendance for five performances of a play was 172, 153, 205, 193, and 182. How many people attended during those performances?

17. Alan bought a Volkswagon with a list price of $8975. He added stereo equipment for $439 and an air conditioner for $615. If he made a down payment of $2450, what balance remained on the car?

In Problems 18 to 20, name the property of addition and/or multiplication that is illustrated.

18. $3 \times (4 \times 7) = (3 \times 4) \times 7$ 19. $7 \times 1 = 7$ 20. $5 \times (2 + 4) = 5 \times 2 + 5 \times 4$

In Problems 21 to 23, perform the indicated operations.

21. 538
    ×703

22. 1372
    × 500

23. Find the product of 27 and 285.

In Problem 24, estimate the product by rounding each factor to the nearest hundred.

24. 1475
    × 418

In Problem 25, solve the application.

25. A classroom is 8 yd wide by 9 yd long. If the room is to be recarpeted with material costing $14 per square yard, find the cost of the carpeting.

In Problems 26 and 27, divide using long division.

26. 48)3259

27. 458)47,350

In Problems 28 to 31, evaluate the expressions.

28. $3 + 5 \times 7$

29. $(3 + 5) \times 7$

30. 4×3^2

31. $2 + 8 \times 3 \div 4$

In Problems 32 and 33, solve the applications.

32. William bought a washer-dryer combination which, with interest charges, cost $841. He paid $145 down and agreed to pay the balance in 12 monthly payments. Find the amount of each payment.

33. Elmer had scores of 89, 71, 93, and 87 on his four mathematics tests. What was his average score?

34. Which of the numbers 5, 9, 13, 17, 22, 27, 31, and 45 are prime numbers? Which are composite numbers?

35. Use the divisibility test to determine which, if any, of the numbers 2, 3, and 5 are factors of 54,204.

36. Find the prime factorization for 264.

In Problems 37 and 38, find the greatest common factor (GCF) for the given numbers.

37. 36 and 84

38. 16, 24, and 72

In Problems 39 and 40, find the least common multiple (LCM) for the given numbers.

39. 12 and 30

40. 6, 15, and 45

PART 2

Fractions

PRETEST for Chapter Five

An Introduction to Fractions

This pretest will point out any difficulties you may be having with fractions.
Do all the problems. Then check your answers on the following page.

1. What fraction names the shaded part of the following diagram? Identify the numerator and denominator.

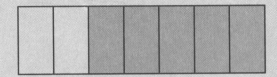

2. From the following group of numbers, list the proper fractions, improper fractions, and mixed numbers:

$$\left\{\frac{5}{6}, \frac{8}{7}, \frac{13}{9}, 2\frac{3}{5}, \frac{3}{8}, 7\frac{2}{9}, \frac{15}{8}, \frac{9}{9}, \frac{20}{21}, 3\frac{2}{7}, \frac{16}{5}, \frac{5}{11}\right\}$$

3. Convert $\frac{42}{5}$ to a mixed number.

4. Convert $7\frac{5}{8}$ to an improper fraction.

5. Are $\frac{8}{12}$ and $\frac{20}{30}$ equivalent fractions?

6. Reduce $\frac{48}{84}$ to lowest terms.

7. Find the missing numerator: $\frac{3}{4} = \frac{?}{32}$

8. Which fraction is larger, $\frac{3}{5}$ or $\frac{4}{7}$?

Write equivalent fractions with a least common denominator.

9. $\frac{5}{6}$ and $\frac{2}{9}$

10. $\frac{3}{8}, \frac{5}{9},$ and $\frac{7}{12}$

ANSWERS TO PRETEST

For help with similar problems, turn to the section indicated.

1. $\dfrac{5}{7}$ ←Numerator ←Denominator (Section 5.1)

2. Proper: $\dfrac{5}{6}, \dfrac{3}{8}, \dfrac{20}{21}, \dfrac{5}{11}$; improper: $\dfrac{8}{7}, \dfrac{13}{9}, \dfrac{15}{8}, \dfrac{9}{9}, \dfrac{16}{5}$; mixed numbers: $2\dfrac{3}{5}, 7\dfrac{2}{9}, 3\dfrac{2}{7}$ (Section 5.2)

3. $8\dfrac{2}{5}$ (Section 5.3)   4. $\dfrac{61}{8}$ (Section 5.3)   5. Yes (Section 5.4)

6. $\dfrac{4}{7}$ (Section 5.5)   7. 24 (Section 5.6)   8. $\dfrac{3}{5}$ (Section 5.6)

9. $\dfrac{15}{18}, \dfrac{4}{18}$ (Section 5.6)   10. $\dfrac{27}{72}, \dfrac{40}{72}, \dfrac{42}{72}$ (Section 5.6)

Chapter Five

An Introduction to Fractions

5.1 The Language of Fractions

OBJECTIVES
1. To identify the numerator and denominator of a fraction
2. To use fractions to name parts of a whole

<small>Our word "fraction" comes from the Latin stem *fractio*, which means "breaking into pieces."</small>

The previous chapters dealt with whole numbers and the operations that are performed on them. We are now ready to consider a new kind of number, a *fraction*. Whenever a unit or a whole quantity is divided into parts, we call those parts *fractions* of the unit.

Example 1

In Figure 1, the whole has been divided into five parts. We use the symbol $\frac{2}{5}$ to represent the shaded portion of the whole.

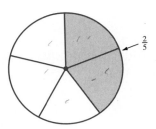

Figure 1

<small>*Common fraction* is technically the correct term. We will normally just use *fraction* in these materials if there is no room for confusion.</small>

The symbol $\frac{2}{5}$ is called a *common fraction*, or more simply a fraction. A fraction is written in the form a/b or $\frac{a}{b}$, where a and b represent whole numbers and b cannot be equal to 0.

We give the numbers a and b special names. The *denominator*, b, is the number on the bottom. This tells us into how many parts the unit or whole has been divided. The *numerator*, a, is the number on the top. This tells us how many parts of the unit are used.

189

Example 2

In Figure 1, the *denominator* is 5; the unit or whole (the circle) has been divided into five parts. The *numerator* is 2. We have taken two parts of the unit.

$$\frac{2}{5} \begin{array}{l} \leftarrow \text{Numerator} \\ \leftarrow \text{Denominator} \end{array}$$

Example 3

The fraction $\frac{4}{7}$ names the shaded part of the rectangle in Figure 2.

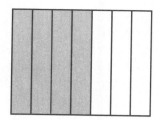

Figure 2

The unit or whole is divided into seven parts, so the denominator is 7. We have shaded four of those parts, and so we have a numerator of 4.

CHECK YOURSELF 1

What fraction names the shaded part of this diagram? Identify the numerator and denominator.

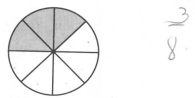

Fractions can also be used to name a part of a collection or a set of objects.

Example 4

The fraction $\frac{5}{6}$ names the shaded part of Figure 3. We have shaded five of the six objects.

Section 5.1: The Language of Fractions 191

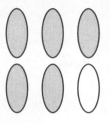

Figure 3

CHECK YOURSELF 2

What fraction names the shaded part of this diagram?

Example 5

Of course, the fraction $\frac{8}{23}$ names the part of the class that are not women.

In a class of 23 students, 15 are women. We can name the part of the class that are women as $\frac{15}{23}$.

CHECK YOURSELF 3

Seven replacement parts out of a shipment of fifty were faulty. What fraction names the portion of the shipment that was faulty?

$\frac{a}{b}$ then names the *quotient* when a is divided by b. Of course, b cannot be 0.

A fraction can also be thought of as indicating division. The symbol $\frac{a}{b}$ also means $a \div b$.

Example 6

$\frac{2}{3}$ names the quotient when 2 is divided by 3. So $\frac{2}{3} = 2 \div 3$.

Note $\frac{2}{3}$ can be read as "two-thirds" or as "two divided by three."

CHECK YOURSELF 4

Using the numbers 5 and 9 write $\frac{5}{9}$ in another way.

CHECK YOURSELF ANSWERS

1. $\dfrac{3}{8}$ ←Numerator ←Denominator   2. $\dfrac{2}{7}$ 3. $\dfrac{7}{50}$ 4. $5 \div 9$

5.1 Exercises

Identify the numerator and denominator of each fraction.

1. $\dfrac{7}{10}$

2. $\dfrac{5}{12}$

3. $\dfrac{3}{11}$

4. $\dfrac{7}{15}$

What fraction names the shaded part of each of the figures below?

5.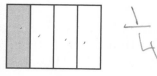

6.

7.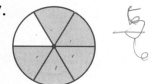

8.

9.

10.

11.

12.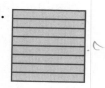

13.

14.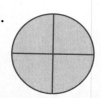

Section 5.1: The Language of Fractions   193

15.

16.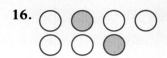

17.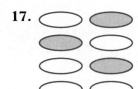

18.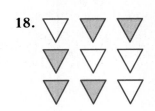

19. You missed three questions on a 20-question test. What fraction names the part you got correct? The part you got wrong?   3/20

20. Two of the five starters on a basketball team fouled out of a game. What fraction names the part of the starting team that fouled out?   2/5

21. A used car dealer sold 13 of the 19 cars that were in stock. What fraction names the portion sold? What fraction names the portion *not* sold?   13/19

22. At lunch, five people out of a group of nine had hamburgers. What fraction names the part of the group that had hamburgers? What fraction names the part that did *not* have hamburgers?   5/9

23. Using the numbers 3 and 4, show another way of writing $\frac{3}{4}$.

24. Using the numbers 4 and 5, show another way of writing $\frac{4}{5}$.

Skillscan (Section 1.6)
Use the symbols < or > to complete the following statements.

a. 2 < 3

b. 7 > 4

c. 5 < 9

d. 10 > 7

e. 11 < 13

f. 18 > 17

Answers
Solutions for the even-numbered exercises are provided in the back of the book.

1. 7 is the numerator; 10 is the denominator   3. 3 is the numerator; 11 is the denominator   5. $\frac{1}{4}$
7. $\frac{5}{6}$ 9. $\frac{5}{5}$ 11. $\frac{11}{12}$ 13. $\frac{7}{12}$ 15. $\frac{4}{5}$ 17. $\frac{3}{8}$ 19. $\frac{17}{20}, \frac{3}{20}$ 21. $\frac{13}{19}, \frac{6}{19}$ 23. $3 \div 4$
a. 2 < 3   b. 7 > 4   c. 5 < 9   d. 10 > 7   e. 11 < 13   f. 18 > 17

5.1 Supplementary Exercises

Identify the numerator and denominator of each fraction.

1. $\dfrac{5}{9}$
2. $\dfrac{3}{20}$

What fraction names the shaded part of each of the figures below?

3.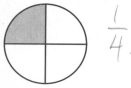

4.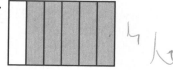

5.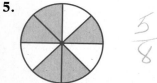

6.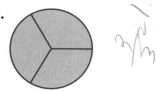

7.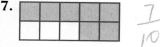

8.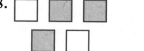

9.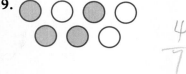

10. On a committee of eight people, five of the members are women. What fraction names the part of the group that are women? What fraction names the part that are not women?

11. A football team won 8 of its games in an 11-game season. What fraction names the portion of games won? What fraction names the portion lost or tied?

12. Using the numbers 7 and 8, show another way of writing $\dfrac{7}{8}$.

5.2 Proper Fractions, Improper Fractions, and Mixed Numbers

OBJECTIVE
To identify proper fractions, improper fractions, and mixed numbers

We can use the relative size of the numerator and denominator of a fraction to separate fractions into different categories.

> If the numerator is *less than* the denominator, the fraction names a number less than 1 and is called a *proper fraction*.

Example 1

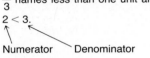

$\frac{2}{3}$ is a proper fraction. The numerator is less than the denominator. $\frac{5}{9}$ is also a proper fraction.

> If the numerator is *greater than or equal to* the denominator, the fraction names a number greater than or equal to 1 and is called an *improper fraction*.

Example 2

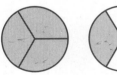

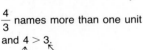

$\frac{4}{3}$ is an improper fraction. The numerator is larger than the denominator. $\frac{9}{7}$ is an improper fraction. $\frac{8}{8}$ is also an improper fraction. It names exactly one unit.

CHECK YOURSELF 1

List the proper fractions and the improper fractions in the following group:

$$\frac{5}{4}, \frac{10}{11}, \frac{3}{4}, \frac{8}{5}, \frac{6}{6}, \frac{13}{10}, \frac{7}{8}, \frac{15}{8}$$

Another way to write a fraction that is larger than 1 is called a *mixed number*.

> A mixed number is the sum of a whole number and a proper fraction.

Example 3

$2\frac{3}{4}$ means $2 + \frac{3}{4}$. In fact, we read the mixed number as "two *and* three-fourths." The addition sign is usually not written.

$2\frac{3}{4}$ is a mixed number. It represents the sum of the whole number 2 and the fraction $\frac{3}{4}$. Look at the diagram below representing $2\frac{3}{4}$.

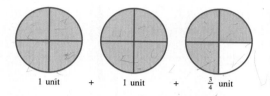

1 unit + 1 unit + $\frac{3}{4}$ unit

CHECK YOURSELF ANSWERS

1. Proper fractions: $\frac{10}{11}, \frac{3}{4}, \frac{7}{8}$ Improper fractions: $\frac{5}{4}, \frac{8}{5}, \frac{6}{6}, \frac{13}{10}, \frac{15}{8}$

5.2 Exercises

Identify each number as a proper fraction, an improper fraction, or a mixed number.

1. $\frac{7}{8}$
2. $\frac{9}{5}$
3. $1\frac{1}{4}$
4. $\frac{9}{10}$

5. $\frac{5}{5}$
6. $1\frac{1}{5}$
7. $\frac{7}{6}$
8. $\frac{8}{8}$

9. $2\frac{1}{3}$
10. $5\frac{3}{7}$
11. $\frac{14}{15}$
12. $\frac{16}{15}$

13. $\frac{21}{20}$
14. $\frac{19}{20}$
15. $7\frac{3}{5}$
16. $\frac{15}{14}$

Give the mixed number that names the shaded portion of each diagram.

17.
18.

19.

20.

Skillscan (Section 3.1)
Divide.

a. $5\overline{)12}$ b. $3\overline{)10}$ c. $7\overline{)16}$

d. $4\overline{)15}$ e. $3\overline{)19}$ f. $6\overline{)17}$

Answers
1. Proper   3. Mixed number   5. Improper   7. Improper   9. Mixed number   11. Proper
13. Improper   15. Mixed number   17. $1\frac{1}{4}$ 19. $3\frac{5}{8}$ a. 2 r2   b. 3 r1   c. 2 r2   d. 3 r3
e. 6 r1   f. 2 r5

5.2 Supplementary Exercises

Identify each number as a proper fraction, an improper fraction, or a mixed number.

1. $\frac{8}{5}$ 2. $\frac{9}{11}$ 3. $3\frac{3}{4}$ 4. $\frac{8}{8}$

5. $\frac{9}{9}$ 6. $2\frac{1}{4}$ 7. $\frac{29}{30}$ 8. $\frac{19}{20}$

Give the mixed number that names the shaded portions of each diagram.

9.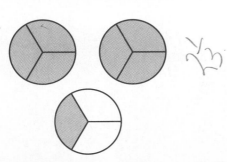

10.

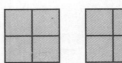

5.3 Converting Improper Fractions and Mixed Numbers

OBJECTIVES
1. To write improper fractions as mixed or whole numbers
2. To write mixed numbers as improper fractions

For our later work it will be important to be able to change back and forth between improper fractions and mixed numbers. Since an improper fraction represents a number that is greater than or equal to 1:

> An improper fraction can always be written as a mixed number or as a whole number.

You can write the fraction $\frac{7}{5}$ as $7 \div 5$. We divide the numerator by the denominator.

To do this, remember that you can think of a fraction as indicating division. The numerator is divided by the denominator. This leads us to the following rule:

> **TO CHANGE AN IMPROPER FRACTION TO A MIXED NUMBER**
>
> STEP 1 Divide the numerator by the denominator.
>
> STEP 2 If there is a remainder, write the remainder over the original denominator.

In step 1 the quotient gives the whole-number portion of the mixed number.

Step 2 gives the fractional portion of the mixed number.

Example 1

To convert $\frac{17}{5}$ to a mixed number, we divide.

$$5\overline{)17} \quad \begin{array}{r} 3 \\ \underline{15} \\ 2 \end{array} \qquad \frac{17}{5} = 3\frac{2}{5} \;\;\leftarrow \text{Remainder}$$

$$\uparrow \text{Quotient}$$

In diagram form:

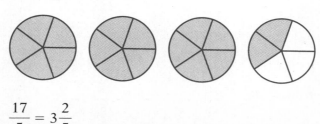

$$\frac{17}{5} = 3\frac{2}{5}$$

Section 5.3: Converting Improper Fractions and Mixed Numbers   199

CHECK YOURSELF 1

Convert $\dfrac{32}{5}$ to a mixed number.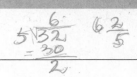

Example 2

To convert $\dfrac{21}{7}$ to a mixed or a whole number, we divide.

If there is *no* remainder, the improper fraction is equal to some whole number; in this case, 3.

$$7\overline{)21} \quad \dfrac{21}{7} = 3$$
$$\underline{21}$$
$$0$$

CHECK YOURSELF 2

Convert $\dfrac{48}{6}$ to a mixed or a whole number.   8

It is also easy to convert mixed numbers to improper fractions. Just use the following rule:

TO CHANGE A MIXED NUMBER TO AN IMPROPER FRACTION

STEP 1 Multiply the denominator of the fraction by the whole-number portion of the mixed number.

STEP 2 Add the numerator of the fraction to that product.

STEP 3 Write that sum over the original denominator to form the improper fraction.

Example 3

To convert $3\dfrac{2}{5}$ to an improper fraction:

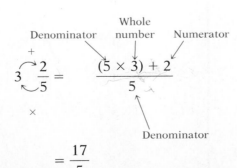

Multiply the denominator by the whole number ($5 \times 3 = 15$).

Add the numerator. We now have 17.

Write 17 over the original denominator.

$$= \dfrac{17}{5}$$

In diagram form:

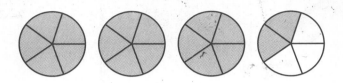

Each of the three units has 5 fifths, so the whole-number portion is 5 × 3 or 15 fifths. Then add the 2 fifths from the fractional portion for 17 fifths.

Example 4

Multiply the denominator, 7, by the whole number, 4, and add the numerator, 5.

To convert $4\frac{5}{7}$ to an improper fraction:

$$4\frac{5}{7} = \frac{(7 \times 4) + 5}{7} = \frac{33}{7}$$

CHECK YOURSELF 3

Convert $5\frac{3}{8}$ to an improper fraction.

One special kind of improper fraction should be mentioned at this point: a fraction with a denominator of 1.

This is the same as our earlier rule for dividing by 1.

> Any fraction with a denominator of 1 is equal to the numerator alone.

Thinking in terms of division, $\frac{5}{1} = 5 \div 1 = 5$

Example 5

$$\frac{5}{1} = 5 \qquad \frac{12}{1} = 12$$

CHECK YOURSELF 4

Convert 9 to an improper fraction.

CHECK YOURSELF ANSWERS

1. $6\frac{2}{5}$ 2. 8   3. $\frac{43}{8}$ 4. $\frac{9}{1}$

5.3 Exercises

Change to a mixed or a whole number.

1. $\dfrac{5}{2}$ 2. $\dfrac{8}{3}$ 3. $\dfrac{5}{4}$ 4. $\dfrac{7}{6}$

5. $\dfrac{17}{7}$ 6. $\dfrac{27}{8}$ 7. $\dfrac{34}{5}$ 8. $\dfrac{25}{6}$

9. $\dfrac{59}{5}$ 10. $\dfrac{58}{7}$ 11. $\dfrac{65}{9}$ 12. $\dfrac{151}{12}$

13. $\dfrac{24}{6}$ 14. $\dfrac{160}{8}$ 15. $\dfrac{9}{1}$ 16. $\dfrac{6}{1}$

Change to an improper fraction.

17. $2\dfrac{1}{5}$ 18. $3\dfrac{1}{4}$ 19. $6\dfrac{6}{1}$ 20. $4\dfrac{5}{8}$

21. $5\dfrac{3}{11}$ 22. 8 23. $3\dfrac{3}{7}$ 24. $2\dfrac{2}{9}$

25. $8\dfrac{5}{12}$ 26. $7\dfrac{3}{10}$ 27. $10\dfrac{2}{5}$ 28. $12\dfrac{1}{4}$

29. $100\dfrac{2}{3}$ 30. $150\dfrac{1}{4}$ 31. $120\dfrac{2}{3}$ 32. $250\dfrac{3}{4}$

Skillscan (Section 2.1)
Multiply and then complete each statement using the symbols = (equal to) or ≠ (not equal to).

a. 2×3 1×6 b. 3×5 4×4 c. 4×5 3×7

d. 6×7 3×14 e. 5×10 25×2 f. 16×3 7×7

Answers

1. $2\frac{1}{2}$ 3. $1\frac{1}{4}$ 5. $2\frac{3}{7}$ 7. $5\overline{)34}$ $\underline{30}$ 4 $\frac{34}{5} = 6\frac{4}{5}$ 9. $11\frac{4}{5}$ 11. $7\frac{2}{9}$

13. 4. There is no remainder in the division, so the result is a whole number.   15. 9   17. $\frac{11}{5}$ 19. $\frac{6}{1}$

21. $\frac{58}{11}$ 23. $3\frac{3}{7} = \frac{(7 \times 3) + 3}{7} = \frac{24}{7}$ 25. $\frac{101}{12}$ 27. $10\frac{2}{5} = \frac{(5 \times 10) + 2}{5} = \frac{52}{5}$ 29. $\frac{302}{3}$ 31. $\frac{362}{3}$

a. =   b. ≠   c. ≠   d. =   e. =   f. ≠

5.3 Supplementary Exercises

Change to a mixed or a whole number.

1. $\frac{4}{3}$ 2. $\frac{6}{5}$ 3. $\frac{44}{7}$ 4. $\frac{51}{8}$

5. $\frac{200}{3}$ 6. $\frac{125}{4}$ 7. $\frac{72}{9}$ 8. $\frac{7}{1}$

Change to an improper fraction.

9. $2\frac{1}{4}$ 10. 4   11. $6\frac{7}{12}$ 12. $5\frac{4}{7}$

13. $10\frac{3}{5}$ 14. $12\frac{1}{4}$ 15. $100\frac{3}{4}$ 16. $200\frac{1}{5}$

5.4 Equivalent Fractions

OBJECTIVE
To determine whether two fractions are equivalent

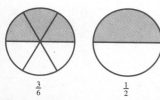

Figure 1

It is possible to represent the same portion of the whole by different fractions. Look at Figure 1, representing $\frac{3}{6}$ and $\frac{1}{2}$. The two fractions are simply different names for the same number. They are called *equivalent fractions* for this reason. Any fraction has a large number of equivalent fractions. Look at our first example.

Section 5.4: Equivalent Fractions

Example 1

The three fractions name the same number.

$\frac{2}{3}$, $\frac{4}{6}$, and $\frac{6}{9}$ are all equivalent fractions, since they name the same part of a unit. This is illustrated in Figure 2.

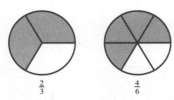

Figure 2

There are many more fractions that are equivalent to $\frac{2}{3}$. All of these names can be used interchangeably.

This has nothing to do with multiplying two fractions. We'll talk about that later.

An easy way to find out if two fractions are equivalent is to use the idea of cross products.

 We call $a \times d$ and $b \times c$ the *cross products*.

TESTING FOR EQUALITY

If the cross products for two fractions are equal, the two fractions are equivalent.

Example 2

Are $\frac{3}{24}$ and $\frac{4}{32}$ equivalent fractions? The cross products are 3×32, or 96, and 24×4, or 96. Since the cross products are equal, the fractions are equivalent.

CHECK YOURSELF 1

Are $\frac{3}{8}$ and $\frac{9}{24}$ equivalent fractions?

Example 3

Are $\frac{2}{5}$ and $\frac{3}{7}$ equivalent fractions? The cross products are 2×7 and 5×3.

$2 \times 7 = 14$ and $5 \times 3 = 15$

Since $14 \neq 15$, the fractions are *not* equivalent.

CHECK YOURSELF 2

Are $\frac{7}{8}$ and $\frac{8}{9}$ equivalent fractions?

CHECK YOURSELF ANSWERS

1. Yes. The cross products are equal.
2. No. The cross products are *not* equal.

5.4 Exercises

Are the pairs of fractions equivalent?

1. $\frac{1}{3}, \frac{3}{10}$
2. $\frac{3}{5}, \frac{6}{10}$
3. $\frac{1}{8}, \frac{3}{24}$
4. $\frac{2}{3}, \frac{3}{5}$

5. $\frac{5}{6}, \frac{15}{18}$
6. $\frac{3}{4}, \frac{16}{20}$
7. $\frac{3}{25}, \frac{2}{16}$
8. $\frac{20}{24}, \frac{5}{6}$

9. $\frac{2}{7}, \frac{3}{11}$
10. $\frac{11}{12}, \frac{22}{24}$
11. $\frac{16}{24}, \frac{40}{60}$
12. $\frac{10}{25}, \frac{20}{48}$

Skillscan (Section 3.1)
Divide.

a. $8 \div 4$
b. $10 \div 2$
c. $9 \div 9$

d. $12 \div 6$
e. $15 \div 3$
f. $20 \div 5$

Answers
1. $1 \times 10 = 10$; $3 \times 3 = 9$. The fractions are not equivalent.   3. Yes   5. Yes   7. No   9. No
11. $16 \times 60 = 960$, and $24 \times 40 = 960$. The fractions are equivalent.   a. 2   b. 5   c. 1   d. 2
e. 5   f. 4

5.4 Supplementary Exercises

Are the pairs of fractions equivalent?

1. $\frac{1}{4}, \frac{6}{25}$
2. $\frac{2}{3}, \frac{10}{15}$
3. $\frac{3}{4}, \frac{18}{24}$

4. $\frac{15}{27}, \frac{10}{18}$
5. $\frac{12}{30}, \frac{15}{38}$
6. $\frac{30}{42}, \frac{21}{27}$

5.5 Simplifying Fractions

OBJECTIVE
To use the fundamental principle to simplify fractions

As we saw in Section 5.4, different fractions can all represent the same value. The fractions

The fractions are all equivalent.

$$\frac{1}{2}, \frac{2}{4}, \frac{3}{6}, \text{ and } \frac{5}{10}$$

all name the same number.
Writing equivalent fractions uses the following important principle.

FUNDAMENTAL PRINCIPLE OF FRACTIONS

For the fraction $\frac{a}{b}$ and any nonzero number c,

$$\frac{a}{b} = \frac{a \times c}{b \times c}$$

We are really multiplying by $\frac{c}{c}$ or 1, and multiplying by 1 does not change the value of a number.

In words, the fundamental principle tells us that we can multiply the numerator and denominator by the same nonzero number. The result will be an equivalent fraction. For instance,

Multiply the numerator and denominator by 2.

$$\frac{1}{2} = \frac{1 \times 2}{2 \times 2} = \frac{2}{4}$$

Multiply the numerator and denominator by 5.

$$\frac{1}{2} = \frac{1 \times 5}{2 \times 5} = \frac{5}{10}$$

For our work in this section, we will use the principle in another way. Since the principle can also be written as

$$\frac{a \div c}{b \div c} = \frac{a}{b} \qquad \text{where } c \text{ is nonzero}$$

we can *divide* the numerator and denominator by the same nonzero number. Let's see how this is applied.
Simplifying a fraction or *reducing a fraction to lower terms* means finding an equivalent fraction with a *smaller* numerator and denominator than those of the original fraction. Dividing the numerator and denominator by the same nonzero number will do exactly that.

Consider the following example.

Example 1

We apply the fundamental principle to divide the numerator and denominator by 5.

(a) $\dfrac{5}{15} = \dfrac{5 \div 5}{15 \div 5} = \dfrac{1}{3}$

$\dfrac{5}{15}$ and $\dfrac{1}{3}$ are equivalent fractions. *Check this using the cross products.*

We divide the numerator and denominator by 2.

(b) $\dfrac{4}{8} = \dfrac{4 \div 2}{8 \div 2} = \dfrac{2}{4}$

$\dfrac{4}{8}$ and $\dfrac{2}{4}$ are equivalent fractions.

CHECK YOURSELF 1

Write two fractions equivalent to $\dfrac{30}{45}$. Divide the numerator and denominator by 5. Divide the numerator and denominator by 15.

We say that a fraction is in *simplest form,* or in *lowest terms,* if the numerator and denominator have no common factors other than 1. This means that the fraction has the smallest possible numerator and denominator.

Example 2

In Example 1:

$\dfrac{1}{3}$ is in simplest form.

In this case the numerator and denominator are *not* as small as possible. The numerator and denominator have a common factor of 2.

$\dfrac{2}{4}$ is *not* in simplest form.   *Do you see that $\dfrac{2}{4}$ can also be written as $\dfrac{1}{2}$?*

This means that the numerator and denominator can have no additional common factors other than 1. The fraction must be in lowest terms.

To write a fraction in simplest form or to *reduce a fraction to lowest terms,* divide the numerator and denominator by their greatest common factor (GCF).

Example 3

From our work in Chapter 4, we know that the greatest common factor of 10 and 15 is 5. To write $\frac{10}{15}$ in simplest form, divide the numerator and denominator by 5.

$$\frac{10}{15} = \frac{10 \div 5}{15 \div 5} = \frac{2}{3}$$

The resulting fraction, $\frac{2}{3}$, is in lowest terms.

CHECK YOURSELF 2

Write $\frac{12}{18}$ in simplest form by dividing the numerator and denominator by the GCF.

This method uses the prime factorization of the numerator and denominator.

Many students prefer another method of reducing fractions. Look at the fraction of Example 3 reduced by this method.

Example 4

To simplify $\frac{10}{15}$, we factor as shown.

$$\frac{10}{15} = \frac{2 \times \cancel{5}}{3 \times \cancel{5}} = \frac{2}{3}$$

We have a common factor of 5 in the original numerator and denominator, so we divide by 5. Note the use of the slash to indicate that division.

CHECK YOURSELF 3

Write $\frac{12}{18}$ in simplest form.

Example 5

To simplify $\frac{24}{42}$, factor.

Divide by the common factors of 2 and 3.

$$\frac{24}{42} = \frac{\cancel{2} \times 2 \times 2 \times \cancel{3}}{\cancel{2} \times \cancel{3} \times 7} = \frac{4}{7}$$

Note The numerator of the simplified fraction is the *product* of the prime factors remaining in the numerator after the division by 2 and 3.

Example 6

To reduce $\dfrac{120}{180}$ to lowest terms, write the prime factorizations of the numerator and denominator. Then divide by any common factors.

$$\frac{120}{180} = \frac{\cancel{2} \times \cancel{2} \times 2 \times \cancel{3} \times \cancel{5}}{\cancel{2} \times \cancel{2} \times \cancel{3} \times 3 \times \cancel{5}} = \frac{2}{3}$$

CHECK YOURSELF 4

Write the fraction $\dfrac{60}{75}$ in simplest form.

There is another way to organize your work in simplifying fractions. It again uses the fundamental principle to divide the numerator and denominator by any common factors. Let's illustrate with the fraction considered in Example 5.

Example 7

$$\frac{24}{42} = \frac{\overset{12}{\cancel{24}}}{\underset{21}{\cancel{42}}} = \frac{\overset{4}{\cancel{12}}}{\underset{7}{\cancel{21}}} = \frac{4}{7}$$

Divide by the common factor of 2. Divide by the common factor of 3.

The original numerator and denominator are divisible by 2, and so we divide by that factor to arrive at $\dfrac{12}{21}$. Our divisibility tests tell us that a common factor of 3 still exists. (Do you remember why?) Divide again for the result $\dfrac{4}{7}$, which is in lowest terms.

Note If we had seen the GCF of 6 at first, we could have divided by 6 and arrived at the same result in one step.

Example 8

$$\frac{42}{63} = \frac{\overset{2}{\overset{\cancel{14}}{\cancel{42}}}}{\underset{3}{\underset{\cancel{21}}{\cancel{63}}}} = \frac{2}{3}$$

Our first step is to divide by the common factor of 3. We then have $\dfrac{14}{21}$. There is still a common factor of 7, so we again divide.

Again we could have removed the GCF of 21 (or 3 × 7) in one step if we had recognized it.

Section 5.5: Simplifying Fractions

CHECK YOURSELF 5

Use the method of Examples 7 and 8 to write $\dfrac{60}{75}$ in simplest form.

To Review In using division to reduce a fraction to lowest terms, we simply continue to divide by common factors until no common factors other than 1 exist in the numerator and denominator.

You have seen several approaches to reducing fractions. All are essentially the same; they are just different ways of writing down the steps. Experiment with the various methods, and use the one you are most comfortable with.

Will you be able to simplify or reduce all fractions? Division requires common factors in the numerator and denominator, and they may or may not exist. Let's look at an example.

Example 9

Try reducing $\dfrac{14}{25}$ to lowest terms.

$$\frac{14}{25} = \frac{2 \times 7}{5 \times 5}$$

Looking at the factors of 14 and 25, we see that no common factors other than 1 exist. The fraction *cannot be reduced*. It is already in lowest terms.

We mentioned earlier that it is very important to remember that the use of the fundamental principle means *dividing* the numerator and denominator by a common *factor*. Compare the two statements below.

Be Careful!

$$\frac{12}{15} = \frac{4 \times \cancel{3}}{5 \times \cancel{3}} = \frac{4}{5}$$

This is correct. We have divided by the *common factor* of 3.

Now look at the following.

$$\frac{4}{5} = \frac{1 + \cancel{3}}{2 + \cancel{3}} \stackrel{?}{=} \frac{1}{2}$$

Since $\dfrac{4}{5}$ is not the same as $\dfrac{1}{2}$, we definitely have a problem now! Do you see what went wrong? In removing the "added 3" we have subtracted 3 in the numerator and denominator. This is *not* a legal step.

CHECK YOURSELF ANSWERS

1. $\dfrac{30}{45}$ is equivalent to $\dfrac{6}{9}$ and $\dfrac{2}{3}$.

2. 6 is the GCF of 12 and 18, so $\dfrac{12}{18} = \dfrac{12 \div 6}{18 \div 6} = \dfrac{2}{3}$.

3. $\dfrac{12}{18} = \dfrac{\cancel{2} \times 2 \times \cancel{3}}{\cancel{2} \times 3 \times \cancel{3}} = \dfrac{2}{3}$ 4. $\dfrac{60}{75} = \dfrac{2 \times 2 \times \cancel{3} \times \cancel{5}}{\cancel{3} \times \cancel{5} \times 5} = \dfrac{4}{5}$

5. Divide by the common factors of 3 and 5. $\dfrac{60}{75} = \dfrac{4}{5}$

5.5 Exercises

Write each fraction in simplest form.

1. $\dfrac{6}{9}$ 2. $\dfrac{8}{10}$ 3. $\dfrac{10}{14}$ 4. $\dfrac{12}{40}$

5. $\dfrac{12}{18}$ 6. $\dfrac{28}{35}$ 7. $\dfrac{25}{30}$ 8. $\dfrac{21}{24}$

9. $\dfrac{9}{36}$ 10. $\dfrac{16}{40}$ 11. $\dfrac{12}{36}$ 12. $\dfrac{9}{24}$

13. $\dfrac{48}{54}$ 14. $\dfrac{30}{50}$ 15. $\dfrac{32}{40}$ 16. $\dfrac{16}{48}$

17. $\dfrac{75}{105}$ 18. $\dfrac{42}{63}$ 19. $\dfrac{60}{75}$ 20. $\dfrac{48}{66}$

21. $\dfrac{84}{108}$ 22. $\dfrac{54}{126}$ 23. $\dfrac{66}{110}$ 24. $\dfrac{210}{240}$

25. $\dfrac{16}{21}$ 26. $\dfrac{21}{32}$ 27. $\dfrac{28}{45}$ 28. $\dfrac{42}{55}$

29. A quarter is what fractional part of a dollar? Simplify your result.

30. A dime is what fractional part of a dollar? Simplify your result.

31. What fractional part of an hour is 15 min? Simplify your result.

32. What fractional part of a day is 3 hours? Simplify your result.

33. A meter is equal to 100 cm. What fractional part of a meter is 40 cm? Simplify your result.

34. A kilometer is equal to 1000 meters. What fractional part of a kilometer is 300 meters? Simplify your result.

Skillscan (Section 4.4)
Find the least common multiple (LCM) for the following numbers.

a. 5 and 6　　　　　**b.** 3 and 8　　　　　**c.** 6 and 8

d. 10 and 15　　　　**e.** 5, 6, and 9　　　　**f.** 12, 15, and 20

Answers
1. $\frac{2}{3}$　 3. $\frac{5}{7}$　 5. $\frac{2}{3}$　 7. $\frac{5}{6}$　 9. $\frac{1}{4}$　 11. $\frac{1}{3}$　 13. $\frac{8}{9}$　 15. $\frac{4}{5}$　 17. $\frac{5}{7}$　 19. $\frac{4}{5}$　 21. $\frac{7}{9}$　 23. $\frac{3}{5}$
25. $\frac{16}{21}$ is already in simplest form.　 27. $\frac{28}{45}$ is already in simplest form.　 29. $\frac{1}{4}$　 31. $\frac{1}{4}$　 33. $\frac{2}{5}$
a. 30　**b.** 24　**c.** 24　**d.** 30　**e.** 90　**f.** 60

5.5　Supplementary Exercises

Write each fraction in simplest form.

1. $\frac{10}{15}$　　　**2.** $\frac{9}{12}$　　　**3.** $\frac{14}{16}$　　　**4.** $\frac{18}{27}$

5. $\frac{50}{70}$　　　**6.** $\frac{30}{36}$　　　**7.** $\frac{48}{56}$　　　**8.** $\frac{42}{48}$

9. $\frac{45}{60}$　　　**10.** $\frac{36}{48}$　　　**11.** $\frac{60}{198}$　　　**12.** $\frac{126}{315}$

13. $\frac{21}{25}$　　　**14.** $\frac{35}{48}$

5.6 Building Fractions

OBJECTIVES
1. To use the fundamental principle to build fractions
2. To write a group of fractions as equivalent fractions with a common denominator
3. To compare the sizes of fractions

Often we must *build* a fraction or raise a fraction to *higher terms*. This means finding an equivalent fraction with a numerator and denominator *larger* than those of the original fraction.

Example 1

$\frac{1}{2}, \frac{2}{4}, \frac{3}{6}, \frac{4}{8}$, and $\frac{5}{10}$ are all equivalent fractions. We say that $\frac{2}{4}$ is in higher terms than $\frac{1}{2}$. We have *multiplied* the numerator and denominator by 2. In the same way, $\frac{3}{6}$ is in higher terms than $\frac{1}{2}$, as is $\frac{4}{8}$, and so on. Of course this uses the fundamental principle of fractions introduced in Section 5.5. It is restated here.

Do you see that this is just the opposite of simplifying a fraction!

> **FUNDAMENTAL PRINCIPLE OF FRACTIONS**
>
> $$\frac{a}{b} = \frac{a \times c}{b \times c}$$
>
> for any nonzero number *c*.
>
> In words: When the numerator and denominator of a fraction are multiplied by the same nonzero number, the result is an equivalent fraction.

Example 2

We can, of course, write as many equivalent fractions as we want by using different numbers as multipliers. For instance,
$\frac{4}{5} = \frac{4 \times 4}{5 \times 4} = \frac{16}{20}$

Write an equivalent fraction for $\frac{4}{5}$. Let's use 3 as our multiplier.

$$\frac{4}{5} = \frac{4 \times 3}{5 \times 3} = \frac{12}{15}$$

$\frac{12}{15}$ is equivalent to $\frac{4}{5}$.

Note Again, multiplying the numerator and denominator by 3 is the same as multiplying the fraction by $\frac{3}{3}$, which is just 1. Since

Section 5.6: Building Fractions

multiplying by 1 does not change the value of the fraction, the result is equivalent to the original fraction.

CHECK YOURSELF 1

Write two fractions equivalent to $\frac{3}{7}$. $\frac{3 \times 4}{7 \times 4} = \frac{12}{28}$

Many applications of our rule will involve building a fraction to an equivalent fraction with some specific denominator.

Example 3

To build $\frac{3}{4}$ to an equivalent fraction with denominator 12, we can write

$$\frac{3 \times 3}{4 \times 3} = \frac{?\ 9}{12}$$

(×3)

Think, "What must 4 be multiplied by to give 12?" The answer is 3, and so we multiply the numerator and denominator by that number.

Applying our rule, we have

$$\frac{3}{4} = \frac{9}{12}$$

(×3)

CHECK YOURSELF 2

Write $\frac{2}{3}$ as an equivalent fraction with a denominator of 6.

Example 4

To build $\frac{5}{8}$ to an equivalent fraction with a denominator of 48, the original denominator must be multiplied by 6.

Think: What must 8 be multiplied by to give 48?

$$\frac{5}{8} = \frac{30}{48}$$

(×6)

Remember: Multiply the numerator by the *same* number as you multiply the denominator by; in this case, 6.

CHECK YOURSELF 3

Write $\frac{3}{5}$ as an equivalent fraction with a denominator of 35.

Thus far we have been working with individual fractions. Much of your work with fractions will require building each of a group of fractions to an equivalent fraction with a common (the same) denominator. When fractions have a common denominator, they are called *like fractions*.

Example 5

$\frac{3}{5}$ and $\frac{4}{5}$ have the common denominator 5.

$\frac{3}{5}$ and $\frac{4}{5}$ (like fractions)

$\frac{2}{3}$ and $\frac{3}{4}$ have different denominators.

$\frac{2}{3}$ and $\frac{3}{4}$ (*not* like fractions)

Suppose you are asked to compare the sizes of the fractions $\frac{3}{7}$ and $\frac{4}{7}$. Since each unit in the diagram is divided into seven parts, it is easy to see that $\frac{4}{7}$ is larger than $\frac{3}{7}$.

Four parts of seven must be a greater portion than three parts.

$\frac{4}{7}$

$\frac{3}{7}$

Now look at the same questions for $\frac{2}{5}$ and $\frac{3}{7}$.

We *cannot* compare fifths with sevenths!

$\frac{2}{5}$

$\frac{3}{7}$

$\frac{2}{5}$ and $\frac{3}{7}$ are *not* like fractions. Since they name different ways of dividing the whole, the question about which fraction is larger is not nearly so easy to answer.

In order to compare the sizes of fractions, we can change them to equivalent fractions having a common denominator. This common denominator must be a multiple of each of the original denominators.

Section 5.6: Building Fractions   215

Example 6

Compare the sizes of $\frac{2}{5}$ and $\frac{3}{7}$.

Solution The original denominators are 5 and 7. Since 35 is a common multiple of 5 and 7, let's use 35 as our common denominator.

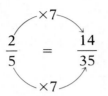

Think, "What must we multiply 5 by to get 35?" The answer is 7. Multiply the numerator and denominator by that number.

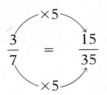

Multiply the numerator and denominator by 5.

15 of 35 parts represents a greater portion of the whole than 14 parts.

Since $\frac{2}{5} = \frac{14}{35}$ and $\frac{3}{7} = \frac{15}{35}$, we see that $\frac{3}{7}$ is larger than $\frac{2}{5}$.

CHECK YOURSELF 4

Which is larger, $\frac{5}{9}$ or $\frac{4}{7}$?

Let's consider an example which uses the inequality notation.

Example 7

Use the symbols < or > to complete the statement below.

$\frac{5}{8}$ _____ $\frac{3}{5}$ $\frac{21}{40}$

Solution Once again we must compare the sizes of the two fractions and this is done by converting the fractions to equivalent fractions with a common denominator. Here we will use 40 as that denominator.

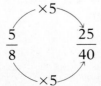

Since $\frac{5}{8}$ (or $\frac{25}{40}$) is larger than $\frac{3}{5}$ (or $\frac{24}{40}$), we write

$$\frac{5}{8} > \frac{3}{5}$$

CHECK YOURSELF 5

Use the symbols < or > to complete the statement

$$\frac{5}{9} \; < \; \frac{6}{11}$$

Comparing sizes is only one reason for changing fractions so that they have common denominators. Common denominators are also necessary in adding and subtracting fractions. You will see this in a later chapter.

Let's look at the process of finding a common denominator in greater detail. The best choice for the common denominator is the least common multiple of the given denominators. Remember from Section 4.4 that this is the *smallest* number that is a multiple of each of the denominators. In our work with fractions this will be called the least common denominator (the LCD).

Using the smallest possible number for the common denominator simply lets us work with the fractions that make the calculation easiest.

To Review How can you find the least common multiple of a group of numbers? Consider the following example.

Example 8

To find the least common multiple of 6, 8, and 12, we write the numbers as a product of their prime factors.

$$\begin{array}{l} 6 = 2 \times 3 \\ 8 = 2 \times 2 \times 2 \\ 12 = 2 \times 2 \times 3 \\ \hline 2 \times 2 \times 2 \times 3 \end{array}$$

Note that 2 appears *three* times as a factor of 8, and so it must appear three times as a factor in forming the LCM.

The least common multiple is $2 \times 2 \times 2 \times 3$, or 24.

A group of fractions with denominators 6, 8, and 12 can be changed to equivalent fractions with denominator 24.

Example 9

Write $\frac{5}{6}$, $\frac{3}{8}$, and $\frac{7}{12}$ as equivalent fractions with the LCD.

$$\frac{5}{6} = \frac{20}{24}$$
(×4 numerator and denominator)

Think, "What must we multiply 6 by to get 24?" The answer is 4, so we multiply the numerator and denominator by 4.

Section 5.6: Building Fractions   217

$$\frac{3}{8} \xrightarrow{\times 3} \frac{9}{24}$$ Multiply the numerator and denominator by 3.

$$\frac{7}{12} \xrightarrow{\times 2} \frac{14}{24}$$ Multiply the numerator and denominator by 2.

$\frac{20}{24}$, $\frac{9}{24}$, and $\frac{14}{24}$ all have the denominator 24, and they are equivalent to the original fractions.

CHECK YOURSELF 6

Write $\frac{2}{3}$, $\frac{3}{4}$, and $\frac{5}{6}$ as equivalent fractions with the least common denominator.

Here is a similar example.

Example 10

To convert $\frac{3}{8}$, $\frac{5}{12}$, and $\frac{7}{15}$ to fractions with the least common denominator, first find the LCD.

Step 1 Find the LCD.

Remember that since 2 appears 3 times as a factor of 8, it must be used three times as a factor in forming the LCD.

$$8 = 2 \times 2 \times 2$$
$$12 = 2 \times 2 \times 3$$
$$15 = 3 \times 5$$
$$\overline{ 2 \times 2 \times 2 \times 3 \times 5} \text{ or 120 is the LCD}$$

Step 2 Write equivalent fractions with that common denominator.

$120 \div 8 = 15$

$$\frac{3}{8} \xrightarrow{\times 15} \frac{45}{120}$$ We must multiply by 15 to change the original denominator, 8, to 120.

$120 \div 12 = 10$

$$\frac{5}{12} \xrightarrow{\times 10} \frac{50}{120}$$ Multiply the numerator and denominator by 10.

Chapter 5: An Introduction to Fractions

$120 \div 15 = 8$

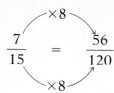

$$\frac{7}{15} = \frac{56}{120}$$

Multiply the numerator and denominator by 8.

$\frac{45}{120}$, $\frac{50}{120}$, and $\frac{56}{120}$ are equivalent to the original fractions, and they now have the least common denominator.

CHECK YOURSELF 7

Convert $\frac{4}{9}$, $\frac{5}{18}$, and $\frac{7}{12}$ to equivalent fractions with the least common denominator.

CHECK YOURSELF ANSWERS

1. Two possible equivalent fractions are $\frac{6}{14}$ (multiply the numerator and denominator by 2) and $\frac{9}{21}$ (multiply by 3). Of course there are many more possibilities.
2. $\frac{4}{6}$ 3. $\frac{21}{35}$
4. $\frac{5}{9} = \frac{35}{63}$ and $\frac{4}{7} = \frac{36}{63}$, so $\frac{4}{7}$ is the larger fraction.
5. $\frac{5}{9} > \frac{6}{11}$ 6. $\frac{8}{12}, \frac{9}{12}$, and $\frac{10}{12}$ 7. $\frac{16}{36}, \frac{10}{36}$, and $\frac{21}{36}$

5.6 Exercises

Find the missing numerators.

1. $\frac{1}{2} = \frac{?}{6}$
2. $\frac{1}{3} = \frac{?}{15}$
3. $\frac{3}{7} = \frac{?}{21}$
4. $\frac{2}{9} = \frac{?}{36}$

5. $\frac{4}{5} = \frac{?}{50}$
6. $\frac{5}{6} = \frac{?}{48}$
7. $\frac{4}{5} = \frac{?}{30}$
8. $\frac{5}{8} = \frac{?}{96}$

9. $\frac{3}{11} = \frac{?}{77}$
10. $\frac{2}{9} = \frac{?}{72}$
11. $\frac{3}{8} = \frac{?}{32}$
12. $\frac{5}{11} = \frac{?}{33}$

13. $\frac{7}{8} = \frac{?}{96}$
14. $\frac{3}{25} = \frac{?}{400}$
15. $\frac{3}{10} = \frac{?}{200}$
16. $\frac{5}{16} = \frac{?}{144}$

Section 5.6: Building Fractions

Arrange the given fractions from smallest to largest.

17. $\dfrac{11}{15}, \dfrac{7}{10}$

18. $\dfrac{4}{9}, \dfrac{5}{11}$

19. $\dfrac{4}{9}, \dfrac{3}{7}$

20. $\dfrac{9}{10}, \dfrac{8}{9}$

21. $\dfrac{3}{8}, \dfrac{1}{3}, \dfrac{1}{4}$

22. $\dfrac{5}{12}, \dfrac{3}{6}, \dfrac{7}{18}$

23. $\dfrac{11}{12}, \dfrac{4}{5}, \dfrac{5}{6}$

24. $\dfrac{3}{8}, \dfrac{11}{32}, \dfrac{7}{16}$

Complete the statements using the symbols < or >.

25. $\dfrac{2}{3} \qquad \dfrac{3}{4}$

26. $\dfrac{5}{6} \qquad \dfrac{11}{12}$

27. $\dfrac{4}{9} \qquad \dfrac{3}{7}$

28. $\dfrac{7}{10} \qquad \dfrac{11}{15}$

29. $\dfrac{9}{20} \qquad \dfrac{11}{25}$

30. $\dfrac{5}{12} \qquad \dfrac{7}{18}$

31. $\dfrac{5}{16} \qquad \dfrac{7}{20}$

32. $\dfrac{7}{12} \qquad \dfrac{9}{15}$

Write as equivalent fractions with the LCD as a common denominator.

33. $\dfrac{3}{5}, \dfrac{3}{4}$

34. $\dfrac{5}{6}, \dfrac{4}{5}$

35. $\dfrac{5}{8}, \dfrac{5}{12}$

36. $\dfrac{7}{12}, \dfrac{5}{18}$

37. $\dfrac{1}{2}, \dfrac{1}{3}, \dfrac{1}{4}$

38. $\dfrac{1}{5}, \dfrac{1}{3}, \dfrac{1}{6}$

39. $\dfrac{3}{25}, \dfrac{7}{10}, \dfrac{5}{6}$

40. $\dfrac{5}{8}, \dfrac{3}{10}, \dfrac{7}{12}$

41. Three drill bits are marked $\dfrac{3}{8}$, $\dfrac{5}{16}$, and $\dfrac{11}{32}$. Which drill bit is largest?

42. Bolts can be purchased with diameters of $\dfrac{3}{8}$ in, $\dfrac{1}{4}$ in, or $\dfrac{3}{16}$ in. Which would be smallest?

43. Plywood comes in $\dfrac{5}{8}$-in, $\dfrac{3}{4}$-in, $\dfrac{1}{2}$-in, and $\dfrac{3}{8}$-in thicknesses. Which size is thickest?

44. Doweling is sold with diameters $\dfrac{1}{2}$ in, $\dfrac{9}{16}$ in, $\dfrac{5}{8}$ in, and $\dfrac{3}{8}$ in. Which size is smallest?

Answers

1. 3   3. 9   5. 40   7. 24   9. 21   11. 12   13. 84   15. 60   17. $\frac{7}{10}, \frac{11}{15}$ 19. $\frac{3}{7}, \frac{4}{9}$
21. $\frac{1}{4}, \frac{1}{3}, \frac{3}{8}$ 23. $\frac{4}{5}, \frac{5}{6}, \frac{11}{12}$ 25. $\frac{2}{3} < \frac{3}{4}$ 27. $\frac{4}{9} > \frac{3}{7}$ 29. $\frac{9}{20} > \frac{11}{25}$ 31. $\frac{5}{16} < \frac{7}{20}$ 33. $\frac{12}{20}, \frac{15}{20}$
35. $\frac{15}{24}, \frac{10}{24}$ 37. $\frac{6}{12}, \frac{4}{12}, \frac{3}{12}$ 39. $\frac{18}{150}, \frac{105}{150}, \frac{125}{150}$ 41. $\frac{3}{8}$ 43. $\frac{3}{4}$ in

5.6 Supplementary Exercises

Find the missing numerators.

1. $\frac{2}{5} = \frac{?}{35}$
2. $\frac{3}{4} = \frac{?}{16}$
3. $\frac{3}{4} = \frac{?}{32}$
4. $\frac{7}{8} = \frac{?}{40}$

5. $\frac{7}{8} = \frac{?}{72}$
6. $\frac{3}{8} = \frac{?}{80}$
7. $\frac{5}{6} = \frac{?}{144}$
8. $\frac{3}{20} = \frac{?}{400}$

Arrange the given fractions from smallest to largest.

9. $\frac{3}{7}, \frac{2}{5}$
10. $\frac{7}{12}, \frac{5}{8}$
11. $\frac{5}{8}, \frac{3}{4}, \frac{2}{3}$
12. $\frac{1}{3}, \frac{2}{9}, \frac{3}{12}$

Complete the statements using the symbols < or >.

13. $\frac{3}{4}$ $\frac{4}{5}$
14. $\frac{9}{16}$ $\frac{7}{12}$
15. $\frac{9}{10}$ $\frac{13}{15}$
16. $\frac{7}{20}$ $\frac{11}{30}$

Write as equivalent fractions with the LCD as a common denominator.

17. $\frac{3}{7}, \frac{4}{5}$
18. $\frac{5}{9}, \frac{1}{12}$
19. $\frac{1}{6}, \frac{3}{8}, \frac{5}{16}$
20. $\frac{3}{4}, \frac{5}{12}, \frac{5}{18}$

SELF-TEST for Chapter Five

The purpose of the Self-Test is to help you check your progress and review for a chapter test in class. Allow yourself about an hour to take the test. When you are done, check your answers in the back of the book. If you missed any problems, be sure to go back and review the appropriate sections in the chapter and do the supplementary exercises provided.

For Problems 1 to 3 what fraction names the shaded part of each diagram? Identify the numerator and denominator.

1.   2.   3.

4. Identify the proper fractions, improper fractions, and mixed numbers in the following group.

$$\left\{\frac{10}{11}, \frac{9}{5}, \frac{7}{7}, \frac{8}{1}, 2\frac{3}{5}, \frac{1}{8}\right\}$$

Proper   Improper   Mixed number

5. Give the mixed number that names the shaded part of the following diagram.

In Problems 6 to 9, convert the fractions to mixed or whole numbers.

6. $\dfrac{13}{5} =$ 7. $\dfrac{65}{9} =$ 8. $\dfrac{18}{6} =$ 9. $\dfrac{5}{1} =$

In Problems 10 to 13, convert the mixed numbers to improper fractions.

10. $3\dfrac{1}{4} =$ 11. $2\dfrac{5}{6} =$ 12. $5\dfrac{2}{7} =$ 13. $9\dfrac{3}{10} =$

In Problems 14 and 15, find out whether the pair of fractions is equivalent.

14. $\dfrac{6}{20}, \dfrac{9}{30}$ 15. $\dfrac{4}{15}, \dfrac{6}{20}$

In Problems 16 to 19, write the fractions in simplest form.

16. $\dfrac{14}{18} =$ 17. $\dfrac{21}{28} =$ 18. $\dfrac{36}{84} =$ 19. $\dfrac{15}{32} =$

In Problems 20 to 22, find the missing numerators.

20. $\dfrac{4}{5} = \dfrac{?}{20}$ 21. $\dfrac{3}{7} = \dfrac{?}{84}$ 22. $\dfrac{7}{8} = \dfrac{?}{120}$

In Problems 23 and 24, arrange the fractions from smallest to largest.

23. $\dfrac{7}{9}, \dfrac{5}{7}$ 24. $\dfrac{2}{5}, \dfrac{1}{3}, \dfrac{4}{15}$

In Problems 25 and 26, complete the statements using the symbols < or >.

25. $\dfrac{5}{6}$ $\dfrac{4}{5}$ 26. $\dfrac{7}{12}$ $\dfrac{11}{18}$

In Problems 27 to 30, write the fractions as equivalent fractions with the LCD as a common denominator.

27. $\dfrac{2}{5}, \dfrac{3}{4}$ 28. $\dfrac{7}{8}, \dfrac{5}{12}$ 29. $\dfrac{3}{8}, \dfrac{1}{4}, \dfrac{5}{6}$ 30. $\dfrac{9}{20}, \dfrac{11}{15}, \dfrac{7}{12}$

PRETEST for Chapter Six

Multiplication and Division of Fractions

This pretest will point out any difficulties you may be having with multiplying and dividing fractions. Do all the problems. Then check your answers on the following page.

1. $\dfrac{3}{8} \times \dfrac{4}{5} =$

2. $7 \times \dfrac{2}{9} =$

3. $2\dfrac{2}{5} \times 1\dfrac{3}{4} =$

4. $\dfrac{15}{18} \times \dfrac{2}{5} =$

5. $3\dfrac{1}{3} \times 4\dfrac{1}{5} =$

6. $\dfrac{6}{25} \div \dfrac{3}{10} =$

7. $\dfrac{1}{3} \div 5 =$

8. $2\dfrac{3}{4} \div 1\dfrac{5}{8} =$

9. A sheet of paper is $5\dfrac{1}{2}$ in long by $3\dfrac{3}{4}$ in wide. What is its area?

10. A piece of wood that is $15\dfrac{3}{4}$ in long is to be cut into blocks $1\dfrac{1}{8}$ in long. How many blocks can be cut?

ANSWERS TO PRETEST

For help with similar problems, turn to the section indicated.

1. $\frac{3}{10}$ (Section 6.1);   2. $1\frac{5}{9}$ (Section 6.1);   3. $4\frac{1}{5}$ (Section 6.2);   4. $\frac{1}{3}$ (Section 6.3);

5. 14 (Section 6.3);   6. $\frac{4}{5}$ (Section 6.4);   7. $\frac{1}{15}$ (Section 6.4);   8. $1\frac{9}{13}$ (Section 6.4);

9. $20\frac{5}{8}$ sq in (Section 6.3);   10. 14 (Section 6.4)

Chapter Six
Multiplication and Division of Fractions

6.1 Multiplying Fractions

OBJECTIVE
To multiply fractions and give the product in simplest form

Multiplication is the easiest of the four operations with fractions. We can illustrate multiplication by picturing fractions as parts of a whole or unit. Using this idea, we show the fractions $\frac{4}{5}$ and $\frac{2}{3}$ in Figure 1.

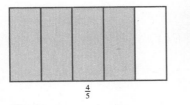

Figure 1

A fraction followed by the word "of" means that we want to multiply by that fraction.

Suppose now that we wish to find $\frac{2}{3}$ of $\frac{4}{5}$. We can combine the diagrams as shown in Figure 2. The part of the whole representing the product $\frac{2}{3} \times \frac{4}{5}$ is the region that has been double-shaded. The unit has been divided into 15 parts and eight of those parts are used, so $\frac{2}{3} \times \frac{4}{5}$ must be $\frac{8}{15}$.

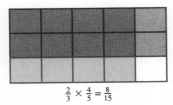

$\frac{2}{3} \times \frac{4}{5} = \frac{8}{15}$

Figure 2

Do you see the rule suggested by the diagrams?

TO MULTIPLY FRACTIONS
STEP 1 Multiply the numerators.
STEP 2 Multiply the denominators.
STEP 3 Simplify the resulting fraction if possible.

This gives the numerator of the product.
This gives the denominator of the product.

In symbols,

$$\frac{a}{b} \times \frac{c}{d} = \frac{a \times c}{b \times d}$$

where b and d are not equal to zero.

Example 1

We multiply fractions this way not because it is easy, but because it works!

$$\frac{2}{3} \times \frac{4}{5} = \frac{2 \times 4}{3 \times 5} = \frac{8}{15}$$

$$\frac{5}{8} \times \frac{7}{9} = \frac{5 \times 7}{8 \times 9} = \frac{35}{72}$$

CHECK YOURSELF 1
Multiply.

$$\frac{7}{8} \times \frac{3}{10}$$

The product of fractions should always be simplified to lowest terms. Consider the following.

Example 2

$$\frac{3}{4} \times \frac{2}{9} = \frac{3 \times 2}{4 \times 9} = \frac{6}{36} = \frac{1}{6}$$

Noting that $\frac{6}{36}$ is not in simplest form, we divide numerator and denominator by 6 to write the product in lowest terms.

CHECK YOURSELF 2
Multiply.

$$\frac{5}{7} \times \frac{3}{10} = \frac{15}{70}$$

Section 6.1: Multiplying Fractions 227

To find the product of a fraction and a whole number, write the whole number as a fraction (the whole number divided by 1) and apply the multiplication rule as before. The following examples illustrate.

Example 3

Remember that $5 = \frac{5}{1}$.

$$5 \times \frac{3}{4} = \frac{5}{1} \times \frac{3}{4} = \frac{5 \times 3}{1 \times 4}$$

$$= \frac{15}{4} = 3\frac{3}{4}$$

We have written the resulting improper fraction as a mixed number.

Example 4

$$\frac{5}{12} \times 6 = \frac{5}{12} \times \frac{6}{1}$$

$$= \frac{5 \times 6}{12 \times 1}$$

$$= \frac{30}{12} = 2\frac{6}{12}$$

$$= 2\frac{1}{2}$$

Write the product as a mixed number, then reduce the fractional portion to simplest form.

CHECK YOURSELF 3

Multiply.

$$\frac{3}{16} \times 8$$

CHECK YOURSELF ANSWERS

1. $\frac{7}{8} \times \frac{3}{10} = \frac{7 \times 3}{8 \times 10} = \frac{21}{80}$ 2. $\frac{3}{14}$ 3. $1\frac{1}{2}$

6.1 Exercises

Multiply. Be sure to simplify each product.

1. $\frac{2}{5} \times \frac{3}{7} = \frac{6}{35}$

2. $\frac{3}{8} \times \frac{7}{11} = \frac{21}{88}$

3. $\frac{3}{4} \times \frac{7}{5} = \frac{21}{20}$

4. $\frac{2}{3} \times \frac{8}{5} = \frac{16}{15}$

5. $\dfrac{3}{5} \times \dfrac{5}{7}$ 6. $\dfrac{7}{8} \times \dfrac{3}{7}$ 7. $\dfrac{4}{7} \times \dfrac{3}{8}$ 8. $\dfrac{5}{9} \times \dfrac{6}{11}$

9. $\dfrac{4}{7} \times \dfrac{5}{12}$ 10. $\dfrac{7}{9} \times \dfrac{3}{5}$ 11. $\dfrac{3}{10} \times \dfrac{5}{9}$ 12. $\dfrac{2}{15} \times \dfrac{5}{8}$

13. $\dfrac{7}{9} \times \dfrac{6}{5}$ 14. $\dfrac{7}{10} \times \dfrac{5}{3}$ 15. $4 \times \dfrac{3}{5}$ 16. $\dfrac{7}{9} \times 5$

17. $\dfrac{3}{7} \times 14$ 18. $9 \times \dfrac{5}{6}$ 19. $12 \times \dfrac{2}{9}$ 20. $\dfrac{7}{8} \times 16$

21. Find $\dfrac{3}{4}$ of $\dfrac{6}{7}$.   22. Find $\dfrac{2}{3}$ of $\dfrac{7}{10}$.

23. What is $\dfrac{3}{5}$ of $\dfrac{5}{6}$?   24. What is $\dfrac{5}{8}$ of $\dfrac{12}{15}$?

Skillscan (Section 5.2)
Convert the mixed numbers to improper fractions.

a. $3\dfrac{1}{5}$ b. $2\dfrac{3}{4}$ c. $5\dfrac{1}{6}$

d. $7\dfrac{3}{8}$ e. $4\dfrac{3}{10}$ f. $6\dfrac{5}{7}$

Answers
Solutions for the even-numbered exercises are provided in the back of the book.

1. $\dfrac{6}{35}$ 3. $\dfrac{3}{4} \times \dfrac{7}{5} = \dfrac{21}{20} = 1\dfrac{1}{20}$ 5. $\dfrac{3}{7}$ 7. $\dfrac{3}{14}$ 9. $\dfrac{5}{21}$ 11. $\dfrac{3}{10} \times \dfrac{5}{9} = \dfrac{15}{90} = \dfrac{1}{6}$ 13. $\dfrac{14}{15}$ 15. $2\dfrac{2}{5}$
17. $\dfrac{3}{7} \times 14 = \dfrac{3}{7} \times \dfrac{14}{1} = \dfrac{3 \times 14}{7 \times 1} = \dfrac{42}{7} = 6$ 19. $2\dfrac{2}{3}$ 21. $\dfrac{3}{4}$ of $\dfrac{6}{7} = \dfrac{3}{4} \times \dfrac{6}{7} = \dfrac{3 \times 6}{4 \times 7} = \dfrac{18}{28} = \dfrac{9}{14}$ 23. $\dfrac{1}{2}$
a. $\dfrac{16}{5}$ b. $\dfrac{11}{4}$ c. $\dfrac{31}{6}$ d. $\dfrac{59}{8}$ e. $\dfrac{43}{10}$ f. $\dfrac{47}{7}$

Section 6.2: Multiplying Mixed Numbers   229

6.1 Supplementary Exercises

Multiply. Be sure to simplify each product.

1. $\dfrac{2}{3} \times \dfrac{5}{7}$

2. $\dfrac{4}{9} \times \dfrac{2}{3}$

3. $\dfrac{5}{8} \times \dfrac{6}{11}$

4. $\dfrac{4}{9} \times \dfrac{7}{12}$

5. $\dfrac{5}{8} \times \dfrac{7}{10}$

6. $\dfrac{3}{5} \times \dfrac{10}{21}$

7. $\dfrac{5}{6} \times \dfrac{9}{20}$

8. $8 \times \dfrac{4}{7}$

9. $\dfrac{5}{8} \times 12$

10. $6 \times \dfrac{5}{8}$

11. Find $\dfrac{3}{5}$ of 9.

12. What is $\dfrac{2}{3}$ of $\dfrac{7}{8}$?

6.2 Multiplying Mixed Numbers

OBJECTIVE
To multiply mixed numbers

When mixed numbers are involved in multiplication, the problem requires an additional step. Change any mixed numbers to improper fractions. Then apply our multiplication rule for fractions.

Example 1

$1\dfrac{1}{2} \times \dfrac{3}{4} = \dfrac{3}{2} \times \dfrac{3}{4}$ Change the mixed number to an improper fraction.

Here $1\dfrac{1}{2} = \dfrac{3}{2}$.

$= \dfrac{3 \times 3}{2 \times 4}$ Multiply as before.

$= \dfrac{9}{8} = 1\dfrac{1}{8}$ The product is usually written in mixed number form.

CHECK YOURSELF 1

Multiply.

$$\frac{5}{8} \times 3\frac{1}{2}$$

If two mixed numbers are involved, change both of the mixed numbers to improper fractions. Our second example illustrates.

Example 2

Change the mixed numbers to improper fractions.

$$3\frac{2}{3} \times 2\frac{1}{2} = \frac{11}{3} \times \frac{5}{2}$$

$$= \frac{11 \times 5}{3 \times 2} = \frac{55}{6} = 9\frac{1}{6}$$

Be Careful! Students sometimes think of

$$3\frac{2}{3} \times 2\frac{1}{2} \quad \text{as} \quad (3 \times 2) + \left(\frac{2}{3} \times \frac{1}{2}\right)$$

This is *not* the correct multiplication pattern. You must first change the mixed numbers to improper fractions.

CHECK YOURSELF 2

Multiply.

$$2\frac{1}{3} \times 3\frac{1}{2}$$

Again, be sure to reduce the product of the mixed numbers to simplest form. Consider our next example.

Example 3

$$5\frac{1}{3} \times 1\frac{1}{4} = \frac{16}{3} \times \frac{5}{4}$$

$$= \frac{16 \times 5}{3 \times 4} = \frac{80}{12} = 6\frac{8}{12} \quad \text{Write the product as a mixed number.}$$

$$= 6\frac{2}{3} \quad \text{Reduce the fractional portion to lowest terms.}$$

Section 6.2: Multiplying Mixed Numbers

CHECK YOURSELF 3

Multiply.

$$3\frac{3}{4} \times 1\frac{1}{5}$$

If a multiplication problem involves a whole number and a mixed number, write the mixed number as an improper fraction. Write the whole number as a fraction (the whole number over 1). Then multiply as before. This is shown in our next example.

Example 4

$$8 \times 5\frac{1}{4} = \frac{8}{1} \times \frac{21}{4}$$

$$= \frac{8 \times 21}{1 \times 4} = \frac{168}{4} = 42$$

CHECK YOURSELF 4

Multiply.

$$5\frac{1}{3} \times 9$$

CHECK YOURSELF ANSWERS

1. $\frac{5}{8} \times 3\frac{1}{2} = \frac{5}{8} \times \frac{7}{2} = \frac{35}{16} = 2\frac{3}{16}$ 2. $8\frac{1}{6}$
3. Write $3\frac{3}{4}$ as $\frac{15}{4}$ and $1\frac{1}{5}$ as $\frac{6}{5}$. Multiply as before. The product is $4\frac{1}{2}$.
4. 48

6.2 Exercises

Multiply. Be sure to simplify each product.

1. $1\frac{1}{3} \times \frac{2}{3}$ 2. $1\frac{1}{5} \times \frac{3}{4}$ 3. $\frac{5}{8} \times 1\frac{3}{4}$

4. $\frac{7}{8} \times 1\frac{3}{5}$ 5. $2\frac{1}{4} \times \frac{5}{6}$ 6. $\frac{2}{3} \times 2\frac{2}{5}$

7. $3\dfrac{1}{3} \times \dfrac{3}{7}$

8. $\dfrac{3}{4} \times 2\dfrac{2}{5}$

9. $1\dfrac{1}{2} \times 1\dfrac{1}{4}$

10. $2\dfrac{1}{3} \times 2\dfrac{1}{2}$

11. $1\dfrac{3}{4} \times 2\dfrac{3}{5}$

12. $3\dfrac{2}{3} \times 1\dfrac{1}{7}$

13. $2\dfrac{2}{3} \times 2\dfrac{3}{4}$

14. $3\dfrac{3}{4} \times 1\dfrac{1}{5}$

15. $1\dfrac{2}{5} \times 1\dfrac{1}{4}$

16. $2\dfrac{1}{7} \times 4\dfrac{1}{5}$

17. $2\dfrac{1}{2} \times 3$

18. $2 \times 3\dfrac{1}{3}$

19. $5 \times 4\dfrac{2}{3}$

20. $5\dfrac{1}{2} \times 6$

21. $1\dfrac{5}{8} \times 4$

22. $10 \times 2\dfrac{4}{5}$

23. $3\dfrac{3}{5} \times 10$

24. $8 \times 5\dfrac{1}{4}$

Skillscan (Section 5.5)
Write each fraction in simplest form.

a. $\dfrac{14}{16}$

b. $\dfrac{27}{30}$

c. $\dfrac{20}{35}$

d. $\dfrac{18}{24}$

e. $\dfrac{32}{40}$

f. $\dfrac{40}{60}$

Answers

1. $\dfrac{8}{9}$ **3.** $1\dfrac{3}{32}$ **5.** $1\dfrac{7}{8}$ **7.** $3\dfrac{1}{3} \times \dfrac{3}{7} = \dfrac{10}{3} \times \dfrac{3}{7} = \dfrac{30}{21} = 1\dfrac{9}{21} = 1\dfrac{3}{7}$ **9.** $1\dfrac{7}{8}$ **11.** $4\dfrac{11}{20}$
13. $2\dfrac{2}{3} \times 2\dfrac{3}{4} = \dfrac{8}{3} \times \dfrac{11}{4} = \dfrac{88}{12} = 7\dfrac{4}{12} = 7\dfrac{1}{3}$ **15.** $1\dfrac{3}{4}$ **17.** $7\dfrac{1}{2}$ **19.** $23\dfrac{1}{3}$ **21.** $6\dfrac{1}{2}$ **23.** 36 **a.** $\dfrac{7}{8}$
b. $\dfrac{9}{10}$ **c.** $\dfrac{4}{7}$ **d.** $\dfrac{3}{4}$ **e.** $\dfrac{4}{5}$ **f.** $\dfrac{2}{3}$

6.2 Supplementary Exercises

Multiply. Be sure to simplify each product.

1. $\dfrac{2}{7} \times 1\dfrac{3}{5}$

2. $2\dfrac{1}{4} \times \dfrac{1}{5}$

3. $1\dfrac{9}{16} \times \dfrac{4}{5}$

4. $\dfrac{3}{8} \times 3\dfrac{3}{7}$

5. $2\dfrac{1}{5} \times 1\dfrac{1}{2}$

6. $3\dfrac{3}{7} \times 1\dfrac{1}{3}$

7. $3\dfrac{1}{2} \times 4\dfrac{2}{3}$

8. $4\dfrac{1}{3} \times 2\dfrac{1}{4}$

9. $2\dfrac{1}{3} \times 4$

10. $5 \times 3\dfrac{1}{2}$

11. $4 \times 6\dfrac{1}{2}$

12. $5\dfrac{2}{3} \times 6$

6.3 Simplifying in Multiplication

OBJECTIVES
1. To use simplification in multiplying fractions
2. To use multiplication in solving word problems

In Sections 6.1 and 6.2, you saw that many of the products were not in lowest terms. In multiplying fractions it is usually easier to simplify, that is, remove any common factors in the numerator and denominator, *before multiplying*. Remember that to simplify means to *divide* by the same common factor.

Example 1

Once again we are applying the fundamental principle to divide the numerator and denominator by 3.

Since we divide by any common factors before we multiply, the resulting product *is in simplest form*.

$$\dfrac{3}{5} \times \dfrac{4}{9} = \dfrac{\overset{1}{\cancel{3}} \times 4}{5 \times \underset{3}{\cancel{9}}}$$

We divide the *numerator* and *denominator* by the common factor 3. Note that $\overset{1}{\cancel{3}}$ means $3 \div 3 = 1$, and $\underset{3}{\cancel{9}}$ means $9 \div 3 = 3$.

$$= \dfrac{1 \times 4}{5 \times 3}$$

$$= \dfrac{4}{15}$$

CHECK YOURSELF 1

Multiply.

$$\frac{7}{8} \times \frac{5}{21}$$

Our work of the previous example leads to the following general rule about simplifying fractions in multiplication.

> In multiplying two or more fractions, we can divide any factor of the numerator and any factor of the denominator by the same nonzero number to simplify the product.

Our next example further illustrates the use of this rule.

Example 2

$$\frac{6}{25} \times \frac{20}{9} = \frac{\overset{2}{\cancel{6}} \times \overset{4}{\cancel{20}}}{\underset{5}{\cancel{25}} \times \underset{3}{\cancel{9}}}$$

We remove the common factor of 3 from 6 and 9 and then the common factor of 5 from 20 and 25.

$$= \frac{2 \times 4}{5 \times 3}$$

$$= \frac{8}{15}$$

As you may have observed, simplifying before you multiply makes the problem much easier. You'll get the same answer if you multiply and then reduce to lowest terms. It's just a lot more work that way.

CHECK YOURSELF 2

Multiply.

$$\frac{5}{12} \times \frac{8}{15}$$

Simplifying is also useful when the multiplication involves whole or mixed numbers as is shown in our following examples.

Section 6.3: Simplifying in Multiplication

Example 3

The whole number, 5, is written as an improper fraction for our first step.

$$5 \times \frac{4}{25} = \frac{5}{1} \times \frac{4}{25} \qquad \text{First write 5 as } \frac{5}{1}.$$

$$= \frac{\overset{1}{\cancel{5}} \times 4}{1 \times \underset{5}{\cancel{25}}} \qquad \text{Now divide by the common factor of 5.}$$

$$= \frac{4}{5}$$

CHECK YOURSELF 3

Multiply.

$$\frac{9}{16} \times 12$$

When mixed numbers are involved, the process is similar. Consider the following.

Example 4

Convert the mixed numbers to improper fractions. This is the same first step we saw earlier.

$$2\frac{2}{3} \times 2\frac{1}{4} = \frac{8}{3} \times \frac{9}{4}$$

$$= \frac{\overset{2}{\cancel{8}} \times \overset{3}{\cancel{9}}}{\underset{1}{\cancel{3}} \times \underset{1}{\cancel{4}}} \qquad \text{Divide by the common factors of 3 and 4.}$$

$$= \frac{2 \times 3}{1 \times 1} \qquad \text{Multiply as before.}$$

$$= \frac{6}{1} = 6$$

CHECK YOURSELF 4

Multiply.

$$3\frac{1}{3} \times 2\frac{2}{5}$$

The ideas of our previous examples will also allow us to find the product of more than two fractions.

Example 5

Remember our earlier rule: We can divide any factor of the numerator and any factor of the denominator by the same nonzero number.

$$\frac{2}{3} \times 1\frac{4}{5} \times \frac{5}{8} = \frac{2}{3} \times \frac{9}{5} \times \frac{5}{8}$$ Write any mixed or whole numbers as improper fractions.

$$= \frac{\overset{1}{\cancel{2}} \times \overset{3}{\cancel{9}} \times \overset{1}{\cancel{5}}}{\underset{1}{\cancel{3}} \times \underset{1}{\cancel{5}} \times \underset{4}{\cancel{8}}}$$ Divide by the common factors in the numerator and denominator.

$$= \frac{3}{4}$$

CHECK YOURSELF 5

Multiply.

$$\frac{5}{8} \times 4\frac{4}{5} \times \frac{1}{6}$$

Let's look at some applications of our work with the multiplication of fractions. In solving these word problems, we will use the same approach we used earlier with whole numbers. Let's review the four-step process introduced in Section 1.9.

SOLVING WORD PROBLEMS

STEP 1 Read the problem carefully to determine the given information and what you are asked to find.

STEP 2 Decide upon the operation or operations to be used.

STEP 3 Write down the complete statement necessary to solve the problem and do the calculations.

STEP 4 Check to make sure that you have answered the question of the problem and that your answer seems reasonable.

Let's work through some examples using these steps.

Example 6

A grocery store survey shows that $\frac{2}{3}$ of the customers will buy meat. Of these, $\frac{3}{4}$ will buy at least one package of beef. What portion of the store's customers will buy beef?

Section 6.3: Simplifying in Multiplication   237

Step 1 We know that $\frac{2}{3}$ of the customers will buy meat and that $\frac{3}{4}$ of these customers will buy beef.

Remember: In this problem, "of" means to multiply.

Step 2 We wish to know $\frac{3}{4}$ of $\frac{2}{3}$. The operation here is multiplication.

Step 3 Multiplying, we have

$$\frac{\cancel{3}^1}{\cancel{4}_2} \times \frac{\cancel{2}^1}{\cancel{3}_1} = \frac{1}{2}$$

Step 4 From step 3 we have the result: One-half of the store's customers will buy beef.

Example 7

A sheet of notepaper is $6\frac{3}{4}$ in wide by $8\frac{2}{3}$ in long. Find the area of the paper.

Recall that the area of a rectangle is the product of its length and its width.

Solution Multiply the given length times the width. This will give the desired area.

$$8\frac{2}{3} \times 6\frac{3}{4} = \frac{\cancel{26}^{13}}{\cancel{3}_1} \times \frac{\cancel{27}^9}{\cancel{4}_2}$$

$$= \frac{117}{2} = 58\frac{1}{2} \text{ sq in}$$

CHECK YOURSELF 6

A window is $4\frac{1}{2}$ ft high by $2\frac{1}{3}$ ft wide. What is its area?

Example 8

A state park contains $38\frac{2}{3}$ acres. According to the plan for the park, $\frac{3}{4}$ of the park is to be left as a wildlife preserve. How many acres will this be?

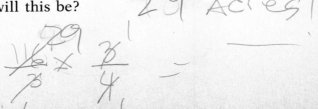

Solution We want to find $\frac{3}{4}$ of $38\frac{2}{3}$ acres. We then multiply as shown:

The word "of" indicates multiplication.

$$\frac{3}{4} \times 38\frac{2}{3} = \frac{\cancel{3}^{1}}{\cancel{4}_{1}} \times \frac{\cancel{116}^{29}}{\cancel{3}_{1}} = 29 \text{ acres}$$

Example 9

Shirley drives at an average speed of 52 mi/h for $3\frac{1}{4}$ h. How far will she have traveled?

Solution

Distance is the product of the speed and the time.

$$52 \times 3\frac{1}{4} = 52 \times \frac{13}{4}$$

Speed (mi/h) Time (hours)

$$= \frac{\cancel{52}^{13} \times 13}{1 \times \cancel{4}_{1}}$$

$$= 169 \text{ mi}$$

CHECK YOURSELF 7

The scale on a map is 1 in equals 60 mi. What is the actual distance between two towns that are $3\frac{1}{2}$ in apart on the map?

CHECK YOURSELF ANSWERS

1. $\frac{\cancel{7}^{1}}{8} \times \frac{5}{\cancel{21}_{3}} = \frac{5}{24}$ 2. $\frac{5}{12} \times \frac{8}{15} = \frac{\cancel{5}^{1} \times \cancel{8}^{2}}{\cancel{12}_{3} \times \cancel{15}_{3}} = \frac{2}{9}$ 3. $6\frac{3}{4}$

4. $3\frac{1}{3} \times 2\frac{2}{5} = \frac{10}{3} \times \frac{12}{5} = \frac{\cancel{10}^{2} \times \cancel{12}^{4}}{\cancel{3}_{1} \times \cancel{5}_{1}} = \frac{8}{1} = 8$

5. $\frac{1}{2}$ 6. $10\frac{1}{2}$ sq ft   7. 210 mi

6.3 Exercises

Multiply.

1. $\dfrac{4}{5} \times \dfrac{1}{8}$

2. $\dfrac{3}{7} \times \dfrac{5}{9}$

3. $\dfrac{7}{15} \times \dfrac{10}{13}$

4. $\dfrac{12}{25} \times \dfrac{11}{18}$

5. $\dfrac{10}{12} \times \dfrac{16}{25}$

6. $\dfrac{14}{15} \times \dfrac{10}{21}$

7. $\dfrac{21}{25} \times \dfrac{30}{7}$

8. $\dfrac{16}{21} \times \dfrac{14}{20}$

9. $5\dfrac{5}{6} \times \dfrac{3}{7}$

10. $\dfrac{4}{9} \times 3\dfrac{3}{5}$

11. $5\dfrac{1}{3} \times \dfrac{7}{8}$

12. $\dfrac{10}{27} \times 3\dfrac{3}{5}$

13. $1\dfrac{1}{2} \times 1\dfrac{1}{6}$

14. $1\dfrac{3}{4} \times 2\dfrac{2}{5}$

15. $3\dfrac{3}{8} \times 1\dfrac{1}{15}$

16. $7\dfrac{1}{5} \times 4\dfrac{1}{6}$

17. $3\dfrac{3}{7} \times 2\dfrac{5}{8}$

18. $4\dfrac{3}{8} \times 1\dfrac{5}{7}$

19. $6 \times 2\dfrac{1}{3}$

20. $1\dfrac{3}{10} \times 5$

21. $3\dfrac{1}{8} \times 4$

22. $6 \times 4\dfrac{2}{3}$

23. $\dfrac{7}{12} \times \dfrac{3}{4} \times \dfrac{8}{15}$

24. $\dfrac{5}{18} \times \dfrac{2}{3} \times \dfrac{9}{10}$

25. $2\dfrac{1}{2} \times \dfrac{4}{5} \times \dfrac{3}{8}$

26. $\dfrac{7}{8} \times 5\dfrac{1}{3} \times \dfrac{5}{14}$

27. $3\dfrac{1}{3} \times \dfrac{4}{5} \times 1\dfrac{1}{8}$

28. $4\dfrac{1}{2} \times 5\dfrac{5}{6} \times \dfrac{8}{15}$

29. Find $\dfrac{3}{5}$ of $\dfrac{5}{9}$.

30. What is $\dfrac{5}{6}$ of $\dfrac{9}{10}$?

31. What is $\dfrac{3}{5}$ of 15?

32. Find $\dfrac{5}{8}$ of 24.

33. Find $\dfrac{3}{4}$ of $2\dfrac{2}{5}$.

34. What is $\dfrac{5}{8}$ of $1\dfrac{5}{7}$?   35. What is $\dfrac{7}{8}$ of $5\dfrac{1}{3}$?   36. Find $\dfrac{4}{9}$ of $3\dfrac{3}{5}$.

37. The scale on a map uses 1 in to represent 200 mi. What actual distance does $\dfrac{3}{8}$ in represent?

38. You make $72 a day on a job. What will you receive for working $\dfrac{3}{4}$ of a day?

39. A lumber yard has a stack of 60 sheets of plywood. If each sheet is $\dfrac{3}{4}$ in thick, how high will the pile be?

40. A family averages $\dfrac{2}{5}$ of its monthly income for housing and utilities. If the family's monthly income is $1750, what is spent for housing and utilities? What amount remains?

41. To find the approximate circumference or distance around a circle, we multiply its diameter by $\dfrac{22}{7}$. What is the circumference of a circle with diameter 14 in?

42. The length of a rectangle is $\dfrac{5}{8}$ yd, and its width is $\dfrac{2}{3}$ yd. What is its area in square yards?

43. Of the eligible voters in an election, $\dfrac{3}{4}$ were registered. Of those registered, $\dfrac{5}{9}$ actually voted. What fraction of those people who were eligible voted?

44. A survey has found that $\dfrac{7}{10}$ of the people in a city own pets. Of those that own pets, $\dfrac{2}{3}$ have dogs. What fraction of those surveyed own dogs?

45. A kitchen has dimensions $3\dfrac{1}{3}$ yd by $3\dfrac{3}{4}$ yd. How many square yards of linoleum must be bought to cover the floor?

46. If you drive at an average speed of 48 mi/h for $2\dfrac{3}{4}$ h, how far will you have traveled?

Section 6.3: Simplifying in Multiplication   241

47. A jet flew at an average speed of 540 mi/h on a $4\frac{2}{3}$-h flight. What was the distance flown?

48. A piece of land which has $11\frac{2}{3}$ acres is being subdivided for home lots. It is estimated that $\frac{2}{7}$ of the area will be used for roads. What amount remains to be used for lots?

The formula for the area of a triangle is

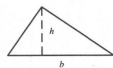

$$A = \frac{1}{2} \times h \times b$$

where h is the height of the triangle and b is the base.

49. Find the area of a triangle with height $1\frac{3}{5}$ in and base $3\frac{3}{4}$ in.

50. Find the area of a triangle with height $1\frac{7}{8}$ in and base $2\frac{2}{5}$ in.

We encountered estimation by rounding in our earlier work with whole numbers. Estimation can also be used to check the "reasonableness" of an answer when working with fractions or mixed numbers. Consider the following product:

$$3\frac{1}{8} \times 5\frac{5}{6}$$

Round each mixed number to the nearest whole number.

$3\frac{1}{8} \to 3$

$5\frac{5}{6} \to 6$

Our estimate of the product is then

$3 \times 6 = 18$

Note The actual product in this case is $18\frac{11}{48}$ which certainly seems reasonable in view of our estimate.

Estimate the following products.

51. $5\frac{1}{4} \times 6\frac{9}{10}$

52. $3\frac{7}{8} \times 6\frac{1}{3}$

53. $9\frac{2}{5} \times 3\frac{5}{6}$

54. $8\frac{1}{3} \times 11\frac{3}{4}$

Skillscan (Sections 6.1 and 6.2)
Multiply.

a. $\frac{2}{3} \times \frac{3}{2}$

b. $\frac{4}{5} \times \frac{5}{4}$

c. $\frac{4}{7} \times \frac{7}{4}$

d. $\frac{3}{10} \times \frac{10}{3}$

e. $5 \times \frac{1}{5}$

f. $9 \times \frac{1}{9}$

Answers

1. $\frac{1}{10}$ **3.** $\frac{7}{\cancel{15}_{3}} \times \frac{\cancel{10}^{2}}{13} = \frac{14}{39}$ **5.** $\frac{8}{15}$ **7.** $3\frac{3}{5}$ **9.** $2\frac{1}{2}$ **11.** $5\frac{1}{3} \times \frac{7}{8} = \frac{\cancel{16}^{2}}{3} \times \frac{7}{\cancel{8}_{1}} = \frac{14}{3} = 4\frac{2}{3}$ **13.** $1\frac{3}{4}$

15. $3\frac{3}{5}$ **17.** 9 **19.** $6 \times 2\frac{1}{3} = \frac{\cancel{6}^{2}}{1} \times \frac{7}{\cancel{3}_{1}} = 14$ **21.** $12\frac{1}{2}$ **23.** $\frac{7}{30}$ **25.** $\frac{3}{4}$ **27.** 3 **29.** $\frac{1}{3}$

31. $\frac{3}{5}$ of 15 is $\frac{3}{5} \times 15 = \frac{3}{\cancel{5}_{1}} \times \frac{\cancel{15}^{3}}{1} = 9$ **33.** $1\frac{4}{5}$ **35.** $4\frac{2}{3}$ **37.** 75 mi **39.** 45 in **41.** 44 in **43.** $\frac{5}{12}$

45. $12\frac{1}{2}$ sq yd **47.** 2520 mi **49.** 3 sq in **51.** 35 **53.** 36 **a.** 1 **b.** 1 **c.** 1 **d.** 1 **e.** 1
f. 1

6.3 Supplementary Exercises

Multiply.

1. $\frac{2}{5} \times \frac{5}{7}$

2. $\frac{4}{9} \times \frac{3}{7}$

3. $\frac{15}{16} \times \frac{7}{20}$

4. $\frac{7}{18} \times \frac{12}{25}$

5. $2\frac{2}{5} \times \frac{3}{4}$

6. $3\frac{3}{5} \times \frac{5}{6}$

Section 6.3: Simplifying in Multiplication

7. $3\frac{1}{5} \times 1\frac{1}{8}$

8. $1\frac{3}{5} \times 2\frac{1}{12}$

9. $1\frac{1}{9} \times 2\frac{7}{10}$

10. $5 \times 3\frac{1}{5}$

11. $2\frac{1}{4} \times 8$

12. $\frac{2}{5} \times \frac{3}{4} \times \frac{1}{6}$

13. $2\frac{1}{2} \times \frac{9}{10} \times 1\frac{1}{3}$

14. $2\frac{2}{5} \times 1\frac{7}{8} \times 2\frac{2}{3}$

15. What is $\frac{2}{3}$ of 27?

16. Find $\frac{3}{10}$ of 25.

17. Find $\frac{2}{5}$ of $3\frac{1}{8}$.

18. What is $\frac{3}{4}$ of $1\frac{7}{9}$?

19. Maria earns $72 per day. If she works $\frac{5}{8}$ of a day, how much will she earn?

20. David drove at an average speed of 55 mi/h for $3\frac{1}{5}$ h. How many miles did he travel?

21. The scale on a map uses 1 in to represent 120 mi. What actual distance does $2\frac{3}{4}$ in on the map represent?

22. At a college, $\frac{2}{5}$ of the students take a science course. Of those, $\frac{1}{4}$ take biology. What fraction of the students take biology?

23. A student survey found that $\frac{5}{6}$ of the students have jobs while going to school. Of those that have jobs, $\frac{2}{3}$ work more than 20 hours per week. What fraction of those surveyed work more than 20 hours per week?

24. A living room has dimensions $5\frac{2}{3}$ yd by $4\frac{1}{2}$ yd. How much carpeting must be purchased to cover the room?

6.4 Dividing Fractions

OBJECTIVES
1. To divide fractions and mixed numbers
2. To use division in solving word problems

We are now ready to look at the operation of division on fractions. First we will need a new concept, the *reciprocal* of a fraction.

In general the reciprocal of the fraction $\frac{a}{b}$ is $\frac{b}{a}$. Neither a nor b can be zero.

> We invert or turn over a fraction to write its reciprocal.

Example 1

The reciprocal of $\frac{3}{4}$ is $\frac{4}{3}$. *Just invert or turn over the fraction.*

The reciprocal of 5 or $\frac{5}{1}$ is $\frac{1}{5}$. *Write 5 as $\frac{5}{1}$ and then turn the fraction over.*

CHECK YOURSELF 1

Write the reciprocal of $\frac{5}{8}$.

To find the reciprocal of a mixed number, first write the mixed number as an improper fraction and then invert that fraction. The following example illustrates.

Example 2

Write $1\frac{2}{3}$ as $\frac{5}{3}$, *then invert.*

The reciprocal of $1\frac{2}{3}$, or $\frac{5}{3}$, is $\frac{3}{5}$.

CHECK YOURSELF 2

What is the reciprocal of $3\frac{1}{4}$?

An important property relating a number and its reciprocal follows.

Section 6.4: Dividing Fractions

> The product of any nonzero number and its reciprocal is 1.

This is shown in the following example.

Example 3

The reciprocal of $\frac{3}{4}$ is $\frac{4}{3}$.

$$\frac{3}{4} \times \frac{4}{3} = 1$$

CHECK YOURSELF 3

Find the product of $\frac{5}{7}$ and its reciprocal.

We are now ready to use the reciprocal to find a rule for dividing fractions. Remember that we can represent the operation of division in several ways. We used the symbol ÷ earlier. A fraction also indicates division. For instance,

$3 \div 5$ and $\frac{3}{5}$ both mean "3 divided by 5."

$$3 \div 5 = \frac{3}{5}$$

In this statement, 5 is called the *divisor*. It follows the division sign.

In the same way, we can rewrite a statement involving fractions and division.

Example 4

$$\frac{2}{3} \div \frac{4}{5} = \frac{\frac{2}{3}}{\frac{4}{5}}$$

$\frac{\frac{2}{3}}{\frac{4}{5}}$ is called a *complex fraction*.

The numerator is $\frac{2}{3}$.

The denominator is $\frac{4}{5}$.

It has a fraction as both its numerator and its denominator.

CHECK YOURSELF 4

Write $\dfrac{2}{5} \div \dfrac{3}{4}$ as a complex fraction.

Let's continue with the same division problem.

Example 5

(1) $\dfrac{2}{5} \div \dfrac{3}{4} = \dfrac{\frac{2}{5}}{\frac{3}{4}}$ Write the original division as a complex fraction.

$= \dfrac{\frac{2}{5} \times \frac{4}{3}}{\frac{3}{4} \times \frac{4}{3}}$ Multiply the numerator and denominator by $\dfrac{4}{3}$, the reciprocal of the denominator. This does *not* change the value of the fraction.

$= \dfrac{\frac{2}{5} \times \frac{4}{3}}{1}$ The denominator becomes 1.

(2) $= \dfrac{2}{5} \times \dfrac{4}{3}$ A number divided by 1 is just that number.

We see from lines (1) and (2) in Example 5 that

Do you see a rule suggested? $\dfrac{2}{5} \div \dfrac{3}{4} = \dfrac{2}{5} \times \dfrac{4}{3}$

We would certainly like to be able to divide fractions easily without all the work of the last example. Look carefully at the example. The following rule is suggested.

TO DIVIDE FRACTIONS

To divide one fraction by another, invert the divisor and multiply. In symbols we write:

$\dfrac{a}{b} \div \dfrac{c}{d} = \dfrac{a}{b} \times \dfrac{d}{c}$ where b, c, and $d \neq 0$.

Section 6.4: Dividing Fractions 247

The following example applies the rule for dividing fractions.

Example 6

Remember: The number inverted is the divisor. It *follows* the division sign.

$$\frac{1}{3} \div \frac{4}{7} = \frac{1}{3} \times \frac{7}{4}$$ We invert the divisor, $\frac{4}{7}$, *and* multiply.

$$= \frac{1 \times 7}{3 \times 4} = \frac{7}{12}$$

CHECK YOURSELF 5
Divide.

$$\frac{2}{5} \div \frac{3}{4} \qquad \frac{2}{5} \times \frac{4}{3} = \frac{8}{15}$$

Let's look at another similar example.

Example 7

$$\frac{5}{8} \div \frac{3}{5} = \frac{5}{8} \times \frac{5}{3} = \frac{5 \times 5}{8 \times 3} = \frac{25}{24} = 1\frac{1}{24}$$ Write the quotient as a mixed number if necessary.

CHECK YOURSELF 6
Divide.

$$\frac{5}{6} \div \frac{3}{8} \qquad \frac{5}{6} \times \frac{8}{3} \quad \frac{40}{18} \quad 2\frac{2}{9}$$

Be careful! We must invert the divisor *before any simplification*.

Simplifying will also be useful in dividing fractions. Consider the next example.

Example 8

$$\frac{3}{5} \div \frac{6}{7} = \frac{3}{5} \times \frac{7}{6}$$ Invert the divisor *first!* Then you can divide by the common factor of 3.

$$= \frac{\overset{1}{3} \times 7}{5 \times \underset{2}{6}} = \frac{7}{10}$$

CHECK YOURSELF 7
Divide.

$$\frac{4}{9} \div \frac{8}{15}$$

(handwritten work: $= \frac{4}{9} \times \frac{15}{8} = \frac{5}{6}$)

When mixed or whole numbers are involved, the process is similar. Simply change the mixed or whole numbers to improper fractions as the first step. Then proceed with the division rule. The following examples illustrate.

Example 9

$$2\frac{3}{8} \div 1\frac{3}{4} = \frac{19}{8} \div \frac{7}{4} \quad \text{Write the mixed numbers as improper fractions.}$$

$$= \frac{19}{\underset{2}{\cancel{8}}} \times \frac{\cancel{4}^{1}}{7} \quad \text{Invert the divisor and multiply as before.}$$

$$= \frac{19}{14} = 1\frac{5}{14}$$

CHECK YOURSELF 8
Divide.

$$3\frac{1}{5} \div 2\frac{2}{5}$$

(handwritten work: $= \frac{16}{5} \div \frac{12}{5} = \frac{16}{5} \times \frac{5}{12} = \frac{4}{3}$)

The following illustrates the division process when a whole number is involved.

Example 10

Write the whole number 6 as $\frac{6}{1}$.

$$1\frac{4}{5} \div 6 = \frac{9}{5} \div \frac{6}{1}$$

$$= \frac{\cancel{9}^{3}}{5} \times \frac{1}{\cancel{6}_{2}} \quad \text{Invert the divisor, then divide by the common factor of 3.}$$

$$= \frac{3}{10}$$

Section 6.4: Dividing Fractions

CHECK YOURSELF 9

Divide.

$$8 \div 4\frac{4}{5}$$

As was the case with multiplication, our work with the division of fractions will be used in the solution of a variety of problems. The steps of the problem-solving process remain the same.

Example 11

Jack traveled 140 km in $2\frac{1}{3}$ h. What was his average speed?

A kilometer, abbreviated km, is a metric unit of distance. It is about $\frac{6}{10}$ of a mile.

Solution

Speed = 140 km (Distance) ÷ $2\frac{1}{3}$ h (Time)

We know the distance traveled and the time for that travel. We want to find the average speed and must use division. Do you remember why?

The important formula is Speed = distance ÷ time

$$= 140 \text{ km} \div \frac{7}{3} \text{ h}$$

$$= \overset{20}{\cancel{140}} \times \frac{3}{\underset{1}{\cancel{7}}} \text{ km/h}$$

km/h is read "kilometers per hour." This is a unit of speed.

$$= 60 \text{ km/h}$$

CHECK YOURSELF 10

A light plane flew 280 mi in $1\frac{3}{4}$ h. What was the average speed?

Example 12

An electrician needs pieces of wire $2\frac{3}{5}$ in long. If she has a $20\frac{4}{5}$-in piece of wire, how many of the shorter pieces can she cut?

Solution

We must divide the length of the longer piece by the desired length of the shorter piece.

$$20\frac{4}{5} \div 2\frac{3}{5} = \frac{104}{5} \div \frac{13}{5}$$

$$= \frac{\overset{8}{\cancel{104}}}{\underset{1}{\cancel{5}}} \times \frac{\overset{1}{\cancel{5}}}{\underset{1}{\cancel{13}}}$$

$$= 8 \text{ pieces}$$

CHECK YOURSELF 11

A piece of plastic water pipe, 63 in long, is to be cut into lengths of $3\frac{1}{2}$ in. How many of the shorter pieces can be cut?

CHECK YOURSELF ANSWERS

1. $\frac{8}{5}$ 2. $3\frac{1}{4}$ is $\frac{13}{4}$, so the reciprocal is $\frac{4}{13}$.   3. $\frac{5}{7} \times \frac{7}{5} = 1$

4. $\dfrac{\frac{2}{5}}{\frac{3}{4}}$ 5. $\frac{8}{15}$ 6. $2\frac{2}{9}$ 7. $\frac{4}{9} \div \frac{8}{15} = \frac{4}{9} \times \frac{15}{8} = \frac{\overset{1}{\cancel{4}} \times \overset{5}{\cancel{15}}}{\underset{3}{\cancel{9}} \times \underset{2}{\cancel{8}}} = \frac{5}{6}$

8. $3\frac{1}{5} \div 2\frac{2}{5} = \frac{16}{5} \div \frac{12}{5} = \frac{\overset{4}{\cancel{16}}}{\underset{1}{\cancel{5}}} \times \frac{\overset{1}{\cancel{5}}}{\underset{3}{\cancel{12}}} = \frac{4}{3} = 1\frac{1}{3}$

9. $1\frac{2}{3}$ 10. 160 mi/h   11. 18 pieces

6.4 Exercises

Divide.

1. $\dfrac{1}{4} \div \dfrac{2}{3}$ 2. $\dfrac{3}{5} \div \dfrac{1}{2}$ 3. $\dfrac{2}{5} \div \dfrac{3}{4}$

4. $\dfrac{5}{8} \div \dfrac{3}{4}$ 5. $\dfrac{7}{10} \div \dfrac{4}{5}$ 6. $\dfrac{5}{9} \div \dfrac{8}{11}$

7. $\dfrac{7}{10} \div \dfrac{5}{9}$ 8. $\dfrac{8}{9} \div \dfrac{11}{15}$ 9. $\dfrac{9}{16} \div \dfrac{3}{8}$

10. $\dfrac{10}{27} \div \dfrac{5}{18}$ 11. $\dfrac{5}{27} \div \dfrac{25}{36}$ 12. $\dfrac{9}{28} \div \dfrac{27}{35}$

13. $\dfrac{3}{5} \div 3$ 14. $32 \div \dfrac{4}{5}$ 15. $12 \div \dfrac{2}{3}$

16. $\dfrac{5}{8} \div 5$ 17. $\dfrac{8}{11} \div 4$ 18. $\dfrac{4}{5} \div 12$

Section 6.4: Dividing Fractions

19. $4 \div \dfrac{6}{7}$

20. $6 \div \dfrac{9}{10}$

21. $4\dfrac{1}{2} \div 6$

22. $5 \div 3\dfrac{1}{3}$

23. $8 \div 2\dfrac{2}{3}$

24. $3\dfrac{1}{5} \div 4$

25. $15 \div 3\dfrac{1}{3}$

26. $2\dfrac{4}{7} \div 12$

27. $1\dfrac{3}{5} \div \dfrac{4}{15}$

28. $\dfrac{7}{10} \div 2\dfrac{4}{5}$

29. $\dfrac{5}{8} \div 3\dfrac{3}{4}$

30. $1\dfrac{3}{8} \div \dfrac{5}{12}$

31. $5\dfrac{3}{5} \div \dfrac{7}{15}$

32. $\dfrac{7}{18} \div 5\dfrac{5}{6}$

33. $1\dfrac{1}{2} \div 1\dfrac{1}{5}$

34. $1\dfrac{1}{3} \div 1\dfrac{3}{5}$

35. $3\dfrac{3}{4} \div 1\dfrac{3}{8}$

36. $5\dfrac{1}{3} \div 2\dfrac{2}{5}$

37. $2\dfrac{2}{3} \div 1\dfrac{7}{9}$

38. $8\dfrac{3}{4} \div 3\dfrac{1}{8}$

39. A wire $5\dfrac{1}{4}$ ft long is to be cut into seven pieces of the same length. How long will each piece be?

40. A potter uses $\dfrac{2}{3}$ lb of clay in making a bowl. How many bowls can be made from 16 lb of clay?

41. Virginia made a trip of 84 mi in $1\dfrac{3}{4}$ h. What was her average speed?

42. A piece of land measures $2\dfrac{1}{2}$ acres and is for sale at $30,000. What is the price per acre?

43. A roast weighs $3\frac{1}{4}$ lb. How many $\frac{1}{4}$-lb servings will this provide?

44. A bookshelf is 45 in long. If the books have an average thickness of $1\frac{1}{4}$ in, how many books can be put on the shelf?

45. A butcher wants to wrap $\frac{3}{4}$-lb packages of ground beef from a cut of meat weighing $17\frac{1}{4}$ lb. How many packages can be prepared?

46. A manufacturer has $45\frac{1}{2}$ yd of imported cotton fabric. A shirt pattern uses $1\frac{3}{4}$ yd. How many shirts can be made?

47. A stack of $\frac{3}{4}$-in-thick plywood is 36 in high. How many sheets of plywood are in the stack?

48. A 24-acre plot of land is to be subdivided into home lots that are each $\frac{3}{8}$ acre. How many lots can be formed?

Answers

1. $\frac{3}{8}$ 3. $\frac{8}{15}$ 5. $\frac{7}{10} \div \frac{4}{5} = \frac{7}{\cancel{10}2} \times \frac{\cancel{5}1}{4} = \frac{7}{8}$ 7. $1\frac{13}{50}$ 9. $\frac{9}{16} \div \frac{3}{8} = \frac{\cancel{9}3}{\cancel{16}2} \times \frac{\cancel{8}1}{\cancel{3}1} = \frac{3}{2} = 1\frac{1}{2}$ 11. $\frac{4}{15}$

13. $\frac{1}{5}$ 15. $12 \div \frac{2}{3} = \frac{\cancel{12}6}{1} \times \frac{3}{\cancel{2}1} = 18$ 17. $\frac{2}{11}$ 19. $4\frac{2}{3}$ 21. $\frac{3}{4}$ 23. $8 \div 2\frac{2}{3} = 8 \div \frac{8}{3} = \frac{\cancel{8}1}{1} \times \frac{3}{\cancel{8}1} = 3$

25. $4\frac{1}{2}$ 27. 6 29. $\frac{5}{8} \div 3\frac{3}{4} = \frac{5}{8} \div \frac{15}{4} = \frac{\cancel{5}1}{\cancel{8}2} \times \frac{\cancel{4}1}{\cancel{15}3} = \frac{1}{6}$ 31. 12 33. $1\frac{1}{4}$

35. $3\frac{3}{4} \div 1\frac{3}{8} = \frac{15}{4} \div \frac{11}{8} = \frac{15}{\cancel{4}1} \times \frac{\cancel{8}2}{11} = \frac{30}{11} = 2\frac{8}{11}$ 37. $1\frac{1}{2}$ 39. $\frac{3}{4}$ ft 41. 48 mi/h 43. 13 servings

45. 23 packages   47. 48 sheets

6.4 Supplementary Exercises

Divide.

1. $\dfrac{3}{7} \div \dfrac{4}{5}$
2. $\dfrac{5}{8} \div \dfrac{6}{7}$
3. $\dfrac{7}{8} \div \dfrac{3}{4}$
4. $\dfrac{3}{10} \div \dfrac{9}{20}$

5. $\dfrac{9}{10} \div \dfrac{6}{25}$
6. $\dfrac{7}{8} \div \dfrac{21}{40}$
7. $16 \div \dfrac{4}{5}$
8. $\dfrac{15}{16} \div 5$

9. $\dfrac{7}{8} \div 14$
10. $30 \div \dfrac{9}{10}$
11. $5\dfrac{1}{3} \div 8$
12. $12 \div 2\dfrac{2}{3}$

13. $6\dfrac{2}{3} \div 5$
14. $1\dfrac{4}{5} \div \dfrac{3}{10}$
15. $\dfrac{5}{9} \div 1\dfrac{2}{3}$
16. $4\dfrac{1}{5} \div \dfrac{28}{15}$

17. $3\dfrac{3}{5} \div 2\dfrac{7}{10}$
18. $2\dfrac{5}{8} \div 1\dfrac{3}{4}$
19. $2\dfrac{5}{6} \div 6\dfrac{3}{8}$

20. A piece of wire $3\dfrac{3}{4}$ ft long is to be cut into five pieces of the same length. How long will each piece be?

21. A blouse pattern requires $1\dfrac{3}{4}$ yd of fabric. How many blouses can be made from a piece of silk that is 21 yd long?

22. If you drive 99 mi in $2\dfrac{1}{4}$ h, what is your average speed?

23. A stack of $\dfrac{5}{8}$-in-thick plywood is 30 in high. How many sheets of plywood are in the stack?

24. An 18-acre plot of land is to be subdivided into home lots that are each $\dfrac{2}{3}$ acre. How many lots will there be in the development?

SELF-TEST for Chapter Six

The purpose of the Self-Test is to help you check your progress and review for a chapter test in class. Allow yourself about an hour to take the test. When you are done, check your answers in the back of the book. If you missed any problems, be sure to go back and review the appropriate sections in the chapter and do the supplementary exercises provided there.

In Problems 1 to 14, multiply.

1. $\dfrac{3}{5} \times \dfrac{4}{7} =$

2. $\dfrac{9}{10} \times \dfrac{5}{8} =$

3. $\dfrac{5}{6} \times \dfrac{4}{15} =$

4. $5 \times \dfrac{3}{7} =$

5. $1\dfrac{1}{3} \times 1\dfrac{4}{5} =$

6. $3\dfrac{5}{6} \times 2\dfrac{2}{5} =$

7. $\dfrac{2}{3} \times \dfrac{9}{10} =$

8. $\dfrac{15}{27} \times \dfrac{18}{25} =$

9. $\dfrac{16}{35} \times \dfrac{14}{24} =$

10. $4\dfrac{4}{5} \times \dfrac{5}{8} =$

11. $4\dfrac{1}{6} \times 1\dfrac{7}{15} =$

12. $6\dfrac{2}{3} \times 1\dfrac{2}{25} =$

13. $8 \times \dfrac{7}{12} =$

14. $\dfrac{3}{5} \times \dfrac{7}{9} \times 1\dfrac{1}{14} =$

In Problems 15 to 18, solve each application.

15. You had $\dfrac{7}{8}$ yd of upholstery material but used only $\dfrac{2}{3}$ of that for a project. How much material did you use?

16. What is the cost of $2\dfrac{3}{4}$ lb of apples if the price per pound is 36¢?

17. A room measures $5\dfrac{1}{3}$ yd by $3\dfrac{3}{4}$ yd. How many square yards of linoleum must be purchased to cover the floor?

18. The scale on a map uses 1 in to represent 80 mi. If two towns are $2\frac{3}{8}$ in apart on the map, what is the actual distance between the towns?

In Problems 19 to 26, divide.

19. $\dfrac{7}{8} \div \dfrac{4}{5} =$

20. $\dfrac{7}{12} \div \dfrac{14}{15} =$

21. $\dfrac{3}{8} \div 4 =$

22. $12 \div \dfrac{3}{4} =$

23. $1\dfrac{3}{7} \div \dfrac{4}{9} =$

24. $\dfrac{7}{8} \div 1\dfrac{5}{16} =$

25. $1\dfrac{2}{3} \div 1\dfrac{1}{9} =$

26. $5\dfrac{3}{5} \div 2\dfrac{1}{10} =$

In Problems 27 to 30, solve each application.

27. A $31\dfrac{1}{3}$-acre piece of land is subdivided into home lots. Each lot is to be $\dfrac{2}{3}$ acre. How many homes can be built on the land?

28. If you drive 99 mi in $1\dfrac{5}{6}$ hours, what is your average speed?

29. A stack of $\dfrac{5}{8}$-in-thick plywood is 40 in high. How many sheets of plywood are in the stack?

30. A bookshelf is 66 in long. If the average thickness of the books on the shelf is $1\dfrac{3}{8}$ in, how many books can be placed on the shelf?

PRETEST for Chapter Seven

Addition and Subtraction of Fractions

This pretest will point out any difficulties you may be having in adding and subtracting fractions. Do all the problems. Then check your answers on the following page.

1. $\dfrac{5}{9} + \dfrac{1}{9} =$

2. Find the least common denominator for fractions with the denominators 9 and 12.

3. $\dfrac{5}{8} + \dfrac{7}{12} =$

4. $\dfrac{4}{5} + \dfrac{7}{8} + \dfrac{9}{20} =$

5. $\dfrac{9}{10} - \dfrac{3}{10} =$

6. $\dfrac{7}{24} - \dfrac{2}{9} =$

7. $3\dfrac{5}{6} + 2\dfrac{7}{8} =$

8. $8\dfrac{3}{16} - 3\dfrac{5}{8} =$

9. A house plan calls for $12\dfrac{3}{4}$ sq yd of carpeting in the living room and $5\dfrac{1}{2}$ sq yd in the hallway. How much carpeting will be needed?

10. A stock is listed at $51\dfrac{3}{4}$ points at the start of a week. By the end of the week it is at $53\dfrac{1}{8}$ points. How much did it gain during the week?

ANSWERS TO PRETEST

For help with similar problems, turn to the section indicated.

1. $\frac{2}{3}$ (Section 7.1); 2. 36 (Section 7.2); 3. $1\frac{5}{24}$ (Section 7.3); 4. $2\frac{1}{8}$ (Section 7.3);
5. $\frac{3}{5}$ (Section 7.4); 6. $\frac{5}{72}$ (Section 7.4); 7. $6\frac{17}{24}$ (Section 7.5); 8. $4\frac{9}{16}$ (Section 7.5);
9. $18\frac{1}{4}$ sq yd (Section 7.6); 10. $1\frac{3}{8}$ points (Section 7.6)

Chapter Seven

Addition and Subtraction of Fractions

7.1 Adding Fractions with a Common Denominator

OBJECTIVE
To add like fractions

For instance, we can add two nickels and three nickels to get five nickels. We *cannot* directly add two nickels and three dimes!

Recall from our work in Chapter 1 that adding can be thought of as combining groups of the *same kinds* of objects. This is also true when you think about adding fractions.

Fractions can be added only if they name a number of the *same parts* of a whole. This means that we can add fractions only when they are *like* fractions; that is, when they have the *same (a common)* denominator.

As long as we are dealing with like fractions, addition is an easy matter. Just use the following rule.

TO ADD LIKE FRACTIONS

STEP 1 Add the numerators.

STEP 2 Place the sum over the common denominator.

STEP 3 Simplify the resulting fraction when necessary.

In symbols we can write our rule as:

$$\frac{a}{c} + \frac{b}{c} = \frac{a+b}{c} \qquad \text{where } c \neq 0.$$

Our first example illustrates the use of this rule.

Example 1

Add.

$$\frac{2}{6} + \frac{3}{6}$$

Step 1 Add the numerators.

2 + 3 = 5

Step 2 Write that sum over the common denominator, 6. We are done at this point because the answer, $\frac{5}{6}$, is in the simplest possible form.

$$\overset{\text{Step 1}}{} \qquad \overset{\text{Step 2}}{}$$

$$\frac{2}{6} + \frac{3}{6} = \frac{2+3}{6} = \frac{5}{6}$$

Let's illustrate with a diagram.

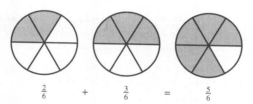

$$\frac{2}{6} \quad + \quad \frac{3}{6} \quad = \quad \frac{5}{6}$$

Combining two of the six parts with three of the six parts gives a total of five of the six equal parts.

CHECK YOURSELF 1

Add.

$$\frac{2}{9} + \frac{5}{9}$$

Step 3 of the addition rule for like fractions tells us to *simplify* the sum. The sum of fractions should always be written in lowest terms. Consider the following.

Example 2

Add.

$$\frac{3}{12} + \frac{5}{12} = \frac{8}{12} \overset{\text{Step 3}}{=} \frac{2}{3}$$

The sum $\frac{8}{12}$ is *not* in lowest terms.
Divide the numerator and denominator by 4 to simplify the result.

Section 7.1: Adding Fractions with a Common Denominator

CHECK YOURSELF 2

Add.

$$\frac{4}{15} + \frac{6}{15}$$

If the sum of two fractions is an improper fraction, we will usually write that sum as a mixed number.

Example 3

Add.

Add as before. Then convert the sum to a mixed number.

$$\frac{5}{9} + \frac{8}{9} = \frac{13}{9} = 1\frac{4}{9} \qquad \text{Write the sum } \frac{13}{9} \text{ as a mixed number.}$$

CHECK YOURSELF 3

Add.

$$\frac{7}{12} + \frac{10}{12}$$

We can also easily extend our addition rule to find the sum of more than two fractions as long as they all have the same denominator. This is shown in the following example.

Example 4

Add.

$$\frac{2}{7} + \frac{3}{7} + \frac{6}{7} = \frac{11}{7} \qquad \text{Add the numerators: } 2 + 3 + 6 = 11$$

$$= 1\frac{4}{7}$$

CHECK YOURSELF 4

Add.

$$\frac{1}{8} + \frac{3}{8} + \frac{5}{8}$$

Be Careful! In multiplying fractions, you *multiply* the numerators and *multiply* the denominators. So

$$\frac{2}{3} \times \frac{4}{5} = \frac{8}{15}$$

In adding fractions, you *cannot* just add the numerators and add the denominators.

$$\frac{2}{15} + \frac{6}{15} \text{ is } not \text{ } \frac{8}{30}$$

The fact that you *do not need* a common denominator when you are multiplying fractions is why most students find multiplication the easiest of the operations on fractions.

In adding fractions, *you must have a common denominator*. Then add the numerators and place that sum over the common denominator. So

$$\frac{2}{15} + \frac{6}{15} = \frac{8}{15}$$

CHECK YOURSELF ANSWERS

1. $\frac{2}{9} + \frac{5}{9} = \frac{2+5}{9} = \frac{7}{9}$ 2. $\frac{2}{3}$ 3. $1\frac{5}{12}$ 4. $1\frac{1}{8}$

7.1 Exercises

Add.

1. $\frac{1}{5} + \frac{2}{5}$
2. $\frac{2}{7} + \frac{3}{7}$
3. $\frac{3}{11} + \frac{5}{11}$
4. $\frac{5}{16} + \frac{4}{16}$

5. $\frac{2}{10} + \frac{3}{10}$
6. $\frac{1}{12} + \frac{7}{12}$
7. $\frac{3}{7} + \frac{4}{7}$
8. $\frac{3}{20} + \frac{7}{20}$

9. $\frac{9}{30} + \frac{11}{30}$
10. $\frac{4}{9} + \frac{5}{9}$
11. $\frac{13}{48} + \frac{23}{48}$
12. $\frac{17}{60} + \frac{31}{60}$

13. $\frac{3}{7} + \frac{6}{7}$
14. $\frac{3}{5} + \frac{4}{5}$
15. $\frac{3}{10} + \frac{9}{10}$
16. $\frac{5}{8} + \frac{7}{8}$

17. $\frac{11}{12} + \frac{10}{12}$
18. $\frac{13}{18} + \frac{11}{18}$
19. $\frac{1}{8} + \frac{1}{8} + \frac{3}{8}$
20. $\frac{1}{10} + \frac{3}{10} + \frac{3}{10}$

21. $\frac{5}{9} + \frac{8}{9} + \frac{5}{9}$
22. $\frac{7}{12} + \frac{11}{12} + \frac{1}{12}$
23. $\frac{3}{10} + \frac{7}{10} + \frac{5}{10}$
24. $\frac{9}{20} + \frac{7}{20} + \frac{11}{20}$

25. You collect 3 dimes, 2 dimes, and then 4 dimes. How much money do you have as a fraction of a dollar?

26. You collect 7 nickels, 4 nickels, and then 5 nickels. How much money do you have as a fraction of a dollar?

27. You work 7 hours one day, 5 hours the second day, and 6 hours the third day. How long did you work, as a fraction of a day?

28. One task takes 7 min, a second task takes 12 min, and a third task takes 21 min. How long did the three tasks take, as a fraction of an hour?

Skillscan (Section 4.2)
Find the prime factorization for each number.

a. 12   b. 20   c. 36

d. 48   e. 60   f. 98

Answers
Solutions for the even-numbered exercises are provided in the back of the book.

1. $\frac{3}{5}$ 3. $\frac{8}{11}$ 5. $\frac{2}{10} + \frac{3}{10} = \frac{5}{10} = \frac{1}{2}$ 7. $\frac{3}{7} + \frac{4}{7} = \frac{7}{7} = 1$ 9. $\frac{2}{3}$ 11. $\frac{3}{4}$ 13. $1\frac{2}{7}$
15. $\frac{3}{10} + \frac{9}{10} = \frac{12}{10} = 1\frac{2}{10} = 1\frac{1}{5}$ 17. $1\frac{3}{4}$ 19. $\frac{5}{8}$ 21. 2   23. $1\frac{1}{2}$ 25. $\frac{9}{10}$ of a dollar
27. $\frac{3}{4}$ of a day   a. $2 \times 2 \times 3$ b. $2 \times 2 \times 5$ c. $2 \times 2 \times 3 \times 3$ d. $2 \times 2 \times 2 \times 2 \times 3$
e. $2 \times 2 \times 3 \times 5$ f. $2 \times 7 \times 7$

7.1 Supplementary Exercises

Add.

1. $\frac{2}{7} + \frac{4}{7}$ 2. $\frac{2}{9} + \frac{3}{9}$ 3. $\frac{5}{8} + \frac{1}{8}$ 4. $\frac{1}{10} + \frac{3}{10}$

5. $\frac{8}{15} + \frac{2}{15}$ 6. $\frac{5}{8} + \frac{3}{8}$ 7. $\frac{6}{11} + \frac{7}{11}$ 8. $\frac{17}{18} + \frac{5}{18}$

9. $\frac{19}{24} + \frac{13}{24}$ 10. $\frac{1}{9} + \frac{2}{9} + \frac{4}{9}$ 11. $\frac{3}{8} + \frac{1}{8} + \frac{7}{8}$ 12. $\frac{4}{15} + \frac{7}{15} + \frac{7}{15}$

7.2 Finding the Least Common Denominator

OBJECTIVE
To find the least common denominator of a group of fractions

In Section 7.1 you dealt with like fractions (fractions with a common denominator). What about a sum such as $\frac{1}{3} + \frac{1}{4}$?

Only *like* fractions can be added.

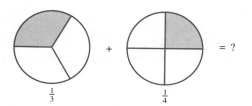

We cannot add the fractions while they are in this form. They have different denominators.

In order to add the fractions, write them as equivalent fractions with a common denominator. In this case let's use 12 as the denominator.

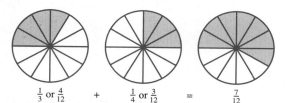

We have chosen 12 because it is a *multiple* of 3 and 4.

$\frac{1}{3}$ is equivalent to $\frac{4}{12}$.

$\frac{1}{4}$ is equivalent to $\frac{3}{12}$.

We can now add because we have like fractions.

Any common multiple of the denominators will work in forming equivalent fractions. For instance, we can write $\frac{1}{3}$ as $\frac{8}{24}$ and $\frac{1}{4}$ as $\frac{6}{24}$. The work is simplest, however, if we use the smallest possible number for the common denominator. This is called the *least common denominator* (LCD).

The LCD is the least common multiple of the denominators of the fractions. This is the *smallest* number that is a multiple of all the denominators. For example, the LCD for $\frac{1}{3}$ and $\frac{1}{4}$ is 12, *not* 24. (See Section 4.5.)

Finding the least common denominator is so important for our work in this chapter that it is worth a review.

Section 7.2: Finding the Least Common Denominator

> **TO FIND THE LEAST COMMON DENOMINATOR**
>
> STEP 1 Write the prime factorization for each of the denominators.
>
> STEP 2 Find all the prime factors that appear in any one of the prime factorizations.
>
> STEP 3 Form the product of those prime factors, using each factor the greatest number of times it occurs in any one factorization.

Example 1

Our first step in adding fractions with denominators 6 and 8 will be to determine the least common denominator. Factor 6 and 8.

$6 = 2 \times 3$
$8 = 2 \times 2 \times 2$

Note that since 2 appears three times as a factor of 8, it is used three times in writing the LCD.

The LCD is $2 \times 2 \times 2 \times 3$, or 24.

CHECK YOURSELF 1

Find the LCD of fractions with denominators 9 and 12.

The process is similar if more than two denominators are involved.

Example 2

To add fractions with denominators 6, 9, and 15, we need to find the LCD. Factor the three numbers.

$6 = 2 \times 3$
$9 = 3 \times 3$
$15 = 3 \times 5$

2 and 5 appear only once in any one factorization.
3 appears twice as a factor of 9.

The LCD is $2 \times 3 \times 3 \times 5$, or 90.

CHECK YOURSELF 2

Find the LCD of fractions with denominators 5, 8, and 20.

Example 3

To find the LCD of fractions with denominators 12, 21, and 28, we again factor the three denominators.

$12 = 2 \times 2 \times 3$
$21 = 3 \times 7$
$28 = 2 \times 2 \times 7$

Do you see that we must use 2 as a factor twice and 3 and 7 each once in forming the LCD?

The LCD is $2 \times 2 \times 3 \times 7$, or 84.

CHECK YOURSELF 3

Find the LCD of fractions with denominators 18, 27, and 12.

CHECK YOURSELF ANSWERS

1. $9 = 3 \times 3$
   $12 = 2 \times 2 \times 3$
   The LCD is $2 \times 2 \times 3 \times 3 = 36$.
2. 40
3. 108

7.2 Exercises

Find the least common denominator (LCD) for fractions with the given denominators.

1. 2 and 3
2. 3 and 5
3. 4 and 8
4. 5 and 10

5. 9 and 27
6. 10 and 30
7. 8 and 12
8. 15 and 40

9. 18 and 27
10. 15 and 20
11. 20 and 30
12. 24 and 36

13. 30 and 50
14. 36 and 48
15. 48 and 80
16. 60 and 84

17. 2, 5, and 7
18. 3, 4, and 5
19. 8, 10, and 15
20. 6, 22, and 33

21. 5, 10, and 25
22. 9, 27, and 54
23. 14, 24, and 28
24. 8, 18, and 30

Skillscan (Section 5.6)
Find the missing numerators.

a. $\dfrac{1}{3} = \dfrac{?}{9}$

b. $\dfrac{1}{5} = \dfrac{?}{20}$

c. $\dfrac{1}{6} = \dfrac{?}{30}$

d. $\dfrac{2}{3} = \dfrac{?}{24}$ e. $\dfrac{3}{8} = \dfrac{?}{32}$ f. $\dfrac{5}{9} = \dfrac{?}{45}$

Answers
1. 6 3. 8 5. 27 7. $8 = 2 \times 2 \times 2$; $12 = 2 \times 2 \times 3$; The LCD is $2 \times 2 \times 2 \times 3 = 24$ 9. 54
11. $20 = 2 \times 2 \times 5$; $30 = 2 \times 3 \times 5$; The LCD is $2 \times 2 \times 3 \times 5 = 60$ 13. 150 15. 240 17. 70
19. 120 21. 50 23. 168 a. 3 b. 4 c. 5 d. 16 e. 12 f. 25

7.2 Supplementary Exercises

Find the least common denominator (LCD) for fractions with the given denominators.

1. 3 and 8
2. 5 and 15
3. 6 and 24

4. 12 and 18
5. 20 and 24
6. 30 and 45

7. 2, 3, and 7
8. 3, 4, and 11
9. 2, 5, and 8

10. 3, 6, and 8
11. 12, 18, and 24
12. 4, 15, and 18

7.3 Adding Fractions with Different Denominators

OBJECTIVES
1. To add any fraction
2. To apply addition in solving word problems

We are now ready to add fractions with different denominators. In this case, the fractions must be renamed as equivalent fractions that have the same denominator. We will use the following rule.

> **TO ADD FRACTIONS WITH DIFFERENT DENOMINATORS**
>
> STEP 1 Find the LCD of the fractions.
>
> STEP 2 Change each fraction to an equivalent fraction with the LCD as a common denominator.
>
> STEP 3 Add the resulting like fractions as before.

Our first example shows the use of this rule.

Example 1

Add the fractions $\frac{1}{6}$ and $\frac{3}{8}$.

Step 1 We find that the LCD of 6 and 8 is 24.

See Section 7.2 if you wish to review how we arrived at 24.

Step 2 Convert the fractions so that they have the denominator 24.

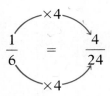

How many 6s are in 24? There are 4. So multiply the numerator and denominator by 4.

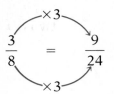

How many 8s are in 24? There are 3. So multiply the numerator and denominator by 3.

Step 3 We can now add the equivalent like fractions.

$$\frac{1}{6} + \frac{3}{8} = \frac{4}{24} + \frac{9}{24} = \frac{13}{24}$$

Add the numerators and place that sum over the common denominator.

CHECK YOURSELF 1

Add.

$$\frac{3}{5} + \frac{1}{3}$$

Here is a similar example. Remember that the sum should always be written in simplest form.

Example 2

Add the fractions $\frac{7}{10}$ and $\frac{2}{15}$.

Step 1 The LCD of 10 and 15 is 30.

Step 2 $\frac{7}{10} = \frac{21}{30}$

$\frac{2}{15} = \frac{4}{30}$

Do you see how the equivalent fractions are formed?

Section 7.3: Adding Fractions with Different Denominators 269

Step 3 $\dfrac{7}{10} + \dfrac{2}{15} = \dfrac{21}{30} + \dfrac{4}{30}$ Add the resulting like fractions. Be sure the sum is in simplest form.

$\phantom{\dfrac{7}{10} + \dfrac{2}{15}} = \dfrac{25}{30} = \dfrac{5}{6}$

CHECK YOURSELF 2

Add.

$\dfrac{1}{6} + \dfrac{7}{12}$

We can easily add more than two fractions using the same procedure. Our final example illustrates.

Example 3

Add $\dfrac{5}{6} + \dfrac{2}{9} + \dfrac{4}{15}$.

Go back and review if you need to.

Step 1 We found the LCD of 90 for these same fractions in Section 7.2.

Step 2 $\dfrac{5}{6} = \dfrac{75}{90}$ Multiply the numerator and denominator by 15.

$\dfrac{2}{9} = \dfrac{20}{90}$ Multiply the numerator and denominator by 10.

$\dfrac{4}{15} = \dfrac{24}{90}$ Multiply the numerator and denominator by 6.

Step 3 $\dfrac{75}{90} + \dfrac{20}{90} + \dfrac{24}{90} = \dfrac{119}{90}$ Now add.

$\phantom{\dfrac{75}{90} + \dfrac{20}{90} + \dfrac{24}{90}} = 1\dfrac{29}{90}$ Remember, if the sum is an improper fraction, it should be changed to a mixed number.

CHECK YOURSELF 3

Add.

$\dfrac{2}{5} + \dfrac{3}{8} + \dfrac{7}{20}$

Hint: You found the LCD in Section 7.2.

Let's look at an application of our work in adding fractions.

Example 4

Jack ran $\frac{1}{2}$ mi on Monday, $\frac{2}{3}$ mi on Wednesday, and $\frac{3}{4}$ mi on Friday. How far did he run during the week?

Solution The three distances that Jack ran are the given information in the problem. We want to find a total distance, so we must add for the solution.

$$\frac{1}{2} + \frac{2}{3} + \frac{3}{4} = \frac{6}{12} + \frac{8}{12} + \frac{9}{12}$$

$$= \frac{23}{12} = 1\frac{11}{12} \text{ mi}$$

Since we have no common denominator, we must convert to equivalent fractions before we can add.

Jack ran $1\frac{11}{12}$ mi during the week.

CHECK YOURSELF 4

Susan is designing an office complex. She needs $\frac{2}{5}$ of an acre for buildings, $\frac{1}{3}$ of an acre for driveways and parking, and $\frac{1}{6}$ of an acre for walks and landscaping. How much land does she need?

CHECK YOURSELF ANSWERS

1. $\frac{14}{15}$ 2. $\frac{1}{6} + \frac{7}{12} = \frac{2}{12} + \frac{7}{12} = \frac{9}{12} = \frac{3}{4}$ 3. $1\frac{1}{8}$ 4. $\frac{9}{10}$ of an acre

7.3 Exercises

Add.

1. $\frac{1}{2} + \frac{1}{5}$ 2. $\frac{3}{5} + \frac{1}{3}$ 3. $\frac{1}{5} + \frac{3}{10}$ 4. $\frac{1}{4} + \frac{1}{12}$

5. $\frac{3}{4} + \frac{1}{8}$ 6. $\frac{4}{5} + \frac{1}{10}$ 7. $\frac{1}{6} + \frac{3}{4}$ 8. $\frac{1}{6} + \frac{2}{15}$

9. $\dfrac{3}{7} + \dfrac{3}{14}$ 10. $\dfrac{3}{10} + \dfrac{7}{30}$ 11. $\dfrac{5}{12} + \dfrac{4}{15}$ 12. $\dfrac{3}{10} + \dfrac{3}{8}$

13. $\dfrac{5}{8} + \dfrac{1}{12}$ 14. $\dfrac{5}{12} + \dfrac{3}{10}$ 15. $\dfrac{5}{6} + \dfrac{7}{9}$ 16. $\dfrac{4}{15} + \dfrac{3}{20}$

17. $\dfrac{5}{12} + \dfrac{7}{18}$ 18. $\dfrac{4}{15} + \dfrac{9}{25}$ 19. $\dfrac{13}{16} + \dfrac{17}{24}$ 20. $\dfrac{17}{30} + \dfrac{14}{25}$

21. $\dfrac{1}{2} + \dfrac{1}{3} + \dfrac{1}{4}$ 22. $\dfrac{1}{3} + \dfrac{1}{4} + \dfrac{1}{5}$ 23. $\dfrac{1}{5} + \dfrac{7}{10} + \dfrac{4}{15}$ 24. $\dfrac{2}{3} + \dfrac{1}{4} + \dfrac{3}{8}$

25. $\dfrac{1}{9} + \dfrac{7}{12} + \dfrac{5}{8}$ 26. $\dfrac{1}{3} + \dfrac{5}{12} + \dfrac{4}{5}$ 27. $\dfrac{5}{12} + \dfrac{2}{21} + \dfrac{11}{28}$ 28. $\dfrac{5}{8} + \dfrac{7}{10} + \dfrac{11}{15}$

29. Paul bought $\dfrac{1}{4}$ lb of peanuts and $\dfrac{1}{8}$ lb of cashews. How many pounds of nuts did he buy?

30. A countertop consists of a board $\dfrac{3}{4}$ in thick and tile $\dfrac{3}{8}$ in thick. What is the overall thickness?

31. Amy budgets $\dfrac{2}{5}$ of her income for housing and $\dfrac{1}{6}$ of her income for food. What fraction of her income is budgeted for these two purposes? What fraction of her income remains?

32. A person spends $\dfrac{3}{8}$ of a day at work and $\dfrac{1}{3}$ of a day sleeping. What fraction of a day do these two activities use? What fraction of the day remains?

33. Jose walked $\dfrac{3}{4}$ mi to the store, $\dfrac{1}{2}$ mi to a friend's house, and then $\dfrac{2}{3}$ mi home. How far did he walk?

34. Find the perimeter of, or the distance around, the accompanying figure.

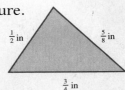

35. A budget guide recommends the following: You should spend $\frac{1}{4}$ of your salary for housing, $\frac{3}{16}$ for food, $\frac{1}{16}$ for clothing, and $\frac{1}{8}$ for transportation. What portion of your salary should these four expenses account for?

36. Deductions from your paycheck are made roughly as follows: $\frac{1}{8}$ for federal tax, $\frac{1}{20}$ for state tax, $\frac{1}{20}$ for social security, and $\frac{1}{40}$ for a savings withholding plan. What portion of your pay is deducted?

Skillscan (Section 1.8)
Find the following differences.

a. $17 - 12$ **b.** $29 - 15$ **c.** $43 - 25$

d. $58 - 39$ **e.** $92 - 77$ **f.** $82 - 76$

Answers
1. $\frac{7}{10}$ **3.** $\frac{1}{2}$ **5.** $\frac{7}{8}$ **7.** $\frac{11}{12}$ **9.** $\frac{9}{14}$ **11.** $\frac{41}{60}$ **13.** $\frac{17}{24}$ **15.** $\frac{5}{6} + \frac{7}{9} = \frac{15}{18} + \frac{14}{18} = \frac{29}{18} = 1\frac{11}{18}$
17. $\frac{29}{36}$ **19.** $1\frac{25}{48}$ **21.** $1\frac{1}{12}$ **23.** $\frac{1}{5} + \frac{7}{10} + \frac{4}{15} = \frac{6}{30} + \frac{21}{30} + \frac{8}{30} = \frac{35}{30} = 1\frac{5}{30} = 1\frac{1}{6}$ **25.** $1\frac{23}{72}$
27. $\frac{19}{21}$ **29.** $\frac{3}{8}$ lb   **31.** $\frac{17}{30}, \frac{13}{30}$ **33.** $1\frac{11}{12}$ mi   **35.** $\frac{5}{8}$ **a.** 5   **b.** 14   **c.** 18   **d.** 19   **e.** 15
f. 6

7.3 Supplementary Exercises

Add.

1. $\frac{2}{5} + \frac{1}{4}$ **2.** $\frac{1}{3} + \frac{1}{10}$ **3.** $\frac{2}{5} + \frac{7}{15}$ **4.** $\frac{3}{10} + \frac{7}{12}$

5. $\frac{3}{8} + \frac{5}{12}$ **6.** $\frac{5}{36} + \frac{7}{24}$ **7.** $\frac{4}{15} + \frac{7}{20}$ **8.** $\frac{9}{14} + \frac{10}{21}$

9. $\frac{8}{15} + \frac{11}{18}$ **10.** $\frac{16}{25} + \frac{13}{30}$ **11.** $\frac{1}{2} + \frac{1}{4} + \frac{1}{8}$ **12.** $\frac{1}{3} + \frac{1}{5} + \frac{1}{10}$

Section 7.4: Subtracting Fractions

13. $\dfrac{3}{8} + \dfrac{5}{12} + \dfrac{7}{18}$ 14. $\dfrac{5}{6} + \dfrac{8}{15} + \dfrac{9}{20}$

15. Marjory spends $\dfrac{1}{4}$ of her income for housing and $\dfrac{1}{5}$ of her income for food. What portion of her income goes for these two expenses? What amount remains?

16. Roberto used $\dfrac{3}{4}$ gal of paint in his living room, $\dfrac{1}{3}$ gal in the dining room, and $\dfrac{1}{2}$ gal in a hallway. How much paint did he use?

17. A sheet of plywood consists of two outer sections that are $\dfrac{3}{16}$ in thick and a center section that is $\dfrac{3}{8}$ in thick. How thick is the plywood overall?

18. A pattern calls for four pieces of fabric with lengths $\dfrac{1}{2}, \dfrac{3}{8}, \dfrac{1}{4},$ and $\dfrac{5}{8}$ yd. How much fabric must be purchased to use the pattern?

7.4 Subtracting Fractions

OBJECTIVES
1. To subtract like fractions
2. To subtract any two fractions
3. To apply subtraction in solving word problems

If a problem involves like fractions, then subtraction, like addition, is not difficult.

Note the similarity to our rule for adding like fractions.

TO SUBTRACT LIKE FRACTIONS

STEP 1 Subtract the numerators.

STEP 2 Place the difference over the common denominator.

STEP 3 Simplify the resulting fraction when necessary.

In symbols we can write our rule as

$$\dfrac{a}{c} - \dfrac{b}{c} = \dfrac{a-b}{c} \quad \text{where } c \neq 0$$

Example 1

Step 1 Step 2

$$\frac{4}{5} - \frac{2}{5} = \frac{4-2}{5} = \frac{2}{5}$$

Subtract the numerators: $4 - 2 = 2$. Write the difference over the common denominator, 5. Step 3 is not necessary because the difference is in simplest form.

Illustrating with a diagram:

Subtracting two of the five parts from four of the five parts leaves two of the five parts.

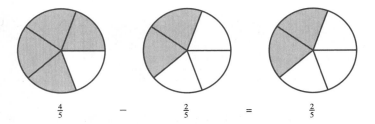

$\frac{4}{5}$ — $\frac{2}{5}$ = $\frac{2}{5}$

Example 2

Always write the result in lowest terms.

$$\frac{5}{8} - \frac{3}{8} = \frac{5-3}{8} = \frac{2}{8} = \frac{1}{4}$$

CHECK YOURSELF 1

Subtract.

$$\frac{11}{12} - \frac{5}{12}$$

To subtract fractions that do not have the same denominator, we have the following rule:

Of course this is the same as our rule for adding fractions. We just subtract instead of add!

> **TO SUBTRACT FRACTIONS WITH DIFFERENT DENOMINATORS**
>
> STEP 1 Find the LCD of the fractions.
>
> STEP 2 Change each fraction to an equivalent fraction with the LCD as a common denominator.
>
> STEP 3 Subtract the resulting like fractions as before.

Example 3

Subtract $\frac{5}{8} - \frac{1}{6}$.

Step 1 The LCD is 24.

Section 7.4: Subtracting Fractions

Step 2 Convert the fractions so that they have the common denominator 24.

$$\frac{5}{8} = \frac{15}{24}$$

$$\frac{1}{6} = \frac{4}{24}$$

The first two steps are exactly the same as if we were adding.

Step 3 Subtract the equivalent like fractions.

$$\frac{5}{8} - \frac{1}{6} = \frac{15}{24} - \frac{4}{24} = \frac{11}{24}$$

Be Careful! You *cannot* subtract the numerators and subtract the denominators.

$$\frac{5}{8} - \frac{1}{6} \text{ is } not \text{ } \frac{4}{2}$$

It works for multiplication, but not for addition or subtraction of fractions.

CHECK YOURSELF 2

Subtract.

$$\frac{7}{10} - \frac{1}{4}$$

The difference of two fractions should always be written in simplest form. The following example illustrates.

Example 4

Subtract $\frac{7}{15} - \frac{3}{10}$.

$$\frac{7}{15} = \frac{14}{30} \quad \text{and} \quad \frac{3}{10} = \frac{9}{30}$$

Convert to fractions with the LCD, 30.

$$\frac{14}{30} - \frac{9}{30} = \frac{5}{30} = \frac{1}{6}$$

Subtract the like fractions and then reduce the difference to simplest form.

CHECK YOURSELF 3

Subtract.

$$\frac{11}{15} - \frac{7}{12}$$

Let's look at an example that applies our work in subtracting fractions.

Example 5

You have $\frac{7}{8}$ yd of a handwoven linen. A pattern for a placemat calls for $\frac{1}{2}$ yd. Will you have enough left for two napkins that will use $\frac{1}{3}$ yd?

Solution First, find out how much fabric is left over after making the placemat.

$$\frac{7}{8} - \frac{1}{2} = \frac{7}{8} - \frac{4}{8} = \frac{3}{8} \text{ yd}$$

Now compare the size of $\frac{1}{3}$ and $\frac{3}{8}$.

Now $\frac{1}{3}$ yd is needed for the two napkins; $\frac{3}{8}$ yd remains.

$$\frac{3}{8} = \frac{9}{24} \text{ yd} \quad \text{and} \quad \frac{1}{3} = \frac{8}{24} \text{ yd}$$

Since $\frac{3}{8}$ yd is *more than* the $\frac{1}{3}$ yd that is needed, there is enough material for the placemat *and* the two napkins.

CHECK YOURSELF 4

A concrete walk will require $\frac{3}{4}$ cu yd of concrete. If you have mixed $\frac{8}{9}$ cu yd will enough concrete remain to do a project which will use $\frac{1}{6}$ cu yd?

Section 7.4: Subtracting Fractions

CHECK YOURSELF ANSWERS

1. $\dfrac{11}{12} - \dfrac{5}{12} = \dfrac{6}{12} = \dfrac{1}{2}$ 2. $\dfrac{9}{20}$

3. The LCD is 60. $\dfrac{11}{15} = \dfrac{44}{60}$ and $\dfrac{7}{12} = \dfrac{35}{60}$ $\dfrac{44}{60} - \dfrac{35}{60} = \dfrac{9}{60} = \dfrac{3}{20}$

4. $\dfrac{5}{36}$ cu yd will remain. You do *not* have enough concrete for both projects.

7.4 Exercises

Subtract.

1. $\dfrac{4}{5} - \dfrac{2}{5}$

2. $\dfrac{5}{7} - \dfrac{2}{7}$

3. $\dfrac{5}{9} - \dfrac{2}{9}$

4. $\dfrac{7}{10} - \dfrac{3}{10}$

5. $\dfrac{13}{20} - \dfrac{3}{20}$

6. $\dfrac{17}{30} - \dfrac{11}{30}$

7. $\dfrac{19}{24} - \dfrac{5}{24}$

8. $\dfrac{25}{36} - \dfrac{13}{36}$

9. $\dfrac{4}{5} - \dfrac{1}{3}$

10. $\dfrac{7}{9} - \dfrac{1}{6}$

11. $\dfrac{7}{15} - \dfrac{2}{5}$

12. $\dfrac{5}{6} - \dfrac{2}{7}$

13. $\dfrac{3}{8} - \dfrac{1}{4}$

14. $\dfrac{7}{10} - \dfrac{2}{5}$

15. $\dfrac{5}{12} - \dfrac{3}{8}$

16. $\dfrac{13}{15} - \dfrac{11}{20}$

17. $\dfrac{13}{25} - \dfrac{2}{15}$

18. $\dfrac{11}{12} - \dfrac{7}{10}$

19. $\dfrac{13}{27} - \dfrac{5}{18}$

20. $\dfrac{7}{20} - \dfrac{4}{15}$

21. $\dfrac{13}{18} - \dfrac{7}{12}$

22. $\dfrac{17}{30} - \dfrac{5}{9}$

23. $\dfrac{33}{40} - \dfrac{7}{24}$

24. $\dfrac{13}{24} - \dfrac{5}{16}$

25. $\dfrac{7}{8} - \dfrac{1}{4} - \dfrac{1}{2}$

26. $\dfrac{9}{10} - \dfrac{1}{5} - \dfrac{1}{2}$

27. $\dfrac{5}{9} + \dfrac{7}{12} - \dfrac{5}{8}$

28. $\dfrac{11}{12} + \dfrac{4}{21} - \dfrac{13}{28}$

29. A pattern calls for $\dfrac{5}{8}$ yd of ribbon. If you have $\dfrac{3}{4}$ yd, how much ribbon will be left over if you use the pattern?

30. Find the missing dimension in the accompanying figure.

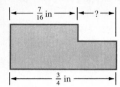

31. Find the missing dimension in the accompanying figure.

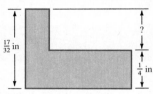

32. A hamburger, which weighed $\frac{1}{4}$ lb before cooking, weighed $\frac{3}{16}$ lb after cooking. How much weight was lost in cooking?

33. Martin owned a $\frac{7}{8}$-acre piece of land. He sold $\frac{1}{3}$ acre. What amount of land remains?

34. On Monday, $\frac{7}{8}$ of a house-painting project remained to be done. John painted $\frac{1}{4}$ of the house on Tuesday and $\frac{3}{16}$ of the house on Wednesday. What portion of the project remained to be done?

35. Geraldo has $\frac{3}{4}$ cup of flour. Biscuits use $\frac{5}{8}$ cup. Will he have enough left over for a small pie crust which requires $\frac{1}{4}$ cup?

36. You have $\frac{5}{6}$ gal of paint. You estimate that one wall will use $\frac{1}{2}$ gal. Can you also finish a smaller wall that will need $\frac{1}{4}$ gal?

Skillscan (Section 5.3)
Write each improper fraction as a mixed or whole number.

a. $\frac{11}{9}$ **b.** $\frac{10}{7}$ **c.** $\frac{16}{15}$ **d.** $\frac{10}{10}$

Write each mixed number as an improper fraction.

e. $1\dfrac{1}{3}$ f. $2\dfrac{3}{5}$ g. $3\dfrac{1}{6}$ h. $5\dfrac{5}{8}$

Answers

1. $\dfrac{2}{5}$ 3. $\dfrac{5}{9} - \dfrac{2}{9} = \dfrac{3}{9} = \dfrac{1}{3}$ 5. $\dfrac{1}{2}$ 7. $\dfrac{7}{12}$ 9. $\dfrac{4}{5} - \dfrac{1}{3} = \dfrac{12}{15} - \dfrac{5}{15} = \dfrac{7}{15}$ 11. $\dfrac{1}{15}$ 13. $\dfrac{1}{8}$ 15. $\dfrac{1}{24}$
17. $\dfrac{29}{75}$ 19. $\dfrac{11}{54}$ 21. $\dfrac{13}{18} - \dfrac{7}{12} = \dfrac{26}{36} - \dfrac{21}{36} = \dfrac{5}{36}$ 23. $\dfrac{33}{40} - \dfrac{7}{24} = \dfrac{99}{120} - \dfrac{35}{120} = \dfrac{64}{120} = \dfrac{8}{15}$ 25. $\dfrac{1}{8}$
27. $\dfrac{37}{72}$ 29. $\dfrac{1}{8}$ yd 31. $\dfrac{9}{32}$ in 33. $\dfrac{13}{24}$ acre 35. No. Only $\dfrac{1}{8}$ cup will be left over. a. $1\dfrac{2}{9}$
b. $1\dfrac{3}{7}$ c. $1\dfrac{1}{15}$ d. 1 e. $\dfrac{4}{3}$ f. $\dfrac{13}{5}$ g. $\dfrac{19}{6}$ h. $\dfrac{45}{8}$

7.4 Supplementary Exercises

Subtract.

1. $\dfrac{7}{9} - \dfrac{5}{9}$
2. $\dfrac{9}{10} - \dfrac{6}{10}$
3. $\dfrac{7}{8} - \dfrac{3}{8}$
4. $\dfrac{11}{12} - \dfrac{7}{12}$

5. $\dfrac{7}{8} - \dfrac{2}{3}$
6. $\dfrac{5}{6} - \dfrac{2}{5}$
7. $\dfrac{11}{18} - \dfrac{2}{9}$
8. $\dfrac{5}{6} - \dfrac{3}{4}$

9. $\dfrac{5}{8} - \dfrac{1}{6}$
10. $\dfrac{13}{18} - \dfrac{5}{12}$
11. $\dfrac{10}{21} - \dfrac{3}{14}$
12. $\dfrac{13}{18} - \dfrac{7}{15}$

13. $\dfrac{11}{12} - \dfrac{1}{4} - \dfrac{1}{3}$
14. $\dfrac{13}{15} + \dfrac{2}{3} - \dfrac{3}{5}$

15. A recipe calls for $\dfrac{1}{3}$ cup of milk. You have $\dfrac{3}{4}$ cup. How much milk will be left over?

16. The outside diameter of a pipe is $\dfrac{7}{8}$ in. The inside diameter is $\dfrac{9}{16}$ in. What is the wall thickness of the pipe? *Hint:* Draw a picture of the pipe.

17. You have $\dfrac{7}{8}$ of a homework assignment left to do over a weekend. On Saturday you do $\dfrac{1}{4}$ of the assignment. What portion remains to be done on Sunday?

18. A concrete divider requires $\frac{1}{2}$ cu yd of concrete. If you have mixed $\frac{7}{8}$ cu yd, will enough be left over for a small project that requires $\frac{1}{4}$ cu yd?

7.5 Adding and Subtracting Mixed Numbers

OBJECTIVES
1. To add mixed numbers
2. To subtract mixed numbers

Once you know how to add fractions, adding mixed numbers should be no problem if you keep in mind that addition involves combining groups of the *same kind* of objects. Since mixed numbers consist of two parts—a whole number and a fraction, we can work with the whole numbers and the fractions separately. Consider the following:

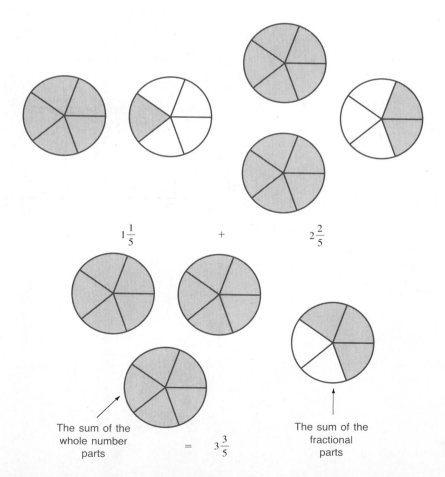

Section 7.5: Adding and Subtracting Mixed Numbers

This suggests the following general rule.

Note: Step 2 requires that the fractional parts have the same denominator.

> **TO ADD MIXED NUMBERS**
> STEP 1 Add the whole-number parts.
> STEP 2 Add the fractional parts.
> STEP 3 Combine the results as a mixed number.

Our first example illustrates the use of this rule.

Example 1

$3 + 4 = 7$ Add the whole numbers.

$$3\frac{1}{5} + 4\frac{2}{5}$$

$$3\frac{1}{5} + 4\frac{2}{5}$$

$\frac{1}{5} + \frac{2}{5} = \frac{3}{5}$ Add the fractional parts.

$7 + \frac{3}{5} = 7\frac{3}{5}$ Now combine the results.

CHECK YOURSELF 1

Add.

$$2\frac{3}{10} + 3\frac{4}{10}$$

Example 2

We add as before, but there is a problem. The fractional portion is an improper fraction.

$$2\frac{5}{7} + 3\frac{6}{7} = 5\frac{11}{7}$$

$$= 5 + 1\frac{4}{7}$$ Write the improper fraction, $\frac{11}{7}$, as a mixed number, $1\frac{4}{7}$.

Add the remaining whole numbers.
$5 + 1 = 6$

$$= 6\frac{4}{7}$$

CHECK YOURSELF 2

Add.

$$4\frac{3}{8} + 2\frac{7}{8}$$

When the fractional portions of the mixed numbers have different denominators, we must rename these fractions as equivalent fractions with the least common denominator in order to perform the addition in step 2 of our rule. Consider our next example.

Example 3

$$3\frac{1}{6} + 2\frac{3}{8} = 3\frac{4}{24} + 2\frac{9}{24}$$ The LCD of the fractions is 24. Rename them with that denominator.

$$= 5\frac{13}{24}$$ Then add as before.

CHECK YOURSELF 3

Add.

$$5\frac{7}{10} + 3\frac{5}{6}$$

If more than two mixed numbers are involved in the addition, use the same procedure.

Example 4

The LCD of the three fractions is 40. So convert the fractions. Take time to make sure you see how the fractions were changed!

$$2\frac{1}{5} + 3\frac{3}{4} + 4\frac{1}{8} = 2\frac{8}{40} + 3\frac{30}{40} + 4\frac{5}{40}$$

$$= 9\frac{43}{40}$$

The fractional portion is an improper fraction. Write it as a mixed number.

$$= 9 + 1\frac{3}{40} = 10\frac{3}{40}$$ Add the whole numbers. $9 + 1 = 10$

CHECK YOURSELF 4

Add.

$$5\frac{1}{2} + 4\frac{2}{3} + 3\frac{3}{4}$$

Section 7.5: Adding and Subtracting Mixed Numbers

We can use a similar technique for *subtracting* mixed numbers if the fractional part being subtracted is the *smaller* of the two fractions. The rule is similar to that stated earlier for adding mixed numbers.

> **TO SUBTRACT MIXED NUMBERS**
> STEP 1 Subtract the whole-number parts.
> STEP 2 Subtract the fractional parts.
> STEP 3 Combine the results as a mixed number.

The following example illustrates the use of this rule.

Example 5

$$5\frac{7}{12} - 3\frac{5}{12} = 2\frac{2}{12} = 2\frac{1}{6}$$

$\{5 - 3\}$ $\left\{\frac{7}{12} - \frac{5}{12}\right\}$ Subtract the whole numbers and then the fractional portions.

CHECK YOURSELF 5

Subtract.

$$8\frac{7}{8} - 5\frac{3}{8}$$

Again, we must rename the fractions if different denominators are involved. This is shown in the next example.

Example 6

$$8\frac{7}{10} - 3\frac{3}{8} = 8\frac{28}{40} - 3\frac{15}{40}$$ Write the fractions with denominator 40.

$$= 5\frac{13}{40}$$ Subtract as before.

CHECK YOURSELF 6

Subtract.

$$7\frac{11}{12} - 3\frac{5}{8}$$

Let's look at an example of subtracting mixed numbers with an important difference.

Example 7

$$4\frac{3}{8} - 2\frac{5}{8} = ?$$

We cannot subtract the fractional portions. $\frac{5}{8}$ is larger than $\frac{3}{8}$. To do the subtraction when the fractional part being subtracted is the *larger* of the fractions, we must *rename* the first mixed number.

$$4\frac{3}{8} = 4 + \frac{3}{8} = 3 + 1 + \frac{3}{8} \quad \text{Borrow 1 from the 4.}$$

$$= 3 + \frac{8}{8} + \frac{3}{8} \quad \text{Think of that 1 as } \frac{8}{8}.$$

$$= 3 + \frac{11}{8} \quad \text{Add } \frac{8}{8} \text{ to the original } \frac{3}{8}.$$

$$= 3\frac{11}{8} \quad \text{We have renamed } 4\frac{3}{8} \text{ as } 3\frac{11}{8}.$$

So

$$4\frac{3}{8} - 2\frac{5}{8} = 3\frac{11}{8} - 2\frac{5}{8} \quad \text{By writing } 4\frac{3}{8} \text{ as } 3\frac{11}{8}, \text{ you can subtract as before.}$$

$$= 1\frac{6}{8} = 1\frac{3}{4}$$

CHECK YOURSELF 7

Subtract.

$$6\frac{3}{5} - 2\frac{4}{5}$$

A second method for handling the subtraction of our last example requires writing the mixed numbers as improper fractions as the first step. This method is shown in Example 8.

Example 8

$$4\frac{3}{8} - 2\frac{5}{8} = \frac{35}{8} - \frac{21}{8}$$ Write the mixed numbers as improper fractions for the first step.

$$= \frac{14}{8}$$ Now you can subtract.

$$= 1\frac{6}{8} = 1\frac{3}{4}$$ Simplify your answer.

CHECK YOURSELF 8

Subtract.

$$6\frac{3}{5} - 2\frac{4}{5}$$

Use the second method and compare your work with that of the last exercise.

Which method do you like better?
Note: You will probably find that the first method is best when the whole numbers in the problem are large.

If the fractions have different denominators, rewrite the fractions so that they have the least common denominator. Either method that we have considered can then be used.

Example 9

Method 1

$$3\frac{1}{9} - 1\frac{5}{6} = 3\frac{2}{18} - 1\frac{15}{18}$$ Write the fractions with the LCD of 18. Do you see that we cannot subtract the way things stand?

$$= 2\frac{20}{18} - 1\frac{15}{18}$$ Borrow 1 from the 3 of the minuend. Think of that 1 as $\frac{18}{18}$ and combine it with the original $\frac{2}{18}$.

$$= 1\frac{5}{18}$$

Method 2

$$3\frac{1}{9} - 1\frac{5}{6} = \frac{28}{9} - \frac{11}{6}$$ Write the mixed numbers as improper fractions.

$$= \frac{56}{18} - \frac{33}{18}$$ Convert the fractions so that they have the denominator 18.

$$= \frac{23}{18} = 1\frac{5}{18}$$

CHECK YOURSELF 9

Subtract.

$$5\frac{1}{6} - 3\frac{3}{4}$$

Try both methods.

To subtract a mixed number from a whole number, we again have two possible approaches.

Example 10

$$6 - 2\frac{3}{4} = ?$$

You can think of 6 as 5 + 1   Borrow 1 from 6.

or $5 + \frac{4}{4}$, or $5\frac{4}{4}$.   Write that 1 as $\frac{4}{4}$.

So

$$6 - 2\frac{3}{4} = 5\frac{4}{4} - 2\frac{3}{4} = 3\frac{1}{4}$$

A second approach to the same problem:

$$6 - 2\frac{3}{4} = \frac{24}{4} - \frac{11}{4}$$ Write both the whole number and the mixed number as improper fractions. *Note:*

$$= \frac{13}{4}$$ $6 = \frac{6}{1} = \frac{24}{4}$

$$= 3\frac{1}{4}$$ (Multiply numerator and denominator by 4 to form a common denominator.)

CHECK YOURSELF 10

Subtract.

$$7 - 3\frac{2}{5}$$

Some students prefer a vertical arrangement when adding or subtracting mixed numbers. Look at the following example.

Section 7.5: Adding and Subtracting Mixed Numbers

Example 11

Subtract $6\frac{3}{8} - 2\frac{5}{6}$.

Arranging vertically,

$$6\frac{3}{8} \longrightarrow 6\frac{9}{24} \longrightarrow 5\frac{33}{24}$$
$$-2\frac{5}{6} \longrightarrow -2\frac{20}{24} \longrightarrow -2\frac{20}{24}$$
$$\overline{\phantom{-2\frac{20}{24}}} 3\frac{13}{24}$$

Rename $6\frac{9}{24}$ as $5\frac{33}{24}$. The subtraction can now be done.

Write the fractional parts as like fractions. We cannot subtract in this form. Do you see why?

This different arrangement is certainly nothing new. It is just another way of writing a problem when adding or subtracting mixed numbers. Experiment with both of the arrangements that have been shown and decide which you like best.

CHECK YOURSELF ANSWERS

1. $5\frac{7}{10}$ 2. $4\frac{3}{8} + 2\frac{7}{8} = 6\frac{10}{8} = 6 + 1\frac{2}{8} = 7\frac{2}{8} = 7\frac{1}{4}$

3. $5\frac{7}{10} + 3\frac{5}{6} = 5\frac{21}{30} + 3\frac{25}{30} = 8\frac{46}{30} = 9\frac{16}{30} = 9\frac{8}{15}$ 4. $13\frac{11}{12}$ 5. $3\frac{1}{2}$

6. $7\frac{11}{12} - 3\frac{5}{8} = 7\frac{22}{24} - 3\frac{15}{24} = 4\frac{7}{24}$ 7. $6\frac{3}{5} - 2\frac{4}{5} = 5\frac{8}{5} - 2\frac{4}{5} = 3\frac{4}{5}$

8. $6\frac{3}{5} - 2\frac{4}{5} = \frac{33}{5} - \frac{14}{5} = \frac{19}{5} = 3\frac{4}{5}$ 9. $1\frac{5}{12}$ 10. $3\frac{3}{5}$

7.5 Exercises

Do the indicated operations.

1. $3\frac{2}{7} + 4\frac{3}{7}$

2. $5\frac{2}{9} + 6\frac{4}{9}$

3. $3\frac{1}{8} + 5\frac{3}{8}$

4. $1\frac{1}{6} + 5\frac{5}{6}$

5. $6\frac{5}{9} + 4\frac{7}{9}$

6. $4\frac{9}{10} + 3\frac{7}{10}$

7. $1\frac{1}{3} + 2\frac{1}{5}$

8. $2\frac{1}{4} + 1\frac{1}{6}$

9. $5\frac{3}{8} + 3\frac{5}{12}$

10. $4\frac{5}{9} + 6\frac{3}{4}$

11. $6\frac{3}{4} + 5\frac{5}{6}$

12. $3\frac{2}{5} + 3\frac{5}{8}$

13. $2\frac{1}{4} + 3\frac{5}{8} + 1\frac{1}{6}$ 14. $3\frac{1}{5} + 2\frac{1}{2} + 5\frac{1}{4}$ 15. $3\frac{3}{5} + 4\frac{1}{4} + 5\frac{3}{10}$ 16. $4\frac{5}{6} + 3\frac{2}{3} + 7\frac{5}{9}$

17. $5\frac{5}{6} + 3\frac{4}{5} + 7\frac{2}{3}$ 18. $5\frac{7}{12} + 2\frac{3}{8} + 4\frac{1}{2}$ 19. $6\frac{5}{9} - 3\frac{2}{9}$ 20. $3\frac{5}{6} - 1\frac{1}{6}$

21. $3\frac{2}{5} - 1\frac{4}{5}$ 22. $6\frac{1}{6} - 4\frac{5}{6}$ 23. $3\frac{2}{3} - 2\frac{1}{4}$ 24. $5\frac{4}{5} - 1\frac{1}{6}$

25. $6\frac{3}{10} - 3\frac{7}{15}$ 26. $8\frac{5}{12} - 4\frac{11}{15}$ 27. $7\frac{5}{12} - 3\frac{11}{18}$ 28. $9\frac{3}{7} - 2\frac{13}{21}$

29. $3 - 1\frac{1}{4}$ 30. $4 - 1\frac{2}{3}$ 31. $8 - 4\frac{7}{8}$ 32. $9 - 5\frac{4}{9}$

33. $3\frac{3}{4} + 5\frac{1}{2} - 2\frac{3}{8}$ 34. $1\frac{5}{6} + 3\frac{5}{12} - 2\frac{1}{4}$ 35. $2\frac{3}{8} + 2\frac{1}{4} - 1\frac{5}{6}$ 36. $1\frac{1}{15} + 3\frac{3}{10} - 2\frac{4}{5}$

Answers

1. $7\frac{5}{7}$ 3. $8\frac{1}{2}$ 5. $11\frac{1}{3}$ 7. $3\frac{8}{15}$ 9. $5\frac{3}{8} + 3\frac{5}{12} = 5\frac{9}{24} + 3\frac{10}{24} = 8\frac{19}{24}$ 11. $12\frac{7}{12}$ 13. $7\frac{1}{24}$

15. $13\frac{3}{20}$ 17. $17\frac{3}{10}$ 19. $3\frac{1}{3}$ 21. $3\frac{2}{5} - 1\frac{4}{5} = 2\frac{7}{5} - 1\frac{4}{5} = 1\frac{3}{5}$ 23. $1\frac{5}{12}$

25. $6\frac{3}{10} - 3\frac{7}{15} = 6\frac{9}{30} - 3\frac{14}{30} = 5\frac{39}{30} - 3\frac{14}{30} = 2\frac{25}{30} = 2\frac{5}{6}$ 27. $3\frac{29}{36}$ 29. $1\frac{3}{4}$

31. $8 - 4\frac{7}{8} = 7\frac{8}{8} - 4\frac{7}{8} = 3\frac{1}{8}$ 33. $6\frac{7}{8}$ 35. $2\frac{19}{24}$

7.5 Supplementary Exercises

Do the indicated operations.

1. $2\frac{2}{9} + 3\frac{4}{9}$ 2. $4\frac{5}{8} + 3\frac{7}{8}$ 3. $6\frac{7}{10} + 3\frac{3}{10}$ 4. $2\frac{1}{3} + 2\frac{1}{4}$

5. $4\frac{5}{6} + 1\frac{3}{8}$ 6. $5\frac{9}{10} + 3\frac{7}{15}$ 7. $1\frac{1}{4} + \frac{1}{8} + 2\frac{1}{2}$ 8. $\frac{1}{5} + 3\frac{1}{2} + 2\frac{1}{10}$

9. $2\frac{1}{4} + 3\frac{5}{8} + 1\frac{1}{6}$ 10. $5\frac{3}{4} - 2\frac{1}{4}$ 11. $9\frac{5}{8} - 3\frac{7}{8}$ 12. $8\frac{3}{8} - 3\frac{5}{16}$

13. $5\frac{5}{6} - 2\frac{1}{9}$ 14. $9\frac{7}{18} - 3\frac{5}{12}$ 15. $9 - 5\frac{3}{4}$ 16. $7 - 2\frac{3}{10}$

17. $3\frac{4}{5} + 2\frac{5}{6} - 1\frac{1}{3}$ 18. $5\frac{1}{4} + 3\frac{1}{5} - 2\frac{9}{10}$

7.6 Applying Fractions

OBJECTIVE
To use fractions and mixed numbers in the solution of word problems

Since many of the measurements you deal with in everyday life involve fractions, there are many applications of your work in this chapter. Let's look at some typical situations.

Example 1

Mr. Goodwin entered a tournament and caught three fish weighing $5\frac{1}{4}$, $3\frac{5}{8}$, and $6\frac{1}{2}$ lb. What was the weight of his total catch?

Solution We need to find the total weight, so we must add.

The abbreviation for pounds is "lb" from the Latin *libra*, if you're curious.

$$5\frac{1}{4} + 3\frac{5}{8} + 6\frac{1}{2} = 5\frac{2}{8} + 3\frac{5}{8} + 6\frac{4}{8}$$

Write each fraction with the denominator 8.

$$= 14\frac{11}{8} = 15\frac{3}{8} \text{ lb}$$

The weight of his total catch was $15\frac{3}{8}$ lb.

CHECK YOURSELF 1

In three fill-ups you purchased $9\frac{7}{10}$, $8\frac{1}{2}$, and $9\frac{3}{10}$ gal of gas. How many gallons did you buy altogether?

Example 2

Linda was $48\frac{1}{4}$ in tall on her sixth birthday. By her seventh year she was $51\frac{5}{8}$ in tall. How much had she grown during the year?

Solution Since we want the difference in height, we must subtract to find the solution.

$$51\frac{5}{8} - 48\frac{1}{4} = 51\frac{5}{8} - 48\frac{2}{8} = 3\frac{3}{8} \text{ in}$$

Linda grew $3\frac{3}{8}$ in during the year.

CHECK YOURSELF 2

You use $4\frac{3}{4}$ yd of fabric from a 50-yd bolt. How much fabric will remain on the bolt?

Often we will have to use more than one operation to find the solution to a problem. Consider the following example.

Example 3

A poster is to have a total length of $12\frac{1}{4}$ in. We want a $1\frac{3}{8}$-in border on the top and a 2-in border on the bottom. What is the length of the printed part of the poster?

Solution First we need to find the total width of the top and bottom borders.

$$1\frac{3}{8} + 2 = 3\frac{3}{8} \text{ in}$$

Now *subtract* that sum (the top and bottom borders) from the total length of the poster.

$$12\frac{1}{4} - 3\frac{3}{8} = 8\frac{7}{8} \text{ in}$$

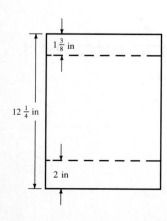

The length of the printed part is $8\frac{7}{8}$ in.

Section 7.6: Applying Fractions   291

CHECK YOURSELF 3

You cut one shelf $3\frac{3}{4}$ ft long and one $4\frac{1}{2}$ ft long from a 12-ft piece of lumber. Can you cut another shelf 4 ft long?

Example 4

Find the diameter of (the distance across) the hole in the washer shown below.

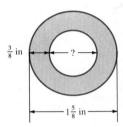

Solution Since the washer itself is $\frac{3}{8}$ in wide, we first add the widths of the two edges of the washer.

$$\frac{3}{8} + \frac{3}{8} = \frac{6}{8} = \frac{3}{4} \text{ in}$$

Now subtract that sum from the outside diameter to find the distance across the hole.

$$1\frac{5}{8} - \frac{3}{4} = \frac{7}{8} \text{ in}$$

The diameter of the hole is $\frac{7}{8}$ in.

CHECK YOURSELF 4

A concrete pipe has an outside diameter of 14 in. The thickness of the wall of the pipe is $\frac{3}{4}$ in. What is the inside diameter of the pipe?

CHECK YOURSELF ANSWERS

1. $27\frac{1}{2}$ gal   2. $45\frac{1}{4}$ yd   3. No, only $3\frac{3}{4}$ ft are "left over."   4. $12\frac{1}{2}$ in

7.6 Exercises

1. A plumber needs pieces of pipe $15\frac{5}{8}$ in and $25\frac{3}{4}$ in long. What is the total length of the pipe that is needed?

2. Marcus has to figure the postage for sending two packages. One weighs $3\frac{7}{8}$ lb and the other weighs $2\frac{3}{4}$ lb. What is the total weight?

3. Franklin worked $2\frac{1}{4}$ hours on Monday, $5\frac{3}{4}$ hours on Wednesday, and $4\frac{1}{2}$ hours on Friday. What was the total number of hours that he worked?

4. Robin ran $5\frac{1}{3}$ mi on Sunday, $2\frac{1}{4}$ mi on Tuesday, and $3\frac{1}{2}$ mi on Friday. How far did she run during the week?

5. Find the perimeter of the figure below.

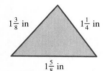

6. Find the perimeter of the figure below.

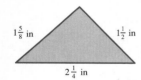

7. Mrs. Selden is working on a project which uses three pieces of fabric with lengths $\frac{3}{4}$, $1\frac{1}{4}$, and $\frac{5}{8}$ yd. How much fabric should she buy? To be safe, allow for $\frac{1}{8}$ yd of waste.

8. The framework of a wall is $3\frac{1}{2}$ in thick. We apply $\frac{5}{8}$-in wall board and $\frac{1}{4}$-in paneling to the inside. Siding that is $\frac{3}{4}$ in thick is applied to the outside. What is the finished thickness of the wall?

Section 7.6: Applying Fractions

9. A stock was listed at $34\frac{3}{8}$ points on Monday. By closing time Friday, it was at $28\frac{3}{4}$. How much did it drop during the week?

10. A roast weighed $4\frac{1}{4}$ lb before cooking and $3\frac{3}{8}$ lb after cooking. How many pounds were lost during the cooking?

11. The interest rate on an auto loan in May was $12\frac{3}{8}$ percent. By September the rate was up to $14\frac{1}{4}$ percent. How much did the interest rate increase over the period?

12. A roll of paper contains $30\frac{1}{4}$ yd. If $16\frac{7}{8}$ yd are cut from the roll, how much paper remains?

13. Find the missing dimension in the figure below.

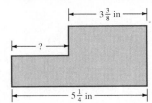

14. A $4\frac{1}{4}$-in bolt is placed through a board that is $3\frac{1}{2}$ in thick. How far does the bolt extend beyond the board?

15. Ben can work 20 hours per week on a part-time job. He works $5\frac{1}{2}$ hours on Monday and $3\frac{3}{4}$ hours on Tuesday. How many more hours can he work during the week?

16. Find the missing dimension in the figure below.

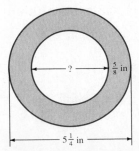

17. The Whites used $20\frac{3}{4}$ sq yd of carpet for their living room, $15\frac{1}{2}$ sq yd for the dining room, and $6\frac{1}{4}$ sq yd for a hallway. How much will remain if a 50-sq yd roll of carpeting is used?

18. A construction company has bids for paving roads of $1\frac{1}{2}$ mi, $\frac{3}{4}$ mi, and $3\frac{1}{3}$ mi for the month of July. With their present equipment, they can pave 8 mi in one month. How much more work can they take on in July?

19. On an 8-hour trip, Jack drives $2\frac{3}{4}$ hours and Pat drives $2\frac{1}{2}$ hours. How many hours are left to drive?

20. A runner has told herself that she will run 20 mi each week. She runs $5\frac{1}{2}$ mi on Sunday, $4\frac{1}{4}$ mi on Tuesday, $4\frac{3}{4}$ mi on Wednesday, and $2\frac{1}{8}$ mi on Friday. How far must she run on Saturday to meet her goal?

Answers

1. $41\frac{3}{8}$ in   3. $12\frac{1}{2}$ hours   5. $4\frac{1}{4}$ in   7. $2\frac{3}{4}$ yd   9. $5\frac{5}{8}$ points   11. $1\frac{7}{8}$ percent   13. $1\frac{7}{8}$ in   15. $10\frac{3}{4}$ hours   17. $7\frac{1}{2}$ sq yd   19. $2\frac{3}{4}$ hours

7.6 Supplementary Exercises

1. Eddie bought two packages of meat weighing $2\frac{7}{16}$ lb and $3\frac{1}{4}$ lb. What was the total weight of his purchase?

2. A recipe calls for $2\frac{1}{3}$ cups of milk and $3\frac{3}{4}$ cups of water. What is the total amount of liquid used?

3. On a trip the Wilsons stopped for gas three times, using $10\frac{1}{2}$, $9\frac{3}{10}$, and $9\frac{9}{10}$ gal of gas. How much gasoline did they use?

Section 7.6: Applying Fractions

4. Michele worked $6\frac{3}{4}$ hours on Monday, $5\frac{1}{2}$ hours on Wednesday, $7\frac{1}{4}$ hours on Thursday, and 6 hours on Friday. How many hours did she work during the week?

5. A post $6\frac{1}{4}$ ft long is to be set $1\frac{1}{2}$ ft into the ground. How much of the post will be above ground?

6. An $18\frac{3}{4}$-in length of wire is cut from a piece 60 in long. How much of the wire remains?

7. If you cut $5\frac{3}{4}$-yd and $6\frac{7}{8}$-yd lengths from a 30-yd roll of wallpaper, how much paper remains on the roll?

8. A poster has an overall length of 15 in. If you want a $1\frac{3}{4}$-in border on the top and a $2\frac{1}{2}$-in border on the bottom, what is the length of the printed portion of the poster?

9. Gene picked $22\frac{3}{4}$ lb, $23\frac{7}{8}$ lb, and $25\frac{1}{2}$ lb of beans in the hours that he worked before noon. He must pick 100 lb during the day to earn a bonus. How many more pounds must he pick to earn the bonus?

10. A house plan calls for $25\frac{1}{2}$ sq yd of carpeting in the living room, $5\frac{2}{3}$ sq yd in a bathroom, and $17\frac{2}{9}$ sq yd in the family room. A floor-covering shop has on hand a 50-sq yd roll of the desired carpet. Is it enough for the house? If so, how much will be left over?

SELF-TEST for Chapter Seven

The purpose of the Self-Test is to help you check your progress and review for a chapter test in class. Allow yourself about an hour to take the test. When you are done, check your answers in the back of the book. If you missed any problems, be sure to go back and review the appropriate sections in the chapter and do the supplementary exercises provided there.

In Problems 1 to 3, add.

1. $\dfrac{2}{7} + \dfrac{4}{7} =$

2. $\dfrac{5}{12} + \dfrac{3}{12} =$

3. $\dfrac{11}{15} + \dfrac{7}{15} =$

In Problems 4 and 5, find the least common denominator for fractions with the given denominators.

4. 12 and 15

5. 3, 4, and 18

In Problems 6 to 10, add.

6. $\dfrac{3}{5} + \dfrac{3}{10} =$

7. $\dfrac{1}{6} + \dfrac{3}{7} =$

8. $\dfrac{3}{8} + \dfrac{5}{12} =$

9. $\dfrac{11}{15} + \dfrac{9}{20} =$

10. $\dfrac{1}{4} + \dfrac{5}{8} + \dfrac{7}{10} =$

11. A recipe calls for $\dfrac{1}{2}$ cup of raisins, $\dfrac{1}{4}$ cup of walnuts, and $\dfrac{2}{3}$ cup of rolled oats. What is the total amount of these ingredients?

In Problems 12 to 15, subtract.

12. $\dfrac{5}{9} - \dfrac{2}{9} =$

13. $\dfrac{7}{12} - \dfrac{1}{6} =$

14. $\dfrac{3}{5} - \dfrac{1}{4} =$

15. $\dfrac{11}{12} - \dfrac{3}{20} =$

16. You have $\dfrac{5}{6}$ hour to take a three-part test. You use $\dfrac{1}{3}$ hour for the first section and $\dfrac{1}{4}$ hour for the second. How much time do you have left to finish the last section of the test?

In Problems 17 to 26, perform the indicated operations.

17. $5\dfrac{3}{10} + 2\dfrac{4}{10} =$

18. $6\dfrac{5}{9} + 3\dfrac{7}{9} =$

19. $4\dfrac{1}{6} + 3\dfrac{3}{4} =$

20. $6\dfrac{3}{8} + 5\dfrac{7}{10} =$

21. $5\dfrac{5}{6} + 3\dfrac{1}{6} + 2\dfrac{5}{6} =$

22. $3\dfrac{1}{2} + 4\dfrac{3}{4} + 5\dfrac{3}{10} =$

23. $7\dfrac{3}{8} - 5\dfrac{5}{8} =$

24. $3\dfrac{5}{6} - 2\dfrac{2}{9} =$

25. $7\dfrac{1}{8} - 3\dfrac{1}{6} = ?$

26. $7 - 5\dfrac{7}{15} =$

In Problems 27 to 30, solve each application.

27. A worker has $2\dfrac{1}{6}$ hours of overtime on Tuesday, $1\dfrac{3}{4}$ hours on Wednesday, and $1\dfrac{5}{6}$ hours on Friday. What is the worker's total overtime for the week?

28. A fence post is $86\dfrac{1}{4}$ in long. If you want $61\dfrac{1}{2}$ in of the post to be above ground, how far should the post be set into the ground?

29. Find the missing dimension in the accompanying figure.

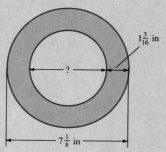

30. You cut two pieces of fabric with length $2\dfrac{3}{4}$ yd and one with length $3\dfrac{2}{3}$ yd from a bolt of fabric containing 40 yd. How much of the fabric remains on the bolt?

SUMMARY for Part Two

The Language of Fractions

Fraction Fractions name a number of equal parts of a unit or whole. A fraction is written in the form $\frac{a}{b}$, where a and b are whole numbers and b cannot be zero.

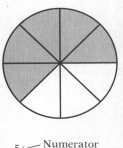

Denominator The number of equal parts the whole is divided into.
Numerator The number of parts of the whole that are used.

$\frac{5}{8}$ ← Numerator
$\phantom{\frac{5}{8}}$ ← Denominator

Proper Fraction A fraction whose numerator is less than its denominator. It names a number less than 1.

$\frac{2}{3}$ and $\frac{11}{15}$ are proper fractions.

Improper Fractions A fraction whose numerator is greater than or equal to its denominator. It names a number greater than or equal to 1.

$\frac{7}{5}$, $\frac{21}{20}$, and $\frac{8}{8}$ are improper fractions.

Mixed Number The sum of a whole number and a proper fraction.

$2\frac{1}{3}$ and $5\frac{7}{8}$ are mixed numbers. Note that $2\frac{1}{3}$ means $2 + \frac{1}{3}$.

Converting Mixed Numbers and Improper Fractions

To Change an Improper Fraction into a Mixed Number

1. Divide the numerator by the denominator. The quotient is the whole-number portion of the mixed number.
2. If there is a remainder, write the remainder over the original denominator. This gives the fractional portion of the mixed number.

To change $\frac{22}{5}$ to a mixed number:

$$5\overline{)22} \quad \text{Quotient} = 4$$
$$20$$
$$2 \quad \text{Remainder}$$

$\frac{22}{5} = 4\frac{2}{5}$

To Change a Mixed Number to an Improper Fraction

1. Multiply the denominator of the fraction by the whole-number portion of the mixed number.
2. Add the numerator of the fraction to that product.
3. Write that sum over the original denominator to form the improper fraction.

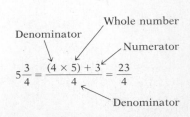

Equivalent Fractions

Equivalent Fractions Two fractions are equivalent (have equal value) if they are different names for the same number.

Cross Products

$$\frac{a}{b} \diagup\!\!\!\!\!\diagdown \frac{c}{d}$$ $a \times d$ and $b \times c$ are called the cross products.

If the cross products for two fractions are equal, the two fractions are equivalent. $\frac{2}{3} = \frac{4}{6}$ because $2 \times 6 = 3 \times 4$.

The Fundamental Principle

For the fraction $\frac{a}{b}$, and any nonzero number c,

$$\frac{a}{b} = \frac{a \times c}{b \times c}$$ $\frac{1}{2} = \frac{1 \times 5}{2 \times 5} = \frac{5}{10}$

In words: We can multiply the numerator and denominator of a fraction by the same nonzero number. The result will be an equivalent fraction. $\frac{1}{2}$ and $\frac{5}{10}$ are equivalent fractions.

Simplest Form A fraction is in simplest form, or in lowest terms, if the numerator and denominator have no common factors other than 1. This means that the fraction has the smallest possible numerator and denominator.

$\frac{2}{3}$ is in simplest form.

$\frac{12}{18}$ is *not* in simplest form. The numerator and denominator have the common factor 6.

To Write a Fraction in Simplest Form

Divide the numerator and denominator by any common factor greater than 1 to reduce a fraction to an equivalent fraction in lower terms. $\frac{10}{15} = \frac{10 \div 5}{15 \div 5} = \frac{2}{3}$

To Build a Fraction

Multiply the numerator and denominator by any whole number greater than 1 to raise a fraction to an equivalent fraction in higher terms. $\frac{3}{4} = \frac{3 \times 2}{4 \times 2} = \frac{6}{8}$

Multiplying Fractions

To Multiply Fractions

1. Multiply numerator by numerator. This gives the numerator of the product.
2. Multiply denominator by denominator. This gives the denominator of the product.
3. Simplify the resulting fraction if possible.

$$\frac{5}{8} \times \frac{3}{7} = \frac{5 \times 3}{8 \times 7} = \frac{15}{56}$$

In multiplying fractions it is usually easiest to divide by any common factors in the numerator and denominator *before* multiplying.

$$\frac{5}{9} \times \frac{3}{10} = \frac{\cancel{5}^1 \times \cancel{3}^1}{\cancel{9}_3 \times \cancel{10}_2} = \frac{1}{6}$$

Dividing Fractions

To Divide Fractions

Invert the divisor and multiply.

$$\frac{3}{7} \div \frac{4}{5} = \frac{3}{7} \times \frac{5}{4} = \frac{15}{28}$$

Multiplying or Dividing Mixed Numbers

To Multiply or Divide Mixed Numbers

Convert any mixed or whole numbers to improper fractions. Then multiply or divide the fractions as before.

$$6\frac{2}{3} \times 3\frac{1}{5} = \frac{20}{3} \times \frac{16}{\cancel{5}_1}^{4} = \frac{64}{3} = 21\frac{1}{3}$$

Finding the Least Common Denominator

To Find the LCD of a Group of Fractions

1. Write the prime factorization for each of the denominators.
2. Find all the prime factors that appear in any one of the prime factorizations.
3. Form the product of those prime factors, using each factor the greatest number of times it occurs in any one factorization.

To find the LCD of fractions with denominators 4, 6, and 15:

$$\begin{array}{r} 4 = 2 \times 2 \\ 6 = 2 \times 3 \\ 15 = 3 \times 5 \\ \hline 2 \times 2 \times 3 \times 5 \end{array}$$

The LCD = $2 \times 2 \times 3 \times 5$, or 60.

Adding Fractions

To Add Like Fractions

1. Add the numerators.
2. Place the sum over the common denominator.
3. Simplify the resulting fraction if necessary.

$$\frac{5}{18} + \frac{7}{18} = \frac{12}{18} = \frac{2}{3}$$

To Add Fractions with Different Denominators

1. Find the LCD of the fractions.
2. Change each fraction to an equivalent fraction with the LCD as a common denominator.
3. Add the resulting like fractions as before.

$$\frac{3}{4} + \frac{7}{10} = \frac{15}{20} + \frac{14}{20} = \frac{29}{20} = 1\frac{9}{20}$$

Subtracting Fractions

To Subtract Like Fractions

1. Subtract the numerators.
2. Place the difference over the common denominator.
3. Simplify the resulting fraction if necessary.

$$\frac{17}{20} - \frac{7}{20} = \frac{10}{20} = \frac{1}{2}$$

To Subtract Fractions with Different Denominators

1. Find the LCD of the fractions.
2. Change each fraction to an equivalent fraction with the LCD as a common denominator.
3. Subtract the resulting like fractions as before.

$$\frac{8}{9} - \frac{5}{6} = \frac{16}{18} - \frac{15}{18} = \frac{1}{18}$$

Adding or Subtracting Mixed Numbers

To Add or Subtract Mixed Numbers

1. Add or subtract the whole-number parts.
2. Add or subtract the fractional parts. *Note:* Subtracting may require renaming the first mixed number.
3. Combine the results as a mixed number.

$$5\frac{1}{2} - 3\frac{3}{4} = 5\frac{2}{4} - 3\frac{3}{4}$$
$$\text{Rename}$$
$$= 4\frac{6}{4} - 3\frac{3}{4} = 1\frac{3}{4}$$
$$\{4 - 3\} \quad \left\{\frac{6}{4} - \frac{3}{4}\right\}$$

SUMMARY EXERCISES for Part Two

You should now be reviewing the material in Part 2 of the text. The following exercises will help in that process. Work all the exercises carefully. Then check your answers in the back of the book. References to the chapter and section for each problem are provided there. If you made an error, go back and review the related material and do the supplementary exercises for that section.

[5.1] Give the fractions that name the shaded portions of the following diagrams. Identify the numerator and the denominator.

1.

   Fraction: 3/8

   Numerator: 3

   Denominator: 8

2.

   Fraction: 5/6

   Numerator: 5

   Denominator: 6

[5.2] 3. From the following group of numbers:

$$\frac{2}{3}, \frac{5}{4}, 2\frac{3}{7}, \frac{45}{8}, \frac{7}{7}, 3\frac{4}{5}, \frac{9}{1}, \frac{7}{10}, \frac{12}{5}, 5\frac{2}{9}$$

List the proper fractions. 2/3, 7/10

List the improper fractions. 5/4, 45/8, 7/7, 9/1, 12/5, 12/5

List the mixed numbers. 2 3/7, 3 4/5, 5 2/9

[5.3] Convert to mixed or whole numbers.

4. $\frac{41}{6}$ 6 5/6

5. $\frac{32}{8}$ 4

[5.3] Convert to improper fractions.

6. $7\frac{5}{8}$ 61/8

7. $4\frac{3}{10}$ 43/10

303

[5.4] Determine whether or not the following pairs of fractions are equivalent.

8. $\dfrac{5}{8}, \dfrac{7}{12}$

9. $\dfrac{8}{15}, \dfrac{32}{60}$

[5.5] Write each fraction in simplest form.

10. $\dfrac{24}{36}$

11. $\dfrac{45}{75}$

12. $\dfrac{140}{180}$

13. $\dfrac{16}{21}$

[5.6] Find the missing numerators.

14. $\dfrac{3}{5} = \dfrac{?}{25}$

15. $\dfrac{4}{5} = \dfrac{?}{40}$

[5.6] Arrange the fractions in order from smallest to largest.

16. $\dfrac{5}{8}, \dfrac{7}{12}$

17. $\dfrac{5}{6}, \dfrac{4}{5}, \dfrac{7}{10}$

[5.6] Complete the following statements using the symbols <, =, or >.

18. $\dfrac{5}{12} \quad \dfrac{3}{8}$

19. $\dfrac{3}{7} \quad \dfrac{9}{21}$

20. $\dfrac{9}{16} \quad \dfrac{7}{12}$

[5.6] Write as equivalent fractions with the LCD as a common denominator.

21. $\dfrac{1}{6}, \dfrac{7}{8}$

22. $\dfrac{3}{10}, \dfrac{5}{8}, \dfrac{7}{12}$

[6.1]
[6.2] Multiply.
[6.3]

23. $\dfrac{7}{15} \times \dfrac{5}{21}$

24. $\dfrac{10}{27} \times \dfrac{9}{20}$

25. $4 \times \dfrac{3}{8}$

26. $3\dfrac{2}{5} \times \dfrac{5}{8}$

27. $5\dfrac{1}{3} \times 1\dfrac{4}{5}$

28. $1\dfrac{5}{12} \times 8$

29. $3\dfrac{1}{5} \times \dfrac{7}{8} \times 2\dfrac{6}{7}$

Summary Exercises for Part Two

[6.4] Divide.

30. $\dfrac{5}{12} \div \dfrac{5}{8}$

31. $\dfrac{7}{15} \div \dfrac{14}{25}$

32. $\dfrac{9}{20} \div 2\dfrac{2}{5}$

33. $3\dfrac{3}{8} \div 2\dfrac{1}{4}$

34. $3\dfrac{3}{7} \div 8$

[6.3]
[6.4] Solve each of the following problems.

35. The scale on a map uses 1 in to represent 80 mi. If two cities are $2\dfrac{3}{4}$ in apart on the map, what is the actual distance between the cities?

36. A kitchen measures $5\dfrac{1}{3}$ yd by $4\dfrac{1}{4}$ yd. If you purchase linoleum which costs \$9 per square yard, what will it cost to cover the floor?

37. If you drive 117 mi in $2\dfrac{1}{4}$ hours, what is your average speed?

38. An 18-acre piece of land is to be subdivided into home lots that are each $\dfrac{3}{8}$ acre. How many lots can be formed?

[7.1]
[7.2] Add.
[7.3]

39. $\dfrac{2}{9} + \dfrac{5}{9}$

40. $\dfrac{7}{10} + \dfrac{9}{10}$

41. $\dfrac{5}{6} + \dfrac{11}{18}$

42. $\dfrac{5}{18} + \dfrac{7}{12}$

43. $\dfrac{3}{5} + \dfrac{1}{4} + \dfrac{5}{6}$

[7.4] Perform the indicated operations.

44. $\dfrac{5}{8} - \dfrac{3}{8}$

45. $\dfrac{11}{18} - \dfrac{2}{9}$

46. $\dfrac{7}{10} - \dfrac{7}{12}$

47. $\dfrac{11}{27} - \dfrac{5}{18}$

48. $\dfrac{4}{9} + \dfrac{5}{12} - \dfrac{3}{8}$

[7.5] Perform the indicated operations.

49. $6\dfrac{5}{7} + 3\dfrac{4}{7}$

50. $5\dfrac{7}{10} + 3\dfrac{11}{12}$

51. $2\dfrac{1}{2} + 3\dfrac{5}{6} + 3\dfrac{3}{8}$

52. $7\dfrac{7}{9} - 3\dfrac{4}{9}$

53. $9\dfrac{1}{6} - 3\dfrac{1}{8}$

54. $6\dfrac{5}{12} - 3\dfrac{5}{8}$

55. $2\dfrac{1}{3} + 5\dfrac{1}{6} - 2\dfrac{4}{5}$

[7.6] Solve each of the following problems.

56. Bradley needs two shelves, one $32\dfrac{3}{8}$ in long, and the other $36\dfrac{11}{16}$ in long. What is the total length of shelving that is needed?

57. Find the perimeter of the figure below.

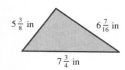

58. At the beginning of a year Miguel was $51\dfrac{3}{4}$ in tall. In June he measured $53\dfrac{1}{8}$ in. How much did he grow during that period?

59. A bookshelf that is $42\dfrac{5}{16}$ in long is cut from a board with a length of 8 ft. If $\dfrac{1}{8}$ in is wasted in the cut, what length board remains?

60. Amelia buys an 8-yd roll of wallpaper on sale. After measuring, she finds that she needs the following amounts of the paper: $2\dfrac{1}{3}$ yd, $1\dfrac{1}{2}$ yd, and $3\dfrac{3}{4}$ yd. Does she have enough for the job? If so, how much will be left over?

CUMULATIVE TEST for Part Two

This test is provided to help you in the process of review over Chapters 5 through 7. Answers are provided in the back of the book. If you missed any problems, be sure to go back and review the appropriate chapter sections.

In Problem 1, give the fraction that names the shaded portion of the diagram. Identify the numerator and denominator.

1.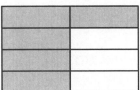

Fraction:

Numerator:

Denominator:

In Problem 2, identify the proper fractions, improper fractions, and mixed numbers from the following group.

$$\frac{7}{12}, \frac{10}{8}, 3\frac{1}{5}, \frac{9}{9}, \frac{7}{1}, \frac{3}{7}, 2\frac{2}{3}$$

2. Proper: Improper:

   Mixed numbers:

In Problems 3 and 4, convert to mixed or whole numbers.

3. $\dfrac{14}{5}$

4. $\dfrac{28}{7}$

In Problems 5 and 6, convert to improper fractions.

5. $4\dfrac{1}{3}$

6. $7\dfrac{7}{8}$

In Problems 7 and 8, find out whether the pair of fractions is equivalent.

7. $\dfrac{7}{21}, \dfrac{8}{24}$

8. $\dfrac{7}{12}, \dfrac{8}{15}$

In Problems 9 and 10, write the fractions in simplest form.

9. $\dfrac{28}{42}$

10. $\dfrac{36}{96}$

In Problems 11 and 12, arrange the fractions in order from smallest to largest.

11. $\dfrac{5}{9}, \dfrac{6}{11}$

12. $\dfrac{7}{10}, \dfrac{3}{5}, \dfrac{8}{15}$

In Problems 13 and 14, write the fractions as equivalent fractions with the LCD as a common denominator.

13. $\dfrac{5}{8}, \dfrac{7}{12}$

14. $\dfrac{2}{3}, \dfrac{5}{9}, \dfrac{3}{4}$

In Problems 15 to 19, multiply.

15. $\dfrac{5}{9} \times \dfrac{8}{15}$

16. $\dfrac{20}{21} \times \dfrac{7}{25}$

17. $1\dfrac{1}{8} \times 4\dfrac{4}{5}$

18. $8 \times 2\dfrac{5}{6}$

19. $\dfrac{2}{3} \times 1\dfrac{4}{5} \times \dfrac{5}{8}$

In Problems 20 to 23, divide.

20. $\dfrac{5}{8} \div \dfrac{15}{32}$

21. $2\dfrac{5}{8} \div \dfrac{7}{12}$

22. $4\dfrac{1}{6} \div 5$

23. $2\dfrac{2}{7} \div 1\dfrac{11}{21}$

In Problems 24 and 25, solve each application.

24. A living room measured $6\dfrac{2}{3}$ yd by $4\dfrac{1}{2}$ yd. If you purchase carpeting at $18 per square yard, what will it cost to carpet the room?

25. If a stack of $\dfrac{5}{8}$-in plywood measures 55 inches in height, how many sheets of plywood are in the stack?

Cumulative Test for Part Two

In Problems 26 to 28, add.

26. $\dfrac{4}{15} + \dfrac{8}{15}$

27. $\dfrac{7}{25} + \dfrac{8}{15}$

28. $\dfrac{2}{5} + \dfrac{3}{4} + \dfrac{5}{8}$

In Problems 29 to 31, perform the indicated operations.

29. $\dfrac{17}{20} - \dfrac{7}{20}$

30. $\dfrac{5}{9} - \dfrac{5}{12}$

31. $\dfrac{5}{18} + \dfrac{4}{9} - \dfrac{1}{6}$

In Problems 32 to 37, perform the indicated operations.

32. $3\dfrac{5}{7} + 2\dfrac{4}{7}$

33. $4\dfrac{7}{8} + 3\dfrac{1}{6}$

34. $8\dfrac{1}{9} - 3\dfrac{5}{9}$

35. $7\dfrac{7}{8} - 3\dfrac{5}{6}$

36. $9 - 5\dfrac{3}{8}$

37. $3\dfrac{1}{6} + 3\dfrac{1}{4} - 2\dfrac{7}{8}$

In Problems 38 to 40, solve each application.

38. In his part-time job, Manuel worked $3\dfrac{5}{6}$ hours on Monday, $4\dfrac{3}{10}$ hours on Wednesday, and $6\dfrac{1}{2}$ hours on Friday. Find the number of hours that he worked during the week?

39. A $6\dfrac{1}{2}$-in bolt is placed through a wall that is $5\dfrac{7}{8}$ in thick. How far does the bolt extend beyond the wall?

40. On a 6-hour trip, Pete drove $1\dfrac{3}{4}$ hours and then Maria drove for another $2\dfrac{1}{3}$ hours. How many hours remained on the trip?

Part 3

Decimals

PRETEST for Chapter Eight

Addition, Subtraction, and Multiplication of Decimals

This pretest will point out any difficulties you may be having in adding, subtracting, or multiplying decimals. Do all the problems. Then check your answers on the following page.

1. Give the place value of 5 in the decimal 2.4857.

2. Write $2\frac{371}{1000}$ in decimal form and in words.

3. 2.43 + 3.17 + 5.82 =

4. 37 + 2.31 + 2.4 + 0.58 =

5. 5.5 − 3.375 =

6. You have $20 in cash and make purchases of $5.36 and $9.25. How much cash do you have left?

7. 0.241 × 3.57 =

8. 0.345 × 1000 =

9. Round 2.35878 to the nearest thousandth.

10. You fill up your car with 9.2 gal of diesel fuel at 93.9¢ per gallon. What is the cost of the fill-up (to the nearest cent)?

ANSWERS TO PRETEST

For help with similar problems, turn to the section indicated.

1. Thousandths (Section 8.1);
2. 2.371, two and three hundred seventy-one thousandths (Section 8.1);
3. 11.42 (Section 8.2);
4. 42.29 (Section 8.2);
5. 2.125 (Section 8.3);
6. $5.39 (Section 8.3);
7. 0.86037 (Section 8.4);
8. 345 (Section 8.5);
9. 2.359 (Section 8.6);
10. $8.64 (Section 8.6)

Chapter Eight

Addition, Subtraction, and Multiplication of Decimals

8.1 Understanding Place Value in Decimal Fractions

OBJECTIVES
1. To identify place values in decimal fractions
2. To write decimal fractions in words
3. To compare the sizes of decimal fractions

Remember that the powers of 10 are 1, 10, 100, 1000, and so on. You might want to review this concept in Section 2.8 before going on.

In Part 2 we looked at common fractions. Let's turn now to a special kind of fraction, a *decimal fraction*. A decimal fraction is a fraction whose denominator is a *power of 10*.

Example 1

$\frac{3}{10}$, $\frac{45}{100}$, and $\frac{231}{1000}$ are examples of decimal fractions.

Earlier we talked about the idea of place value. Recall that in our decimal place-value system, each place has *one-tenth* the value of the place to its left.

Example 2

Label the place values for 538.

$$\underset{\text{Hundreds}}{5} \quad \underset{\text{Tens}}{3} \quad \underset{\text{Ones}}{8}$$

The ones place value is one-tenth of the tens place value.

The tens place value is one-tenth of the hundreds place value.

CHECK YOURSELF 1

Label the place values for 2793.

We now want to extend this idea *to the right* of the ones place. Write a period to the *right* of the ones place. This is called the *decimal point*. Each digit to the right of that decimal point will represent a fraction whose denominator is a power of 10. The first place to the right of the decimal point is the tenths place $\left(0.1 = \frac{1}{10}\right)$.

The decimal point separates the whole-number part and the fractional part of a decimal fraction.

Example 3

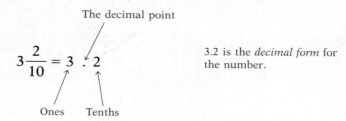

3.2 is the *decimal form* for the number.

CHECK YOURSELF 2

Write $5\frac{3}{10}$ in decimal form. 5.3.

As you move further to the *right*, each place value must be one-tenth of the value before it. The second place value is hundredths $\left(0.01 = \frac{1}{100}\right)$. The next place is thousandths, the next position is the ten-thousandths place, and so on. Figure 1 illustrates the value of each position as we move to the right of the decimal point.

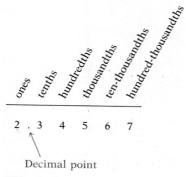

Figure 1

Example 4

For convenience we will shorten the term *decimal fraction* to *decimal* from this point on.

For the decimal 2.34567, the place value of 4 is hundredths and the place value of 6 is ten-thousandths.

CHECK YOURSELF 3

What is the place value of 5 in the decimal of Example 4?

Understanding place values will allow you to read and write decimals using the following steps.

Section 8.1: Understanding Place Value in Decimal Fractions 317

> **READING AND WRITING DECIMALS**
>
> STEP 1 The digits *to the left* of the decimal point are read as a whole number.
>
> STEP 2 The word "and" indicates the decimal point.
>
> STEP 3 The digits *to the right* of the decimal point are read as a whole number followed by the place value of the rightmost digit.

If there are *no* nonzero digits to the left of the decimal point, start directly with step 3.

Example 5

5.03 is read "five and three hundredths"

Hundredths — The rightmost digit, 3, is in the hundredths position.

12.057 is read "twelve and fifty-seven thousandths"

Thousandths — The rightmost digit, 7, is in the thousandths position.

Example 6

0.5321 is read

"five thousand three hundred twenty-one ten-thousandths."

When the decimal has no whole-number part, we have chosen to write a 0 to the left of the decimal point. This simply makes sure that you don't miss the decimal point. However, both 0.5321 and .5321 are correct.

An informal way of reading decimals is to simply read the digits in order and use the word "point" to indicate the decimal point. 2.58 can be read "two point five eight." 0.689 can be read as "point six eight nine."

CHECK YOURSELF 4

Write 2.58 in words. *two and fifty eight Hundredths*

The number of digits to the right of the decimal point is called the number of *decimal places*.

Example 7

3.45 has two decimal places.

12.368 has three decimal places.

One quick way to write a decimal as a common fraction is to remember that the number of decimal places must be the same as the number of zeros in the denominator of the common fraction.

Example 8

$$0.35 = \frac{35}{100}$$

Two places ↑ (under 0.35)
Two zeros ↑ (under 100)

The same method can be used with decimals that are greater than 1. Here the result will be a mixed number. Our next example illustrates.

Example 9

The zero to the right of the decimal point is a "placeholder" which is not needed in the common fraction form.

$$2.058 = 2\frac{58}{1000}$$

Three places ↑
Three zeros ↑

CHECK YOURSELF 5

Write as common fractions or mixed numbers.

(1) 0.528
(2) 5.08

It is often useful to compare the sizes of two decimal fractions. One approach to comparing decimals uses the following fact.

Remember: By the fundamental principle of fractions, multiplying the numerator and denominator of a fraction by the same nonzero number does not change the value of the fraction.

> Adding zeros to the right *does not change* the value of a decimal. 0.53 is the same as 0.530. Look at the fractional form:
>
> $$\frac{53}{100} = \frac{530}{1000}$$
>
> The fractions are equivalent. We have multiplied the numerator and denominator by 10.

Let's see how this is used to compare decimals in our final example.

Example 10

Which is larger?

0.84   or   0.842

Section 8.1: Understanding Place Value in Decimal Fractions   319

Write 0.84 as 0.840. Then we see that 0.842 (or 842 thousandths) is greater than 0.840 (or 840 thousandths), and we can write

0.842 > 0.84

CHECK YOURSELF 6

Complete the statement below using the symbols < or >.

0.588 __<__ 0.59

CHECK YOURSELF ANSWERS

1. 2 7 9 3 — Thousands, Hundreds, Tens, Ones   2. $5\frac{3}{10} = 5.3$.   3. Thousandths.

4. Two and fifty-eight hundredths.   5. (1) $\frac{528}{1000}$; (2) $5\frac{8}{100}$.   6. $0.588 < 0.59$.

8.1 Exercises

For the decimal 8.57932:

1. What is the place value of 7?

2. What is the place value of 5?

3. What is the place value of 3?

4. What is the place value of 2?

Write in decimal form.

5. $\frac{23}{100}$ 0.23

6. $\frac{371}{1000}$ 0.371

7. $\frac{209}{10,000}$ 0.0209

8. $3\frac{5}{10}$

9. $23\frac{56}{1000}$ 23.056

10. $7\frac{431}{10,000}$ 7.0431

Write in words.

11. 0.23

12. 0.371

13. 0.071

14. 0.0251 ten thousn.

15. 12.07   and 7 hundredths

16. 23.056   23 and 56 thousandth

Write in decimal form.

17. Fifty-one thousandths .051

18. Two hundred fifty-three ten-thousandths .0253

19. Seven and three tenths 7.3

20. Twelve and two hundred forty-five thousandths 12.245

Write as common fractions or mixed numbers.

21. 0.65 $\frac{65}{100}$

22. 0.00765 $\frac{765}{100,000}$

23. 5.231 $5\frac{231}{1000}$

24. 4.0171 $4\frac{171}{10,000}$

Complete each of the following statements using the symbols <, =, or >.

25. 0.69 > 0.689

26. 0.75 < 0.752

27. 1.230 = 1.230

28. 2.451 > 2.450

29. 10 > 9.9

30. 4.98 < 5

31. 1.459 < 1.46

32. 0.235   0.2350

Skillscan (Section 1.5)
Add.

a. 43
     +58

b. 79
     +37

c. 584
     +749

d. 675
     +985

e. 29
      58
     +43

f. 129
      538
     +794

Answers
Solutions for the even-numbered exercises are provided in the back of the book. **1.** hundredths **3.** Ten-thousandths **5.** 0.23 **7.** 0.0209 **9.** 23.056 **11.** Twenty-three hundredths **13.** Seventy-one thousandths **15.** Twelve and seven hundredths **17.** 0.051 **19.** 7.3 **21.** $\frac{65}{100}$ (or $\frac{13}{20}$) **23.** $5\frac{231}{1000}$ **25.** 0.69 > 0.689 **27.** 1.23 = 1.230 **29.** 10 > 9.9 **31.** 1.459 < 1.46 **a.** 101 **b.** 116 **c.** 1333 **d.** 1660 **e.** 130 **f.** 1461

8.1 Supplementary Exercises

For the decimal 9.63584:

1. What is the place value of 3?
2. What is the place value of 8?

Write in decimal form.

3. $\frac{47}{100}$ 0.47

4. $\frac{47}{1000}$ 0.047

5. $12\frac{251}{10,000}$ 12.0251

Write in words.

6. 0.47
7. 0.419
8. 2.0043

Write in decimal form.

9. Two hundred forty-five ten-thousandths
10. Seven and thirty-two thousandths

Write as common fractions or mixed numbers.

11. 0.0067
12. 21.857

Complete each of the following statements using the symbols <, =, or >.

13. 0.78   0.778
14. 0.53   0.532
15. 0.27   0.270
16. 2.31   2.308

8.2 Adding Decimals

OBJECTIVES
1. To add decimals
2. To use addition in solving word problems

Working with decimals rather than common fractions makes the basic operations much easier. Let's start by looking at addition. One method for adding decimals is to write the decimals as common fractions, add, and then change the sum back to a decimal.

Example 1

$$0.34 + 0.52 = \frac{34}{100} + \frac{52}{100} = \frac{86}{100} = 0.86$$

It is much more efficient to leave the numbers in decimal form and perform the addition in the same way as we did with whole numbers. You can use the following rule:

TO ADD DECIMALS

STEP 1 Write the numbers being added in column form *with their decimal points in a vertical line.*

STEP 2 Add just as you would with whole numbers.

STEP 3 Place the decimal point of the sum in line with the decimal points of the addends.

Our next example illustrates the use of this rule.

Example 2

To add 0.13, 0.42, and 0.31, write:

Placing the decimal points in a vertical line assures that we are adding digits of the same place value.

```
  0.13
  0.42
 +0.31
  ----
  0.86
```

CHECK YOURSELF 1
Add.

0.23, 0.15, and 0.41

In adding decimals you can use the *carrying process* just as you did in adding whole numbers. Consider the following.

Example 3

To add 0.35, 1.58, and 0.67, write:

```
  1 2    ← Carries
  0.35
  1.58
 +0.67
  ----
  2.60
```

In the hundredths column:
5 + 8 + 7 = 20
Write 0 and carry 2 to the tenths column.

In the tenths column:
2 + 3 + 5 + 6 = 16
Write 6 and carry 1 to the ones column.

Section 8.2: Adding Decimals

Note The carrying process works with decimals, just as it did with whole numbers, because each place value is again *one-tenth* the value of the place to its left.

CHECK YOURSELF 2

Add.

23.546, 0.489, 2.312, and 6.135

In adding decimals, the numbers may not have the same number of decimal places. Just fill in as many zeros as needed so that all the numbers added have the same number of decimal places.

Recall that adding zeros to the right *does not change* the value of a decimal. 0.53 is the same as 0.530.

Let's see how this is used in our next example.

Example 4

To add 0.53, 4, 2.7, and 3.234, write:

Be sure that the decimal points are in a vertical line.

```
  0.53
  4.         Note that for a whole number, the decimal
  2.7        is understood to be to its right. So 4 = 4.
+3.234
```

Now fill in the missing zeros, and add as before.

```
   0.530
   4.000     Now all the numbers being added
   2.700     have three decimal places.
+  3.234
  10.464
```

CHECK YOURSELF 3

Add.

6, 2.583, 4.7, and 2.54

Many applied problems require working with decimals. For instance, filling up at a gas station means reading decimal amounts.

Example 5

On a trip the Jacksons kept track of their gas purchases. If they bought 12.3, 14.2, 10.7, and 13.8 gal, how much gas did they use on the trip?

Since we want a total amount, addition is used for the solution.

Solution

```
  12.3
  14.2
  10.7
+ 13.8
------
  51.0 gal
```

Every day you deal with amounts of money. Since our system of money is a decimal system most problems involving money will also involve operations with decimals.

Example 6

Mr. Black makes deposits of $3.24, $15.73, $50, $28.79, and $124.38 during the month of May. What is his total deposit for the month?

Solution

```
$   3.24
   15.73     Simply add the amounts of money
   50.00     deposited as decimals. Note that
   28.79     we write $50 as $50.00.
+ 124.38
--------
$ 222.14  ← The total deposit for May
```

CHECK YOURSELF 4

Your textbooks for the fall term cost $33.50, $18.95, $23.15, $12, and $5.85. What was the total cost of texts for the term?

CHECK YOURSELF ANSWERS

1. 0.79. 2. 32.482. 3.
```
   6.000
   2.583
   4.700
 + 2.540
 -------
  15.823
```
4. $93.45.

8.2 Exercises

Add.

1. 0.28
   +0.79

2. 2.59
   +0.63

3. 1.045
   +0.23

4. 2.485
   +1.25

5. 0.62
   4.23
   +12.5

6. 0.50
   2.99
   +24.8

7. 5.28
   +19.455

8. 23.845
   + 7.29

Section 8.2: Adding Decimals

9. 13.58
    7.239
  + 1.5

10. 8.625
     1.38
   +12.6

11. 25.3582
      6.5
      1.898
   + 0.69

12. 3.459
    15.6857
     7.9
   + 0.85

13. 0.43 + 0.8 + 0.561

14. 1.25 + 0.7 + 0.259

15. 5 + 23.7 + 8.7 + 9.85

16. 28.3 + 6 + 8.76 + 3.8

17. 25.83 + 1.7 + 3.92

18. 4.8 + 32.59 + 3.87

19. 42.731 + 1.058 + 103.24

20. 27.4 + 213.321 + 39.38

21. Add: twenty-three hundredths, five tenths, and two hundred sixty-eight thousandths.

22. Add: seven tenths, four hundred fifty-eight thousandths, and fifty-six hundredths.

23. Add: five and three tenths, seventy-five hundredths, twenty and thirteen hundredths, and twelve and seven tenths.

24. Add: thirty-eight and nine tenths, five and fifty-eight hundredths, seven, and fifteen and eight tenths.

25. On a 3-day trip Tom bought 12.7, 15.9, and 13.8 gal of gas. How many gallons of gas did he buy?

26. Jan ran 2.7 mi on Monday, 1.9 mi on Wednesday, and 3.6 mi on Friday. How far did she run during the week?

27. Rainfall was recorded during the winter months as follows: December, 5.38 cm; January, 3.2 cm; and February, 4.79 cm. How much rain fell during those months?

28. A metal fitting has three sections with lengths 2.5, 1.775, and 1.45 in. What is the total length of the fitting?

29. Steve had the following expenses on a business trip: gas, $45.69; food, $123; lodging, $95.60; and parking and tolls, $8.65. What were his total expenses during the trip?

30. Cheryl's textbooks for one quarter cost $4.95, $25, $37.25, $22.95, and $10. What was her total cost for books?

31. Bruce wrote checks of $50, $11.38, $112.57, and $9.73 during a single week. What was the total amount of the checks that were written?

32. Joyce made deposits of $75.35, $58, $7.89, and $100 to her checking account in a single month. What was the total amount of her deposits?

Skillscan (Section 1.8)
Subtract.

a. 43	b. 58	c. 247	d. 923	e. 4283	f. 5324
−37	−29	−158	−257	−1059	−2568

Answers

1. 1.07   3. 1.275   5. 17.35   7. 24.735   9. 22.319   11. 34.4462   13. 1.791   15. 5.00
                                                                                         23.70
                                                                                          8.70
                                                                                       + 9.85
                                                                                         47.25
17. 31.45   19. 147.029   21. 0.998

23. 5.30
    0.75
    20.13
    +12.70
    38.88

25. 42.4 gal   27. 13.37 cm   29. $272.94   31. $183.68

a. 6   b. 29   c. 89   d. 666   e. 3224   f. 2756

8.2 Supplementary Exercises

Add.

1. 0.57	2. 2.875	3. 13.86
+0.93	+0.95	3.8
		+ 0.59

4. 4.23	5. 29.358	6. 36.8954
+97.984	0.24	8.7
	+ 6.7	1.789
		+ 0.78

7. 23.5 + 3.26 + 0.96

8. 528.271 + 1.85 + 18.9

9. 13.675 + 9 + 0.87 + 235.8

10. 27.2 + 0.815 + 0.3484 + 7

11. Add: five tenths, seventy-three hundredths, and two hundred forty-five thousandths.

12. Add: thirty-five and six tenths, twenty-seven hundredths, four and five tenths, and five hundred thirty-seven thousandths.

13. Tim worked 6.7 hours on Monday, 5.9 hours on Wednesday, and 7.2 hours on Friday. How many hours did he work during the week?

14. Your bill from a service station includes $7.85 for oil, $5.25 for parts, and $9.75 for labor. What is the total charge?

15. On a vacation trip the Harts bought the following amounts of gasoline: 12.3, 15, 9.8, and 13.7 gal. How much gasoline did they purchase during the trip?

16. Rainfall amounts during a week were 2.38, 0.45, 1.5, and 2.19 cm. What was the total amount of rain during the week?

8.3 Subtracting Decimals

OBJECTIVES
1. To subtract decimals
2. To use subtraction in solving word problems

Much of what we said in the last section about adding decimals is also true of subtraction. To subtract decimals, we use the following rule:

> **TO SUBTRACT DECIMALS**
>
> STEP 1 Write the numbers being subtracted in column form *with their decimal points in a vertical line.*
>
> STEP 2 Subtract just as you would with whole numbers.
>
> STEP 3 Place the decimal point of the difference in line with the decimal points of the numbers being subtracted.

Our first example illustrates the use of this rule.

Example 1

To subtract 1.23 from 3.58, write

$$\begin{array}{r} 3.58 \\ -1.23 \\ \hline 2.35 \end{array}$$

Subtract in the hundredths, the tenths, and then the ones columns.

CHECK YOURSELF 1

Subtract.

$9.87 - 5.45$

Since each place value is one-tenth the value of the place to its left, borrowing, when subtracting decimals, works just as it did in subtracting whole numbers.

Example 2

To subtract 1.86 from 6.54, write

$$\begin{array}{r} {}^{5}\,{}^{1}4\,1 \\ \cancel{6}.\cancel{5}4 \\ -1.86 \\ \hline 4.68 \end{array}$$

Here, borrow in the tenths and ones places to do the subtraction.

CHECK YOURSELF 2

Subtract.

$35.35 - 13.89$

In subtracting decimals, as in adding, we can add zeros to the right of the decimal point so that both decimals have the same number of decimal places.

Example 3

To subtract 2.36 from 7.5, write

$$\begin{array}{r} {}^{4}\,{}^{1} \\ 7.\cancel{5}0 \\ -2.36 \\ \hline 5.14 \end{array}$$

We have added a 0 to 7.5. Next, borrow one tenth from the five tenths in the minuend.

Example 4

To subtract 3.657 from 9, write

$$\begin{array}{r}{}^{8}{}^{9}{}^{9}{}^{9}{}_{1}\\ \cancel{9}.000\\ -3.657\\ \hline 5.343\end{array}$$

In this case, move left to the ones place to begin the borrowing process.

CHECK YOURSELF 3

Subtract.

$5 - 2.345$

In solving word problems involving decimals, we may have to apply the same subtraction methods.

Example 5

John was 98.3 cm tall on his sixth birthday. On his seventh birthday he was 104.2 cm. How much did he grow during the year?

Solution

We want to find the difference between the two measurements so we subtract.

$$\begin{array}{r}104.2 \text{ cm}\\ -98.3 \text{ cm}\\ \hline 5.9 \text{ cm}\end{array}$$

John grew 5.9 cm during the year.

CHECK YOURSELF 4

A car's highway mileage before a tune-up was 28.8 mi/gal. After the tune-up it measured 30.1 mi/gal. What was the increase in mileage?

Example 6

Sally buys a roast at the grocery store which is marked $12.37. She pays for her purchase with a $20 bill. How much change does she get?

Solution

Sally's change will be the *difference* between the price of the roast and the $20 paid. We must use subtraction for the solution.

$$\begin{array}{r}\$20.00\\ -12.37\\ \hline \$7.63\end{array}$$

Add 0s to write $20 as $20.00. Then subtract as before.

Her change

Sally will receive $7.63 in change after her purchase.

CHECK YOURSELF 5

A stereo system which normally sells for $549.50 is discounted (or marked down) to $499.95 for a sale. What is the savings?

CHECK YOURSELF ANSWERS

1. 4.42.   2. 21.46.   3. 2.655.   4. 1.3 mi/gal.   5. $49.55.

8.3 Exercises

Subtract.

1. 0.85
   −0.59

2. 5.68
   −2.65

3. 23.81
   − 6.57

4. 48.73
   −19.95

5. 17.134
   − 3.502

6. 40.092
   −21.595

7. 35.8
   − 7.45

8. 7.83
   −5.2

9. 3.82
   −1.565

10. 8.59
    −5.6

11. 7.32
    −4.7

12. 45.6
    − 8.75

13. 12
    − 5.35

14. 15
    − 8.85

15. 15.02
    − 2.545

16. 36.05
    − 3.675

17. 28
    −24.725

18. 40
    −13.875

19. Subtract 2.87 from 6.84.

20. Subtract 3.69 from 10.57.

21. Subtract 7.75 from 9.4.

22. Subtract 5.82 from 12.

23. Subtract 0.24 from 5.

24. Subtract 8.7 from 16.32.

25. A television set selling for $399.50 is discounted (or marked down) to $365.75. What is the saving?

26. The outer radius of a piece of tubing is 2.8325 in. The inner radius is 2.775 in. What is the thickness of the wall of the tubing?

Section 8.3: Subtracting Decimals 331

27. If normal body temperature is 98.6° and a person is running a temperature of 101.3°, how much is that temperature above normal?

28. You pay your hotel bill of $84.58 with two $50 travelers checks. What change will you receive?

29. Given the following figure, find dimension *a*.

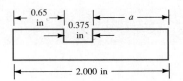

30. You make charges of $37.25, $8.78, and $53.45 on a credit card. If you make a payment of $73.50, how much do you still owe?

31. At the start of a trip, Laura noted that her odometer (mileage indicator) read 15,785.3 mi. Upon finishing the trip it read 16,479.8 mi. How far did she drive?

32. Rainfall for the first 3 months of 1988 was recorded at 2.73, 1.41, and 1.48 in. If the normal rainfall for that period was 6.51 in, was the 1988 amount above or below normal? By what amount?

Recall that a magic square is one in which the sum of every row, column, and diagonal is the same. Complete the magic squares below.

33.

1.6		1.2
	1	
0.8		

34.

2.4		7.2
10.8		
4.8		

Skillscan (Section 6.1)
Multiply.

a. $\dfrac{3}{10} \times \dfrac{7}{10}$

b. $\dfrac{7}{10} \times \dfrac{9}{10}$

c. $\dfrac{3}{10} \times \dfrac{23}{100}$

d. $\dfrac{7}{10} \times \dfrac{33}{100}$

e. $\dfrac{53}{100} \times \dfrac{7}{100}$

f. $\dfrac{21}{100} \times \dfrac{17}{100}$

Answers

1. 0.26   3. $\overset{11\ 7\ 1}{\cancel{23.81}}$ 5. 13.632   7. 28.35   9. 2.255   11. 2.62   13. $\begin{array}{r}12.00\\-\ 5.35\\\hline 6.65\end{array}$ 15. 12.475
   $\quad\quad\quad\begin{array}{r}-\ 6.57\\\hline 17.24\end{array}$

17. 3.275   19. 3.97   21. 1.65   23. 4.76   25. $33.75   27. 2.7°   29. 0.975 in   31. 694.5 mi

33.

1.6	0.2	1.2
0.6	1	1.4
0.8	1.8	0.4

a. $\dfrac{21}{100}$ b. $\dfrac{63}{100}$ c. $\dfrac{69}{1000}$ d. $\dfrac{231}{1000}$ e. $\dfrac{371}{10,000}$ f. $\dfrac{357}{10,000}$

8.3 Supplementary Exercises

Subtract.

1. $\begin{array}{r}5.97\\-0.45\end{array}$ 2. $\begin{array}{r}27.53\\-14.98\end{array}$ 3. $\begin{array}{r}20.235\\-12.857\end{array}$ 4. $\begin{array}{r}27.3\\-\ 8.55\end{array}$ 5. $\begin{array}{r}4.59\\-1.8\end{array}$

6. $\begin{array}{r}8.235\\-0.87\end{array}$ 7. $\begin{array}{r}5.25\\-1.345\end{array}$ 8. $\begin{array}{r}15\\-\ 8.35\end{array}$ 9. $\begin{array}{r}20\\-\ 7.875\end{array}$

10. Subtract 2.75 from 9.   11. Subtract 0.82 from 3.   12. Subtract 2.358 from 8.65.

13. A stereo which lists for $498.50 is discounted (or marked down) to $429.75. What is the saving?

14. At the beginning of a trip, your odometer read 12,583.7 mi. At the end of the trip, the reading was 14,271.2 mi. How many miles did you drive? 1,687.5

15. You pay for purchases of $2.45, $9.78, and $.59 with a $20 bill. How much money do you have left?

16. Find dimension *a* in the figure below.

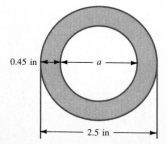

Section 8.3: Subtracting Decimals 333

Using Your Calculator

Remember, the reason for this book is to help you review the basic skills of arithmetic. We are using these calculator sections to show you how the calculator can be helpful as a tool. You should be using your calculator *only on the problems in these special sections.*

Entering decimals in your calculator is similar to entering whole numbers. There is just one difference.

The decimal point key ⦁ is used to place the decimal point as you enter the number.

Example 1

To enter 12.345, press

1 2 ⦁ 3 4 5

Display 12.345

Example 2

To enter 0.678, press

You don't have to press the 0 key for the digit 0 to the left of the decimal point.

⦁ 6 7 8

Display 0.678

The process of adding and subtracting decimals on your calculator is the same as we saw earlier in the sections about adding and subtracting whole numbers.

Example 3

You don't need to worry about the fact that the decimals don't have the same number of places.
If a whole number is involved, just enter that whole number. The decimal point key is not necessary.

To add 2.567 + 0.89 + 5, enter

2.567 + 0.89 + 5 =

Display 8.457

Subtraction of decimals on the calculator is similar.

Example 4

To subtract 4.2 − 2.875, enter

4.2 − 2.875 =

Display 1.325

Often both addition and subtraction are involved in a calculation. In this case, just enter the decimals and the operation signs, + or −, as they appear in the problem.

Example 5

To find 23.7 − 5.2 + 3.87 − 2.341, enter

23.7 $\boxed{-}$ 5.2 $\boxed{+}$ 3.87 $\boxed{-}$ 2.341 $\boxed{=}$

Display (20.029)

Again there are differences in the operation of various calculators. Try this problem on yours to check its operation.

Exercises Using Your Calculator

Compute.

1. 5.87 + 3.6 + 9.25

2. 3.456 + 10 + 2.8 + 5.62

3. 28.21 + 387.6 + 3935.21

4. 10,345.2 + 2308.35 + 153.58

5. 4.59 − 2.389

6. 19.375 − 14.2

7. 27.85 − 3.45 − 2.8

8. 8.8 − 4.59 − 2.325 + 8.5

9. 14 + 3.2 − 9.35 − 3.375

10. 8.7675 + 2.8 − 3.375 − 6

11. Your checking account has a balance of $532.89. You write checks of $50, $27.54, and $134.75 and make a deposit of $50. What is your ending balance?

12. Your checking account has a balance of $278.45. You make deposits of $200 and $135.46. Checks are written for $389.34, $249, and $53.21. What is your ending balance? Be careful with this problem. A negative balance means an overdrawn account.

13. You purchase a car with a sticker price of $9548. If you buy additional options for $85.75, $236, and $95.50 and make a down payment of $1500, how much do you owe on the car?

14. A small store makes a profit of $934.20 in the first week of a given month, $1238.34 in the second week, and $853 in the third week. If their goal is a profit of $4000 for the month, what profit must they make during the remainder of the month?

Answers

1. 18.72   3. 4351.02   5. 2.201   7. 21.6   9. 4.475   11. $370.60   13. $8465.25

8.4 Multiplying Decimals

OBJECTIVES
1. To multiply decimals
2. To use multiplication in solving word problems

To start our discussion of the multiplication of decimals, let's write the decimals in common fraction form and then multiply.

Example 1

$$0.32 \times 0.2 = \frac{32}{100} \times \frac{2}{10} = \frac{64}{1000} = 0.064$$

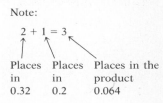

In Example 1, 0.32 has *two* decimal places and 0.2 has *one* decimal place. The product 0.064 has *three* decimal places.

Let's look at another example.

Example 2

$$0.14 \times 0.054 = \frac{14}{100} \times \frac{54}{1000}$$

$$= \frac{756}{100{,}000} = 0.00756$$

Note: $2 + 3 = 5$

The decimals being multiplied in Example 2 have *two* and *three* decimal places. The product has *five* decimal places.

You don't need to write decimals as common fractions to multiply. Examples 1 and 2 suggest the following rule:

TO MULTIPLY DECIMALS

STEP 1 Multiply the decimals as though they were whole numbers.

STEP 2 Add the number of decimal places in the numbers being multiplied.

STEP 3 Place the decimal point in the product so that the number of decimal places in the product is the sum of the number of decimal places in the factors.

The following example illustrates.

Example 3

```
    0.23   ← Two places
  × 0.7    ← One place
  ─────
    0.161  ← Three places
```

CHECK YOURSELF 1

Multiply.

0.36 × 1.52

You may have to affix zeros to the left in the product in order to place the decimal point. Consider our next example.

Example 4

```
    0.136   ← Three places
  × 0.28    ← Two places         3 + 2 = 5
    1088
    272
    0.03808 ← Five places        Insert a 0 to mark off five
    ↑                            decimal places.
  Insert zero
```

CHECK YOURSELF 2

Multiply.

0.234 × 0.24

Let's look at some applications of our work in multiplying decimals.

Example 5

A sheet of paper has dimensions 27.5 cm by 21.5 cm. What is its area?

Solution We multiply to find the required area.

```
    27.5 cm
  × 21.5 cm
    137 5
    275
    550
  591.25 sq cm
```

The area of the paper is 591.25 sq cm.

Recall that area is length times width, so multiplication is the necessary operation.

CHECK YOURSELF 3

If 1 kg is 2.2 lb, how many pounds equal 5.3 kg?

Sometimes we will have to use more than one operation for a solution. Look at our final example.

Example 6

Steve purchased a television set with a price of $299.50. He agreed to pay for the set by making payments of $27.70 for 12 months. How much extra did he pay on the installment plan?

Solution First we multiply to find the amount actually paid.

$$\begin{array}{r} \$\ 27.70 \\ \times\ \ \ \ 12 \\ \hline 55\ 40 \\ 277\ 0\ \ \\ \hline \$332.40 \end{array} \leftarrow \text{Amount paid}$$

Now subtract the listed price. The difference will give the extra amount Steve paid.

$$\begin{array}{r} \$332.40 \\ -299.50 \\ \hline \$\ \ 32.90 \end{array} \leftarrow \text{Extra amount}$$

Steve will pay an additional $32.90 on the installment plan.

CHECK YOURSELF 4

Sandy's new car had a list price of $7385. She paid $1500 down and will pay $205.35 per month for 36 months on the balance. How much extra will she pay with this loan arrangement?

CHECK YOURSELF ANSWERS

1. 0.5472.   **2.** 0.05616.   **3.** 11.66 lb.   **4.** $1507.60.

8.4 Exercises

Multiply.

1. 2.3
   × 3.4

2. 6.5
   × 4.3

3. 8.4
   × 5.2

4. 9.2
   × 4.6

5. 2.56
   ×   72

6. 56.7
   ×   35

7. 0.78
   × 2.3

8. 9.5
   × 0.45

9.	15.7	10.	53.8	11.	0.354	12.	1.34
	× 2.35		× 0.86		× 0.8		× 0.57

13.	3.28	14.	0.582	15.	5.238	16.	0.372
	× 5.07		× 6.3		× 0.48		× 58

17.	1.053	18.	2.375	19.	0.0056	20.	1.008
	× 0.552		× 0.28		× 0.082		× 0.046

21. 0.8 × 2.376   22. 3.52 × 58   23. 0.3085 × 4.5   24. 0.028 × 0.685

25. John bought four shirts on sale for $9.98 each. What was the total cost of the purchase?

26. Dan makes monthly payments of $75.45 on his car. What will he pay in 1 year?

27. If 1 gal of water weighs 8.34 lb, how much will 2.5 gal weigh?

28. Tony worked 37.4 hours in 1 week. If his hourly rate of pay is $6.75, what was his pay for the week?

29. To find the interest on a loan at $9\frac{1}{2}\%$, we have to multiply the amount of the loan by 0.095. Find the interest on a $1500 loan for 1 year.

30. A beef roast weighing 5.8 lb costs $3.25 per pound. What is the cost of the roast?

31. To find the circumference of a circle, you multiply the diameter of the circle by 3.14. Find the circumference of a circle with a diameter of 7.15 in.

32. Cynthia earns $6.40 per hour. For overtime (each hour over 40 hours) she earns $9.60. If she worked 48.5 hours in a week, what pay should she receive?

33. A sheet of typing paper has dimensions 21.6 cm by 28 cm. What is its area?

34. A rental car costs $24 per day plus 18¢ per mile. If you rent a car for 5 days and drive 785 mi, what will the total car rental bill be?

35. You buy a stereo with a list price of $429.90 by agreeing to make monthly payments of $21.40 for 2 years. How much extra are you paying on this installment plan?

36. Jack cuts 12 pieces of length 7.35 cm from a piece of tubing 100 cm long. How much of the tubing will be left over?

Skillscan (Section 2.5)
Multiply.

a. 27
   × 10

b. 59
   × 10

c. 43
   × 100

d. 971
   × 100

e. 523
   × 1000

f. 498
   × 1000

Answers
1. 7.82   3. 43.68   5. 184.32   7. 1.794   9. 36.895   11. 0.2832   13. 16.6296   15. 2.51424
17. 0.581256   19. 0.0004592   21. 1.9008   23. 1.38825   25. $39.92   27. 20.85 lb   29. $142.50
31. 22.451 in   33. 604.8 sq cm   35. $83.70   a. 270   b. 590   c. 4300   d. 97,100   e. 523,000
f. 498,000

8.4 Supplementary Exercises

Multiply.

1. 5.8
   × 3.4

2. 7.3
   × 8

3. 0.42
   × 4.8

4. 2.17
   × 2.9

5. 4.31
   × 8.6

6. 3.85
   × 0.57

7. 2.53
   × 0.45

8. 0.482
   × 0.9

9. 4.537
   × 0.24

10. 0.0048
    × 0.28

11. 12.36 × 0.005

12. 0.054 × 0.0048

13. Alan bought 12 pens on sale at $1.49 each. How much did he pay for the 12 pens?

14. A salesperson is allowed 22¢ per mile by her company as a business expense. If she drove 425 mi during a week, how much should she claim for mileage expense?

15. How much will 7.5 yd of fabric cost if the cost per yard is $2.98?

16. What amount will you pay for 13.6 gal of gasoline if the cost per gallon is 97.5¢?

17. The cost of a rental car is $28.95 per day plus 27¢ per mile. What will it cost you to rent a car for a single day if you drive 140 mi?

18. John purchases a television set and agrees to make monthly payments of $39.60 for 12 months. If the list price of the set was $395.50, how much extra is John paying on this installment plan?

Using Your Calculator

The steps for finding the product of decimals on a calculator are similar to the ones we used for multiplying whole numbers.

Example 1

To multiply 34.2 × 1.387, enter

34.2 × 1.387 =

Display   47.4354

To find the product of more than two decimals, just extend the process.

Example 2

To multiply 2.8 × 3.45 × 3.725, enter

2.8 × 3.45 × 3.725 =

Display   35.9835

You can also easily find powers of decimals with your calculator using a procedure similar to that of Example 2.

Example 3

Find $(2.35)^3$.

Solution Enter

2.35 × 2.35 × 2.35 =

Display   12.977875

Remember: $(2.35)^3 = 2.35 \times 2.35 \times 2.35$

Some calculators have keys that will find powers more quickly. Look for keys marked x^2 or y^x and check your manual for details.

Exercises Using Your Calculator

Compute.

1. 0.08×7.375

2. 21.34×0.005

3. 21.38×13.75

4. 58.05×13.02

5. $127.85 \times 0.055 \times 15.84$

6. $18.28 \times 143.45 \times 0.075$

7. $(2.65)^2$

8. $(0.08)^3$

9. $(3.95)^3$

10. $(0.521)^2$

11. Find the area of a rectangle with length 3.75 in and width 2.35 in.

12. Mark works 38.4 hours in a given week. If his hourly rate of pay is $5.85, what will he be paid for the week?

13. If fuel oil costs 87.5¢ per gallon, what will 150.4 gal cost?

14. To find the interest on a loan for one year at 12.5%, multiply the amount of the loan by 0.125. What interest will you pay on a loan of $1458 at 12.5% for 1 year?

Answers
1. 0.59   3. 293.975   5. 111.38292   7. 7.0225   9. 61.629875   11. 8.8125 sq in   13. $131.60

8.5 Multiplying Decimals by Powers of 10

OBJECTIVE
To multiply a decimal by a power of 10

The rule will be used to multiply by 10, 100, 1000, and so on.

There are enough applications involving multiplication by the powers of 10 to make it worthwhile to develop a special rule in order that you can do such operations quickly and easily. Look at the patterns of the following examples.

Example 1

```
  0.679          23.58
×   10         ×    10
  6.790, or 6.79  235.80, or 235.8
```

The digits remain the same. Only the *position* of the decimal point is changed.

Do you see that multiplying by 10 has moved the decimal point *one place to the right?* Now let's look at what happens when we multiply by 100.

Example 2

```
  0.892          5.74
×   100        ×  100
  89.200, or 89.2  574.00, or 574
```

Multiplying by 100 shifts the decimal point *two places to the right.* The pattern of the past examples gives us the following rule:

TO MULTIPLY BY A POWER OF 10

Move the decimal point to the right the same number of places as there are zeros in the power of 10.

Multiplying by 10, 100, or any other larger power of 10, makes the number *larger.* Move the decimal point *to the right.*

Example 3

$2.356 \times 10 = 23.56$

One zero → One place to the right

$34.788 \times 100 = 3478.8$

Two zeros → Two places to the right

We have had to add a zero to place the decimal point correctly.

$3.67 \times 1000 = 3670.$

Three zeros → Three places to the right

Remember that 10^5 is just a 1 followed by five zeros.

$0.005672 \times 10^5 = 567.2$

Five zeros → Five places to the right

CHECK YOURSELF 1

Multiply.

(1) 43.875×100 (2) 0.0083×10^3

The following is just one of many applications which require multiplying by a power of 10.

Example 4

To convert from kilometers to meters, multiply by 1000. Find the number of meters in 2.45 km.

Solution

2.45 km = 2450 m   Just move the decimal point three places right to make the conversion.

If the result is a whole number, there is no need to write the decimal point.

CHECK YOURSELF 2

To convert from kilograms to grams, multiply by 1000. Find the number of grams in 5.23 kg.

CHECK YOURSELF ANSWERS

1. (1) 4387.5; (2) 8.3.   2. 5230 grams.

8.5 Exercises

Multiply.

1. 5.89×10

2. 0.895×100

3. 23.79×100

4. 2.41×10

5. 0.045×10

6. 5.8×100

7. 0.431×100

8. 0.025×10

9. 0.471×100

10. $0.95 \times 10,000$

11. 0.7125×1000

12. 23.42×1000

13. 4.25×10^2

14. 0.36×10^3

15. 3.45×10^4

16. 0.058×10^5

17. A store purchases 100 items at a cost of $1.38 each. Find the total cost of the order.

18. To convert from meters to centimeters, multiply by 100. How many centimeters are there in 5.3 meters?

19. How many grams are there in 2.2 kg? Multiply by 1000 to make the conversion.

20. An office purchases 1000 pens at a cost of 17.8¢ each. What is the cost of the purchase in dollars?

Skillscan (Section 1.6)
Round each number to the indicated place value.

a. 5378 (nearest thousand)

b. 25,189 (nearest hundred)

c. 219,473 (nearest ten-thousand)

d. 3,438,000 (nearest hundred-thousand)

e. 351,098 (nearest ten-thousand)

f. 5,298,500 (nearest hundred-thousand)

Answers
1. 58.9   3. 2379   5. 0.45   7. 43.1   9. 47.1   11. 712.5   13. 425   15. 34,500   17. $138
19. 2200 grams   a. 5000   b. 25,200   c. 220,000   d. 3,400,000   e. 350,000   f. 5,300,000

8.5 Supplementary Exercises

Multiply.

1. 4.82×10

2. 3.57×100

3. 0.785×100

4. 14.85×100

5. 45.71×1000

6. 0.0573×1000
   57.3

7. 0.247×10^5
   24,700

8. 1.53×10^4

9. A builder bought 100 lighting fixtures from a supplier at $24.39 each. What was the total cost of the order?

10. How many meters are there in 5.8 km? Multiply by 1000 to make the conversion.

8.6 Rounding Decimals

OBJECTIVES
1. To round a decimal to a specified decimal place
2. To apply rounding in the solution of word problems

Whenever a decimal represents a measurement made by some instrument (a rule, a scale, and so on), the decimals are not exact. They are accurate only to a certain number of places and are called *approximate numbers*. Often we will want to make all decimals in a particular problem accurate to some specified tolerance. This will require *rounding* the decimals.

We can picture the process on a number line.

Example 1

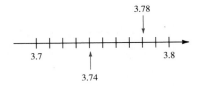

3.74 is rounded to the nearest tenth, 3.7.   3.74 is closer to 3.7 than it is to 3.8.

3.78 is rounded to 3.8.   3.78 is closer to 3.8.

Rather than using the number line, we can apply the following rule:

TO ROUND A DECIMAL

STEP 1 Find the place the decimal is to be rounded to.

STEP 2 If the next digit to the right is 5 or more, increase the digit in the place you are rounding to by 1. Discard any remaining digits to the right.

STEP 3 If the next digit to the right is less than 5, just discard that digit and any remaining digits to the right.

Example 2

Round 34.58 to the nearest tenth.

Many students find it easiest to mark this digit with an arrow.

34.58   Locate the digit you are rounding to.
 ↑
Tenths

Since the next digit to the right (8) is 5 or more, increase the tenths digit by 1. Then discard the remaining digits.

34.58 is rounded to 34.6.

CHECK YOURSELF 1

Round 48.82 to the nearest tenth.

Example 3

Round 5.673 to the nearest hundredth.

5.673
  ↑
  Hundredths

The next digit to the right (3) is less than 5. Leave the hundredths digit as it is and discard the remaining digits to the right.

5.673 is rounded to 5.67.

CHECK YOURSELF 2

Round 29.247 to the nearest hundredth.

Example 4

Round 3.14159 to four decimal places.

3.14159
    ↑
    Ten-thousandths

The fourth place to the right of the decimal point is the ten-thousandths place.

The next digit to the right (9) is 5 or more, so increase the digit you are rounding to by 1. Discard the remaining digits to the right.

3.14159 is rounded to 3.1416.

CHECK YOURSELF 3

Round 0.8235 to three decimal places.

Let's look at an application of our work with decimals in which we will have to round our result.

Example 5

Usually in problems dealing with money we round the result to the nearest cent (hundredth of a dollar).

Jack buys 8.7 gal of gas at 98.9 cents per gallon. Find the cost of the gas.

Solution We multiply the cost per gallon by the number of gallons. Then round the result to the nearest cent.

```
    98.9
  ×  8.7
   69 23
  791 2
  860.43
```

The product 860.43 (cents) is rounded to 860 (cents), or $8.60.

The cost of Jack's gas will be $8.60.

CHECK YOURSELF 4

A liter is approximately 0.265 gal. On a trip to Europe, the Bernard's purchased 88.4 liters of gas for their rental car. How many gallons of gas was this, to the nearest tenth of a gallon?

CHECK YOURSELF ANSWERS

1. 48.8.   **2.** 29.25.   **3.** 0.824.   **4.** 23.4 gal.

8.6 Exercises

Round to the indicated place.

1. 53.48 (tenth)

2. 6.785 (hundredth)

3. 21.534 (hundredth)

4. 5.842 (tenth)

5. 0.342 (hundredth)

6. 2.3576 (thousandth)

7. 2.71828 (thousandth)

8. 1.543 (tenth)

9. 0.5475 (thousandth)

10. 0.85356 (ten-thousandth)

11. 4.85344 (ten-thousandth)

12. 52.8728 (thousandth)

13. 6.734 (to two decimal places)

14. 12.5467 (to three decimal places)   15. 6.58739 (to four decimal places)

16. 503.824 (to two decimal places)

17. An inch is approximately 2.54 cm. How many centimeters would 5.3 in be equal to? Give your answer to the nearest hundredth of a centimeter.

18. A light plane uses 5.8 gal of fuel per hour. How much fuel is used on a flight of 3.2 hours? Give your answer to the nearest tenth of a gallon.

19. The Hallstons select a carpet costing $15.49 per square yard. If they need 7.8 sq yd of carpet, what is the cost to the nearest cent?

20. We can find the circumference of a circle by multiplying the diameter of a circle by 3.14 (an approximation for the number pi). To the nearest hundredth of an inch, what is the circumference of a circle whose diameter is 4.7 in?

Estimation can be a useful tool when working with decimal fractions. To estimate a sum, one approach is to round the addends to the nearest whole number and add for your estimate. For instance, to estimate the sum below:

```
        Round
 19.8    →      20
  3.5            4
 24.2           24
+10.4          +10
               ―――   Add for
                58   the estimate.
```

21. Your restaurant bill includes $18.25 for dinners, $6.80 for salads, $8.75 for wine, $7.40 for dessert, and $1.70 for coffee. Estimate your bill by rounding each amount to the nearest dollar.

22. Your bill for a car tune-up includes $7.80 for oil, $5.90 for a filter, $3.40 for spark plugs, $4.10 for points, and $28.70 for labor. Estimate your total cost.

Estimation is also helpful in multiplying decimals. For instance, to estimate the product below:

```
       Round
 24.3   →      24
× 5.8         × 6
              ―――   Multiply
              144   for the estimate.
```

23. A classroom is 7.9 meters wide and 11.2 meters long. Estimate its area.

24. You buy a 6.2-lb roast which costs $3.89 per pound. Estimate the cost of the roast.

Answers
1. 53.5   3. 21.53   5. 0.34   7. 2.718   9. 0.548   11. 4.8534   13. 6.73   15. 6.5874
17. 13.46 cm   19. $120.82   21. $43   23. 88 sq m

8.6 Supplementary Exercises

Round to the indicated place.

1. 0.738   (tenth)

2. 23.454   (hundredth)

3. 5.8796   (thousandth)

4. 5.853   (tenth)

5. 5.87194   (ten-thousandth)

6. 27.32178   (ten-thousandth)

7. 4.8281   (to two decimal places)

8. 0.8257   (to three decimal places)

9. A room is 15.3 ft long by 9.5 ft wide. What is its area to the nearest tenth of a square foot?

10. To find the amount deducted from your paycheck for social security, multiply your salary by 0.076. If your weekly pay is $507, what amount will be deducted for social security to the nearest cent?

SELF-TEST for Chapter Eight

The purpose of the Self-Test is to help you check your progress and review for a chapter test in class. Allow yourself about an hour to take the test. When you are done, check your answers in the back of the book. If you missed any problems, be sure to go back and review the appropriate sections in the chapter and do the supplementary exercises provided there.

1. Write $\dfrac{431}{1000}$ in decimal form.

2. Write $5\dfrac{13}{100}$ in decimal form.

3. Write 0.431 in words.

4. Write 5.13 in words.

In Problems 5 and 6, complete the statements using the symbols $<$ or $>$.

5. 5.93 _____ 5.928

6. 2.149 _____ 2.15

In Problems 7 to 10, add.

7. 1.238
   +0.97

8. 2.581
   0.24
   +0.7

9. 31.7, 6, 2.81, and 0.254.

10. Three and four-tenths, four-hundred-five thousandths, and seven.

In Problems 11 and 12, solve the applications.

11. The Dawsons purchased 7, 12.7, 10, 11.3, and 9.8 gal of gas on a vacation trip. How much gas did they buy on the trip?

12. Find the perimeter (the distance around) for the figure below.

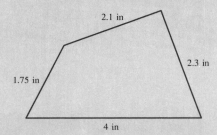

In Problems 13 to 15, subtract.

13. 5.47
    −2.89

14. 8.75
    −3.875

15. Subtract 5.485 from 12.

In Problems 16 to 18, solve the applications.

16. Wally's car odometer read 8534.8 mi at the beginning of a trip. If it read 9472.1 mi upon his return, how far did he drive?

17. Marion earns $360.40 per week. If $27.38 is deducted for social security taxes and $53.45 for federal tax withholding, what is her take-home pay?

18. Find dimension a in the following drawing:

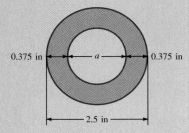

In Problems 19 to 21, multiply.

19. 5.8
    $\times 3.7$

20. 2.71
    $\times 0.58$

21. 0.235 by 0.04.

In Problems 22 and 23, solve the applications.

22. Find the area of a rectangle that has dimensions 8.7 cm by 4.3 cm.

23. A washer-dryer combination has an advertised price of $535.90. You buy the set and agree to make payments of $20.50 for 36 months. How much extra are you paying for this installment plan?

In Problems 24 and 25, multiply.

24. 0.054×100

25. 8.432×10^4

In Problem 26, solve the application.

26. A store buys 1000 items costing $0.47 per item. What is the total cost of the order?

In Problems 27 and 28, round to the indicated place.

27. 2.571   (tenth)

28. 23.3448   (to two decimal places)

In Problems 29 and 30, solve the applications.

29. One mile is equal to 1.609 km. How many kilometers (to the nearest tenth of a kilometer) are equal to 25 mi?

30. To find the amount deducted from your pay for social security taxes, you multiply the amount of your pay by 0.076. If your monthly pay is $1458, what amount should be deducted from your pay per month (to the nearest cent)?

PRETEST for Chapter Nine

Division of Decimals

This pretest will point out any difficulties you may be having in dividing decimals. Do all the problems. Then check your answers on the following page.

1. Divide $38\overline{)123.5}$

2. Divide $108.59 \div 46$. Give the quotient to the nearest hundredth.

3. A shipment of 75 items cost $250.50. What is the cost per item?

4. Divide $0.994 \div 0.35$

5. Divide $2.18\overline{)3.587}$. Give the quotient to the nearest thousandth.

6. John worked 27.5 hours a week and earned $148.50. What was his hourly rate of pay?

7. Divide $24.7 \div 1000$

8. Find the decimal equivalent of $\frac{5}{8}$.

9. Find the decimal equivalent of $\frac{7}{24}$ (to the nearest hundredth).

10. Write 0.78 as a common fraction.

ANSWERS TO PRETEST

For help with similar problems, turn to the section indicated.

1. 3.25 (Section 9.1);
2. 2.36 (Section 9.1);
3. $3.34 (Section 9.1);
4. 2.84 (Section 9.2);
5. 1.645 (Section 9.2);
6. $5.40 (Section 9.2);
7. 0.0247 (Section 9.3);
8. 0.625 (Section 9.4);
9. 0.29 (Section 9.4);
10. $\frac{39}{50}$ (Section 9.5)

Chapter Nine
Division of Decimals

9.1 Dividing Decimals by Whole Numbers

OBJECTIVES
1. To divide a decimal by a whole number
2. To apply division to the solution of word problems

The division of decimals is very similar to our earlier work with dividing whole numbers. The only difference is in learning to place the decimal point in the quotient. Let's start with the case of dividing a decimal by a whole number. Here, placing the decimal point is easy. You can apply the following rule.

TO DIVIDE A DECIMAL BY A WHOLE NUMBER

STEP 1 Place the decimal point in the quotient *directly above* the decimal point of the dividend.

STEP 2 Divide as you would with whole numbers.

Example 1

To divide 29.21 by 23, write

Do the division just as if you were dealing with whole numbers. Just remember to place the decimal point in the quotient directly above *the one in the dividend.*

```
         1.27
    23)29.21
       23
       ‾‾
        6 2
        4 6
        ‾‾‾
        1 61
        1 61
        ‾‾‾‾
           0
```

The quotient is 1.27

CHECK YOURSELF 1

Divide.

80.24 by 34

Let's look at another example of dividing a decimal by a whole number.

Example 2

To divide 122.2 by 52, write

Again place the decimal point of the quotient above that of the dividend.

```
       2.3
52)122.2
    104
     18 2
     15 6
      2 6
```

We normally do not use a remainder when dealing with decimals. Add a zero to the dividend and continue.

Remember that adding a zero does not change the value of the dividend. It simply allows us to complete the division process in this case.

```
       2.35
52)122.20   ← Add a zero.
    104
     18 2
     15 6
      2 60
      2 60
         0
```

So 122.2 ÷ 52 = 2.35. The quotient is 2.35.

CHECK YOURSELF 2

Divide.

234.6 by 68

Often you will be asked to give a quotient to a certain place value. In this case, continue the division process to *one digit past* the indicated place value. Then round the result back to the desired accuracy.

In dealing with money, for instance, we normally would give the quotient to the nearest hundredth of a dollar (the nearest cent). This would mean carrying the division out to the thousandths place and then rounding back.

Example 3

Find the quotient of 25.75 ÷ 15 to the nearest hundredth.

Find the quotient to one place past the desired place and then round the result.

```
        1.716
    15)25.750
       15
       10 7
       10 5
          25
          15
          100
           90
           10
```

Add a zero to carry the division to the thousandths place.

So 25.75 ÷ 15 = 1.72 (to the nearest hundredth).

CHECK YOURSELF 3

Find.

99.26 ÷ 35 to the nearest hundredth.

As we mentioned, problems similar to the one in Example 3 will occur often when you are dealing with money. Our next example is one of the many applications of this type of division.

Example 4

A carton of 144 items costs $56.10. What is the price per item to the nearest cent?

Solution To find the price per item, divide the total price by 144.

```
          .389
     144)56.100
         43 2
         12 90
         11 52
          1 380
          1 296
             84
```

Carry the division to the thousandths place and then round back.

The cost per item is rounded to $0.39 or 39¢.

CHECK YOURSELF 4

An office paid $26.55 for 72 pens. What was the cost per pen to the nearest cent?

Example 5

Ed's grade points were 2.37 in fall term, 2.58 in winter, and 2.87 in spring. What was his grade-point average for this year?

Solution Do you remember the method for finding an average? We add the numbers and then divide by the number of items.

$$2.37 + 2.58 + 2.87 = 7.82$$

Divide the total by 3, the number of items.

Now divide by 3:

```
    2.606
3)7.820
    6
    1 8
    1 8
      020
       18
        2
```

Now round 2.606 to 2.61.

We choose to round the average to the decimal place (hundredths) of the given grade points.

Ed's grade-point average is 2.61 (rounded to the nearest hundredth). Note that we have carried the division to the thousandths place and then rounded back.

CHECK YOURSELF 5

Whitney's new car had mileage of 38.9, 41.2, and 40.7 mi/gal in three readings. What was the average mileage for the period? Give your answer to the nearest tenth of a mile per gallon.

CHECK YOURSELF ANSWERS

1. 2.36.   **2.** 3.45.   **3.** 2.84.   **4.** $0.37 or 37¢.   **5.** 40.3 mi/gal.

9.1 Exercises

Divide:

1. $16.68 \div 6$ **2.** $43.92 \div 8$ **3.** $1.92 \div 4$ **4.** $5.52 \div 6$

5. $5.48 \div 8$ **6.** $2.76 \div 8$ **7.** $13.89 \div 6$ **8.** $21.92 \div 5$

Section 9.1: Dividing Decimals by Whole Numbers

9. 185.6 ÷ 32   10. 165.6 ÷ 36   11. 79.9 ÷ 34   12. 179.3 ÷ 55

13. 13.78 ÷ 52   14. 26.22 ÷ 76   15. 144.63 ÷ 45   16. 183.04 ÷ 65

Divide and round the quotient to the indicated decimal place.

17. 23.8 ÷ 9 (to the nearest tenth)   18. 5.27 ÷ 8 (to the nearest hundredth)

19. 38.48 ÷ 46 (to the nearest hundredth)   20. 3.36 ÷ 36 (to the nearest thousandth)

21. 125.4 ÷ 52 (to the nearest tenth)   22. 2.563 ÷ 54 (to the nearest thousandth)

23. 0.927 ÷ 28 (to the nearest thousandth)   24. 5.8 ÷ 65 (to the nearest hundredth)

25. Marv paid $13.47 for three records on sale. What was the cost per record?

26. Seven employees of an office donated $172.06 during a charity drive. What was the average donation?

27. A shipment of 72 paperback books costs a store $190.25. What was the average cost per book to the nearest cent?

28. A restaurant bought 50 glasses at a cost of $39.90. What was the cost per glass to the nearest cent?

29. The cost of a case of 48 items is $28.20. What is the cost of an individual item to the nearest cent?

30. An office bought 18 hand-held calculators for $284. What was the cost per calculator to the nearest cent?

31. Al purchased a new refrigerator which cost $736.12 with interest included. He paid $100 as a down payment and agreed to pay the remainder in 18 monthly payments. What amount will he be paying per month?

32. The cost of a television set with interest is $490.64. If you make a down payment of $50 and agree to pay the balance in 12 monthly payments, what will the amount of each monthly payment be?

33. In five readings Fred's gas mileage was 32.3, 31.6, 29.5, 27.3, and 33.4 mi/gal. What was his average mileage to the nearest tenth of a mile per gallon?

34. Pollution index readings were 53.3, 47.8, 41.9, 55.8, 43.7, 41.7, and 52.3 for a 7-day period. What was the average reading (to the nearest tenth) for the seven days?

35. Jeremy's grade points for eight semesters were 2.81, 3.05, 3.62, 2.95, 3.15, 2.79, 3.45, and 3.53. Find his grade-point average to the nearest hundredth of a point.

36. The ratings for a weekly television program over a 6-week period were 19.7, 15.2, 18.5, 17.8, 16.3, and 18.6. What was the average rating per show to the nearest tenth of a point?

Skillscan (Section 8.6)
Round each decimal to the indicated place.

a. 5.43 (tenth)   **b.** 6.87 (tenth)   **c.** 27.428 (hundredth)

d. 30.583 (hundredth)   **e.** 0.0587 (thousandth)   **f.** 0.12545 (thousandth)

Answers
Solutions for the even-numbered exercises are provided in the back of the book.

1. $6\overline{)16.68} = 2.78$ 3. 0.48   5. $8\overline{)5.480} = .685$ 7. 2.315   9. 5.8   11. 2.35   13. $52\overline{)13.780} = .265$ 15. 3.214

17. 2.6   19. 0.84   21. $52\overline{)125.40} = 2.41$; $125.4 \div 52 = 2.4$ (to the nearest tenth)   23. 0.033

25. $4.49   27. $2.64   29. $0.59 or 59¢   31. $35.34   33. 30.8 mi/gal   35. 3.17   a. 5.4   b. 6.9
c. 27.43   d. 30.58   e. 0.059   f. 0.125

9.1 Supplementary Exercises

Divide.

1. 4.68 ÷ 6   **2.** 34.65 ÷ 9   **3.** 3.745 ÷ 7   **4.** 4.95 ÷ 6

5. 12.555 ÷ 27   **6.** 15.98 ÷ 68   **7.** 3.712 ÷ 64   **8.** 4.437 ÷ 58

Divide and round the quotient to the indicated decimal place.

9. 43.76 ÷ 8 (to the nearest tenth)

10. 2.106 ÷ 9 (to the nearest hundredth)

11. 98.23 ÷ 64 (to the nearest hundredth)

12. 27.81 ÷ 52 (to the nearest thousandth)

13. Marc bought three T-shirts on sale for $19.47. What was the price per shirt?

14. A group of eight people shared expenses of $66.88 for a party. What amount should each person pay?

15. Phil buys a stereo system costing $945.76 with interest. He makes a $100 down payment and agrees to pay the balance in 24 monthly payments. What will he pay per month?

16. Kathy bought a used car that cost, with the interest included, $1936.30. She paid $400 down and agreed to pay off the balance in equal monthly payments for 18 months. What will she pay per month?

17. Jack's grade points for six semesters were 2.74, 3.15, 2.86, 3.21, 2.84, and 3.04. What was his grade-point average to the nearest hundredth of a point?

18. In four fill-ups, Jan's gas mileage was 31.5, 32.8, 37.3, and 32.2 mi/gal. What was her average mileage to the nearest tenth of a mile per gallon?

9.2 Dividing by Decimals

OBJECTIVE
To divide by a decimal

We want now to look at division *by* decimals. Here is an example using a fractional form.

Example 1

$$2.57 \div 3.4 = \frac{2.57}{3.4}$$

Write the division as a fraction.

$$= \frac{2.57 \times 10}{3.4 \times 10}$$

Multiplying the numerator and denominator by 10 *does not change* the value of the fraction.

$$= \frac{25.7}{34}$$

Shift the decimal point in the numerator and denominator *one place to the right* to multiply by 10.

$$= 25.7 \div 34$$

So

$2.57 \div 3.4 = 25.7 \div 34$

Dividing by a whole number makes it easy to place the decimal point in the quotient.

Look closely at the original division problem in Example 1. After we multiply the numerator and denominator by 10, we see that $2.57 \div 3.4$ is the same as $25.7 \div 34$. This is very useful because we would rather divide by a whole number. Let's look at a second example.

Example 2

$$14.835 \div 2.14 = \frac{14.835}{2.14}$$

$$= \frac{14.835 \times 100}{2.14 \times 100}$$

$$= \frac{1483.5}{214} \quad \text{Shift the decimal } \textit{two places right} \text{ to multiply by 100.}$$

$$= 1483.5 \div 214 \quad \text{Now we are dividing by a whole number.}$$

So,

$14.835 \div 2.14 = 1483.5 \div 214$

Of course, multiplying by any whole-number power of 10, greater than one, is just a matter of shifting the decimal point to the right.

Do you see the rule suggested by these examples? In Example 1 we multiplied the numerator and the denominator (the dividend and the divisor) by 10. In Example 2 we multiplied by 100. In each case this made the divisor a whole number without altering the actual digits involved. All we did was shift the decimal point in the divisor and dividend the same number of places. This leads us to the rule:

TO DIVIDE BY A DECIMAL

STEP 1 Move the decimal point in the divisor *to the right*, making the divisor a whole number.

STEP 2 Move the decimal point in the dividend to the right *the same number of places*. Add zeros if necessary.

STEP 3 Place the decimal point in the quotient directly above the decimal point of the dividend.

STEP 4 Divide as you would with whole numbers.

Example 3

$2.3\overline{)15.85} \longrightarrow 23.\overline{)158.5}$ Shift the decimal points one place to the right. The divisor is now the whole number 23.

$4.53\overline{)12.40} \longrightarrow 453.\overline{)1240.}$ Shift the decimal points two places to the right.

$3.245\overline{)34.5} \longrightarrow 3245.\overline{)34500.}$ Shift the decimal points three places to the right. As you can see, you may have to add zeros in the dividend.

$0.34\overline{)58} \longrightarrow 34.\overline{)5800.}$ The decimal point is assumed to be to the right of 58!

58 = 58.

CHECK YOURSELF 1

Rewrite the division problems so that the divisor is a whole number.

(1) $3.7\overline{)5.93}$ (2) $2.58\overline{)125.7}$

Let's look at one complete example of the use of our division rule.

Example 4

Divide 1.573 by 0.48. (Give the quotient to the nearest tenth.)
Write

$0.48\overline{)1.57\,3}$ Shift the decimal points two places to the right to make the divisor a whole number.

Now divide:

Once the division statement is rewritten, place the decimal point in the quotient above that in the dividend.

```
         3.27
    48)157.30
        144
         13 3
          9 6
          3 70
          3 36
            34
```

Note that we add a zero to carry the division to the hundredths place. In this case, we want to find the quotient to the nearest tenth.

Round 3.27 to 3.3. So,

1.573 ÷ 0.48 = 3.3 (to the nearest tenth)

CHECK YOURSELF 2

Divide.

3.4 ÷ 1.24

Give the quotient to the nearest tenth.

Let's look at some applications of our work in dividing by decimals.

Example 5

Andrea worked 41.5 hours in a week and earned $239.87. What was her hourly rate of pay?

Solution To find the hourly rate of pay we must use division. We divide the number of hours worked into the total pay.

Note that we must add a zero to the dividend to complete the division process.

```
            5.78
   41.5 )239.8 70
         207 5
          32 3 7
          29 0 5
           3 3 20
           3 3 20
                0
```

Andrea's hourly rate of pay is $5.78.

CHECK YOURSELF 3

A developer wants to subdivide a 12.6-acre piece of land into 0.45-acre lots. How many lots are possible?

We may also have to round the quotient in some problems. Our final example illustrates.

Example 6

Jesse drove 185 mi in 3.5 hours. What was his average speed (to the nearest mile per hour)?

Solution To find the average speed (miles per hour) we divide the distance by the time.

Section 9.2: Dividing by Decimals

Remember: We assume that the decimal point is to the right of 185.

185 = 185.0

Then shift the decimal points.

```
            5 2.8
    3.5 )185.0 0
          175
          ---
          10 0
           7 0
           ---
           3 0 0
           2 8 0
           -----
             2 0
```

Once again we use

Speed = distance ÷ time

Carry the division to the tenths place and then round 52.8 to 53. So Jesse's average speed was 53 mi/h.

CHECK YOURSELF 4

To convert from centimeters to inches, you can divide by 2.54. Find the number of inches in 25 cm (to the nearest hundredth of an inch).

CHECK YOURSELF ANSWERS

1. (1) 3.7)5.9 3 ; (2) 2.58)125.70 .
2. 2.7.   3. 28 lots.   4. 9.84 in.

9.2 Exercises

Divide.

1. 0.6)11.07

2. 0.8)10.84

3. 3.8)7.22

4. 2.9)13.34

5. 5.2)11.622

6. 6.4)3.616

7. 0.27)1.8495

8. 0.038)0.8132

9. 0.046)1.587

10. 0.52)3.2318

11. 0.658 ÷ 2.8

12. 0.882 ÷ 0.36

13. 3.275 ÷ 0.524

14. 0.6837 ÷ 3.18

Find the quotients to the indicated place.

15. 0.7)1.642   (hundredth)

16. 0.6)7.695   (tenth)

17. 4.5)8.415   (tenth)

18. 5.8)16   (hundredth)

19. $3.12\overline{)4.75}$ (hundredth)

20. $64.2\overline{)16.3}$ (thousandth)

21. $5.38\overline{)0.205}$ (thousandth)

22. $0.347\overline{)0.8193}$ (hundredth)

23. $2.42\overline{)1.3}$ (hundredth)

24. $96.3\overline{)1.753}$ (thousandth)

25. 0.99 ÷ 0.624 (thousandth)

26. 3.75 ÷ 1.58 (hundredth)

27. 0.125 ÷ 2.135 (ten-thousandth)

28. 0.428 ÷ 1.452 (thousandth)

29. We have 91.25 in of plastic labeling tape and wish to make labels that are 1.25 in long. How many can be made?

30. Perry worked 32.5 hours, earning $204.10. How much did he make per hour?

31. A roast weighing 5.3 lb sold for $14.89. Find the cost per pound to the nearest cent.

32. One nail weighs 0.025 oz. How many nails are there in 1 lb? (1 lb is 16 oz)

33. A family drove 1390 mi, stopping for gas three times. If they purchased 15.5, 16.2, and 10.8 gal of gas, find the number of miles per gallon (the mileage) to the nearest tenth of a mile per gallon.

34. The water in an aquarium weighs 1025 lb. If water weighs 62.5 lb per cubic foot, how many cubic feet of water does the aquarium hold?

35. To convert from millimeters to inches, we can divide by 25.4. If film is 35 mm wide, find the width to the nearest hundredth of an inch.

36. To convert from centimeters to inches, we can divide by 2.54. The rainfall in Paris was 11.8 cm during 1 week. What was that rainfall to the nearest hundredth of an inch?

Skillscan (Section 8.5)
Multiply.

a. 2.3 × 10

b. 0.452 × 10

c. 1.58 × 100

d. 0.248 × 100

e. 1.4579 × 1000

f. 0.0427 × 10,000

Section 9.2: Dividing by Decimals

Answers

1. 18.45   3.
$$3.8\overline{)7.2.2}\begin{array}{r}1.9\\\hline\\3\ 8\\\hline 3\ 4\ 2\\3\ 4\ 2\\\hline 0\end{array}$$
5. 2.235   7. 6.85   9. 34.5   11. 0.235   13. 6.25   15. 2.35

17. 1.9   19. 1.52   21. 0.038   23. $2.42\overline{)1.30,000}\begin{array}{r}.537\\\hline 1\ 21\ 0\\9\ 00\\7\ 26\\\hline 1\ 740\\1\ 694\\\hline 46\end{array}$ 1.3 ÷ 2.42 = 0.54 (to the nearest hundredth)

25. 1.587   27. 0.0585   29. 73 labels   31. $2.81   33. 32.7 mi/gal   35. 1.38 in   a. 23   b. 4.52
c. 158   d. 24.8   e. 1457.9   f. 427

9.2 Supplementary Exercises

Divide.

1. $0.8\overline{)4.28}$
2. $2.4\overline{)12.48}$
3. $3.8\overline{)9.766}$

4. $0.58\overline{)2.262}$
5. $0.518\overline{)1.1137}$
6. 1.971 ÷ 0.45

7. 2.3625 ÷ 0.525

Find the quotients to the indicated place.

8. $0.8\overline{)2.534}$ (hundredth)
9. $2.8\overline{)15}$ (tenth)

10. $4.8\overline{)25.07}$ (tenth)
11. $1.3\overline{)27.48}$ (hundredth)

12. $0.59\overline{)1.698}$ (hundredth)
13. 0.1398 ÷ 0.578 (ten-thousandth)

14. 5.342 ÷ 27.8 (thousandth)

15. You received $220.80 in pay during a week. If your hourly rate of pay was $5.75, how many hours did you work?

16. A piece of bubble gum weighs 0.25 oz. How many pieces of gum are there in 2 lb (32 oz)?

17. At the start of a trip, an odometer read 27,458. At the end of the trip, it read 28,808 and 38.7 gal of gas had been used. Find the number of miles per gallon (gas mileage) to the nearest tenth.

18. Carlos drove 224 mi in 4.5 hours. What was his average speed to the nearest mile per hour?

9.3 Dividing Decimals by Powers of 10

OBJECTIVE
To divide a decimal by a power of 10

Recall that you can multiply decimals by powers of 10 by simply shifting the decimal point to the right. A similar approach will work for division by powers of 10. Let's look at some examples.

Example 1

```
        3.53
   10)35.30
       30
        5 3
        5 0
          30
          30
           0
```

The dividend is 35.3. The quotient is 3.53. The decimal point has been shifted *one place to the left*. Note also that the divisor, 10, has *one* zero.

Let's try it again!

Example 2

```
          3.785
  100)378.500
      300
       78 5
       70 0
        8 50
        8 00
          500
          500
            0
```

Here the dividend is 378.5 while the quotient is 3.785. The decimal point is now shifted *two places to the left*. In this case the divisor, 100, has *two* zeros.

Section 9.3: Dividing Decimals by Powers of 10

The two examples suggest that division by powers of 10 is just as easy as multiplication. We have the following rule:

> **TO DIVIDE A DECIMAL BY A POWER OF 10**
>
> Move the decimal point *to the left* the same number of places as there are zeros in the power of 10.

The following example illustrates the use of this rule.

Example 3

As you can see, we may have to add zeros to correctly place the decimal point.

Remember, 10^4 is a 1 followed by *four* zeros.

$27.3 \div 10 = 2.73$	Shift one place to the left
$57.53 \div 100 = 0.5753$	Shift two places to the left
$39.75 \div 1000 = 0.03975$	Shift three places to the left
$85 \div 1000 = 0.085$	Note that $85 = 85.$
$235.72 \div 10^4 = 0.023572$	Shift four places to the left.

CHECK YOURSELF 1

(1) $3.84 \div 10 =$
(2) $27.3 \div 1000 =$

Let's look at an application of our work in dividing by powers of 10.

Example 4

To convert from millimeters to meters, we divide by 1000. How many meters are 3450 mm?

Solution

$3450 \text{ mm} = 3.450 \text{ m}$ Shift three places to the left to divide by 1000

CHECK YOURSELF 2

A shipment of 1000 pens cost a stationery store $658. What was the cost per pen to the nearest cent?

CHECK YOURSELF ANSWERS

1. (1) 0.384; (2) 0.0273.   2. 66¢.

9.3 Exercises

Divide.

1. $5.8 \div 10$
2. $2.3 \div 100$
3. $4.568 \div 100$
4. $0.672 \div 10$

5. $24.39 \div 1000$
6. $5.92 \div 100$
7. $6.9 \div 1000$
8. $48 \div 1000$

9. $7.8 \div 10^2$
10. $2.43 \div 10^3$
11. $45.2 \div 10^5$
12. $237.1 \div 10^4$

13. The cost of a street-lighting project, $4850, will be shared by 10 homeowners in a neighborhood. What will each homeowner pay?

14. A road-paving project will cost $23,500. If the cost is to be shared by 100 families, how much will each family pay?

15. A builder ordered 100 spotlight fixtures at a cost of $2780. What was the cost per fixture?

16. A school ordered 10 new television monitors at a cost of $2890. What was the cost per monitor?

17. To convert from milligrams to grams, we divide by 1000. A tablet is 250 mg. What is its weight in grams?

18. To convert from milliliters to liters, we divide by 1000. If a bottle of wine holds 750 mL, what is its volume in liters?

19. A shipment of 100 calculators cost a store $593.88. Find the cost per calculator (to the nearest cent).

20. A shipment of 1000 writing tablets cost an office supplier $756.80. Find the cost per tablet (to the nearest cent).

Skillscan (Section 5.1)
Write each of the common fractions as a division statement.

a. $\dfrac{2}{3}$
b. $\dfrac{3}{8}$
c. $\dfrac{7}{10}$

d. $\dfrac{5}{7}$ e. $\dfrac{7}{9}$ f. $\dfrac{1}{4}$

Answers
1. 0.58   3. 0.04568   5. 0.02439   7. 0.0069   9. 0.078   11. 0.000452   13. $485   15. $27.80
17. 0.25 gram   19. $5.94   a. $2 \div 3$ b. $3 \div 8$ c. $7 \div 10$ d. $5 \div 7$ e. $7 \div 9$ f. $1 \div 4$

9.3 Supplementary Exercises

Divide.

1. $4.93 \div 10$
2. $157.9 \div 100$
3. $5.23 \div 1000$

4. $0.953 \div 10$
5. $27.1 \div 10^4$
6. $523.8 \div 10^5$

7. A sewer project costing $44,350 will be paid for by 100 families. What will the cost per family be?

8. To convert from centiliters (cL) to liters, divide by 100. If a glass holds 30 cL, what is its volume in liters?

9. A shipment of 1000 items costs a store $438.75. What is the cost per item to the nearest cent?

10. A desk top is 750 mm wide. You divide by 1000 to convert millimeters to meters. What is its width in meters?

Using Your Calculator

Now that you have had a chance to work with the division of decimals by hand, let's look at the process of division on a calculator. We use the same steps as we used earlier to divide whole numbers.

Example 1

To divide 345.12 by 8.37, follow the steps shown:

Step 1   Enter the dividend   345.12

Step 2   Press the divide key $\boxed{\div}$

> Don't worry about shifting decimal points. The calculator takes care of all that.

Step 3 Enter the divisor   8.37

Step 4 Press the equals key $\boxed{=}$

Display $\boxed{41.232975}$ The calculator shows the quotient to its eight-digit capacity.

To use this result, we will normally want to state the quotient to some specified decimal place. So we must round the quotient.

Example 1 (continued)

345.12 ÷ 8.37 = 41.23   (to the nearest hundredth)

Round 41.232975 to the nearest hundredth.

Remember that a division problem can be checked by multiplying the quotient and the divisor.

Example 2

To divide 60.9352 by 25.82, follow the steps:

60.9352 $\boxed{\div}$ 25.82 $\boxed{=}$ $\boxed{2.36}$

Display

To check:

2.36 $\boxed{\times}$ 25.82 $\boxed{=}$ $\boxed{60.9352}$ We have checked the division, since the product is the original dividend.

You may run into problems, however, if the quotient exceeds the capacity of your calculator's display.

Example 3

Divide.

21.8 $\boxed{\div}$ 7.4 $\boxed{=}$ $\boxed{2.9459459}$

Display

To check:

2.9459459 $\boxed{\times}$ 7.4 $\boxed{=}$ $\boxed{21.799999}$

> You may not find this happening at all if your calculator rounds its results. Try the example and find out!

This result, 21.799999 instead of 21.8, is due to the capacity of the display. It is still close enough to verify the division.

Exercises Using Your Calculator

Divide and check.

1. 8.901 ÷ 2.58
2. 16.848 ÷ 0.288
3. 99.705 ÷ 34.5

4. 171.25 ÷ 2.74
5. 0.01372 ÷ 0.056
6. 0.200754 ÷ 0.00855

7. 2.546 ÷ 1.38   (to the nearest hundredth)

8. 45.8 ÷ 9.4   (to the nearest tenth)

9. 0.5782 ÷ 1.236   (to the nearest thousandth)

10. 1.25 ÷ 0.785   (to the nearest hundredth)

11. 1.34 ÷ 2.63   (to two decimal places)

12. 12.364 ÷ 4.361   (to three decimal places)

13. In one week Tom earned $178.30 by working 36.25 hours. What was his hourly rate of pay to the nearest cent?

14. An 80.5-acre piece of land is being subdivided into lots of 0.35 acre. How many lots are possible in the subdivision?

15. You buy 18.7 gal of gas for $17.30. What is the cost per gallon to the nearest tenth of a cent?

16. On a trip you traveled 1030 mi and used 32.8 gal of gas. What was your average mileage for the trip to the nearest tenth of a mile per gallon?

Answers
1. 3.45   3. 2.89   5. 0.245   7. 1.84   9. 0.468   11. 0.51   13. $4.92   15. 92.5¢

9.4 Converting Common Fractions to Decimals

OBJECTIVE
To convert a common fraction to a decimal

Since a common fraction can be interpreted as division, to convert a common fraction to a decimal, you can divide the numerator of the common fraction by its denominator.

Example 1

Write $\frac{5}{8}$ as a decimal.

Solution Since $\frac{5}{8}$ means $5 \div 8$, divide 8 into 5.

Remember that 5 can be written as 5.0, 5.00, 5.000, and so on. In this case we continue the division by adding zeros to the dividend until a zero remainder is reached.

```
     .625
  8)5.000
    4 8
    ---
      20
      16
      ---
       40
       40
       ---
        0
```

We see that $\frac{5}{8} = 0.625$. We call 0.625 the *decimal equivalent* of $\frac{5}{8}$.

CHECK YOURSELF 1

Find the decimal equivalent of $\frac{3}{8}$.

Some fractions are used so often that we have listed their decimal equivalents for your reference.

SOME COMMON DECIMAL EQUIVALENTS

$\frac{1}{2} = 0.5$	$\frac{1}{4} = 0.25$	$\frac{1}{5} = 0.2$	$\frac{1}{8} = 0.125$
	$\frac{3}{4} = 0.75$	$\frac{2}{5} = 0.4$	$\frac{3}{8} = 0.375$
		$\frac{3}{5} = 0.6$	$\frac{5}{8} = 0.625$
		$\frac{4}{5} = 0.8$	$\frac{7}{8} = 0.875$

Section 9.4: Converting Common Fractions to Decimals 375

The division used to find these decimal equivalents stops when a zero remainder is reached. The equivalents are called *terminating decimals*.

If a decimal equivalent does not terminate, you may round the result to approximate the fraction to some specified number of decimal places. Consider the following.

Example 2

To write $\frac{3}{7}$ as a decimal, divide.

```
    .4285
 7)3.0000       In this example we are choosing to
    2 8         round to three decimal places, so we
    ――          must add enough zeros to carry the
     20         division to four decimal places.
     14
     ――
      60
      56
      ――
       40
       35
       ――
        5
```

So $\frac{3}{7}$ = 0.429 (to the nearest thousandth).

CHECK YOURSELF 2

Find the decimal equivalent of $\frac{5}{11}$ to the nearest thousandth.

If a decimal equivalent does *not* terminate, it will *repeat* a sequence of digits. These decimals are called *repeating decimals*.

Example 3

To write $\frac{1}{3}$ as a decimal, divide.

```
    .333        The digit 3 will just repeat itself
 3)1.000        indefinitely, since each new remainder
    9           will be 1.
    ―
    10
     9
    ―
    10
     9          Adding more zeros and going on will
    ―           simply lead to more 3s in the quotient.
     1
```

The three dots mean "and so on" and tell us that 3 will repeat itself indefinitely.

We can say: $\frac{1}{3}$ = 0.333 . . . .

Example 4

To write $\frac{5}{12}$ as a decimal, divide:

$$\begin{array}{r} .4166\ldots \\ 12\overline{)5.0000} \\ \underline{4\ 8} \\ 20 \\ \underline{12} \\ 80 \\ \underline{72} \\ 80 \\ \underline{72} \\ 8 \end{array}$$

In this example, the digit 6 will just repeat itself, since the remainder, 8, will keep occurring if we add more zeros and continue the division.

CHECK YOURSELF 3

Find the decimal equivalent of $\frac{7}{12}$.

Some important decimal equivalents (rounded to the nearest thousandth) are shown below for reference.

$\frac{1}{3} = 0.333$ $\qquad$ $\frac{1}{6} = 0.167$

$\frac{2}{3} = 0.667$ $\qquad$ $\frac{5}{6} = 0.833$

Another way to write a repeating decimal is with a bar placed over the digit or digits that repeat.

Example 5

We can write $0.37373737\ldots$ as $0.\overline{37}$.

The bar placed over the digits indicates that "37" repeats indefinitely.

Example 6

$0.234\overline{567}$ means $0.234567567567\ldots$

CHECK YOURSELF 4

Write $2.6252525\ldots$ using the bar notation.

When long division leads to a repeating decimal, we can use the bar notation to write the decimal equivalent.

Example 7

To write $\frac{5}{11}$ as a decimal, divide.

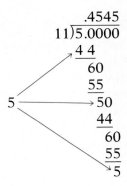

As soon as a remainder repeats itself, as 5 does here, the pattern of digits will repeat in the quotient.
Write.

$$\frac{5}{11} = 0.\overline{45}$$

CHECK YOURSELF 5

Use the bar notation to write the decimal equivalent of $\frac{5}{7}$. (Be patient. You'll have to divide for a while to find the repeating pattern.)

You can find the decimal equivalents for mixed numbers in a similar way. Find the decimal equivalent of the fractional part of the mixed number, and then combine that with the whole number part. Our final example illustrates.

Example 8

Find the decimal equivalent of $3\frac{5}{16}$.

Solution First find the equivalent of $\frac{5}{16}$ by division.

$$\frac{5}{16} = 0.3125$$

Add 3 to the result.

$$3\frac{5}{16} = 3.3125$$

CHECK YOURSELF 6

Find the decimal equivalent of $2\frac{5}{8}$.

We have learned something important in this section. To find the decimal equivalent of a fraction, we use long division. Since the remainder must be less than the divisor, the remainder must *either repeat or become 0*. So *every common fraction* will have a *repeating* or a *terminating* decimal as its decimal equivalent.

CHECK YOURSELF ANSWERS

1. 0.375. 2. $\frac{5}{11} = 0.455$ (to the nearest thousandth).
3. $\frac{7}{12} = 0.583\ldots$ The digit 3 will continue indefinitely. 4. $2.6\overline{25}$.
5. $\frac{5}{7} = 0.\overline{714285}$. 6. 2.625.

9.4 Exercises

Find the decimal equivalents for the following fractions.

1. $\frac{3}{4}$ 2. $\frac{4}{5}$ 3. $\frac{9}{20}$ 4. $\frac{3}{10}$

5. $\frac{1}{5}$ 6. $\frac{7}{8}$ 7. $\frac{5}{16}$ 8. $\frac{11}{20}$

9. $\frac{7}{10}$ 10. $\frac{7}{16}$ 11. $\frac{27}{40}$ 12. $\frac{17}{32}$

13. $\frac{13}{25}$ 14. $\frac{9}{16}$ 15. $\frac{7}{32}$ 16. $\frac{11}{25}$

Find the decimal equivalents rounded to the indicated place.

17. $\frac{5}{6}$ (thousandth)   18. $\frac{7}{12}$ (hundredth)   19. $\frac{4}{15}$ (thousandth)   20. $\frac{5}{24}$ (thousandth)

Section 9.4: Converting Common Fractions to Decimals

Write the decimal equivalents using the bar notation.

21. $\dfrac{1}{18}$ **22.** $\dfrac{4}{9}$ **23.** $\dfrac{3}{11}$ **24.** $\dfrac{7}{15}$

Find the decimal equivalents for the following mixed numbers.

25. $5\dfrac{3}{5}$ **26.** $7\dfrac{3}{4}$ **27.** $3\dfrac{5}{8}$ **28.** $4\dfrac{7}{16}$

Comparing the sizes of common fractions and decimals requires finding the decimal equivalent of the common fraction and then comparing the resulting decimals. For instance: Which is larger, $\dfrac{3}{8}$ or 0.38?

Write the decimal equivalent of $\dfrac{3}{8}$. That decimal is 0.375. Now comparing 0.375 and 0.38, we see that 0.38 is the larger of the numbers.

Complete each of the following statements using the symbols < or >.

29. $\dfrac{3}{4}$ 0.8   **30.** $\dfrac{7}{8}$ 0.87   **31.** $\dfrac{5}{16}$ 0.313   **32.** $\dfrac{9}{25}$ 0.4

Skillscan (Section 5.5)
Reduce each fraction to simplest form.

a. $\dfrac{8}{10}$ **b.** $\dfrac{16}{100}$ **c.** $\dfrac{35}{100}$

d. $\dfrac{225}{1000}$ **e.** $\dfrac{450}{1000}$ **f.** $\dfrac{625}{10{,}000}$

Answers
1. 0.75 **3.** 0.45 **5.** 0.2 **7.** 0.3125 **9.** 0.7 **11.** 0.675 **13.** 0.52 **15.** 0.21875 **17.** 0.833
19. 0.267 **21.** $0.0\overline{5}$ **23.** $0.\overline{27}$ **25.** 5.6 **27.** 3.625 **29.** $\dfrac{3}{4} < 0.8$ **31.** $\dfrac{5}{16} < 0.313$ **a.** $\dfrac{4}{5}$
b. $\dfrac{4}{25}$ **c.** $\dfrac{7}{20}$ **d.** $\dfrac{9}{40}$ **e.** $\dfrac{9}{20}$ **f.** $\dfrac{1}{16}$

9.4 Supplementary Exercises

Find the decimal equivalents for the following fractions.

1. $\frac{3}{5}$
2. $\frac{7}{20}$
3. $\frac{5}{8}$
4. $\frac{3}{16}$

5. $\frac{11}{20}$
6. $\frac{11}{32}$
7. $\frac{17}{25}$
8. $\frac{13}{40}$

Find the decimal equivalents rounded to the indicated place.

9. $\frac{5}{12}$ (thousandth)
10. $\frac{4}{9}$ (hundredth)

Write the decimal equivalents using the bar notation.

11. $\frac{7}{9}$
12. $\frac{4}{11}$

Find the decimal equivalents for the following mixed numbers.

13. $4\frac{3}{8}$
14. $5\frac{5}{16}$

Using Your Calculator

A calculator is very useful in converting common fractions to decimals. Just divide the numerator by the denominator, and the decimal equivalent will be in the display.

Example 1

To find the decimal equivalent of $\frac{7}{16}$:

7 ÷ 16 = 0.4375   0.4375 is the decimal equivalent of $\frac{7}{16}$.

Display

You may have to round the result in the display.

Section 9.4: Converting Common Fractions to Decimals

Example 2

To find the decimal equivalent of $\frac{5}{24}$ to the nearest hundredth:

Some calculators show the 0 to the left of the decimal point and seven digits to the right. Others omit the 0 and show eight digits to the right. Check yours with this example.

5 [÷] 24 [=] (0.2083333)

Display

$\frac{5}{24} = 0.21$ (to the nearest hundredth)

To find the decimal equivalent of a mixed number, use the following sequence.

Example 3

For $7\frac{5}{8}$:

There are several ways to do this, depending on the calculator you are using. For example,

7 [+] 5 [÷] 8 [=]

will work on most scientific calculators. Try it on yours.

5 [÷] 8 [+] 7 [=] (7.625)

Display

Other variations between types of calculators lead to the following:

Example 4

For $\frac{5}{9}$.

Depending on the machine you are using, the result may be rounded at its last displayed digit.

5 [÷] 9 [=] (0.5555556)

or

(0.5555555)

Exercises Using Your Calculator

Find the decimal equivalents.

1. $\frac{7}{8}$

2. $\frac{11}{16}$

3. $\frac{9}{16}$

4. $\frac{7}{24}$ (to the nearest hundredth)

5. $\frac{5}{32}$ (to the nearest thousandth)

6. $\dfrac{11}{75}$ (to the nearest thousandth)

7. $\dfrac{3}{11}$ (use the bar notation)

8. $\dfrac{7}{11}$ (use the bar notation)

9. $\dfrac{16}{33}$ (use the bar notation)

10. $3\dfrac{4}{5}$

11. $3\dfrac{7}{8}$

12. $8\dfrac{3}{16}$

Answers

1. 0.875   3. 0.5625   5. 0.156   7. $0.\overline{27}$ 9. $0.\overline{48}$ 11. 3.875

9.5 Converting Decimals to Common Fractions

OBJECTIVE
To convert a decimal to a common fraction

Using what we have learned about place values, you can easily write decimals as common fractions. The following rule is used.

> **TO CONVERT DECIMALS TO COMMON FRACTIONS**
>
> For a terminating decimal that is less than 1:
>
> STEP 1 Write the digits of the decimal without the decimal point. This will be the numerator of the common fraction.
>
> STEP 2 The denominator of the fraction is a 1 followed by as many zeros as there are places in the decimal.

Our first example illustrates.

Example 1

$0.7 = \dfrac{7}{10}$
One place / One zero

$0.09 = \dfrac{9}{100}$
Two places / Two zeros

$0.13 = \dfrac{13}{100}$
Two places / Two zeros

$0.257 = \dfrac{257}{1000}$
Three places / Three zeros

Section 9.5: Converting Decimals to Common Fractions

CHECK YOURSELF 1

Write as common fractions.

(1) 0.3   (2) 0.311

When converting a decimal to a common fraction, the common fraction that results should be written in lowest terms.

Example 2

$$0.395 = \frac{395}{1000} = \frac{79}{200}$$

Divide the numerator and denominator by 5.

CHECK YOURSELF 2

Write 0.275 as a common fraction.

If the decimal has a whole-number portion, write the digits to the right of the decimal point as a proper fraction and then form a mixed number for your result. This is illustrated in the following example.

Example 3

To write 12.277 as a mixed number,

$$0.277 = \frac{277}{1000}$$

so

$$12.277 = 12\frac{277}{1000}$$

Repeating decimals can also be written as common fractions, although the process is more complicated. We will limit ourselves to the conversion of terminating decimals in these materials.

CHECK YOURSELF ANSWERS

1. (1) $\frac{3}{10}$; (2) $\frac{311}{1000}$.   **2.** $0.275 = \frac{11}{40}$.

9.5 Exercises

Write the following as common fractions or mixed numbers.

1. 0.9
2. 0.3
3. 0.8
4. 0.6

5. 0.37
6. 0.97
7. 0.587
8. 0.379

9. 0.48
10. 0.75
11. 0.58
12. 0.65

13. 0.425
14. 0.116
15. 0.375
16. 0.225

17. 0.136
18. 0.575
19. 0.059
20. 0.067

21. 0.0625
22. 0.0425
23. 6.3
24. 5.7

25. 2.17
26. 3.31
27. 5.28
28. 15.35

Answers

1. $\frac{9}{10}$ 3. $\frac{4}{5}$ 5. $\frac{37}{100}$ 7. $\frac{587}{1000}$ 9. $\frac{12}{25}$ 11. $\frac{29}{50}$ 13. $\frac{17}{40}$ 15. $\frac{3}{8}$ 17. $\frac{17}{125}$ 19. $\frac{59}{1000}$
21. $\frac{1}{16}$ 23. $6\frac{3}{10}$ 25. $2\frac{17}{100}$ 27. $5\frac{7}{25}$

9.5 Supplementary Exercises

Write the following as common fractions or mixed numbers.

1. 0.7
2. 0.6
3. 0.41
4. 0.863

5. 0.35
6. 0.36
7. 0.325
8. 0.248

9. 0.875
10. 0.037
11. 0.0225
12. 4.7

13. 6.8
14. 5.32

SELF-TEST for Chapter Nine

The purpose of the Self-Test is to help you check your progress and review for a chapter test in class. Allow yourself about an hour to take the test. When you are done, check your answers in the back of the book. If you missed any problems, be sure to go back and review the appropriate sections in the chapter and do the supplementary exercises provided there.

In Problems 1 to 4, divide.

1. $9\overline{)24.75}$

2. $138.33 \div 58$

3. $28\overline{)12.85}$ (to the nearest hundredth)

4. $8.97 \div 62$ (to the nearest thousandth)

In Problems 5 and 6, solve the application.

5. Alice buys a television set and, with interest charges included, owes $558.72. If she agrees to make equal monthly payments for 2 years, what is the amount of those monthly payments?

6. Rainfall amounts at an airport weather station were measured at 1.12, 0.68, 0.04, 0.2, 1.31, 0.5, and 0.72 in for a 7-day period. What was the average rainfall (to the nearest hundredth of an inch) for the 7 days?

In Problems 7 to 12, divide.

7. $5.8\overline{)38.86}$

8. $0.24\overline{)0.774}$

9. $0.075\overline{)0.213}$

10. $5.3\overline{)29.284}$ (to the nearest hundredth)

11. $0.258\overline{)0.5218}$ (to the nearest hundredth)

12. $2 \div 3.7$ (to the nearest thousandth)

In Problems 13 to 15, solve the applications.

13. An automobile travels 627 mi on 20.7 gal of gas. What is the average gas mileage for that period (to the nearest tenth of a mile per gallon)?

14. A shirt pattern requires 2.25 yd of fabric. If a tailor has 65.25 yd of the fabric available, how many shirts can be made?

15. A 21-acre piece of land is being divided for a housing project. Roads will use 4.8 acres, and the remainder will be divided into 0.36-acre lots. How many lots can be formed?

In Problems 16 to 20, divide.

16. $385.7 \div 100 =$

17. $28.47 \div 1000 =$

18. $37.95 \div 10^4 =$

19. A shipment of 100 items costs a store $537. What is the cost of an individual item?

20. To convert from milliliters to liters, we divide by 1000. A bottle of soft drink contains 828 milliliters. What is its volume in liters?

In Problems 21 to 25, find the decimal equivalents of the given fractions.

21. $\dfrac{7}{8} =$

22. $\dfrac{7}{32} =$

23. $\dfrac{6}{7} =$ (to the nearest thousandth)

24. $\dfrac{7}{11} =$ (use the bar notation)

25. $2\dfrac{9}{16} =$

In Problems 26 to 30, write the decimals as common fractions or mixed numbers.

26. $0.29 =$

27. $0.56 =$

28. $0.355 =$

29. $0.7825 =$

30. $2.76 =$

SUMMARY for Part Three

The Language of Decimals

Decimal Fraction   A fraction whose denominator is a power of 10. We call decimal fractions *decimals*.

$\frac{7}{10}$ and $\frac{47}{100}$ are decimal fractions.

Decimal Place   The position of a digit to the right of the decimal point. The position gives the value of the digit.

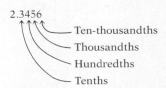

The number of decimal places is the number of digits to the right of the decimal point.

2.3456 has four decimal places.

Decimal Form   Using place values to the right of the decimal point to write decimal fractions.

0.7 is the decimal form of $\frac{7}{10}$.

0.47 is the decimal form of $\frac{47}{100}$.

Reading and Writing Decimals in Words

To Read or Write a Decimal

1. The digits *to the left* of the decimal point are read as a whole number.
2. The word *and* indicates the decimal point.
3. The digits *to the right* of the decimal point are read as a whole number followed by the place value of the rightmost digit.

Note: If there are *no* nonzero digits to the left of the decimal point, start directly with step 3.

8.15 is read "eight and fifteen hundredths."

Adding Decimals

To Add Decimals

1. Write the numbers being added in column form with their decimal points in a vertical line.
2. Add just as you would with whole numbers.
3. Place the decimal point of the sum in line with the decimal points of the addends.

To add 2.7, 3.15, and 0.48:

```
  2.7
  3.15
+0.48
─────
  6.33
```

Subtracting Decimals

To Subtract Decimals

1. Write the numbers being subtracted in column form with their decimal points in a vertical line. You may have to place zeros to the right of the existing digits.
2. Subtract just as you would with whole numbers.
3. Place the decimal point of the difference in line with the decimal points of the numbers being subtracted.

To subtract 5.875 from 8.5:

```
  8.500
 -5.875
  2.625
```

Multiplying Decimals

To Multiply Decimals

1. Multiply the decimals as though they were whole numbers.
2. Add the number of decimal places in the numbers being multiplied.
3. Place the decimal point in the product so that the number of decimal places in the product is the sum of the number of decimal places in the factors.

To multiply 2.85×0.045:

```
   2.85    ← Two places
 ×0.045    ← Three places
  1425
  1140
 0.12825   ← Five places
```

Multiplying by Powers of 10

To Multiply by a Power of 10

Move the decimal point to the right the same number of places as there are zeros in the power of 10.

$2.37 \times 10 = 23.7$

$0.567 \times 1000 = 567.$

Rounding Decimals

To Round a Decimal

1. Find the place the decimal is to be rounded to.
2. If the next digit to the right is 5 or more, increase the digit in the place you are rounding to by 1. Discard any remaining digits to the right.
3. If the next digit to the right is less than 5, just discard that digit and any remaining digits to the right.

To round 5.87 to the nearest tenth:

5.87 is rounded to 5.9

To round 12.3454 to the nearest thousandth:

12.3454 is rounded to 12.345

Dividing Decimals

To Divide by a Whole Number

1. Place the decimal point in the quotient directly above the decimal point of the dividend.
2. Divide as you would with whole numbers.

To divide 63.18 by 27:

```
        2.34
    27)63.18
       54
        9 1
        8 1
        1 08
        1 08
            0
```

To Divide by a Decimal

1. Move the decimal point to the right, making the divisor a whole number.
2. Move the decimal point in the dividend to the right the same number of places. Add zeros if necessary.
3. Place the decimal point in the quotient directly above the decimal point of the dividend.
4. Divide as you would with whole numbers.

To divide 2.3147 by 0.395, write

```
0.395)2.3147
```

Move the decimal points:

```
             5.86
   0.395_)2.314_70
          1 975
            339 7
            316 0
             23 70
             23 70
                 0
```

To Divide by a Power of 10

Move the decimal point to the left the same number of places as there are zeros in the power of 10.

$25.8 \div 10 = 2\_5.8 = 2.58$

$34.789 \div 1000 = 0\_034\_789$
$= 0.034789$

Converting Common Fractions and Decimals

To Convert a Common Fraction to a Decimal

1. Divide the numerator of the common fraction by its denominator.
2. The quotient is the decimal equivalent of the common fraction.

To convert $\frac{5}{32}$ to a decimal:

```
         .15625
    32)5.00000
       3 2
       1 80
       1 60
         200
         192
          80
          64
          160
          160
            0
```

$\frac{5}{32} = 0.15625$

To Convert a Decimal to a Common Fraction

For a terminating decimal that is less than 1:

1. Write the digits of the decimal without the decimal point. This will be the numerator of the common fraction.
2. The denominator of the fraction is a 1 followed by as many zeros as there are places in the decimal.

To convert 0.275 to a common fraction:

$$0.275 = \frac{275}{1000} = \frac{11}{40}$$

SUMMARY EXERCISES for Part Three

You should now be reviewing the material in Part 3 of the text. The following exercises will help in that process. Work all the exercises carefully. Then check your answers in the back of the book. References are provided to the chapter and section for each problem. If you made an error, go back and review the related material and do the supplementary exercises for that section.

[8.1] In Problems 1 and 2, find the indicated place values.

1. 7 in 3.5742
2. 3 in 0.5273

[8.1] In Problems 3 and 4, write the fractions in decimal form.

3. $\dfrac{37}{100}$
4. $\dfrac{307}{10,000}$

[8.1] In Problems 5 and 6, write the decimals in words.

5. 0.071
6. 12.39

[8.1] In Problems 7 and 8, write the fractions in decimal form.

7. Four and five tenths

8. Four hundred and thirty-seven thousandths.

[8.1] In Problems 9 to 12, complete each statement using the symbols <, =, or >.

9. 0.79   0.785   10. 1.25   1.250   11. 12.8   13   12. 0.832   0.83

[8.2] In Problems 13 to 16, add:

13. 2.58
    +0.89

14. 3.14
     0.8
     2.912
    +12

15. 1.3, 25, 5.27, and 6.158

16. Add: eight, forty-three thousandths, five and nineteen-hundredths, and seven and three-tenths.

391

[8.2] In Problems 17 and 18, solve the applications.

17. Janice ran 4.8 mi on Sunday, 5.3 mi on Tuesday, 3.9 mi on Thursday, and 8.2 mi on Saturday. How far did she run during the week?

18. Find dimension a in the following figure. (Dimensions are in centimeters.)

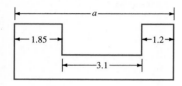

[8.3] In Problems 19 to 22, subtract.

19. 29.21
       − 5.75

20. 6.73
       −2.485

21. 1.735 from 2.81

22. 12.38 from 19

[8.3] In Problems 23 and 24, solve the applications.

23. A stereo system which normally sells for $499.50 is discounted (or marked down) to $437.75 for a sale. Find the savings.

24. If you cash a $50 check and make purchases of $8.71, $12.53, and $9.83, how much money do you have left?

[8.4] In Problems 25 to 28, multiply.

25. 22.8
       ×0.72

26. 0.0045
       ×0.058

27. 1.24 × 56

28. 0.0025 × 0.491

[8.4] In Problems 29 to 31, solve the applications.

29. Neal worked for 37.4 hours during a week. If his hourly rate of pay was $7.25, how much did he earn?

30. To find the interest on a loan at $11\frac{1}{2}$% for 1 year, we must multiply the amount of the loan by 0.115. Find the interest on a $2500 loan at $11\frac{1}{2}$% for 1 year.

31. A television set has an advertised price of $499.50. You buy the set and agree to make payments of $27.15 for 2 years. How much extra are you paying by buying on this installment plan?

Summary Exercises for Part Three

[8.5] In Problems 32 and 33, multiply.

32. 0.052×1000

33. 0.045×10^4

[8.5] In Problem 34, solve the application.

34. A stereo dealer buys 100 portable radios for a promotion sale. If she pays $57.42 per radio, what is her total cost?

[8.6] In Problems 35 to 37, round to the indicated place.

35. 5.837 (hundredth)

36. 9.5723 (thousandth)

37. 4.87625 (to three decimal places)

[8.6] In Problem 38, solve the application.

38. Find the area (to the nearest hundredth of a square centimeter) of a rectangle which has dimensions 5.25 cm by 8.75 cm.

[9.1] In Problems 39 to 41, divide.

39. $8 \overline{)3.08}$

40. $58 \overline{)269.7}$

41. $55 \overline{)17.69}$ (to the nearest thousandth)

[9.1] In Problems 42 and 43, solve the applications.

42. Thirty-seven employees of a company donated a total of $867.65 during a charity fund-raising drive. What was the average donation per employee?

43. In six readings, Faith's gas mileage was 38.9, 35.3, 39.0, 41.2, 40.5, and 40.8 mi/gal. What was her average mileage to the nearest tenth of a mile per gallon?

[9.2] In Problems 44 to 47, divide.

44. $0.7 \overline{)1.862}$

45. $3.042 \div 0.36$

46. $5.3 \overline{)6.748}$ (to the nearest tenth)

47. $0.2549 \div 2.87$ (to three decimal places)

[9.2] In Problems 48 and 49, solve the applications.

48. A developer is planning to subdivide an 18.5-acre piece of land. She estimates that 5 acres will be used for roads and wants individual lots of 0.25 acre. How many lots are possible?

49. Paul drives 949 mi using 31.8 gal of gas. What is his mileage for the trip (to the nearest tenth of a mile per gallon)?

[9.3] In Problems 50 to 52, divide.

50. $7.6 \div 10$ **51.** $80.7 \div 1000$ **52.** $457 \div 10^4$

[9.3] In Problem 53, solve the application.

53. A shipment of 1000 VHS-video tapes cost a dealer $7090. What was the cost per tape to the dealer?

[9.4] In Problems 54 to 57, find the decimal equivalents.

54. $\dfrac{7}{16}$ **55.** $\dfrac{3}{7}$ (to the nearest thousandth)

56. $\dfrac{4}{15}$ (use the bar notation) **57.** $3\dfrac{3}{4}$

[9.5] In Problems 58 to 60, write as common fractions or mixed numbers.

58. 0.21 **59.** 0.084 **60.** 5.28

CUMULATIVE TEST for Part Three

This test is provided to help you in the process of review over Chapters 8 and 9. Answers are provided in the back of the book. If you missed any problems, be sure to go back and review the appropriate chapter sections.

1. Find the place value of 8 in 0.5248.

2. Write $\dfrac{49}{1000}$ in decimal form.

3. Write 2.53 in words.

4. Write twelve and seventeen-thousandths in decimal form.

In Problems 5 and 6, complete the statement using the symbols < or >.

5. 0.889   0.89

6. 0.531   0.53

In Problems 7 to 9, add.

7. 3.45
   0.6
   +12.59

8. 2.4, 35, 4.73, and 5.123.

9. Seven, seventy-nine hundredths, and five and thirteen-thousandths.

10. On a business trip, Martin bought the following amounts of gasoline: 14.4, 12, 13.8, and 10 gal. How much gasoline did he purchase on the trip?

In Problems 11 to 13, subtract.

11. 18.32
    − 7.78

12. 40
    −15.625

13. 1.742 from 5.63

14. You pay for purchases of $13.99, $18.75, $9.20, and $5 with a $50 check. How much cash will you have left?

In Problems 15 to 17, multiply.

15. 32.9
    ×0.53

16. 0.049
    × 0.57

17. 2.75 × 0.53

18. Find the area of a rectangle with length 3.5 in and width 2.15 in.

In Problems 19 and 20, multiply.

19. 0.735×1000

20. 1.257×10^4

21. A college bookstore purchases 1000 pens at a cost of 54.3¢ per pen. Find the total cost of the order in dollars.

In Problems 22 and 23, round to the indicated place.

22. 0.5977 (to the nearest thousandth)

23. 23.5724 (to two decimal places)

24. We find the circumference of a circle by multiplying the diameter of the circle by 3.14. If a circle has a diameter of 3.2 ft, find its circumference, to the nearest hundredth of a foot.

In Problems 25 to 27, divide.

25. $8\overline{)3.72}$

26. $27\overline{)63.45}$

27. $2.72 \div 53$ (to the nearest thousandth)

28. The Michael's expenses for home heating were $56.79, $67.20, $56.89, $98.45, and $105.45 over a 5-month period. Find their average monthly expense for that period, to the nearest cent.

In Problems 29 to 31, divide.

29. $0.6\overline{)1.431}$

30. $3.969 \div 0.54$

31. $0.263 \div 3.91$ (to three decimal places)

32. A 14-acre piece of land is being developed into home lots. If 2.8 acres of land will be used for roads, and each home site is to be 0.35 acre, how many lots can be formed?

In Problems 33 and 34, divide.

33. $4.983 \div 1000$

34. $523 \div 10^5$

35. A street improvement project will cost $57,340 and that cost is to be divided among the 100 families in the area. What will be the cost to each individual family?

In Problems 36 to 38, find the decimal equivalents of the common fractions.

36. $\dfrac{7}{16}$

37. $\dfrac{3}{7}$ (to the nearest thousandth)

38. $\dfrac{7}{11}$ (use the bar notation)

In Problems 39 and 40, write the decimals as common fractions or mixed numbers.

39. 0.072

40. 4.44

PART 4

Ratios, Proportions, and Percents

PRETEST for Chapter Ten

This pretest will point out any difficulties you may be having with ratios and proportions. Do all the problems. Then check your answers on the following page.

1. Write the ratio of 8 to 11.

2. Write the ratio of 20 to 15 in lowest terms.

3. Is $\dfrac{4}{7} = \dfrac{8}{14}$ a true proportion?

4. Is $\dfrac{5}{9} = \dfrac{9}{16}$ a true proportion?

5. Solve for x: $\dfrac{x}{4} = \dfrac{9}{12}$

6. Solve for a: $\dfrac{5}{a} = \dfrac{6}{18}$

7. Solve for n: $\dfrac{3}{16} = \dfrac{9}{n}$

8. Cans of tomato juice are marked 2 for 95¢. At this price, what will a dozen cans cost?

9. The ratio of compact cars to larger model cars sold during a month was 8 to 5. If 96 compact cars were sold during that period, how many of the larger cars were sold?

10. If 2 gal of paint will cover 450 sq ft how many gallons will be needed to paint a room with 2025 ft² of wall surface?

ANSWERS TO PRETEST

For help with similar problems, turn to the section indicated.

1. $\frac{8}{11}$ (Section 10.1);
2. $\frac{4}{3}$ (Section 10.1);
3. Yes (Section 10.2);
4. No (Section 10.2);
5. 3 (Section 10.3);
6. 15 (Section 10.3);
7. 48 (Section 10.3);
8. $5.70 (Section 10.4);
9. 60 (Section 10.4);
10. 9 gal (Section 10.4)

Chapter Ten
Ratio and Proportion

10.1 Using Ratios

OBJECTIVES
1. To write the ratio of two numbers in simplest form
2. To write the ratio of two quantities in simplest form

In earlier chapters you have seen two meanings for a fraction:

First Meaning A fraction can name a certain number of parts of a whole.

$\frac{3}{5}$ names 3 parts of a whole that has been divided into 5 parts.

Second Meaning A fraction can indicate division.

$\frac{3}{5}$ can be thought of as $3 \div 5$.

We now want to turn to a third meaning for a fraction:

Third Meaning A fraction can be a ratio. A *ratio* is a means of comparing two numbers or quantities.

Note: Another way of writing the ratio of 3 to 5 is 3:5. We have chosen to stay with the fraction notation for a ratio in these materials.

Example 1

To compare 3 to 5, we write the ratio of 3 to 5 as $\frac{3}{5}$.

So $\frac{3}{5}$ also means "the ratio of 3 to 5."

CHECK YOURSELF 1

Write the ratio of 7 to 12 as a fraction.

Comparing like quantities means comparing "fish to fish" or "inches to inches."

Our second example illustrates the use of a ratio in comparing *like quantities*.

Example 2

The width of a rectangle is 7 cm and its length is 19 cm. The ratio of its width to its length is 7 cm to 19 cm, or

$$\frac{7 \text{ cm}}{19 \text{ cm}} = \frac{7}{19}$$ We are comparing centimeters to centimeters, so we don't need to write the units.

The ratio of its length to its width is

$$\frac{19 \text{ cm}}{7 \text{ cm}} = \frac{19}{7}$$

A ratio fraction can be greater than 1.
Note: In this case the ratio is never written as a mixed number. It is left as an improper fraction.

CHECK YOURSELF 2

A basketball team wins 17 of its 29 games in a season.

(1) Write the ratio of wins to games played.
(2) Write the ratio of wins to losses.

Since a ratio is a fraction, we can reduce a ratio to simplest form. Consider the following.

Example 3

The ratio of 20 to 30 is

$$\frac{20}{30} \text{ or } \frac{2}{3}$$ Divide the numerator and denominator by the common factor of 10.

CHECK YOURSELF 3

Write the ratio of 24 to 32 in lowest terms.

Example 4

A medium-sized car has a mileage rating of 18 mi/gal. A compact car has a rating of 30 mi/gal. The ratio of the mileage rating of the medium-sized to that of the compact car is

$$\frac{18}{30} = \frac{3}{5}$$ Divide the numerator and denominator by 6.

Section 10.1: Using Ratios   403

Example 5

The cost of 32 oz of orange juice is $1.20. The ratio of ounces to cost (in cents) is

$$\frac{32 \text{ oz}}{120¢} = \frac{4 \text{ oz}}{15¢}$$ Divide the numerator and denominator by 8.

CHECK YOURSELF 4

A baseball player has 12 hits in 42 times at bat. Write the ratio of hits to times at bat.

Sometimes we can convert the identifying units to simplify a ratio. This is shown in the following examples.

Example 6

The ratio of 3 min to 2 hours is

1 hour is 60 min, so 2 hours is 120 min.

$$\frac{3 \text{ min}}{2 \text{ hours}} = \frac{3 \text{ min}}{120 \text{ min}} = \frac{3}{120} = \frac{1}{40}$$

Write 2 hours as 120 min. Then divide the like units and reduce the ratio fraction to lowest terms.

CHECK YOURSELF 5

Write the ratio of 3 in to 2 ft.

Example 7

The ratio of 8 oz to 3 lb is

1 lb is 16 oz, so write 3 lb as 48 oz.

$$\frac{8 \text{ oz}}{3 \text{ lb}} = \frac{8 \text{ oz}}{48 \text{ oz}} = \frac{1}{6}$$

Example 8

The ratio of 5 quarters to 3 dollars is

$$\frac{5 \text{ quarters}}{3 \text{ dollars}} = \frac{125¢}{300¢} = \frac{5}{12}$$ Convert both terms of the ratio to cents. Then reduce the fraction.

CHECK YOURSELF 6

Write the ratio of 4 dimes to 2 dollars.

CHECK YOURSELF ANSWERS

1. $\frac{7}{12}$.   2. (1) $\frac{17}{29}$; (2) $\frac{17}{12}$ (the team lost 12 games).
3. $\frac{3}{4}$.   4. $\frac{2}{7}$ Did you reduce to lowest terms?   5. $\frac{1}{8}$.   6. $\frac{1}{5}$.

10.1 Exercises

Write the following ratios in simplest form.

1. The ratio of 9 to 13.

2. The ratio of 5 to 4.

3. The ratio of 7 to 5.

4. The ratio of 5 to 12.

5. The ratio of 10 to 15.

6. The ratio of 12 to 8.

7. The ratio of 24 to 16.

8. The ratio of 25 to 40.

9. The ratio of 17 in to 30 in.

10. The ratio of 23 lb to 36 lb.

11. The ratio of 8 mi to 12 mi.

12. The ratio of 50 cm to 40 cm.

13. The ratio of 90 yd to 75 yd.

14. The ratio of 12 oz to 18 oz.

15. The ratio of $48 to $42.

16. The ratio of 20 ft to 24 ft.

17. The ratio of 7 min to 2 hours.

18. The ratio of 9 oz to 2 lb.

19. The ratio of 4 nickels to 5 dimes.

20. The ratio of 8 in to 3 ft.

21. The ratio of 2 days to 10 hours.

22. The ratio of 4 ft to 4 yd.

23. The ratio of 2 gal to 3 quarts.

24. The ratio of 5 dimes to 5 quarters.

Section 10.1: Using Ratios

25. An algebra class has 11 men and 15 women. Write the ratio of men to women. Write the ratio of women to men.

26. A French bread recipe calls for 7 cups of flour for 4 loaves of bread. Write the ratio of cups of flour to loaves of bread.

27. A football team wins 8 of its 11 games with no ties. Write the ratio of wins to games played. Write the ratio of wins to losses.

28. Rick makes $53 in an 8-hour day. Write the ratio of dollars to hours.

29. In a school election 4500 yes votes were cast and 3000 no votes were cast. Write the ratio of yes to no votes.

30. A basketball player made 48 of the 80 shots taken in a tournament. Write the ratio of shots made to shots taken.

31. A 32-oz bottle of dishwashing liquid costs $1.92. Write the ratio of cents to ounces.

32. A new compact automobile travels 384 mi on 8 gal of gasoline. Write the ratio of miles to gallons.

Skillscan (Section 5.4)
Determine whether or not the following pairs of fractions are equivalent.

a. $\dfrac{1}{3}, \dfrac{3}{9}$

b. $\dfrac{2}{5}, \dfrac{3}{7}$

c. $\dfrac{5}{6}, \dfrac{6}{7}$

d. $\dfrac{5}{6}, \dfrac{25}{30}$

e. $\dfrac{3}{10}, \dfrac{2}{7}$

f. $\dfrac{4}{15}, \dfrac{12}{45}$

Answers
Solutions for the even-numbered exercises are provided in the back of the book.

1. $\dfrac{9}{13}$ 3. $\dfrac{7}{5}$ 5. $\dfrac{2}{3}$ 7. $\dfrac{3}{2}$ 9. $\dfrac{17}{30}$ 11. $\dfrac{2}{3}$ 13. $\dfrac{6}{5}$ 15. $\dfrac{8}{7}$ 17. $\dfrac{7}{120}$ 19. $\dfrac{2}{5}$ 21. $\dfrac{24}{5}$
23. $\dfrac{8}{3}$ 25. $\dfrac{11}{15}, \dfrac{15}{11}$ 27. $\dfrac{8}{11}, \dfrac{8}{3}$ 29. $\dfrac{3}{2}$ 31. $\dfrac{6¢}{1\text{ oz}}$ a. Yes   b. No   c. No   d. Yes   e. No
f. Yes

10.1 Supplementary Exercises

Write the following ratios in simplest form.

1. The ratio of 7 to 11.
2. The ratio of 8 to 5.
3. The ratio of 20 to 15.
4. The ratio of 8 to 12.
5. The ratio of 14 lb to 19 lb.
6. The ratio of $12 to $7.
7. The ratio of 28 oz to 21 oz.
8. The ratio of 100 mi to 120 mi.
9. The ratio of 7 oz to 3 lb.
10. The ratio of 5 in to 2 ft.
11. The ratio of 5 nickels to 4 dimes.
12. The ratio of 15 min to 2 hours.
13. A Toyota travels 600 mi on 17 gal of gas. Write the ratio of miles to gallons.
14. A basketball team wins 25 of its 35 games. Write the ratio of wins to games played. Write the ratio of wins to losses.
15. A compact car weighs 2400 lb, while a medium-sized car weighs 3600 lb. Write the ratio of the weight of the smaller car to that of the larger.
16. An election on a city budget resulted in 5500 no votes and 4400 yes votes. Write the ratio of no votes to yes votes.

10.2 The Language of Proportions

OBJECTIVES
1. To use the language of proportions
2. To determine whether or not a proportion is a true statement

PROPORTIONS

A statement that two ratios are equal is called a *proportion.*

Section 10.2: The Language of Proportions

Example 1

Since the ratio of 1 to 3 is equal to the ratio of 2 to 6, we can write the proportion

This is the same as saying the fractions are equivalent. They name the same number.

$$\frac{1}{3} = \frac{2}{6}$$

The proportion $\frac{a}{b} = \frac{c}{d}$ is read "a is to b as c is to d."

Example 2

We read the proportion $\frac{1}{3} = \frac{2}{6}$ as "one is to three as two is to six."

We can number the terms of a proportion and give them special names. The terms are numbered as shown.

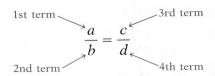

1st term → a (numerator, left) ← 3rd term c (numerator, right)
2nd term → b (denominator, left) ← 4th term d (denominator, right)

$$\frac{a}{b} = \frac{c}{d}$$

The 2nd and 3rd terms are called the *means* of the proportion. The 1st and 4th terms are the *extremes* of the proportion. So in the proportion $\frac{a}{b} = \frac{c}{d}$, b and c are the means and a and d are the extremes.

Example 3

In the proportion $\frac{1}{3} = \frac{2}{6}$, 3 and 2 are the means and 1 and 6 are the extremes.

Example 4

In the proportion $\frac{a}{5} = \frac{3}{15}$, 5 and 3 are the means and a and 15 are the extremes.

CHECK YOURSELF 1

List the means and extremes in the proportion $\frac{5}{8} = \frac{20}{32}$.

A useful property holds for the terms of a true proportion. Let's look at another example and calculate the product of the means and the product of the extremes.

Example 5

In the proportion

$$\frac{3}{8} = \frac{9}{24}$$

the means are 8 and 9, so the product of the means is $8 \times 9 = 72$. The extremes are 3 and 24, so the product of the extremes is $3 \times 24 = 72$. The product of the means is equal to the product of the extremes.

Since $8 \times 9 = 3 \times 24 = 72$

The property shown in Example 5 will hold for any true proportion.

> **THE PROPORTION RULE**
>
> The product of the means is equal to the product of the extremes.

Using symbols in the proportion:

$$\frac{a}{b} = \frac{c}{d}$$

$$b \times c = a \times d$$

$\uparrow$ The product of the means $\uparrow$ The product of the extremes

This is sometimes called cross multiplication.

CHECK YOURSELF 2

Show that the product of the means is equal to the product of the extremes in the following proportion.

$$\frac{5}{8} = \frac{20}{32}$$

You may wish to review Section 5.4 at this point.

Do you see that the proportion rule is the same as our earlier rule for testing equivalent fractions? If the cross products are equal, the two fractions are equivalent and a true proportion is formed.

The proportion rule can be used to find out whether a given proportion is a true statement. Look at the following examples.

Section 10.2: The Language of Proportions   409

Example 6

Is $\dfrac{4}{5} = \dfrac{20}{25}$ a true proportion?

$5 \times 20 = 100$ The product of the means is 100.
$4 \times 25 = 100$ The product of the extremes is also 100.

Since the products are equal, the statement is a true proportion.

Example 7

Is $\dfrac{2}{3} = \dfrac{15}{20}$ a true proportion?

$3 \times 15 = 45$ The product of the means is 45.
$2 \times 20 = 40$ The product of the extremes is 40.

Since the products are not equal, the statement is not a true proportion.

CHECK YOURSELF 3

Are the following true proportions?

(1) $\dfrac{3}{8} = \dfrac{12}{30}$ (2) $\dfrac{5}{9} = \dfrac{15}{27}$

Later we will be using proportions with terms that are fractions or decimals. Our final examples illustrate.

Example 8

Is $\dfrac{3}{\frac{1}{2}} = \dfrac{30}{5}$ a true proportion?

$\dfrac{1}{2} \times 30 = 15$ The product of the means is 15.

$3 \times 5 = 15$ The product of the extremes is 15.

Since the products are equal, the statement is a true proportion.

CHECK YOURSELF 4

Is $\dfrac{\frac{1}{4}}{6} = \dfrac{3}{80}$ a true proportion?

Example 9

Is $\dfrac{0.4}{20} = \dfrac{3}{100}$ a true proportion?

$20 \times 3 = 60$ The product of the means is 60.

$0.4 \times 100 = 40$ The product of the extremes is 40.

Since the products are *not* equal, the statement is not a true proportion.

CHECK YOURSELF 5

Is $\dfrac{0.5}{8} = \dfrac{3}{48}$ a true proportion?

CHECK YOURSELF ANSWERS

1. The means are 8 and 20. The extremes are 5 and 32.
2. $8 \times 20 = 160$.
   $5 \times 32 = 160$.   The products are equal.
3. (1) No; (2) yes.   4. No.   5. Yes.

10.2 Exercises

List the means and the extremes for the following proportions.

1. $\dfrac{2}{3} = \dfrac{6}{9}$ 2. $\dfrac{2}{5} = \dfrac{4}{10}$ 3. $\dfrac{7}{9} = \dfrac{21}{27}$

4. $\dfrac{5}{x} = \dfrac{20}{24}$ 5. $\dfrac{4}{7} = \dfrac{a}{35}$ 6. $\dfrac{3}{8} = \dfrac{15}{40}$

7. $\dfrac{x}{6} = \dfrac{5}{30}$ 8. $\dfrac{7}{9} = \dfrac{n}{45}$

Which of the following are true proportions?

9. $\dfrac{1}{4} = \dfrac{2}{7}$ 10. $\dfrac{2}{3} = \dfrac{6}{9}$ 11. $\dfrac{2}{5} = \dfrac{8}{20}$ 12. $\dfrac{3}{4} = \dfrac{15}{20}$

13. $\dfrac{3}{5} = \dfrac{6}{10}$ 14. $\dfrac{5}{6} = \dfrac{7}{8}$ 15. $\dfrac{4}{7} = \dfrac{3}{5}$ 16. $\dfrac{8}{3} = \dfrac{24}{9}$

17. $\dfrac{5}{8} = \dfrac{15}{24}$ 18. $\dfrac{5}{9} = \dfrac{6}{11}$ 19. $\dfrac{5}{12} = \dfrac{8}{20}$ 20. $\dfrac{7}{16} = \dfrac{21}{48}$

21. $\dfrac{7}{9} = \dfrac{20}{27}$ 22. $\dfrac{10}{3} = \dfrac{150}{50}$ 23. $\dfrac{5}{8} = \dfrac{75}{120}$ 24. $\dfrac{5}{16} = \dfrac{20}{64}$

25. $\dfrac{12}{7} = \dfrac{96}{50}$ 26. $\dfrac{7}{15} = \dfrac{84}{180}$ 27. $\dfrac{84}{48} = \dfrac{14}{8}$ 28. $\dfrac{60}{36} = \dfrac{25}{15}$

29. $\dfrac{\frac{1}{2}}{4} = \dfrac{5}{40}$ 30. $\dfrac{6}{\frac{1}{4}} = \dfrac{40}{2}$ 31. $\dfrac{\frac{2}{3}}{6} = \dfrac{1}{12}$ 32. $\dfrac{\frac{3}{4}}{12} = \dfrac{1}{16}$

33. $\dfrac{0.2}{5} = \dfrac{1}{30}$ 34. $\dfrac{3}{60} = \dfrac{0.3}{6}$ 35. $\dfrac{0.4}{20} = \dfrac{1}{50}$ 36. $\dfrac{0.6}{15} = \dfrac{2}{75}$

Skillscan (Sections 6.4, 9.2)
Divide.

a. $16 \div \dfrac{1}{3}$ b. $24 \div \dfrac{1}{4}$ c. $25 \div \dfrac{5}{6}$ d. $32 \div \dfrac{4}{3}$

e. $20 \div 0.5$ f. $18 \div 0.3$ g. $48 \div 0.8$ h. $60 \div 1.2$

Answers
1. 3 and 6 are the means, 2 and 9 are the extremes.   3. 9 and 21 are the means, 7 and 27 are the extremes.
5. 7 and a are the means, 4 and 35 are the extremes.   7. 6 and 5 are the means, x and 30 are the extremes.
9. F   11. T   13. T; $5 \times 6 = 3 \times 10$ 15. F   17. T   19. F; $12 \times 8 \neq 5 \times 20$ 21. F   23. T
25. F   27. T   29. $4 \times 5 = 20$; $\dfrac{1}{2} \times 40 = 20$. The products are equal, and so the proportion is a true statement.   31. F   33. F   35. T   a. 48   b. 96   c. 30   d. 24   e. 40   f. 60   g. 60   h. 50

10.2 Supplementary Exercises

List the means and the extremes for the following proportions.

1. $\dfrac{4}{5} = \dfrac{8}{10}$ 2. $\dfrac{a}{5} = \dfrac{12}{30}$

3. $\dfrac{6}{x} = \dfrac{18}{24}$ 4. $\dfrac{9}{5} = \dfrac{36}{20}$

Which of the following are true proportions?

5. $\dfrac{3}{4} = \dfrac{6}{9}$

6. $\dfrac{3}{5} = \dfrac{6}{10}$

7. $\dfrac{4}{5} = \dfrac{8}{10}$

8. $\dfrac{2}{7} = \dfrac{3}{10}$

9. $\dfrac{12}{7} = \dfrac{60}{35}$

10. $\dfrac{5}{8} = \dfrac{15}{24}$

11. $\dfrac{3}{14} = \dfrac{5}{23}$

12. $\dfrac{7}{9} = \dfrac{35}{50}$

13. $\dfrac{4}{30} = \dfrac{20}{150}$

14. $\dfrac{28}{12} = \dfrac{35}{15}$

15. $\dfrac{\frac{1}{3}}{8} = \dfrac{2}{48}$

16. $\dfrac{5}{\frac{1}{2}} = \dfrac{28}{3}$

17. $\dfrac{0.3}{8} = \dfrac{2}{50}$

18. $\dfrac{5}{0.2} = \dfrac{100}{4}$

10.3 Solving Proportions

OBJECTIVE
Solving a proportion for an unknown term

$\dfrac{?}{3} = \dfrac{10}{15}$ is a proportion in which the first term is unknown. Our work in this section will be learning how to find that unknown value.

In solving applied problems later in this chapter, we will be using proportions in which one of the four terms is *missing* or *unknown*. If three of the four terms of a proportion are known, you can always find the missing or unknown term.

Example 1

Remember that we call a letter representing an unknown value a *variable*. Here *a* is a variable. We could have chosen any other letter.

In the proportion $\dfrac{a}{3} = \dfrac{10}{15}$ the first term is unknown. We have chosen to represent the unknown value with the letter *a*.

Since the product of the means is equal to the product of the extremes, we can proceed as follows.

$$\dfrac{a}{3} = \dfrac{10}{15}$$

Section 10.3: Solving Proportions

Use the raised dot symbol (·) for multiplication rather than the cross symbol (×). This is so that the cross symbol won't be confused with the letter x.

The product of the extremes is $15 \cdot a$.
The product of the means is $3 \cdot 10$.

Since the products of the means and extremes must be equal, we have

$$15 \cdot a = 3 \cdot 10$$

The equals sign tells us that $15 \cdot a$ and $3 \cdot 10$ are just different names for the same number. This type of statement is called an equation.

> An *equation* is a statement that two expressions are equal.

CHECK YOURSELF 1

Write an equation from the proportion.

$$\frac{x}{2} = \frac{15}{6}$$

Let's go on with the example.

Example 1 (continued)

The proportion

$$\frac{a}{3} = \frac{10}{15}$$

has led to the equation

$$15 \cdot a = 3 \cdot 10$$

or

$$15 \cdot a = 30$$

One important property of an equation is that we can divide both sides by the same nonzero number. Here let's divide by 15.

$$15 \cdot a = 30$$

$$\frac{15 \cdot a}{15} = \frac{30}{15}$$

We will always divide by the number multiplying the variable. This is called the *coefficient*.

$$\frac{\cancel{15} \cdot a}{\cancel{15}} = \frac{\overset{2}{\cancel{30}}}{\underset{1}{\cancel{15}}}$$

Divide by the like factors.
Do you see why we divided by 15? It leaves our unknown a by itself in the left term.

$$a = 2$$

We have found a value of 2 for a, the missing term.

CHECK YOURSELF 2

The proportion $\frac{x}{2} = \frac{15}{6}$ has led to the equation $6 \cdot x = 2 \cdot 15$. Can you find the value for x?

You should always check your result. It is easy in this case. Returning to our earlier example, we found a value of 2 for a. Replace the unknown a with that value. Then cross-multiply to verify that the proportion is true.

Example 1 (continued)

We started with $\frac{a}{3} = \frac{10}{15}$ and found a value of 2 for a. So write

Replace a with 2 and cross-multiply.

$$3 \cdot 10 = 2 \cdot 15$$
$$30 = 30$$

The value of 2 for a checks.

CHECK YOURSELF 3

In the proportion $\frac{x}{2} = \frac{15}{6}$ we found a value of 5 for x. Check this result.

Let's summarize the steps for solving a proportion:

Section 10.3: Solving Proportions

> **TO SOLVE A PROPORTION**
>
> **STEP 1** Set the product of the means equal to the product of the extremes.
>
> **STEP 2** Divide both terms of the resulting equation by the coefficient of the variable.
>
> **STEP 3** Use the value found to replace the unknown in the original proportion. Cross-multiply to check that the proportion is true.

This gives us the unknown value.

Now check the result.

Example 2

Find the unknown value:

$$\frac{8}{x} = \frac{6}{9}$$

You are really using algebra to solve these proportions. In algebra we write the product $6 \cdot x$ as $6x$, omitting the dot. Multiplication of the number and the variable is understood and doesn't need to be written.

Step 1 Set the product of the means equal to the product of the extremes:

$$6 \cdot x = 8 \cdot 9$$

or $\quad 6x = 72$

Step 2 Locate the coefficient, 6, and divide both sides of the equation by that coefficient.

$$\frac{6x}{6} = \frac{72}{6}$$

$$x = 12$$

Step 3 To check: Replace x with 12 in the original proportion.

$$\frac{8}{12} = \frac{6}{9}$$

Cross-multiply:

$12 \cdot 6 = 8 \cdot 9$

$72 = 72$ The value of 12 for x checks.

CHECK YOURSELF 4

Solve the proportion for n. Check your result.

$$\frac{4}{5} = \frac{n}{25}$$

In solving for a missing term in a proportion we may find an equation involving fractions or decimals. Our final examples involve finding the unknown value in such cases.

Example 3

Solve the proportion for x.

$$\frac{\frac{1}{4}}{3} = \frac{4}{x}$$

$\frac{1}{4}x = 12$ The product of the extremes equals the product of the means.

$\dfrac{\frac{1}{4}x}{\frac{1}{4}} = \dfrac{12}{\frac{1}{4}}$ We divide by the coefficient of x. In this case it is $\frac{1}{4}$.

$x = \dfrac{12}{\frac{1}{4}}$ Remember: $\dfrac{12}{\frac{1}{4}}$ is $12 \div \frac{1}{4}$.

$x = 48$ Invert the divisor and multiply. We have

$$x = 12 \cdot \frac{4}{1} = 48$$

To check, replace x with 48 in the original proportion.

$$\frac{\frac{1}{4}}{3} = \frac{4}{48}$$

$$3 \cdot 4 = \frac{1}{4} \cdot 48$$

$$12 = 12$$

CHECK YOURSELF 5

Solve for a.

$$\frac{\frac{1}{2}}{5} = \frac{3}{a}$$

Example 4

$$\frac{0.5}{2} = \frac{3}{a}$$

$0.5a = 6$

Section 10.3: Solving Proportions

Divide by the coefficient, 0.5.

$$\frac{0.5a}{0.5} = \frac{6}{0.5}$$

$$a = 12$$

Here we must divide 6 by 0.5 to find the unknown value. The steps of that division are shown below for review.

```
        1 2.
0.5 )6.0
      5
      1 0
      1 0
          0
```

CHECK YOURSELF 6

Solve for x.

$$\frac{0.4}{x} = \frac{2}{30}$$

CHECK YOURSELF ANSWERS

1. $6 \cdot x = 2 \cdot 15$
2. $6 \cdot x = 2 \cdot 15$ or $6 \cdot x = 30$

$$\frac{\cancel{6} \cdot x}{\cancel{6}} = \frac{\overset{5}{\cancel{30}}}{\cancel{6}}$$

$$x = 5$$

3. Replace x with 5:

$$\frac{5}{2} = \frac{15}{6}$$

$$2 \cdot 15 = 5 \cdot 6$$

$$30 = 30$$

4. $5n = 100$ To check:
$$\frac{5n}{5} = \frac{100}{5}$$
$$n = 20$$

$$\frac{4}{5} = \frac{20}{25}$$
$$5 \cdot 20 = 4 \cdot 25$$

5. $a = 30$ 6. $x = 6$

10.3 Exercises

Solve for the unknown in each of the following proportions.

1. $\dfrac{x}{3} = \dfrac{6}{9}$ 2. $\dfrac{x}{8} = \dfrac{5}{20}$ 3. $\dfrac{5}{n} = \dfrac{15}{12}$

4. $\dfrac{4}{3} = \dfrac{8}{n}$ 5. $\dfrac{4}{7} = \dfrac{y}{14}$ 6. $\dfrac{6}{m} = \dfrac{18}{15}$

7. $\dfrac{3}{4} = \dfrac{a}{12}$ 8. $\dfrac{5}{7} = \dfrac{x}{35}$ 9. $\dfrac{8}{p} = \dfrac{6}{3}$

10. $\dfrac{4}{15} = \dfrac{8}{n}$

11. $\dfrac{8}{a} = \dfrac{5}{15}$

12. $\dfrac{5}{x} = \dfrac{15}{9}$

13. $\dfrac{35}{40} = \dfrac{7}{n}$

14. $\dfrac{a}{9} = \dfrac{10}{18}$

15. $\dfrac{a}{42} = \dfrac{5}{7}$

16. $\dfrac{7}{12} = \dfrac{m}{24}$

17. $\dfrac{18}{12} = \dfrac{12}{p}$

18. $\dfrac{x}{32} = \dfrac{7}{8}$

19. $\dfrac{x}{10} = \dfrac{40}{25}$

20. $\dfrac{20}{15} = \dfrac{100}{a}$

21. $\dfrac{6}{n} = \dfrac{75}{100}$

22. $\dfrac{36}{x} = \dfrac{8}{6}$

23. $\dfrac{3}{21} = \dfrac{a}{28}$

24. $\dfrac{20}{24} = \dfrac{p}{18}$

25. $\dfrac{12}{100} = \dfrac{3}{x}$

26. $\dfrac{b}{7} = \dfrac{21}{49}$

27. $\dfrac{p}{24} = \dfrac{25}{120}$

28. $\dfrac{6}{x} = \dfrac{15}{55}$

29. $\dfrac{\frac{1}{2}}{2} = \dfrac{3}{a}$

30. $\dfrac{x}{5} = \dfrac{2}{\frac{1}{3}}$

31. $\dfrac{\frac{1}{4}}{12} = \dfrac{m}{96}$

32. $\dfrac{10}{\frac{1}{5}} = \dfrac{150}{y}$

33. $\dfrac{\frac{3}{4}}{6} = \dfrac{3}{n}$

34. $\dfrac{4}{a} = \dfrac{\frac{2}{5}}{10}$

35. $\dfrac{0.2}{2} = \dfrac{1.2}{a}$

36. $\dfrac{n}{3} = \dfrac{6}{0.5}$

37. $\dfrac{p}{4} = \dfrac{8}{0.4}$

38. $\dfrac{y}{12} = \dfrac{5}{0.6}$

39. $\dfrac{x}{1.5} = \dfrac{2.5}{7.5}$

40. $\dfrac{0.5}{a} = \dfrac{1.25}{5}$

An important use of proportions is in solving problems involving *similar* geometric figures. These are figures that have the same shape and their corresponding sides are proportional.
  For instance in the similar triangles shown below,

 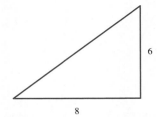

a proportion involving corresponding sides is

$$\frac{3}{4} = \frac{6}{8}$$

Use a proportion to find the unknown side, labeled *x*, in each of the following pairs of similar figures.

41.

42.

43.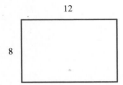

44.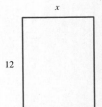

Skillscan (Section 10.1)
Write each ratio in simplest form.

a. 5 in to 2 ft

b. 18 min to 2 hours

c. 2 lb to 7 oz

d. 5 gal to 3 qt

e. 7 ft to 3 yd

f. 2 days to 10 hours

Answers

1. $9x = 18$; $\frac{9x}{9} = \frac{18}{9}$; $x = 2$ 3. 4   5. 8   7. 9   9. 4   11. 24   13. $35n = 280$; $\frac{35n}{35} = \frac{280}{35}$; $n = 8$
15. 30   17. $18p = 144$; $\frac{18p}{18} = \frac{144}{18}$; $p = 8$ 19. 16   21. 8   23. $21a = 84$; $\frac{21a}{21} = \frac{84}{21}$; $a = 4$ 25. 25
27. 5   29. 12   31. 2   33. 24   35. 12   37. 80   39. 0.5   41. 3   43. 6   a. $\frac{5}{24}$ b. $\frac{3}{20}$
c. $\frac{32}{7}$ d. $\frac{20}{3}$ e. $\frac{7}{9}$ f. $\frac{24}{5}$

10.3 Supplementary Exercises

Solve for the unknown in each of the following proportions.

1. $\frac{x}{4} = \frac{15}{20}$ 2. $\frac{3}{2} = \frac{n}{4}$ 3. $\frac{5}{n} = \frac{10}{16}$

4. $\frac{8}{10} = \frac{x}{5}$ 5. $\frac{3}{10} = \frac{9}{y}$ 6. $\frac{4}{24} = \frac{6}{p}$

7. $\frac{8}{s} = \frac{4}{9}$ 8. $\frac{5}{7} = \frac{x}{35}$ 9. $\frac{5}{12} = \frac{15}{t}$

10. $\frac{3}{s} = \frac{15}{25}$ 11. $\frac{7}{b} = \frac{28}{48}$ 12. $\frac{x}{22} = \frac{24}{33}$

13. $\frac{p}{25} = \frac{18}{15}$ 14. $\frac{45}{a} = \frac{27}{21}$ 15. $\frac{\frac{1}{2}}{5} = \frac{3}{x}$

16. $\frac{a}{6} = \frac{3}{\frac{1}{4}}$ 17. $\frac{\frac{2}{3}}{5} = \frac{6}{y}$ 18. $\frac{0.3}{6} = \frac{2}{x}$

19. $\frac{n}{5} = \frac{4}{0.2}$ 20. $\frac{3}{x} = \frac{7.5}{1.5}$

10.4 Using Proportions to Solve Word Problems

OBJECTIVE
To apply proportions to the solution of word problems

Now that you have learned how to find an unknown value in a proportion, let's see how this can be used in the solution of word problems.

> **APPLYING PROPORTIONS**
>
> STEP 1 Read the problem carefully to determine the given information.
>
> STEP 2 Write the proportion necessary to solve the problem. Use a letter to represent the unknown quantity. Be sure to include the units in writing the proportion.
>
> STEP 3 Solve, answer the question of the original problem, and check the proportion as before.

Example 1

If six tickets to a play cost $15, how much will eight tickets cost?

Solution Several proportions can be written for this solution. For example, since the ratio of the number of tickets to the cost must remain the same, we can write:

$$\frac{6 \text{ tickets}}{\$15} = \frac{8 \text{ tickets}}{\$x}$$

We have decided to let x be the cost (in dollars) of the eight tickets.

Cross-multiply as before.

$6x = 120$ You can drop the labels in solving the proportion.

$\dfrac{6x}{6} = \dfrac{120}{6}$ Divide by the coefficient, 6.

$x = 20$ (dollars)

The eight tickets will cost $20.

Note that each ratio must have the same units in the numerators and in the denominators (here tickets and dollars). This helps check that the proportion is written correctly.

CHECK YOURSELF 1

If the ratio of women to men in a class is $\dfrac{4 \text{ women}}{5 \text{ men}}$, how many women will there be in a class with 25 men?

Example 2

In a shipment of 400 parts, 14 are found to be defective. How many defective parts should be expected in a shipment of 1000?

Solution Assume that the ratio of defective parts to the total number remains the same.

$$\frac{14 \text{ defective}}{400 \text{ total}} = \frac{x \text{ defective}}{1000 \text{ total}}$$ We have decided to let x be the unknown number of defective parts.

Cross-multiply:

$400x = 14{,}000$

Divide by the coefficient, 400.

$x = 35$

35 defective parts should be expected in the shipment.

CHECK YOURSELF 2

An investment of $3000 earned $330 for 1 year. How much will an investment of $10,000 earn at the same rate for 1 year?

Let's look at an application involving fractions in the proportion.

Example 3

The scale on a map is given as $\frac{1}{4}$ in = 3 mi. The distance between two towns is 4 in on the map. How far apart are they in miles?

Solution For this solution we use the fact that the ratio of inches (on the map) to miles remains the same.

We solved this proportion in Section 10.3. Go back and check the details of the solution if you need to.

$$\frac{\frac{1}{4} \text{ in}}{3 \text{ mi}} = \frac{4 \text{ in}}{x \text{ mi}}$$

$x = 48$ (mi)

CHECK YOURSELF 3

Jack drives 125 mi in $2\frac{1}{2}$ hours. At the same rate, how far will he be able to travel in 4 hours?

(*Hint:* Write $2\frac{1}{2}$ as an improper fraction.)

We may also find decimals in the solution of a word problem.

Example 4

Jill works 4.2 hours and receives $21 pay. How much will she receive if she works 10 hours?

Solution The ratio of the number of hours worked to the amount of pay remains the same.

$$\frac{4.2 \text{ hours}}{\$21} = \frac{10 \text{ hours}}{\$a} \quad \text{Let } a \text{ be the unknown amount of pay.}$$

$$4.2a = 210$$

$$\frac{4.2a}{4.2} = \frac{210}{4.2} \quad \text{Divide both terms by 4.2.}$$

$$a = \$50$$

CHECK YOURSELF 4

A piece of cable 8.5 cm long weighs 68 grams. What will a 10-cm length of the same cable weigh?

Be careful that the ratios used in a proportion are comparing the *same units*. In the following example we must convert the units stated in the problem. Consider our final example.

Example 5

A machine can produce 15 tin cans in 2 min. At this rate how many cans can it make in an 8-hour time period?

Solution In writing a proportion for this problem, we must write the times involved in terms of the same units.

$$\frac{15 \text{ cans}}{2 \text{ min}} = \frac{x \text{ cans}}{480 \text{ min}} \quad \text{Since 1 hour is 60 min, convert 8 hours to 480 min.}$$

$$2x = 15 \cdot 480$$

or $\quad 2x = 7200$

$$x = 3600 \text{ cans}$$

CHECK YOURSELF 5

Instructions on a can of film developer call for 2 oz of concentrate to 1 qt of water. How much of the concentrate is needed to mix with 1 gal of water?

CHECK YOURSELF ANSWERS

1. The ratio of women to men remains the same, and so you can write

$$\frac{4 \text{ women}}{5 \text{ men}} = \frac{n \text{ women}}{25 \text{ men}}$$

You solved this proportion in the last section and found that n has the value of 20, the number of women.

2. $1100

3. $\dfrac{125 \text{ mi}}{\frac{5}{2} \text{ h}} = \dfrac{x \text{ mi}}{4 \text{ h}}$

$$\frac{5}{2} x = 500$$

Divide both members by $\dfrac{5}{2}$.

$x = 200$ (mi)

4. 80 grams
5. 8 oz

10.4 Exercises

1. If 12 books are purchased for $30, how much will you pay for 16 books at the same rate?

2. If an 8-ft two-by-four costs 96¢, what should a 12-ft two-by-four cost?

3. A box of 18 tea bags is marked 90¢. At that price, what should a box of 48 tea bags cost?

4. Cans of orange juice are marked 2 for 79¢. What should be the price of a case of 24 cans?

5. A worker can complete the assembly of 12 tape players in 5 hours. At this rate, how many can the worker complete in a 40-hour work week?

6. If 3 lb of apples cost 75¢, what will 8 lb cost?

7. The ratio of yes to no votes in an election was 3 to 2. How many no votes were cast if there were 2880 yes votes?

8. The ratio of men to women at a college is 7 to 5. How many women students are there if there are 3500 men?

9. A photograph that is 5 in wide by 6 in high is to be enlarged so that the new width is 15 in. What will be the height of the enlargement?

10. Jason's job is assembling lawn chairs. He can put together 45 chairs in 2 hours. At this rate, how many chairs can he assemble in an 8-h shift?

11. Christy can travel 110 mi in her new car on 5 gal of gas. How far can she travel on a full tank, which has 12 usable gal?

12. The Kelloggs purchased a $40,000 home and the property taxes were $800. If they make improvements and the house is now valued at $60,000, what will the new property tax be?

13. A car travels 165 mi in 3 hours. How far will it travel in 8 hours if it continues at the same speed?

14. If two cities on a map are 7 in apart, and the actual distance between the cities is 420 mi, find the distance between two cities that are 4 in apart on the map.

15. The ratio of teeth on a smaller gear to those on a larger gear is 3 to 7. If the smaller gear has 15 teeth, how many teeth does the larger gear have?

16. A store has T-shirts on sale at "2 for $4.90." At this rate, what will five shirts cost?

17. An inspection finds 30 defective parts in a shipment of 500. How many defective parts should be expected in a shipment of 1200?

18. You invest $2000 in a stock that pays a $240 dividend in 1 year. At the same rate, how much will you need to invest to earn $600?

19. A football back ran 212 yd in the first two games of the season. If he continues at the same pace, how many yards should he gain in the 11-game season?

20. A 6-lb roast will serve 14 people. What size roast is needed to serve 21 people?

21. A 2-lb box of grass seed is supposed to cover 1500 sq ft of lawn. How much seed will you need for 3750 sq ft of lawn?

22. A 6-ft fence post casts a 9-ft shadow. How tall is a nearby pole that casts a 15-ft shadow?

23. A 9-ft light pole casts a 15-ft shadow. Find the height of a nearby tree which is casting a 40-ft shadow.

24. On the blueprint of the Wilson's new home, the scale is 3 in equals 5 ft. What will be the actual length of the living room if it measures 12 in long on the blueprint?

25. The scale on a map uses $\frac{1}{2}$ in to represent 50 mi. If the distance between towns on the map is 6 in, how far apart are they in miles?

26. A metal bar expands $\frac{1}{4}$ in for each 12° rise in temperature. How much will it expand if the temperature rises 48°?

27. Your car burns $2\frac{1}{2}$ qt of oil on a trip of 3000 mi. How many quarts should you expect to use in driving 4800 mi?

28. A 6-ft man casts a $7\frac{1}{2}$-ft shadow. If the shadow of a nearby pole is 30 ft long, how tall is the pole?

29. A piece of tubing 10.5 cm long weighs 35 grams. What is the weight of a piece of the same tubing that is 15 cm long?

30. Jane works 6.25 hours and receives $25 pay. What will she receive at the same rate if she works 12 hours?

31. The sales tax on an item costing $80 is $3.60. What will the tax be for an item costing $150?

32. If 8 km is approximately 4.8 mi, how many kilometers will equal 12 mi?

33. You find that your watch gains 2 min in 8 hours. How much will it gain in 2 days?

34. If 2 qt of paint will cover 225 sq ft, how many square feet will 2 gal cover? (1 gal = 4 qt)

35. Directions on a box of 4 cups of wallpaper paste are to mix the contents with 5 qt of water. To mix a smaller batch using 1 cup of paste, how much water (in ounces) should be added? (1 qt = 32 oz)

36. A film processing machine can develop three rolls of film every 8 min. At this rate, how many rolls can be developed in a 4-hour period?

Answers

1. $40 3. $\dfrac{18 \text{ tea bags}}{90¢} = \dfrac{48 \text{ tea bags}}{x¢}$; $18x = 4320$; $x = 240¢$ or $2.40 5. 96 players 7. 1920 no votes
9. 18 in 11. $\dfrac{5 \text{ gal}}{110 \text{ mi}} = \dfrac{12 \text{ gal}}{x \text{ mi}}$; $5x = 1320$; $x = 264$ mi 13. 440 mi 15. 35 teeth
17. 72 defective parts 19. $\dfrac{212 \text{ yds}}{2 \text{ games}} = \dfrac{x \text{ yd}}{11 \text{ games}}$; $x = 1166$ yd 21. 5 lb 23. 24 ft 25. 600 mi
27. $\left(\text{Write } 2\dfrac{1}{2} \text{ as } \dfrac{5}{2}\right)$ $\dfrac{\frac{5}{2} \text{ qt}}{1250 \text{ mi}} = \dfrac{x \text{ qt}}{2000 \text{ mi}}$; $1250x = 5000$; $x = 4$ qt 29. 50 grams 31. $6.75
33. $\left(\text{Write 2 days as 48 hours}\right)$ $\dfrac{2 \text{ min}}{8 \text{ hours}} = \dfrac{x \text{ min}}{48 \text{ hours}}$; $x = 12$ min 35. 40 oz

10.4 Supplementary Exercises

1. A bread recipe calls for 3 cups of flour for 2 loaves of bread. How many cups of flour will be needed for 6 loaves of bread?

2. A painter uses 2 gal of paint in doing 3 rooms. At this rate, how many gallons will be needed to paint 15 similar rooms?

3. Cans of tomato soup are marked "2 for 59¢." How much will a case of 24 cans cost at the same rate?

4. Dick drives 256 mi using 8 gal of gasoline. If the gas tank holds 11 gal, how far can he travel on a full tank?

5. You want to enlarge a color print which is 3 in wide by 5 in long. If the new width is to be 9 in, find the length of the enlargement.

6. A baseball team won 14 of its first 20 games. At this rate, how many games should it win in the 160-game season?

7. In a sample of 250 parts taken from a shipment, 12 are found to be improperly assembled. At this rate, how many faulty parts can be expected in a shipment of 1500?

8. An investment of $5000 earns $600 in 1 year. What will an investment of $8000 earn at the same rate?

9. The shadow cast by a 30-ft telephone pole is 18 ft long. How tall is a nearby tree which is casting a 15-ft shadow?

10. The scale on a map is 2 in to 100 mi. How many miles is it between two cities that are 7 in apart on the map?

11. If 5 lb of fertilizer will cover 1000 sq ft, what amount will be needed to cover 1600 sq ft?

12. A plane can fly 480 mi in 3 hours. At this rate, how long will it take to travel 800 mi?

13. A medicine label calls for $\frac{1}{16}$ oz of medicine for 20 lb of body weight. If a patient weighs 160 lb, how much of the medicine should be prescribed?

14. A $3\frac{1}{2}$-lb roast costs $14. At the same price, what will a $5\frac{1}{2}$-lb roast cost?

15. If 3 mi is approximately the same distance as 4.8 km, how many miles equal 8 km?

16. George works 4.25 hours and receives $51 pay. What will he make if he works 8 hours?

17. Instructions on a bottle of concentrated wall cleaner call for 2 oz of cleaner to 1 qt of water. If you wish to use 2 gal of water, how much of the concentrated cleaner should you add?

18. A traffic counter registers 25 vehicles in 5 min. At the same rate, how many vehicles will pass the same point in 8 hours?

Using Your Calculator

When the numbers involved in a proportion are large, your calculator can be very useful. Consider the following.

Example 1

Solve the proportion for n.

$$\frac{n}{43} = \frac{105}{1505}$$

Start the solution as before.

$1505n = 43 \cdot 105$ Set the product of the means equal to the product of the extremes.

$\dfrac{\cancel{1505}n}{\cancel{1505}} = \dfrac{43 \cdot 105}{1505}$ Divide both terms by the coefficient, 1505.

To find n, using the calculator,

43 $\boxed{\times}$ 105 $\boxed{\div}$ 1505 $\boxed{=}$

Display (3) n has the value 3. To check, replace n with 3 and use your calculator to cross-multiply.

In practical applications you may have to round the result after using your calculator in the solution of a proportion. The following example shows such a situation.

Example 2

Micki drives 278 mi on 13.6 gal of gas. If the gas tank of her car holds 21 gal, how far can she travel on a full tank of gas?

Solution   We can write the proportion

$$\frac{278 \text{ mi}}{13.6 \text{ gal}} = \frac{x \text{ mi}}{21 \text{ gal}}$$

Cross-multiply.

$13.6x = 278 \cdot 21$

Now divide both terms by 13.6.

$$\frac{\cancel{13.6}x}{\cancel{13.6}} = \frac{278 \cdot 21}{13.6}$$

Now to find x we must multiply 278 by 21 and then divide by 13.6.

On the calculator,

278 ⊠ 21 ⊡ 13.6 ⊟ (429.26471)

Let's round the result to the nearest mile; Micki can drive 429 mi on a full tank of gas.

Unit Pricing

Your calculator can be very handy for comparing prices at the grocery store.

When items are sold in different-sized containers, it is often difficult to determine which is the better buy. Using your calculator to find your own unit prices may help you save some money.

> A unit price is a price *per* unit.

| A *unit price* is a ratio of price to some unit. |

> The unit used may be ounces, pints, pounds, or some other measure.

To find the unit price, just divide the cost of the item by the number of units.

Example 3

A dishwashing liquid comes in three sizes.

(a) 12 oz for 77¢
(b) 22 oz for $1.33
(c) 32 oz for $1.85

Which is the best buy?

Solution For each size, let's find the unit price in cents per ounce.

(a) $\dfrac{77¢}{12 \text{ oz}}$

Using your calculator, divide.

77 ⊡ 12 ⊟ (6.4166667)

$\dfrac{77¢}{12 \text{ oz}} \approx 6.4¢$ per ounce   We have chosen to round to the nearest tenth of a cent.

(b) To find the unit price for the second-size container, divide again.

> Note that we consider $1.33 as 133¢ to find the ratio "cents per ounce."

133 ⊡ 22 ⊟ (6.0454545)   Treat $1.33 as 133¢.

$\dfrac{\$1.33}{22 \text{ oz}} \approx 6¢$ per ounce   Again round to the nearest tenth of a cent.

(c) For the third-size container:

185 ÷ 32 = 5.78125

$$\frac{\$1.85}{32 \text{ oz}} \approx 5.8¢ \text{ per ounce}$$

Comparing the three unit prices, we see that the 32-oz size of dishwashing liquid is the best buy at 5.8¢ per ounce.
    All the ratios used must be in terms of the same units. If quantities involve different units, they must be converted.

Example 4

Vegetable oil is sold in the following quantities:

(a) 16 oz for $1.27
(b) 1 pt 8 oz for $1.79
(c) 1 qt 6 oz for $2.89

Which is the best buy?

(a) $\dfrac{\$1.27}{16 \text{ oz}} \approx 7.9¢$ per ounce

(b) Since 1 pt is 16 oz, 1 pt 8 oz is (16 + 8) oz, or 24 oz. So write 1 pt 8 oz as 24 oz.

$$\frac{\$1.79}{24 \text{ oz}} \approx 7.5¢ \text{ per ounce}$$

(c) Since 1 qt is 32 oz, 1 qt 6 oz is (32 + 6) oz, or 38 oz. Write 1 qt 6 oz as 38 oz.

$$\frac{\$2.89}{38 \text{ oz}} \approx 7.6¢ \text{ per ounce}$$

In this case, by comparing the unit prices we see that the 1-pt 8-oz size is the best buy.

Exercises Using Your Calculator

Solve for the unknowns.

1. $\dfrac{630}{1365} = \dfrac{15}{a}$

2. $\dfrac{770}{1988} = \dfrac{n}{71}$

3. $\dfrac{x}{5.8} = \dfrac{10.9}{7.3}$ (to the nearest tenth)

4. $\dfrac{12.8}{9.5} = \dfrac{n}{8.7}$ (to the nearest hundredth)

5. $\dfrac{2.7}{3.8} = \dfrac{5.9}{n}$ (to the nearest tenth)

6. $\dfrac{12.2}{0.042} = \dfrac{x}{0.08}$ (to the nearest hundredth)

7. Bill earns $198.45 for working 31.5 hours. How much will he receive if he works at the same pay rate for 34.25 hours?

8. Construction-grade lumber costs $384.50 per thousand board feet. What will be the cost of 686 board feet?

9. A speed of 88 ft/s is equal to a speed of 60 mi/h. If the speed of sound is 750 mi/h, what is the speed of sound in feet per second?

10. A shipment of 68 parts is inspected, and 4 are found to be faulty. At the same rate, how many defective parts should be found in a shipment of 139? Round your result to the nearest whole number.

11. The property tax on a $58,750 home is $1253. At the same rate, what will be the tax on a home valued at $75,350? Round your result to the nearest dollar.

12. A machine produces 158 items in 12 min. At the same rate, how many items will it produce in 8 hours?

13. Sally travels 407 mi using 12.8 gal of gas. How many gallons of gas will she need for a trip of 1300 mi? Give your answer to the nearest tenth of a gallon.

14. A 55-ft steel cable expands 1.85 inches in length for each 10° rise in temperature. If a similar cable is 37 ft long, how much will it expand (to the nearest hundredth of an inch) with a 10° temperature rise?

Find the best buy in each of the following problems.

15. A dishwashing liquid comes in the following size bottles.
    (*a*) 12 fl oz for 79¢
    (*b*) 22 fl oz for $1.29

16. Canned corn comes in the following quantities.
    (*a*) 10 oz for 21¢
    (*b*) 17 oz for 39¢

17. Syrup comes in the following size bottles:
    (*a*) 12 oz for 99¢
    (*b*) 24 oz for $1.59
    (*c*) 36 oz for $2.19

18. Shampoo is sold as follows:
    (*a*) 4 oz for $1.16
    (*b*) 7 oz for $1.52
    (*c*) 15 oz for $3.39

19. Salad oil is sold in the following size containers (1 qt is 32 oz):
    (*a*) 18 oz for 89¢
    (*b*) 1 qt for $1.39
    (*c*) 1 qt 16 oz for $2.19

20. Tomato juice is sold in cans with the following sizes (1 pt is 16 oz):
    (*a*) 8 oz for 37¢
    (*b*) 1 pt 10 oz for $1.19
    (*c*) 1 qt 14 oz for $1.99

21. Peanut butter comes in jars of the following sizes (1 lb is 16 oz):
    (*a*) 12 oz for $1.25
    (*b*) 18 oz for $1.72
    (*c*) 1 lb 12 oz for $2.59
    (*d*) 2 lb 8 oz for $3.76

22. Laundry detergent comes in the following size boxes:
    (*a*) 1 lb 2 oz for $1.99
    (*b*) 1 lb 12 oz for $2.89
    (*c*) 2 lb 8 oz for $4.19
    (*d*) 5 lb for $7.99

Answers
1. 32.5   3. 8.7   5. 8.3   7. $215.78   9. 1100 ft/s   11. $1607   13. 40.9 gal
15. The 22-oz size is the best buy.   17. The 36-oz size is the best buy.   19. The 1-qt size is the best buy.
21. The 1-lb 12-oz size is the best buy.

SELF-TEST for Chapter Ten

The purpose of the Self-Test is to help you check your progress and review for a chapter test in class. Allow yourself about an hour to take the test. When you are done, check your answers in the back of the book. If you missed any problems, be sure to go back and review the appropriate sections in the chapter and do the supplementary exercises provided there.

In Problems 1 to 5, write the ratios in lowest terms.

1. The ratio of 5 to 14.

2. The ratio of 25 to 10.

3. The ratio of 4 ft to 6 yd.

4. The ratio of 3 hours to 2 days.

5. A basketball team wins 23 of its 27 games during a season. What is the ratio of wins to games played? What is the ratio of wins to losses?

In Problems 6 and 7, list the means and extremes for the proportions.

6. $\dfrac{5}{12} = \dfrac{15}{36}$.

7. $\dfrac{5}{a} = \dfrac{15}{21}$

In Problems 8 to 11, which are true proportions?

8. $\dfrac{5}{10} = \dfrac{6}{12}$

9. $\dfrac{3}{8} = \dfrac{7}{18}$

10. $\dfrac{9}{10} = \dfrac{27}{30}$

11. $\dfrac{\frac{1}{2}}{5} = \dfrac{2}{18}$

In Problems 12 to 21, solve for the unknowns.

12. $\dfrac{a}{6} = \dfrac{5}{15}$

13. $\dfrac{7}{x} = \dfrac{21}{36}$

14. $\dfrac{10}{14} = \dfrac{n}{35}$

15. $\dfrac{8}{15} = \dfrac{24}{t}$

16. $\dfrac{45}{75} = \dfrac{12}{x}$

17. $\dfrac{a}{24} = \dfrac{45}{60}$

18. $\dfrac{\frac{1}{2}}{p} = \dfrac{5}{30}$

19. $\dfrac{\frac{3}{4}}{12} = \dfrac{3}{a}$

20. $\dfrac{x}{0.3} = \dfrac{60}{3}$

21. $\dfrac{2}{m} = \dfrac{0.6}{4.8}$

In Problems 22 to 30, solve each application using a proportion.

22. If ballpoint pens are marked 5 for 95¢, how much will a dozen cost?

23. A basketball player scores 230 points in her first 10 games. At the same rate, how many points will she score in the 28-game season?

24. Your new compact car travels 324 mi on 9 gal of gas. If the tank holds 16 usable gallons, how far can you drive on a tankful of gas?

25. The ratio of yes to no votes in an election was 5 to 4. How many no votes were cast if 3600 people voted yes?

26. Two towns that are 120 mi apart are 2 in apart on a map. What is the actual distance between two towns that are 7 in apart on the map?

27. You are using a photocopier to reduce an article which was 15 in long by 9 in wide. If the new width is 6 in, find the length of the reduced article.

28. If the scale of a drawing is $\frac{1}{4}$ in equals 1 ft, what actual length is represented by 3 in on the drawing?

29. An assembly line can install six car mufflers in 5 min. At this rate, how many mufflers can be installed in an 8-hour shift?

30. Instructions on a package of concentrated plant food call for 2 teaspoons to a quart of water. We have a large job and wish to use 3 gal of water. How much of the plant food concentrate should be added to the 3 gal of water?

PRETEST
for
Chapter Eleven

Percent

This pretest will point out any difficulties you may be having with percents.
Do all the problems. Then check your answers on the following page.

1. Write 9% as a fraction.

2. Write 37% as a decimal.

3. Write 0.045 as a percent.

4. Write $\frac{4}{5}$ as a percent.

5. What is 25% of 184?

6. What percent of 400 is 36?

7. 15% of a number is 75. What is the number?

8. What interest will you pay on a $2000 loan for 1 year if the interest rate is 12%?

9. A salesperson earns a $400 commission on sales of $8000. What is the commission rate?

10. A salary increase of 8% amounts to a $96 monthly raise. What was the monthly salary before the increase?

ANSWERS TO PRETEST

For help with similar problems, turn to the section indicated.

1. $\frac{9}{100}$ (Section 11.2);
2. 0.37 (Section 11.2);
3. $4\frac{1}{2}\%$ (Section 11.3);
4. 80% (Section 11.3);
5. 46 (Section 11.5);
6. 9% (Section 11.5);
7. 500 (Section 11.5);
8. $240 (Section 11.6);
9. 5% (Section 11.6);
10. $1200 (Section 11.6)

Chapter Eleven

Percent

11.1 The Meaning of Percent

OBJECTIVE
To use the percent notation

When we considered parts of a whole in earlier chapters, we used fractions and decimals. The idea of *percent* is another useful way of naming parts of a whole. We can think of percents as ratios whose denominators are 100.

Example 1

We can say:

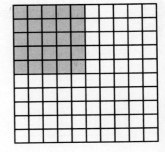

1. $\frac{1}{4}$ of the drawing is shaded, or
2. 0.25 of the drawing is shaded, or
3. 25 percent of the drawing is shaded.

Since
$\frac{1}{4} = \frac{25}{100} = 0.25$

25 percent means the ratio of 25 to 100.

The word *percent* means per hundredths. So $\frac{25}{100}$ is the same as 25 percent. We use the symbol % for percent. It means parts per hundred. So we write 25 percent as 25% and we read 25% as "twenty-five percent."

The symbol % comes from an arrangement of the digits one, zero, zero.

You will find the idea of percent used in a great variety of situations.

Example 2

(*a*) Four out of five geography students passed their midterm exams. Write this statement using the percent notation.

The ratio of students passing to students taking the class is $\frac{4}{5}$.

$$\frac{4}{5} = \frac{80}{100} = 80\%$$

To obtain a denominator of 100, multiply the numerator and denominator of the original fraction by 20.

So we can say that 80% of the geography students passed.

439

The ratio of compact cars to all cars is $\frac{35}{50}$

(b) Thirty-five of fifty automobiles sold by a dealer in 1 month were compact cars. Write this statement using the percent notation.

$$\frac{35}{50} = \frac{70}{100} = 70\%$$

We can say that 70% of the cars sold were compact cars.

CHECK YOURSELF 1

Rewrite the following statement using the percent notation: 4 of the 50 parts were defective.

CHECK YOURSELF ANSWER

1. 8% were defective.

11.1 Exercises

Use percents to name the shaded portion of each drawing.

1.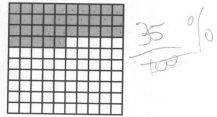

2.

3.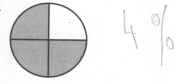

4.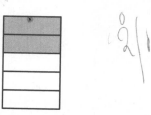

Rewrite each statement using the percent notation.

5. 41 out of every 100 eligible people voted in a recent election.

6. You receive $6 in interest for every $100 saved for 1 year.

Section 11.1: The Meaning of Percent

7. Out of every 100 entering students, 61 register for English composition.

8. 29 of 100 people surveyed watched a particular sports event on television.

9. 3 out of 10 voters in a state are registered as independents.

10. A dealer sold 7 of the 20 cars that were available during a 1-day sale.

11. 37 of the 50 houses in a development are sold.

12. 9 of the 25 employees of a company are part-time.

13. 19 out of 50 people surveyed prefer decaffeinated coffee.

14. 17 out of 20 college students work on part-time jobs.

15. 5 of the 20 students in an algebra class receive a grade of A.

16. 38 of the 50 families in a neighborhood have children attending public schools.

Skillscan (Sections 5.5 and 9.3)
Write each fraction in simplest form.

a. $\dfrac{6}{100}$
b. $\dfrac{8}{100}$
c. $\dfrac{35}{100}$
d. $\dfrac{64}{100}$

Divide.

e. $7 \div 100$
f. $15 \div 100$
g. $42 \div 100$
h. $120 \div 100$

Answers
Solutions for the even-numbered exercises are provided in the back of the book.
1. 35%   3. 75%   5. 41% of the eligible people voted.   7. 61% registered for English composition.
9. 30% are registered as independents.   11. 74% of the houses are sold.
13. 38% prefer decaffeinated coffee.   15. $\dfrac{5}{20} = \dfrac{25}{100} = 25\%$; 25% of the students received A's.

a. $\dfrac{3}{50}$ b. $\dfrac{2}{25}$ c. $\dfrac{7}{20}$ d. $\dfrac{16}{25}$ e. 0.07   f. 0.15   g. 0.42   h. 1.2

11.1 Supplementary Exercises

Use percents to name the shaded portions of each drawing.

1.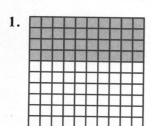

2.

Rewrite each statement using the percent notation.

3. 57 of 100 people surveyed prefer brand A of a product.

4. 38 of the 100 cars in a parking lot are imported cars.

5. 3 out of 10 students at a college are freshmen.

6. 9 out of 20 people working downtown ride a bus to work.

7. 4 out of the 50 parts in a shipment were defective.

8. 8 of the 25 students in a psychology class receive a grade of B.

11.2 Changing a Percent to a Fraction or a Decimal

OBJECTIVES
1. To change a percent to a common fraction or a mixed number
2. To change a percent to a decimal

Since there are different ways of naming the parts of a whole, you need to know how to change from one of these ways to another. First let's look at changing a percent to a fraction. Since a percent is a fraction or a ratio with denominator 100, we can use the following rule.

Section 11.2: Changing a Percent to a Fraction or a Decimal 443

> **CHANGING A PERCENT TO A FRACTION**
>
> To change a percent to a common fraction, remove the percent symbol and write the number over 100.

The use of this rule is shown in our first example.

Example 1

(a) $7\% = \dfrac{7}{100}$

You can choose to reduce $\dfrac{25}{100}$ to simplest form.

(b) $25\% = \dfrac{25}{100} = \dfrac{1}{4}$

CHECK YOURSELF 1

Write 12% as a fraction.

If a percent is *greater than 100*, the resulting fraction will be *greater than 1*. This is shown below.

Example 2

$150\% = \dfrac{150}{100} = 1\dfrac{50}{100} = 1\dfrac{1}{2}$

CHECK YOURSELF 2

Write 125% as a fraction.

The fractional equivalents of certain percents should be memorized.

$33\dfrac{1}{3}\% = \dfrac{1}{3}$ $66\dfrac{2}{3}\% = \dfrac{2}{3}$

To see how these values are formed, look at $33\dfrac{1}{3}\%$:

$$33\dfrac{1}{3}\% = \dfrac{33\dfrac{1}{3}}{100} = \dfrac{\dfrac{100}{3}}{\dfrac{100}{1}} \quad \text{Convert to improper fractions}$$

$$= \dfrac{100}{3} \times \dfrac{1}{100} = \dfrac{1}{3}$$

↑ Invert the divisor and multiply

Instead of working this hard, it's better just to try to remember these fractional equivalents.

Earlier in this section, we wrote percents as fractions by removing the percent sign and dividing by 100. How do we divide by 100 when we are working with decimals? Just move the decimal point two places to the left. This gives us a second rule for converting percents.

> **CHANGING A PERCENT TO A DECIMAL**
>
> To change a percent to a decimal, remove the percent symbol and move the decimal point *two* places to the *left*.

Example 3

$25\% = \dfrac{25}{100}$

$ = 0.25$

Note: A percent greater than 100 gives a decimal greater than 1.

$25\% = 0.25$

$8\% = 0.08$

$130\% = 1.30$

The decimal point is understood to be after the 5.
We must add a zero to move the decimal point.

CHECK YOURSELF 3

Write as decimals.

(1) 5%   (2) 32%   (3) 115%

 0.05   0.32   1.15

Look at the following examples involving fractions of a percent. In this case decimal fractions are involved.

Example 4

$4.5\% = 0.045$

$0.5\% = 0.005$

CHECK YOURSELF 4

Write as decimals

(1) 8.5%   (2) 0.3%

You will also find applications where common fractions are involved in a percent. Our final example illustrates.

Section 11.2: Changing a Percent to a Fraction or a Decimal 445

Example 5

Write the common fractions as decimals. Then remove the percent symbol by our earlier rule.

$9\frac{1}{2}\% = 9.5\% = 0.095$

$\frac{3}{4}\% = 0.75\% = 0.0075$

CHECK YOURSELF 5

Write as decimals.

(1) $7\frac{1}{2}\%$

(2) $\frac{1}{2}\%$

CHECK YOURSELF ANSWERS

1. $12\% = \frac{12}{100} = \frac{3}{25}$. 2. $1\frac{1}{4}$. 3. (1) 0.05; (2) 0.32; (3) 1.15.
4. (1) 0.085; (2) 0.003. 5. (1) 0.075; (2) 0.005.

11.2 Exercises

Write as fractions or mixed numbers.

1. 7%	**2.** 13%	**3.** 75%	**4.** 20%
5. 15%	**6.** 24%	**7.** 50%	**8.** 52%
9. 46%	**10.** 35%	**11.** 72%	**12.** 48%
13. 125%	**14.** 140%	**15.** $133\frac{1}{3}\%$	**16.** $166\frac{2}{3}\%$

Write as decimals.

17. 10%	**18.** 70%	**19.** 25%	**20.** 75%
21. 39%	**22.** 23%	**23.** 3%	**24.** 7%

25. 115%	26. 250%	27. 240%	28. 110%
29. 18.5%	30. 10.5%	31. 7.5%	32. 3.5%
33. 0.4%	34. 0.5%	35. $5\frac{1}{2}$%	36. $8\frac{1}{4}$%

Skillscan (Sections 8.5 and 9.4)
Multiply.

a. 0.05×100 b. 0.15×100 c. 0.45×100 d. 1.40×100

Find the decimal equivalents for each of the following.

e. $\frac{2}{5}$ f. $\frac{3}{4}$ g. $1\frac{1}{2}$ h. $2\frac{4}{5}$

Answers

1. $\frac{7}{100}$ 3. $\frac{3}{4}$ 5. $\frac{3}{20}$ 7. $\frac{1}{2}$ 9. $\frac{23}{50}$ 11. $\frac{18}{25}$ 13. $1\frac{1}{4}$ 15. $1\frac{1}{3}$ 17. 0.1 19. 0.25
21. 0.39 23. 0.03 25. 1.15 27. 2.4 29. 0.185 31. 0.075 33. 0.004 35. 0.055
a. 5 b. 15 c. 45 d. 140 e. 0.4 f. 0.75 g. 1.5 h. 2.8

11.2 Supplementary Exercises

Write as fractions or as mixed numbers.

1. 9%   2. 17%   3. 30%   4. 45%

5. 36%   6. 56%   7. 32%   8. 175%

Write as decimals.

9. 40%   10. 53%   11. 85%   12. 2%

13. 120%   14. 350%   15. 12.5%   16. 4.5%

17. 0.1%   18. $3\frac{1}{4}$%

11.3 Changing a Decimal or a Fraction to a Percent

OBJECTIVES
1. To change a decimal to a percent
2. To change a fraction to a percent

As you might expect, changing a decimal to a percent is the opposite of changing from a percent to a decimal. So we reverse the process of Section 11.2. Here is the rule that will be needed:

> **CHANGING A DECIMAL TO A PERCENT**
>
> To change a decimal to a percent, move the decimal point *two* places to the *right* and attach the percent symbol.

In changing 0.18 to a percent, we move the decimal point two places to the right. The decimal point is no longer necessary. It is understood to be after the 8.

Example 1

0.18 = 18%

CHECK YOURSELF 1

Write 0.27 as a percent.   27%

Example 2

0.03 = 3%

CHECK YOURSELF 2

Write 0.05 as a percent.   5%

Let's look at an example of converting a decimal, greater than 1, to a percent.

Example 3

1.25 = 125%

A decimal greater than 1 always gives a percent greater than 100.

CHECK YOURSELF 3

Write 1.3 as a percent.   130%

If the percent still includes a decimal after the decimal point is moved two places to the right, the fractional portion can be written as a decimal or as a fraction.

Example 4

(a) $0.045 = 4.5\%$ or $4\frac{1}{2}\%$

(b) $0.003 = 0.3\%$ or $\frac{3}{10}\%$

CHECK YOURSELF 4

Write 0.075 as a percent.

The following rule will allow us to change fractions to percents.

You may want to review Section 9.4 on writing decimal equivalents.

CHANGING A FRACTION TO A PERCENT

To change a fraction to a percent, write the decimal equivalent of the fraction. Then use the previous rule to change the decimal to a percent.

Example 5

Write $\frac{3}{5}$ as a percent.

Solution First write the decimal equivalent.

$\frac{3}{5} = 0.60$ To find the decimal equivalent, just divide the denominator into the numerator.

Now write the percent.

Now move the decimal point two places to the right and attach the percent symbol.

$\frac{3}{5} = 0.60 = 60\%$

CHECK YOURSELF 5

Write $\frac{3}{4}$ as a percent.

Section 11.3: Changing a Decimal or a Fraction to a Percent 449

Again, you will find both decimals and fractions used in writing percents. Consider the following.

Example 6

$$\frac{1}{8} = 0.125 = 12.5\% \text{ or } 12\frac{1}{2}\%$$

CHECK YOURSELF 6

Write $\frac{3}{8}$ as a percent.

To write a mixed number as a percent, we use exactly the same steps.

Example 7

Note that the resulting percent must be greater than 100 because the original mixed number was greater than 1.

$$1\frac{1}{4} = 1.25 = 125\%$$

CHECK YOURSELF 7

Write $1\frac{2}{5}$ as a percent.

Recall that some fractions have repeating decimal equivalents. In writing these as percents, we will either round to some indicated place or use a fractional remainder form.

Example 8

Here we use the fractional remainder in writing the decimal equivalent.

$$\frac{1}{3} = 0.33\frac{1}{3} = 33\frac{1}{3}\%$$

CHECK YOURSELF 8

Write $\frac{2}{3}$ as a percent.

Example 9

In this case, we round the decimal equivalent. Then write the percent.

$\frac{5}{7} = 0.714$ (rounded to the nearest thousandth)

$= 71.4\%$ (to the nearest tenth of a percent)

CHECK YOURSELF 9

Write $\frac{2}{9}$ to the nearest tenth of a percent.

CHECK YOURSELF ANSWERS

1. 27%.   2. 0.05 = 5%.   3. 130%.   4. 7.5% or $7\frac{1}{2}\%$.
5. $\frac{3}{4} = 0.75 = 75\%$.   6. 37.5% or $37\frac{1}{2}\%$.   7. $1\frac{2}{5} = 1.4 = 140\%$.
8. $\frac{2}{3} = 66\frac{2}{3}\%$.   9. 22.2%.

11.3 Exercises

Write each decimal as a percent.

1. 0.07
2. 0.09
3. 0.03
4. 0.05

5. 0.18
6. 0.76
7. 0.95
8. 0.45

9. 0.2
10. 0.3
11. 0.7
12. 0.8

13. 1.10
14. 1.50
15. 2.20
16. 3.75

17. 0.065
18. 0.095
19. 0.025
20. 0.075

21. 0.002
22. 0.006
23. 0.004
24. 0.001

Write each fraction as a percent.

25. $\frac{1}{4}$
26. $\frac{3}{4}$
27. $\frac{2}{5}$
28. $\frac{1}{2}$

29. $\frac{1}{5}$
30. $\frac{4}{5}$
31. $\frac{3}{8}$
32. $\frac{7}{8}$

33. $\frac{5}{8}$

34. $1\frac{1}{5}$

35. $2\frac{1}{2}$

36. $\frac{2}{3}$

37. $\frac{1}{6}$

38. $\frac{3}{16}$

39. $\frac{5}{9}$ (to the nearest tenth of a percent)

40. $\frac{3}{7}$ (to the nearest tenth of a percent)

Answers

1. 7%   3. 3%   5. 18%   7. 95%   9. 20%   11. 70%   13. 110%   15. 220%   17. 6.5% or $6\frac{1}{2}$%
19. 2.5% or $2\frac{1}{2}$%   21. 0.2% or $\frac{1}{5}$%   23. 0.4% or $\frac{2}{5}$%   25. $\frac{1}{4} = 0.25 = 25\%$ 27. 40%   29. 20%
31. $\frac{3}{8} = 0.375 = 37.5\%$ or $37\frac{1}{2}$%   33. 62.5% or $62\frac{1}{2}$%   35. 250%   37. $16\frac{2}{3}$%   39. 55.6%

11.3 Supplementary Exercises

Write as percents:

1. 0.06
2. 0.08
3. 0.24
4. 0.72

5. 0.5
6. 0.6
7. 1.25
8. 2.65

9. 0.085
10. 0.055
11. 0.008
12. 0.005

13. $\frac{3}{4}$
14. $\frac{3}{5}$
15. $\frac{4}{5}$
16. $\frac{1}{8}$

17. $\frac{7}{16}$
18. $1\frac{1}{4}$
19. $\frac{5}{6}$

20. $\frac{4}{9}$ (to the nearest tenth of a percent)

11.4 Identifying the Rate, Base, and Amount

OBJECTIVE
To identify the rate, the base, and the amount in an application

There are many practical applications of our work with percents. All of these problems have three basic parts which need to be identified. Let's look at some definitions which will help with that process.

> The *base* is the whole in a problem. It is the standard used for comparison.
>
> The *amount* is the part of the whole being compared to the base.
>
> The *rate* is the ratio of the amount to the base. It is written as a percent.

Let's look at some examples of determining the parts of a percent problem.

Example 1

The *rate* (which we will label $R\%$) is the easiest of the terms to identify. The rate is written with the percent symbol (%) or the word "percent."

What is **15%** of 200?
   ↑
  $R\%$ Here 15% is the rate because it has the percent symbol attached.

25% of what number is 50?
 ↑
$R\%$ 25% is the rate.

20 is **what percent** of 40?
        ↑
       $R\%$ Here the rate is unknown.

CHECK YOURSELF 1

Identify the rate.

(1) 15% of what number is 75?
(2) What is 8.5% of 200?
(3) 200 is what percent of 500?

The *base* (labeled B) is the whole, or 100%, in the problem. The base will often follow the word "of." Look at our next example.

Example 2

What is 15% of 200?   200 is the base. It follows the word "of."
↑
B

25% of what number is 50?   Here the base is the unknown.
↑
B

20 is what percent of 40?   40 is the base.
↑
B

CHECK YOURSELF 2

Identify the base.

(1) 70 is what percent of 350?
(2) What is 25% of 300?
(3) 14% of what number is 280?

The *amount* (labeled A) will be the part of the problem remaining once the rate and the base have been identified.

In many applications, the amount is found with the word "is."

Example 3

What is 15% of 200?   Here the amount is the unknown part of the problem. Note that the word "is" follows.
↑
A

25% of what number is 50?   Here the amount, 50, follows the word "is."
↑
A

20 is what percent of 40?
↑
A   Again the amount, here 20, can be found with the word "is."

CHECK YOURSELF 3

Identify the amount.

(1) 30 is what percent of 600?
(2) What is 12% of 5000?
(3) 24% of what number is 96?

Let's look at another example of identifying the three parts in a percent problem.

Example 4

Determine the rate, base, and amount in the problem:

12% of 800 is what number?

Solution Finding the *rate* is not difficult. Just look for the percent symbol or the word "percent." 12% is the rate in this problem.

The *base* is the whole. Here it follows the word "of." 800 is the whole or the base.

The *amount* remains when the rate and the base have been found. Here the amount is the unknown. It follows the word "is." "What number" asks for the unknown amount.

CHECK YOURSELF 4

Find the rate, the base, and the amount in the following statements or questions.

(1) 75 is 25% of 300.   (2) 20% of what number is 50?

We will use percents to solve a variety of applied problems. In all these situations you will have to identify the three parts of the problem. Let's work through some examples intended to help you build that skill.

Example 5

Determine the rate, base, and amount in the following application.

In an algebra class of 35 students, 7 received a grade of A. What percent of the class received a grade of A?

Solution The *base* is the whole in the problem, or the number of students in the class. 35 is the base.

The *amount* is the portion of the base, here the number of students that receive the A grade. 7 is the amount.

The *rate* is the unknown in this example. "What percent" asks for the unknown rate.

CHECK YOURSELF 5

Determine the rate, base, and amount in the following application: In a shipment of 150 parts, 9 of the parts were defective. What percent were defective?

Example 6

Determine the rate, base, and amount in the following application:

Doyle borrows $2000 for 1 year. If the interest rate is 12%, how much interest will he pay?

Solution The base is again the whole, the size of the loan in this example. $2000 is the base.
 The rate is of course the interest rate. 12% is the rate.
 The amount is the quantity left once the base and rate have been identified. Here the amount is the amount of interest that Doyle must pay. The amount is the unknown in this example.

CHECK YOURSELF 6

Determine the rate, base, and amount in the following application: Robert earned $120 interest from a savings account paying 8% interest. What amount did he have invested?

CHECK YOURSELF ANSWERS

1. (1) 15%; (2) 8.5%; (3) what percent (the unknown).
2. (1) 350; (2) 300; (3) what number (the unknown).
3. (1) 30; (2) "what is" (the unknown); (3) 96.
4. (1) $R\% = 25\%$; $B = 300$; $A = 75$. (2) $R\% = 20\%$; $B = $ "what number"; $A = 50$.
5. $B = 150$; $A = 9$; $R\% = $ "what percent" (the unknown).
6. $R\% = 8\%$; $A = \$120$; $B = $ "what amount" (the unknown).

11.4 Exercises

Identify the rate, the base, and the amount in each statement or question. *Do not solve* the problem at this point.

1. 15% of 200 is 30.

2. 150 is 20% of 750.

3. 40% of 600 is 240.

4. 90 is 30% of 300.

5. What is 8% of 150?

6. 80 is what percent of 400?

7. 12% of what number is 48?

8. What percent of 150 is 30?

9. 200 is 40% of what number?

10. What is 60% of 250?

11. What percent of 80 is 20?

12. 150 is 75% of what number?

13. Jan has a 3% commission rate on all her sales. If she sells $60,000 worth of merchandise in 1 month, what commission will she earn?

14. 22% of Shirley's monthly salary is deducted for withholding. If those deductions total $209, what is her salary?

15. In a chemistry class of 50 students, 8 received a grade of A. What percent of the students received A's?

16. A can of mixed nuts contains 80% peanuts. If the can holds 16 oz, how many ounces of peanuts does it contain?

17. The sales tax rate in a state is 5.5%. If you pay a tax of $3.30 on an item that you purchase, what was its selling price?

18. In a shipment of 500 parts, 20 were found to be defective. What percent of the parts were faulty?

19. A college had 8000 students at the start of a school year. If there is an enrollment increase of 4% by the beginning of the next year, how many additional students were there?

20. Paul invested $5000 in a time deposit. What interest will he earn for 1 year if the interest rate is 8.5%?

Skillscan (Section 10.3)
Solve each of the following proportions.

a. $\dfrac{x}{150} = \dfrac{20}{100}$

b. $\dfrac{36}{y} = \dfrac{30}{100}$

c. $\dfrac{45}{180} = \dfrac{m}{100}$

d. $\dfrac{21}{p} = \dfrac{35}{100}$

e. $\dfrac{a}{80} = \dfrac{45}{100}$

f. $\dfrac{150}{120} = \dfrac{r}{100}$

Answers

1. 15% of 200 is 30. (R%, B, A)
3. 40% of 600 is 240. (R%, B, A)
5. What is 8% of 150? (A, R%, B)
7. 12% of what number is 48? (R%, B, A)
9. 200 is 40% of what number? (A, R%, B)
11. What percent of 80 is 20? (R%, B, A)
13. $60,000 is the base. 3% is the rate. Her commission, the unknown, is the amount.

15. 50 is the base. 8 is the amount. The unknown percent is the rate.
17. 5.5% is the rate. The tax, $3.30, is the amount. The unknown selling price is the base.
19. The base is 8000. The rate is 4%. The unknown number of additional students is the amount.
a. 30 **b.** 120 **c.** 25 **d.** 60 **e.** 36 **f.** 125

11.4 Supplementary Exercises

Identify the rate, the base, and the amount in each statement or question. *Do not solve* the problem at this point.

1. 80 is 20% of 400.

2. 45% of 200 is 90.

3. What is 9% of 300?

4. 70 is 14% of what number?

5. 1200 is what percent of 2000?

6. What is 150% of 1200?

7. On a test, Alice had 80% of the problems right. If she did 20 problems correctly, how many questions were on the test?

8. In a shipment of 250 parts, 40 are found to be defective. What percent of the parts are faulty?

9. There are 88 grams of acid in 800 grams of a solution. What percent of the solution is acid?

10. 82% of the students in a psychology class passed their midterm examination. If 123 students passed, how many students were in the class?

11.5 The Three Types of Percent Problems

OBJECTIVES
1. To find the amount in a percent problem
2. To find the rate in a percent problem
3. To find the base in a percent problem

From your work in Section 11.4, you may have observed that there are three basic types of percent problems. These depend on which of the three parts—the amount, the rate, or the base—are missing in the problem statement. The solution for each type of problem depends on the *percent relationship*

> Amount = Rate × Base
>
> or
>
> Rate × Base = Amount

We will illustrate the solution of each type of problem in the following examples. Let's start with a problem in which we want to find the amount.

Example 1

What is 18% of 300?

This is a type 1 problem. We know the rate, 18%, the base, 300, and the amount is the unknown.
    Using the percent relationship we can translate the problem to an equation.

Amount = 0.18 × 300
        = 54

(Rate = 0.18, Base = 300)

Write 18% as the decimal 0.18 by the rule of Section 11.2. Then multiply to find the amount.

So 54 is 18% of 300.

CHECK YOURSELF 1

Find 65% of 200.

Note

1. If the rate is *less than* 100%, the amount will be *less than* the base.

   25 is 50% of 50 and 25 < 50

2. If the rate is *greater than* 100%, the amount will be *greater than* the base.

   75 is 150% of 50 and 75 > 50

Let's proceed to consider the second type of percent problem.

Section 11.5: The Three Types of Percent Problems

Example 2

30 is what percent of 150?

This is a type 2 problem. We know the amount, 30, the base, 150, and the rate (what percent) is the unknown.
  Again using the percent relationship to translate to an equation we have

Rate × $\overset{\text{Base}}{150}$ = $\overset{\text{Amount}}{30}$

This will leave the rate alone *on the left.*

Now we must *divide* both sides by 150 to find the rate.

$$\text{Rate} = \frac{30}{150} = \frac{1}{5} = 0.20$$

Remember that 0.20 = 20% by the rule of Section 11.3.

Converting 0.20 to a percent we see that

30 is 20% of 150

CHECK YOURSELF 2

75 is what percent of 300?

Note

1. If the amount is *less than* the base, the rate will be *less than* 100%.
2. If the amount is *greater than* the base, the rate will be greater than 100%.

Let's look at the final type of percent problem in our next example.

Example 3

28 is 40% of what number?

This is a type 3 problem. We know the amount, 28, the rate, 40%. The base (what number) is the unknown.
  From the percent relationship we have

Note that 40% is written as 0.40.

$\overset{\text{Rate}}{0.40}$ × Base = $\overset{\text{Amount}}{28}$

Here we must *divide* both sides by 0.40 to find the base.

$$\text{Base} = \frac{28}{0.40} = 70$$

So 28 is 40% of 70.

CHECK YOURSELF 3

70 is 35% of what number?

We have now seen solution methods for the three basic types of percent problems: finding the amount, finding the rate, and finding the base. As you will see in the remainder of this section, our work in Chapter 10, with proportions will allow us to solve each type of problem in an identical fashion. In fact, many students find percent problems easier to approach with the proportion method.

First, we will write what is called the *percent proportion*.

$$\frac{\text{Amount}}{\text{Base}} = \frac{R}{100}$$

In symbols,

On the right, $\frac{R}{100}$ is the rate, and this proportion is equivalent to our earlier percent relationship.

$$\frac{A}{B} = \frac{R}{100}$$

Since in any percent problem we know two of the three quantities (A, B, or R), we can always solve for the unknown term. Consider our first example of the use of the percent proportion.

Example 4

This is a *type 1 problem*. We know the rate and the base, and we want to find the amount.

_____ is 30% of 150.
  ↑   ↑   ↑
  A R B

Substitute the values into the percent proportion.

A, the amount, is the unknown term of the proportion.

$$\frac{A}{150} = \frac{30}{100}$$

where $B \to 150$ and $R \to 30$.

We solve the proportion with the methods of Section 10.3.

$100A = 150 \cdot 30$

Divide by the coefficient, 100.

$$\frac{\cancel{100}A}{\cancel{100}} = \frac{4500}{100}$$

$$A = 45$$

The amount is 45. This means that 45 is 30% of 150.

Section 11.5: The Three Types of Percent Problems   461

CHECK YOURSELF 4

Use the percent proportion to answer the question, What is 24% of 300?

The same percent proportion will work if you want to find the rate. Look at our next example.

Example 5

This is a type 2 problem. We know the base and the amount, and we want to find the rate.

_____% of 400 is 72.

Substitute the known values into the percent proportion.

$$\underset{B}{\overset{A}{}}\,\frac{72}{400} = \frac{R}{100}$$

R, the rate, is the unknown term in this case.

Solving,

$$400R = 7200$$

$$\frac{400R}{400} = \frac{7200}{400}$$

$$R = 18$$

The rate is 18%. So 18% of 400 is 72.

CHECK YOURSELF 5

Use the percent proportion to answer the question, What percent of 50 is 12.5?

Finally, we use the same proportion to find an unknown base.

Example 6

This is a type 3 problem. We know the rate and the amount, and we want to find the base.

40% of _____ is 200.

Substitute the known values into the percent proportion.

$$\overset{A}{}\,\frac{200}{B} = \frac{40}{100}\,\overset{R}{}$$

In this case B, the base, is the missing term of the proportion.

Solving,

$$40B = 20{,}000$$

$$\frac{40B}{40} = \frac{20{,}000}{40}$$

$$B = 500$$

The base is 500 and 40% of 500 is 200.

CHECK YOURSELF 6

Use the percent proportion to answer the following question: 288 is 60% of what number?

Remember that a percent (the rate) can be greater than 100. Let's see what happens in that case.

Example 7

The rate is 125%. The base is 300.

What is 125% of 300?

In the percent proportion we have

When the rate is greater than 100%, the amount will be greater than the base.

$$\frac{A}{300} = \frac{125}{100}$$

So $100A = 300 \cdot 125$.
 Dividing by 100 yields

$$A = \frac{37{,}500}{100} = 375$$

So 375 is 125% of 300.

CHECK YOURSELF 7

Find 150% of 500.

In the following example we want to find a rate where the amount is greater than the base.

Example 8

92 is what percent of 80?

In the percent proportion we have

$$\frac{92}{80} = \frac{R}{100}$$

So $80R = 92 \cdot 100$.
   Now solving for R we divide by 80

$$R = \frac{9200}{80} = 115$$

So 92 is 115% of 80.

The amount is 92, the base is 80.

CHECK YOURSELF 8
What percent of 120 is 156?

We will conclude this section by looking at two examples of solving percent problems involving fractions of a percent.

Example 9

34 is 8.5% of what number?

Using the percent proportion yields

$$\frac{34}{B} = \frac{8.5}{100}$$

The amount is 34, the rate is 8.5%. We want to find the base.

Solving we have

$$8.5B = 34 \cdot 100$$

or

$$B = \frac{3400}{8.5} = 400$$

Divide by 8.5.

So 34 is 8.5% of 400.

CHECK YOURSELF 9
12.5% of what number is 75?

Example 10

40 is what percent of 120?

With the percent proportion we have

Here we know the amount, 40, and the base, 120.

$$\frac{40}{120} = \frac{R}{100}$$

or

$$120R = 40 \cdot 100$$

Now solving for R:

We have chosen to solve for R by dividing numerator and denominator by 40. We then write the result as a mixed number.

$$R = \frac{4000}{120} = \frac{100}{3} = 33\frac{1}{3}$$

and 40 is $33\frac{1}{3}\%$ of 120.

CHECK YOURSELF 10

80 is what percent of 120?

CHECK YOURSELF ANSWERS

1. 130.   2. 25%.   3. 200.
4. $\frac{A}{300} = \frac{24}{100}$ (A from numerator, R from 24)
   $100A = 7200; A = 72.$
5. $\frac{12.5}{50} = \frac{R}{100}$ (A from 12.5, B from 50)
   $50R = 1250; R = 25\ (\%).$
6. $\frac{288}{B} = \frac{60}{100}$ (A from 288, R from 60)
   $60B = 28,800; B = 480.$
7. 750.   8. 130%.   9. 600.   10. $66\frac{2}{3}\%.$

11.5 Exercises

Solve each of the following problems involving percent.

1. What is 25% of 500?

2. 20% of 400 is what number?

3. 45% of 200 is what number?

4. What is 60% of 800?

Section 11.5: The Three Types of Percent Problems

5. Find 30% of 1500.

6. What is 75% of 120?

7. What percent of 50 is 4?

8. 45 is what percent of 750?

9. What percent of 300 is 27?

10. 14 is what percent of 200?

11. What percent of 200 is 340?

12. 210 is what percent of 1500?

13. 24 is 8% of what number?

14. 7% of what number is 42?

15. Find the base if 11% of the base is 55.

16. 16% of what number is 128?

17. 58.5 is 13% of what number?

18. 21% of what number is 73.5?

19. Find 130% of 500.

20. What is 125% of 400?

21. What is 108% of 4000?

22. Find 160% of 2000.

23. 180 is what percent of 150?

24. What percent of 40 is 52?

25. 360 is what percent of 200?

26. What percent of 5000 is 5300?

27. 625 is 125% of what number?

28. 140% of what number is 350?

29. Find the base if 120% of the base is 900.

30. 130% of what number is 1170?

31. Find 8.5% of 300.

32. $7\frac{1}{2}$% of 600 is what number?

33. Find $12\frac{1}{2}$% of 2000.

34. What is 3.5% of 500?

35. What is 5.25% of 3000?

36. What is 8.25% of 6000?

37. 34 is what percent of 400?

38. 500 is what percent of 1500?

39. What percent of 180 is 120?

40. What percent of 400 is 39?

41. What percent of 2400 is 1500?

42. 68 is what percent of 800?

43. 10.5% of what number is 315?

44. Find the base if $11\frac{1}{2}$% of the base is 46.

45. 58.5 is 13% of what number?

46. 7.5% of what number is 150?

47. 285 is 9.5% of what number?

48. 21% of what number is 73.5?

Estimation is a very useful skill when working with percents. For example, to see whether an amount seems reasonable, one approach is to round the rate to a "convenient" value and then use that rounded rate to estimate the amount. As an example, to find 19.3% of 500, round the rate to 20% $\left(\text{as a fraction}, \frac{1}{5}\right)$. An estimate of the amount is then

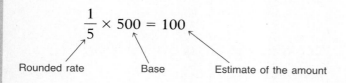

Rounded rate   Base   Estimate of the amount

Use this approach to find an estimate of the amount in each of the following problems.

49. Find 25.8% of 4000.

50. What is 48.3% of 1500?

51. 74.7% of 600 is what number?

52. 9.8% of 1200 is what number?

53. Find 152% of 400.

54. What is 118% of 5000?

Answers

1. 125 **3.** 90 **5.** 450 **7.** 8% **9.** 9% **11.** 170% **13.** 300 **15.** 500 **17.** 450 **19.** 650
21. 4320 **23.** 120% **25.** 180% **27.** 500 **29.** 750 **31.** 25.5 **33.** 250 **35.** 157.5 **37.** 8.5%
39. $66\frac{2}{3}$% **41.** 62.5% **43.** 3000 **45.** 450 **47.** 3000 **49.** 1000 **51.** 450 **53.** 600

11.5 Supplementary Exercises

Solve each of the following problems involving percent.

1. What is 20% of 400?

2. Find 15% of 300.

3. 45% of 2000 is what number?

4. What percent of 800 is 48?

5. 60 is what percent of 500?

6. What percent of 800 is 120?

7. 9% of what number is 27?

8. Find the base if 12% of the base is 96.

9. 59.5 is 17% of what number?

10. What is 160% of 800?

11. Find 108% of 2000.

12. What percent of 480 is 600?

13. 960 is what percent of 800?

14. Find the base if 125% of the base is 750.

15. 1560 is 130% of what number?

16. Find 7.5% of 2000.

17. What is 6.25% of 500?

18. Find $8\frac{1}{2}$% of 600.

19. 225 is what percent of 600?

20. 1050 is what percent of 1200?

21. What percent of 1500 is 1000?

22. 102 is 8.5% of what number?

23. Find the base if 10.5% of the base is 42.

24. 148.5 is 27% of what number?

11.6 Applications of Percent

OBJECTIVES
1. To solve percent applications for an amount
2. To solve percent applications for a rate
3. To solve percent applications for a base

The concept of percent is perhaps the most frequently encountered arithmetic idea that we will consider in this text. In this section we will show some of the many applications of percent and the special terms that are used in these applications.

A *rate*, *base*, and *amount* will appear in *all* problems involving percents.

In order to use percents in problem solving, you should always read the problem carefully to determine the rate, the base, and the amount in the problem. This is illustrated in our first example.

Example 1

A student needs 70% to pass an examination containing 50 questions. How many questions must she get right?

Solution The *rate* is 70%. The *base* is the number of questions on the test, here 50. The *amount* is the number of questions that must be correct.

To find the amount, we will use the percent proportion.

Substitute 50 for B and 70 for R.

$$\frac{A}{50} = \frac{70}{100} \quad \text{so} \quad 100A = 50 \cdot 70$$

Dividing by 100,

$$A = \frac{3500}{100} = 35$$

She must have 35 questions correct to pass.

CHECK YOURSELF 1

Generally, 72% of the students in a chemistry class pass the course. If there are 150 students in the class, how many can you expect to pass?

The money borrowed or saved is called the *principal*.

As we said earlier, there are many applications of percent to daily life. One that almost all of us encounter is *interest*. When you borrow money, you pay interest. When you place money in a sav-

Section 11.6: Applications of Percent

ings account, you earn interest. Interest is a percent of the whole (in this case the *principal*), and the percent is called the *interest rate*.

Example 2

Find the interest you must pay if you borrow $2000 for one year with an interest rate of $9\frac{1}{2}\%$.

Solution The base (the principal) is $2000, the rate is $9\frac{1}{2}\%$, and we want to find the interest (the amount). Using the percent proportion,

Remember: $9\frac{1}{2}\% = 9.5\%$

$$\frac{A}{2000} = \frac{9.5}{100}$$

so

$$100A = 2000 \cdot 9.5$$

or

$$A = \frac{19,000}{100} = 190$$

The interest (amount) is $190.

CHECK YOURSELF 2

You invest $5000 for 1 year at $8\frac{1}{2}\%$. How much interest will you earn?

Let's look at some applications of finding the rate in an interest problem.

Example 3

You borrow $2000 from a bank for 1 year and are charged $150 interest. What is the interest rate?

Solution The base is the amount of the loan (the principal). The amount is the interest paid. To find the interest rate we again use the percent proportion.

$$\frac{150}{2000} = \frac{R}{100}$$

Then

$$2000R = 150 \cdot 100$$

$$R = \frac{15{,}000}{2000} = 7.5$$

The interest rate is 7.5%.

CHECK YOURSELF 3

Sam borrowed $3200 and was charged $352 in interest for 1 year. What was the interest rate?

Now here is an application of finding the base in an interest problem.

Example 4

Mr. Hobson agrees to pay 11% interest on a loan for his new automobile. He is charged $550 interest on a loan for 1 year. How much did he borrow?

Solution The rate is 11%. The amount, or interest, is $550. We want to find the base, which is the principal, or the size of the loan. To solve the problem we have

$$\frac{550}{B} = \frac{11}{100}$$

or

$$11B = 550 \cdot 100$$

$$B = \frac{55{,}000}{11} = 5000$$

He borrowed $5000.

CHECK YOURSELF 4

Sue pays $210 interest for a 1-year loan at 10.5%. What was the size of her loan? $2,000

Percents are used in too many ways for us to list. Look at the variety in the following examples, which illustrate some additional situations in which you will find percents.

Section 11.6: Applications of Percent

Example 5

A *commission* is the amount that a person is paid for a sale. That commission is the amount of the total sold.

A salesman sells a car for $9500. His commission rate is 4%. What will his commission be for the sale?

Solution The base is the total of the sale, in this problem $9500. The rate is 4%, and we want to find the commission. This is the amount. By the percent proportion

$$\frac{A}{9500} = \frac{4}{100}$$

so

$$100A = 4 \cdot 9500$$

and

$$A = \frac{38,000}{100} = 380$$

The salesman's commission is $380.

CHECK YOURSELF 5

Jenny sells a $12,000 building lot. If her real estate commission rate is 5%, what commission will she receive for the sale? $600

Example 6

A clerk sold $3500 in merchandise during 1 week. If he received a commission of $140, what was the commission rate?

Solution The base is $3500, and the amount is the commission of $140. Using the percent proportion we have

$$\frac{140}{3500} = \frac{R}{100}$$

or

$$3500R = 140 \cdot 100$$

$$R = \frac{14,000}{3500}$$

$$= 4$$

The commission rate is 4%.

CHECK YOURSELF 6

On a purchase of $500 you pay sales tax of $21. What is the tax rate? 4.2

Our final example involving a commission shows how to find a base (the total sold).

Example 7

A salesman has a commission rate of 3.5%. To earn $280, how much must he sell?

Solution The rate is 3.5%. The amount is the commission, $280. We want to find the base. In this case this is the amount that the salesman needs to sell.
By the percent proportion:

$$\frac{280}{B} = \frac{3.5}{100}$$

or

$$3.5B = 280 \cdot 100$$

Solving for B:

$$B = \frac{28,000}{3.5} = 8000$$

The salesman must sell $8000 to earn $280 in commissions.

CHECK YOURSELF 7

Edward works with a commission rate of 5.5%. If he wants to earn $825 in commissions, find the total sales that must be made.

Another common application of percents involves tax rates. Consider the following example.

Example 8

A state taxes sales at 5.5%. How much sales tax will you pay on a purchase of $48?

Section 11.6: Applications of Percent

Solution The tax you pay is the amount (the part of the whole). Here the base is the purchase price, $48, the rate is the tax rate, 5.5%.

In an application involving taxes, the tax paid is always the amount.

$$\frac{A}{48} = \frac{5.5}{100} \quad \text{or} \quad 100A = 48 \cdot 5.5$$

Now

$$A = \frac{264}{100} = 2.64$$

The sales tax paid is $2.64.

CHECK YOURSELF 8

Suppose that a state has a sales tax rate of $4\frac{1}{2}\%$. If you buy a used car for $1200, how much sales tax must you pay?

Percents are also used to deal with markups or discounts. Consider the following.

Example 9

A store marks up items to make a 30% profit. If an item cost $2.50 from the supplier, what will the selling price be?

This problem can be done in one step. We'll look at that method in the calculator section later in this chapter.

Solution The base is the cost of the item, $2.50, and the rate is 30%. In the percent proportion, the markup is the amount in this application.

$$\frac{A}{2.50} = \frac{30}{100} \quad \text{or} \quad 100A = 30 \cdot 2.50$$

Then

$$A = \frac{75}{100} = 0.75$$

The markup is $0.75. Finally we have

Selling price = $2.50 + $0.75 = $3.25   *Add the cost and the markup to find the selling price.*

CHECK YOURSELF 9

A store wants to discount (or mark down) an item by 25% for a sale. If the original price of the item was $45, find the sale price. *Hint:* Find the discount (the amount the item will be marked down), and subtract that from the original price.

Increases and decreases are often stated in terms of percents. Our final examples illustrate.

Example 10

The population of a town increased 15% in a 3-year period. If the original population was 12,000, what was the population at the end of the period?

Solution First we find the increase in the population. That increase is the amount in the problem.

$$\frac{A}{12,000} = \frac{15}{100}$$

so

$$100A = 15 \cdot 12,000$$

or $A = \dfrac{180,000}{100}$

$= 1800$

To find the population at the end of the period, we add

$$12,000 + 1800 = 13,800$$

Original population — Increase — New population

CHECK YOURSELF 10

A school's enrollment decreased by 8% from a given year to the next. If the enrollment was 550 students the first year, how many students were enrolled the second year?

Example 11

Enrollment at a school increased from 800 to 888 students from a given year to the next. What was the rate of increase?

Section 11.6: Applications of Percent

Solution First we must subtract to find the actual increase.

Increase: 888 − 800 = 88 students

Now to find the rate we have

Note: We use the *original* enrollment, 800, as our base.

$$\frac{88}{800} = \frac{R}{100}$$

so

$$800R = 88 \cdot 100$$

or

$$R = \frac{8800}{800} = 11$$

The enrollment has increased at a rate of 11%.

CHECK YOURSELF 11

Car sales at a dealership decreased from 350 units one year to 322 units the next. What was the rate of decrease?

Example 12

A company hired 18 new employees in 1 year. If this was a 15% increase, how many employees did the company have before the increase?

Solution The rate is 15%. The amount is 18, the number of new employees. The base in this problem is the number of employees *before the increase.* So

$$\frac{18}{B} = \frac{15}{100}$$

$$15B = 18 \cdot 100 \quad \text{or} \quad B = \frac{1800}{15} = 120$$

The company had 120 employees before the increase.

CHECK YOURSELF 12

A school had 54 new students in one term. If this was a 12% increase over the previous term, how many students were there before the increase?

CHECK YOURSELF ANSWERS

1. 108.   2. $425.   3. 11%.   4. $2000.   5. $600.   6. 4.2%.
7. $15,000.   8. $54.   9. $33.75.   10. 506.   11. 8%.   12. 450

11.6 Exercises

Solve each of the following applications.

1. What interest will you pay on a $2500 loan for 1 year if the interest rate is 11%?

2. A chemist has 300 mL of solution which is 18% acid. How many milliliters of acid are in the solution?

3. Anthony has 22% of his pay withheld for deductions. If he earns $350 per week, what amount is withheld?

4. A real estate agent's commission rate is 6%. What will be the amount of the commission on the sale of a $85,000 home?

5. If a salesman is paid a $140 commission on the sale of a $2800 sailboat, what is his commission rate?

6. Mr. Jordan has been given a loan of $1500 for 1 year. If the interest charged is $135, what is the interest rate on the loan?

7. Joan was charged $18 interest for 1 month on a $1200 credit card balance. What was the monthly interest rate?

8. There are 88 grams of acid in 800 grams of a solution of acid and water. What percent of the solution is acid?

9. On a test, Alice had 80% of the problems right. If she had 20 problems correct, how many questions were on the test?

10. A state sales tax rate is 3.5%. If the tax on a purchase is $7, what was the amount of the purchase?

11. Patty pays $285 interest for a 1-year loan at 9.5%. What was the amount of her loan?

Section 11.6: Applications of Percent   477

12. A salesperson is working on a 4% commission basis. If she wants to make $1400 in 1 month, how much must she sell?

13. A state sales tax is levied at a rate of 4.2%. How much tax would one pay on a purchase of $50?

14. Betty must make a $9\frac{1}{2}\%$ down payment on the purchase of a $2000 motorcycle. How much must she pay down?

15. If a house sells for $80,000 and the commission rate is $5\frac{1}{2}\%$, how much will the salesperson make for the sale?

16. Marla needs 70% on a final test to receive a "C" for a course. If the exam has 80 questions, how many questions must she answer correctly?

17. A study has shown that 102 of the 1200 people in the work force of a small town are unemployed. What is the town's unemployment rate?

18. A survey of 400 people found that 66 were left-handed. What percent of those surveyed were left-handed?

19. Of 40 people that start a training program, 35 complete the course. What is the dropout rate?

20. In a shipment of 250 parts, 40 are found to be defective. What percent of the parts are faulty?

21. In a recent survey, 65% of those responding were in favor of a freeway improvement project. If 780 people were in favor of the project, how many people responded to the survey?

22. A college finds that 38% of the students taking a foreign language are enrolled in Spanish. If 570 students are taking Spanish, how many foreign language students are there?

23. 22% of Samuel's monthly salary is deducted for withholding. If those deductions total $209, what is his salary?

24. The Townsends budget 24% of their monthly income for food. If they spend $432 on food, what is their monthly income?

25. An appliance dealer marks up refrigerators 22% (based on cost). If the cost of one model was $600, what will its selling price be?

26. A school had 900 students at the start of a school year. If there is an enrollment increase of 7% by the beginning of the next year, what is the new enrollment?

27. A home lot purchased for $18,000 increased in value by 15% over 3 years. What was the lot's value at the end of the period?

28. New cars depreciate an average of 28% in their first year of use. What will a $9000 car be worth after 1 year?

29. A school's enrollment was up from 850 students in 1 year to 918 students in the next. What was the rate of increase?

30. Under a new contract, the salary for a position increases from $9000 to $10,035. What rate of increase does this represent?

31. A stereo system is marked down from $450 to $382.50. What is the discount rate?

32. The electricity costs of a business decrease from $12,000 one year to $10,920 the next. What is the rate of decrease?

33. The price of a new van has increased $1500 which amounts to a 12% increase. What was the price of the van before the increase?

34. A television set is marked down $75 to be placed on sale. If this is a 12.5% decrease from the original price, what was the selling price before the sale?

35. A company had 66 fewer employees in July of 1988 than in July of 1987. If this represents a 5.5% decrease, how many employees did the company have in July of 1987?

36. Linda received a monthly raise of $127.50. If this represented an 8.5% increase, what was her monthly salary before the raise?

37. Mr. White buys stock for $15,000. At the end of 6 months its value has decreased 7.5%. What is it worth at the end of the period?

38. The population of a town increases 18% in 2 years. If the population was 8000 originally, what is the population after the increase?

39. A store marks up merchandise 25% to allow for profit. If an item costs the store $9, what will its selling price be?

40. John's pay is $350 per week. If deductions from his paycheck average 25%, what is the amount of his weekly paycheck (after deductions)? $262.50

Suppose that you invest $1000 at 5% in a savings account for 1 year.

For year 1, the interest is 5% of $1000, or $0.05 \times \$1000 = \$50$. At the end of year 1, you will have $1050 in the account.

$1000 $\xrightarrow{\text{At 5\%}}$ $1050
Start   Year 1

Now if you leave that amount in the account for a second year, the interest will be calculated on the original principal, $1000, plus the first year's interest, $50. This is called *compound interest*.

For year 2, the interest is 5% of $1050, or $0.05 \times \$1050 = \$52.50$. At the end of year 2, you will have $1102.50 in the account.

$1000 $\xrightarrow{\text{At 5\%}}$ $1050 $\xrightarrow{\text{At 5\%}}$ $1102.50
Start   Year 1   Year 2

In each of the following problems, assume the interest is compounded annually (at the end of each year), and find the amount in an account with the given interest rate and principal.

41. $2000, 6%, 2 years

42. $3000, 7%, 2 years

43. $4000, 5%, 3 years

44. $5000, 6%, 3 years

Answers
1. $275 3. $77 5. 5% 7. 1.5% 9. 25 questions 11. $3000 13. $2.10 15. $4400
17. 8.5% 19. 12.5% 21. 1200 people 23. $950 25. $732 27. $20,700 29. 8% 31. 15%
33. $12,500 35. 1200 employees 37. $13,875 39. $11.25 41. $2247.20 43. $4630.50

11.6 Supplementary Exercises

Solve each of the following applications.

1. A can of mixed nuts contains 80% peanuts. If the can holds 16 oz, how many ounces of peanuts does it contain?

2. A sales tax rate is $5\frac{1}{2}\%$. If Susan buys a television set for $450, how much sales tax must she pay?

3. You are charged $420 in interest on a $3000 loan for 1 year. What is the interest rate?

4. Mr. Moore made a $180 commission on the sale of a $6000 pickup truck. What was his commission rate?

5. Cynthia makes a 4% commission on all her sales. She earned $1000 in commissions during 1 month. What were her gross sales for the month?

6. Nat places 6% of his salary in a payroll savings plan. If he saves $900 in 1 year, what is his annual salary?

7. You are told that if you install storm windows, you will save 12% on your heating costs. If your annual cost for heating was $750 before the windows were installed, what will the annual savings be?

8. On his 50-question biology examination, Bruce answered 78% correctly. How many questions did he answer correctly?

9. The sales tax on a $400 tape deck is $18. What is the sales tax rate?

10. The utility costs for a business increased from $5000 to $5450 in 1 year. What was the rate of increase?

11. 82% of the students in a psychology class passed the midterm examination. If 123 students passed, how many students were in the class?

12. Students entering college are asked to take a math placement exam. On the average, 35% of those taking the test place in beginning algebra. If 875 students were placed in beginning algebra this year, how many students took the examination?

13. Cheryl earned $12,000 one year and then received an 11.5% raise. What is her new yearly salary?

14. A school's enrollment increases 8.2% in a given year. If the enrollment was 1500 students originally, how many students are there after the increase?

15. The enrollment of a 2-year college decreased from 4200 to 3906 in 1 year. What was the rate of decrease?

16. A living room furniture set is discounted from $800 to $716 during a sale. What is the rate of the discount?

17. Roger receives a raise of $112.50 per month. If this represents a 12.5% increase, what was his salary before the raise?

18. A losing football team found its attendance down 5100 people per game. If this was an 8.5% decrease, what was the attendance before the decrease?

19. Bob's salary is $1500 per month. If deductions average 23%, what is his take-home pay?

20. A car was rated at 32 mi/gal. A new model is advertised as having a 7.5% increase in mileage. What mileage should the new model have?

Using Your Calculator

In many everyday applications of percent the computations required become quite lengthy, and so your calculator can be a great help. Let's look at some examples.

Example 1

In a test, 41 of 720 light bulbs burn out before their advertised life of 700 hours. What percent of the bulbs fail to last the advertised life?

Solution This is a type 2 problem. We know the amount and base and want to find the percent (a rate).

Let's use the percent proportion for the solution.

$$\underset{B}{\overset{A}{}} \quad \frac{41}{720} = \frac{R}{100}$$

$$720R = 4100$$

$$\frac{\cancel{720}R}{\cancel{720}} = \frac{4100}{720}$$

Now use your calculator to divide

4100 ÷ 720 = 5.6944444

5.7% of the light bulbs fail.   We round the result to the nearest tenth.

Example 2

The price of a particular model of sofa has increased $48.20. If this represents an increase of 9.65%, what was the price before the increase?

Solution This is a type 3 problem. We want to find the base (the original price). Again let's use the percent proportion for the solution.

$$\underset{}{\overset{A}{}} \quad \frac{\$48.20}{B} = \frac{9.65}{100} \overset{R}{}$$

$$9.65B = 4820$$

$$\frac{\cancel{9.65}B}{\cancel{9.65}} = \frac{4820}{9.65}$$

Using the calculator,

4820 ÷ 9.65 = 499.48187   Round to the nearest cent.

The original price was $499.48.

Earlier we mentioned an alternative method for doing percent problems when increases or decreases are involved. Let's look at an example that uses this second approach.

Example 3

A store marks up items 22.5% to allow for profit. If an item costs a store $36.40, what will the selling price be?

Let's diagram the problem:

Cost	Markup
100%	22.5%
$36.40	$?

Before, we did a similar example in two steps, finding the markup and then adding that amount to the cost. This method allows you to do the problem in one step.

Selling price
122.5%

Now the base is $36.40, the rate is 122.5% and we want to find the amount (the selling price).

This approach may lead to time-consuming hand calculations, but using a calculator reduces the amount of work involved.

$$\frac{A}{36.40} = \frac{122.5}{100}$$

so

$$A = \frac{122.5 \times 36.40}{100} = \$44.59$$

The selling price should be $44.59.

CHECK YOURSELF 1

A car that cost a dealer $7700 is marked up 28% to allow for overhead, commissions, and profit. What will the selling price of the car be?

A similar approach will allow us to solve problems that involve a decrease in one step.

Example 4

Paul invests $5250 in a piece of property. At the end of a 6-month period, the value has decreased 7.5%. What is the property worth at the end of the period?

Earlier we did a problem like this by finding the decrease and then subtracting from the original value. Again, using this method requires just one step.

Solution Again let's diagram the problem.

```
        Original value
            100%
          or $5250
```

Decrease   Ending value
  7.5%   100% − 7.5% = 92.5%

So the amount (ending value) is found as

$$\frac{A}{5250} = \frac{92.5}{100}$$

so

$$A = \frac{92.5 \times \$5250}{100} = \$4856.25$$

The ending value is $4856.25.

CHECK YOURSELF ANSWER

1. $A = \dfrac{\$7700 \times 128}{100} = \$9856$

Exercises Using Your Calculator

Solve each of the following percent problems.

1. What percent is 1240 of 9920?

2. 53.1875 is 9.25% of what number?

3. Find 8.75% of 225.

4. 18.7 is what percent (to the nearest tenth) of 73.6?

5. Find the base if 18.2% of the base is 101.01.

Section 11.6: Applications of Percent   485

6. What is 3.52% of 2450?

7. What percent (to the nearest tenth) of 1375 is 148?

8. 37.5% of what number is 5430?

Solve each of the following applications.

9. What were Eric's total sales in a given month if he earned a commission of $2458 at a commission rate of 1.6%?

10. A retirement plan calls for a 3.18% deduction from your salary. What amount (to the nearest cent) will be deducted from your pay if your monthly salary is $1675?

11. You receive an 11.8% salary increase. If your salary was $975 per month before the raise, how much will your raise be?

12. In a shipment of 558 parts, 49 are found to be defective. What percent (to the nearest tenth of a percent) of the parts are faulty?

13. Statistics show that an average of 42.4% of the students entering a 2-year program will complete their course work. If 588 students completed the program, how many students started?

14. A time-deposit savings plan gives an interest rate of 8.78% on deposits. If the interest on an account for 1 year was $658.50, how much was deposited?

15. The property taxes on a home increased from $832.10 to $957.70 in 1 year. What was the rate of increase (to the nearest tenth of a percent)?

16. A dealer marks down the last year's model appliances 22.5% for a sale. If the regular price of an air conditioner was $279.95, how much will it be discounted (to the nearest cent)?

17. The population of a town increases 8.4% in 1 year. If the original population was 9750, what is the population after the increase?

18. A store marks up items 42.5% to allow for profit. If an item cost a store $24.40, what will its selling price be?

19. An item which originally sold for $98.50 is marked down by 12.5% for a sale. Find its sale price (to the nearest cent).

20. Jerry earned $18,500 one year and then received a 10.5% raise. What was his new yearly salary?

21. Carolyn's salary is $1820 per month. If deductions average 23.5%, what is her take-home pay?

22. John made a $6400 investment at the beginning of a year. By the end of the year, the value of the investment had decreased by 8.2%. What was its value at the end of the year?

Answers

1. 12.5%   **3.** 19.6875   **5.** 555   **7.** 10.8%   **9.** $153,625   **11.** $115.05   **13.** 1387 students
15. 15.1%   **17.** 10,569   **19.** $86.19   **21.** $1392.30

SELF-TEST for Chapter Eleven

The purpose of the Self-Test is to help you check your progress and review for a chapter test in class. Allow yourself about an hour to take the test. When you are done, check your answers in the back of the book. If you missed any problems, be sure to go back and review the appropriate sections in the chapter and do the supplementary exercises provided there.

1. Use a percent to name the shaded portion of the diagram below.

In Problems 2 and 3, write as fractions.

2. 9% **3.** 48%

In Problems 4 to 6, write as decimals.

4. 35%   **5.** 2%   **6.** 130%

In Problems 7 to 10, write as percents.

7. 0.05   **8.** 0.075   **9.** $\frac{3}{5}$ **10.** $\frac{5}{8}$

In Problems 11 to 13, identify the rate, the base, and the amount. Do *not* solve at this point.

11. 50 is 25% of 200.   **12.** What is 8% of 500?

13. A state sales tax rate is 4%. If the tax on a purchase is $20, what is the amount of the purchase?

In Problems 14 to 21, solve the percent problems.

14. What is 7.5% of 250?   **15.** $33\frac{1}{3}$% of 1500 is what number?

16. Find 135% of 400.   **17.** What percent of 150 is 30?

18. 4.5 is what percent of 60?

19. 375 is what percent of 300?

20. 36 is 9% of what number?

21. 8.5% of what number is 25.5?

In Problems 22 to 30, solve the applications.

22. A state taxes sales at 4.8%. What tax will you pay on an item which cost $55?

23. You receive a grade of 75% on a test of 80 questions. How many questions did you have correct?

24. An item which cost a store $48 is marked up 30% (based on cost). Find its selling price.

25. Mr. Sanford pays $300 in interest on a $2500 loan for 1 year. What is the interest rate for the loan?

26. Otto's monthly salary is $2200. If the deductions for taxes from his monthly paycheck are $528, what percent of his salary goes for these deductions?

27. A car is marked down $1410 from its original selling price of $9400. What is the discount rate?

28. Sarah earns $720 in commissions in 1 month. If her commission rate is 4%, what were her total sales?

29. A community college has 480 more students in Fall 1988 than in Fall 1987. If this is a 7.5% increase, what was the Fall 1987 enrollment?

30. Shawn arranges financing for his new car. The interest rate for the financing plan is 12% and he will pay $1020 interest for 1 year. How much money did he borrow to finance the car?

SUMMARY for Part Four

Ratios, Proportions, and Percents

Ratios and Proportions

Ratio   A means of comparing two numbers or quantities. A ratio can be written as a fraction.

$\frac{4}{7}$ can be thought of as "the ratio of 4 to 7."

Proportion   A statement that two ratios are equal.

$\frac{3}{5} = \frac{6}{10}$ is a proportion read "3 is to 5 as 6 is to 10."

Extremes   The first and fourth terms of a proportion.

3 and 10 are the extremes.

Means   The second and third terms of a proportion.

5 and 6 are the means.

The Proportion Rule   In a true proportion, the product of the means is equal to the product of the extremes.

$5 \cdot 6 = 3 \cdot 10$

Solving Proportions for Unknown Values

To Solve a Proportion

1. Set the product of the means equal to the product of the extremes.
2. Divide both terms of the resulting equation by the coefficient of the unknown.
3. Use the value found to replace the unknown in the original proportion. Cross-multiply to check that the proportion is true.

To solve: $\frac{x}{5} = \frac{16}{20}$

$20x = 5 \cdot 16$

$20x = 80$

$\frac{20x}{20} = \frac{80}{20}$

$x = 4$

Check:

$\frac{4}{5} = \frac{16}{20}$

Applying Proportions

To Solve a Problem Using Proportions

1. Read the problem carefully to determine the given information.
2. Write the proportion necessary to solve the problem, using a letter to represent the unknown quantity. Be sure to include the units in writing the proportion.
3. Solve, answer the question of the original problem, and check the proportion as before.

Percents

Percent Another way of naming parts of a whole. Percent means per hundredths.

Fractions and decimals are other ways of naming parts of a whole.

$$21\% = \frac{21}{100} = 0.21$$

Converting between Fractions, Decimals, and Percents

To Convert a Percent to a Fraction

Remove the percent symbol and write the number over 100.

$37\% = \frac{37}{100}$

To Convert a Percent to a Decimal

Remove the percent symbol and move the decimal point two places to the left.

$37\% = 0.37$

To Convert a Decimal to a Percent

Move the decimal point two places to the right and attach the percent symbol.

$0.58 = 58\%$

To Convert a Fraction to a Percent

Write the decimal equivalent of the fraction and then change that decimal to a percent.

$\frac{3}{5} = 0.60 = 60\%$

Percent Problems

Every percent problem has the following three parts:

1. *The base.* This is the whole in the problem. It is the standard used for comparison. Label the base B.
2. *The amount.* This is the part of the whole being compared to the base. Label the amount A.
3. *The rate.* This is the ratio of the amount to the base. The rate is written as a percent. Label the rate $R\%$.

45 is 30% of 150
↑   ↑   ↑
A $R\%$ B

These quantities are related by the equation:

Amount = rate × base

To Solve a Percent Problem

There Are Three Basic Types of Percent Problems.

Type 1 Problems
Finding the amount when the rate and base are known.

Type 2 Problems
Finding the rate when the amount and the base are known.

Type 3 Problems
Finding the base when the rate and the amount are known.

Using the Percent Proportion
The percent proportion is

$$\frac{A}{B} = \frac{R}{100}$$

To solve a percent problem using this proportion:

1. Substitute the two known values into the proportion.
2. Solve the proportion as before to find the unknown value.

SUMMARY EXERCISES for Part Four

You should now be reviewing the material in Part 4 of the text. The following exercises will help in that process. Work all the exercises carefully. Then check your answers in the back of the book. References are provided to the chapter and section for each problem. If you made an error, go back and review the related material and do the supplementary exercises for that section.

[10.1] In Problems 1 to 5, write the following ratios in simplest form.

1. The ratio of 5 to 13.

2. The ratio of 28 to 42.

3. The ratio of 5 dimes to 5 quarters.

4. The ratio of 5 in to 2 ft.

5. The ratio of 48 hours to 3 days.

[10.2] In Problems 6 to 11, determine which of the statements are true proportions.

6. $\dfrac{5}{12} = \dfrac{6}{14}$

7. $\dfrac{8}{11} = \dfrac{24}{33}$

8. $\dfrac{9}{24} = \dfrac{12}{32}$

9. $\dfrac{7}{18} = \dfrac{35}{80}$

10. $\dfrac{5}{\frac{1}{3}} = \dfrac{45}{3}$

11. $\dfrac{0.8}{4} = \dfrac{12}{50}$

[10.3] In Problems 12 to 20, solve for the unknown in each proportion.

12. $\dfrac{8}{12} = \dfrac{m}{3}$

13. $\dfrac{12}{a} = \dfrac{9}{15}$

14. $\dfrac{14}{35} = \dfrac{t}{10}$

15. $\dfrac{y}{22} = \dfrac{15}{55}$

16. $\dfrac{55}{88} = \dfrac{10}{p}$

17. $\dfrac{\frac{1}{4}}{10} = \dfrac{3}{w}$

18. $\dfrac{\frac{3}{2}}{9} = \dfrac{5}{a}$

19. $\dfrac{3}{x} = \dfrac{0.4}{8}$

20. $\dfrac{s}{2.5} = \dfrac{1.5}{7.5}$

[10.4] In Problems 21 to 29, solve each application using a proportion.

21. If 6 tickets to a civic theater performance cost $33, what will the price for 8 tickets be?

22. The ratio of freshmen to sophomores at a school is 8 to 7. If there are 224 sophomores, how many freshmen are there?

23. A photograph that is 5 in wide by 7 in tall is to be enlarged so that the new height will be 21 in. What will be the width of the enlargement?

24. Marcia assembles disk drives for a computer manufacturer. If she can assemble 9 drives in 2 hours, how many can she assemble in a work week (40 hours)?

25. A firm finds 18 defective parts in a shipment of 200. How many defective parts can be expected in a shipment of 500 parts?

26. The scale on a map uses $\frac{1}{4}$ in to represent 10 mi. How many miles apart are two towns that are 3 in apart on the map?

27. A piece of tubing that is 16.5 cm long weighs 55 grams. What is the weight of the same tubing that is 21 cm long?

28. If 1 qt of paint will cover 120 sq ft, how many square feet will 2 gal cover? (1 gal = 4 qt).

29. Instructions for a film developer are to use 6 oz of powder to make 16 oz of the solution. How much powder is needed to make 2 qt of the developer? Remember that 1 qt is 32 oz.

[11.1] In Problem 30, use a percent to name the shaded portion of the diagram.

30.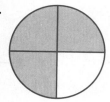

[11.2] In Problems 31 to 33, write the percent as a common fraction or a mixed number.

31. 3%   **32.** 15%   **33.** 130%

Summary Exercises for Part Four

[1.2] In Problems 34 to 36, write the percents as decimals.

34. 45% **35.** 12.5% **36.** 225%

[1.3] In Problems 37 to 41, write as percents.

37. 0.08 **38.** 0.125 **39.** 2.4

40. $\dfrac{2}{5}$ **41.** $1\dfrac{1}{3}$

[1.5] In Problems 42 to 50, find the unknown.

42. 80 is 4% of what number?

43. 35 is what percent of 25?

44. 12% of 2000 is what number?

45. 24 is what percent of 192?

46. Find the base if 12.5% of the base is 625.

47. 90 is 120% of what number?

48. What is 11.5% of 500?

49. Find 150% of 30.

50. Find the base if 130% of the base is 780.

[1.6] In Problems 51 to 60, solve the applications.

51. Joan works on a 5% commission basis. She sold $35,000 in merchandise during 1 month. What was the amount of her commission?

52. David buys a dishwasher that is marked down $77 from its original price of $350. What is the discount rate?

53. A chemist prepares a 400-mL acid and water solution. If the solution contains 30 mL of acid, what percent of the solution is acid?

54. The price of a new compact car has increased $308 over the previous year. If this amounts to a 3.5% increase, what was the price of the car before the increase?

55. A store advertises, "Buy the red-tagged items at 25% off their listed price." If you buy a coat marked $124, what will you pay for the coat during the sale?

56. Tom has 6% of his salary deducted for a retirement plan. If that deduction is $78, what is his monthly salary?

57. A college finds that 45% of its science students take biology. If there are 306 biology students, how many science students are there altogether?

58. A company finds that its advertising costs increased from $72,000 to $76,680 in 1 year. What was the rate of increase?

59. A savings bank offers 7.25% on 1-year time deposits. If you place $5000 in an account, how much will you have at the end of the year?

60. Maria's company offers her a 9% pay raise. This will amount to a $99 per month increase in her salary. What is her monthly salary before and after the raise?

CUMULATIVE TEST for Part Four

This test is provided to help you in the process of review over Chapters 10 and 11. Answers are provided in the back of the book. If you missed any problems, be sure to go back and review the appropriate chapter sections.

In Problems 1 to 3, write the ratios in simplest form.

1. The ratio of 32 to 48.

2. The ratio of 6 dimes to 4 quarters.

3. The ratio of 9 in to 2 ft.

In Problems 4 to 6, determine which of the statements are true proportions.

4. $\dfrac{3}{10} = \dfrac{7}{23}$

5. $\dfrac{7}{24} = \dfrac{21}{72}$

6. $\dfrac{11}{18} = \dfrac{33}{54}$

In Problems 7 to 10, solve for the unknown in each proportion.

7. $\dfrac{10}{y} = \dfrac{2}{3}$

8. $\dfrac{24}{40} = \dfrac{n}{5}$

9. $\dfrac{\frac{2}{3}}{8} = \dfrac{5}{a}$

10. $\dfrac{5}{m} = \dfrac{0.3}{6}$

In Problems 11 to 17, solve each application using a proportion.

11. If 4 tickets to a theater cost $22, what will 14 tickets to the same performance cost?

12. Jeffrey drove 222 mi using 6 gal of gasoline. At the same rate, how far can he travel on a tankful (15 gal) of gas?

13. A company finds that in one branch office 4 of the 15 employees sign up for a new health plan. If the company has 120 employees overall, how many should be expected to sign up for the new plan?

14. You are using a photocopy machine to reduce an advertisement which is 14 in wide by 21 in long. If the new width is to be 8 in, what will the new length be?

15. If the scale on a map is $\frac{1}{3}$ in equals 10 mi, how far apart are two towns that are 4 in apart on the map?

16. Diane worked 23.5 hours on a part-time job and was paid $131.60. She is asked to work 30 hours the next week at the same rate of pay. What salary will she receive?

17. Instructions for a powdered plant fertilizer call for 3 oz of powder to be mixed with water to form 16 oz of solution. How much of the fertilizer should be used to make 3 qt of solution? Remember that 1 qt is 32 oz.

In Problem 18, use a percent to name the shaded portion of the diagram.

18.

In Problems 19 and 20, write the percent as a common fraction or a mixed number.

19. 35%

20. 175%

In Problems 21 and 22, write the percents as decimals.

21. 55%

22. 18.5%

In Problems 23 to 26, write as percents.

23. 0.125

24. 0.001

25. $\frac{3}{5}$

26. $\frac{5}{8}$

In Problems 27 to 32, solve the percent problem.

27. 72 is 12% of what number?

28. Find 7.5% of 400.

29. 27.5 is what percent of 250?

30. $8\frac{1}{2}$% of 3000 is what number?

31. 120% of what number is 180?

32. What percent of 60 is 105?

In Problems 33 to 40, solve the percent application.

33. Anthony takes out a $2000 loan for 1 year to pay for a remodeling project. If he will pay $190 in interest, what is the interest rate on the loan?

34. Jackie works on an 8% commission basis. If she wishes to earn $1400 in commissions in 1 month, how much must she sell during that period?

35. Marilyn has calculated that she must score at least 68% in her last examination to receive a C in her chemistry course. She answered 84 of the 120 questions on the test correctly. Will she receive the C?

36. A college predicts that 66% of their entering freshmen will take English composition. If 2500 freshmen students are enrolled, how many composition students should be expected?

37. Martin averages 27% deductions from his paycheck for taxes and insurance. If in 1 month those deductions were $432, what was his pay before the deductions?

38. A home that was purchased for $75,000 increased in value by 14% over a 3-year period. What was its value at the end of that period?

39. The number of employees of a business increased from 440 at the end of one fiscal year to 473 at the end of the next. What was the rate of increase for that period?

40. A stereo dealer decides to discount or mark down one model of video recorder for a sale. If the recorder's original price was $425, and it is to be discounted by 24%, find its sale price.

Part 5

Measurement

PRETEST for Chapter Twelve

The English System of Measurement

This pretest will point out any difficulties you may be having with measurement. Do all the problems. Then check your answers on the following page.

1. 7 ft = _____ in.

2. Simplify: 7 min 80 seconds.

3. Add: 5 gal 3 qt and 7 gal 2 qt.

In Problems 4 and 5, find the perimeter or circumference.

4. (trapezoid: 25 in top, 22 in left, 20 in right, 35 in bottom)

5. (circle with radius 3 yd) Use 3.14 for π.

In Problems 6 to 8, find the area.

6. (rectangle: 7 in by 5 in)

7. (circle with radius 7 in) Use $\frac{22}{7}$ for π.

8. (triangle: height 3 in, base 5 in)

In Problems 9 and 10, find the volume.

9. (rectangular box: 4 in by 4 in by 5 in)

10. (cylinder: height 4 ft, radius 2 ft) Use 3.14 for π.

ANSWERS TO PRETEST

For help with similar problems, turn to the section indicated.

1. 84 (Section 12.1);
2. 8 min 20 seconds (Section 12.2);
3. 13 gal 1 qt (Section 12.2);
4. 102 in (Section 12.3);
5. 18.84 yd (Section 12.3);
6. 35 in² (Section 12.4);
7. 154 in² (Section 12.4);
8. $7\frac{1}{2}$ in² (Section 12.4);
9. 80 in³ (Section 12.5);
10. 50.24 ft³ (Section 12.5)

Chapter Twelve

The English System of Measurement

12.1 The Units of the English System

OBJECTIVE
To be able to convert between various English units of measure

Many arithmetic problems involve *units of measure*. When we measure an object, we give it a number and some unit. For instance, we might say a board is 6 feet long, a container holds 4 quarts, or a package weighs 5 pounds. Feet, quarts, and pounds are the units of measure.

The system you are probably most familiar with is called the *English* system of measurement. This system is used in the United States and a few other countries.

The following table will list the units of measurement you should be familiar with.

Don't let the name mislead you. The English system is no longer used in England as that country has converted to the metric system, which we will look at in Chapter 13.

ENGLISH UNITS OF MEASURE AND EQUIVALENTS

Length	Weight
1 foot (ft) = 12 inches (in)	1 pound (lb) = 16 ounces (oz)
1 yard (yd) = 3 ft	1 ton = 2000 lb
1 mile (mi) = 5280 ft	
Volume	**Time**
1 pint (pt) = 16 fluid ounces (fl oz)	1 minute (min) = 60 seconds (s)
	1 hour (h) = 60 min
1 quart (qt) = 2 pt	1 day = 24 h
1 gallon (gal) = 4 qt	1 week = 7 days

You may want to use the equivalencies shown in the table to change from one unit to another. Let's look at one approach.

> To change from one unit to another, replace the unit of measure with the appropriate equivalent measure and multiply.

Example 1

We write 5 ft as 5 (1 ft) and then change 1 ft to 12 in.

$$5 \text{ ft} = 5(1 \text{ ft}) = 5(12 \text{ in}) = 60 \text{ in}$$ Replace 1 ft with 12 in.
$$3 \text{ lb} = 3(1 \text{ lb}) = 3(16 \text{ oz}) = 48 \text{ oz}$$ Replace 1 lb with 16 oz.
$$6 \text{ gal} = 6(1 \text{ gal}) = 6(4 \text{ qt}) = 24 \text{ qt}$$ Replace 1 gal with 4 qt.

$$48 \text{ in} = 48(1 \text{ in}) = 48\left(\frac{1}{12} \text{ ft}\right) = 4 \text{ ft}$$ Since 12 in equals 1 ft, replace 1 in with $\frac{1}{12}$ ft.

$$180 \text{ min} = 180(1 \text{ min}) = 180\left(\frac{1}{60} \text{ h}\right) = 3 \text{ h}$$ Replace 1 min with $\frac{1}{60}$ h.

CHECK YOURSELF 1

Complete the following statements.

(1) 4 ft = _____ in
(2) 12 qt = _____ pt
(3) 48 fl oz = _____ pt
(4) 240 s = _____ min

If you had any difficulty with the exercises, here is another idea that may help. You can use a *unit ratio* to convert from one unit to another. A *unit ratio* is a fraction whose value is 1.

Example 2

To convert from feet to inches, you can multiply by the ratio 12 in/1 ft. So, to convert 5 ft to inches, write:

$\frac{12 \text{ in}}{1 \text{ ft}}$ is our first example of a unit ratio. It can be reduced to 1.

$$5 \text{ ft} = 5 \text{ ft} \left(\frac{12 \text{ in}}{1 \text{ ft}}\right)$$ We are multiplying by 1, and so the value of the expression is not changed.

$$= 60 \text{ in}$$ Note that we can divide out units just as we do numbers.

To decide which unit ratio to use, just choose one with the unit you *want* in the numerator (inches in the example) and the unit you *want to remove* in the denominator (feet in the example).

Example 3

To convert ounces to pounds, you can multiply by the unit ratio:

$$\frac{1 \text{ lb}}{16 \text{ oz}}$$ Use the unit you want (pounds) in the numerator and the unit you want to remove (ounces) in the denominator.

So to convert 96 oz to pounds, write:

$$96 \text{ oz} = \overset{6}{\cancel{96 \text{ oz}}} \left(\frac{1 \text{ lb}}{\underset{1}{\cancel{16 \text{ oz}}}}\right) = 6 \text{ lb}$$

CHECK YOURSELF 2

Use a unit ratio to complete this statement:

240 min = _____ h

You have now had a chance to use two different methods for converting from one unit of measurement to another. You should use whichever approach seems best to you for the exercises at the end of this section.

From our work so far, it should be clear that one big disadvantage of the English system is that the relationships between units are all different. One foot is 12 in, 1 lb is 16 oz, and so on. We shall see later that this problem doesn't exist in the metric system.

Historically units were associated with various things. A foot was the length of a foot, of course. The yard was the distance from the end of a nose to the fingertips of an outstretched arm. Objects were weighed by comparing them with grains of barley.

CHECK YOURSELF ANSWERS

1. (1) 48; (2) 24; (3) 3; (4) 4. **2.** $240 \text{ min} \left(\dfrac{1 \text{ h}}{60 \text{ min}} \right) = 4 \text{ h}$.

12.1 Exercises

Complete the following statements.

1. 6 ft = _____ in
2. 8 gal = _____ qt
3. 3 lb = _____ oz

4. 300 s = _____ min
5. 240 min = _____ h
6. 6 lb = _____ oz

7. 4 days = _____ h
8. 3 pt = _____ fl oz
9. 16 qt = _____ gal

10. 6 h = _____ min
11. 6000 lb = _____ tons
12. 7 min = _____ s

13. 20 pt = _____ qt
14. 3 mi = _____ ft
15. 64 oz = _____ lb

16. 64 fl oz = _____ pt
17. 4 yd = _____ ft
18. 360 min = _____ h

19. 21 ft = _____ yd
20. 24 qt = _____ gal
21. 3 h = _____ min

22. 12 ft = _____ in
23. 80 fl oz = _____ pt
24. 6 yd = _____ ft

25. 120 s =   min   **26.** 120 h =   days   **27.** 7 gal =   qt

28. 12 pt =   qt   **29.** 2 mi =   ft   **30.** 80 oz =   lb

31. 4 min =   s   **32.** 18 qt =   pt   **33.** 144 h =   days

34. 4000 lb =   tons   **35.** 12 qt =   pt   **36.** 7 days =   h

Answers

Solutions for the even-numbered exercises are provided in the back of the book.

1. 72   **3.** 48   **5.** 4   **7.** 96   **9.** 4   **11.** 3   **13.** 10   **15.** 4   **17.** 12   **19.** 7   **21.** 180   **23.** 5
25. 2   **27.** 28   **29.** 10,560   **31.** 240   **33.** 6   **35.** 24

12.1 Supplementary Exercises

Complete the following statements.

1. 8 ft =   in   **2.** 4 yd =   ft   **3.** 4 gal =   qt

4. 300 min =   h   **5.** 8 lb =   oz   **6.** 84 in =   ft

7. 180 s =   min   **8.** 48 h =   days   **9.** 48 fl oz =   pt

10. 6 pt =   fl oz   **11.** 2 mi =   ft   **12.** 64 oz =   lb

13. 12 qt =   gal   **14.** 5 min =   s   **15.** 3 days =   h

16. 8000 lb =   tons   **17.** 16 pt =   qt   **18.** 4 h =   min

12.2 Denominate Numbers

OBJECTIVES
1. To simplify denominate numbers
2. To perform operations with denominate numbers

Numbers used to give measurements are called *denominate numbers*. A denominate number is a number with a unit or name attached.

Section 12.2: Denominate Numbers

Example 1

3 ft
4 yd
5 lb
6 qt

These are all denominate numbers.

Numbers without units attached are *abstract numbers*.
   A denominate number may involve two or more different units. We regularly combine feet and inches, pounds and ounces, and so on. The measures 5 ft 10 in or 6 lb 7 oz are examples. In combining units, you should let the larger unit represent *as much as is possible:* The denominate number should always be simplified. For example, 7 ft 3 in is in simplest form, while 4 ft 18 in is *not* in simplest form; 18 in can be written as a combination of feet and inches.
   Our next example shows the steps of simplifying a denominate number.

Example 2

Simplify 4 ft 18 in.

Solution

4 ft 18 in = 4 ft + $\underbrace{1\text{ ft} + 6\text{ in}}_{18\text{ in}}$ Write 18 in as 1 ft 6 in since 12 in is 1 ft.

= 5 ft 6 in

CHECK YOURSELF 1
Simplify 7 ft 20 in.

Example 3

Simplify 5 h 75 min.   Write 75 min as 1 h 15 min since 1 h is 60 min.

5 h 75 min = 5 h + $\underbrace{1\text{ h} + 15\text{ min}}_{75\text{ min}}$

= 6 h 15 min

CHECK YOURSELF 2
Simplify 5 lb 24 oz.

Denominate numbers with the same units are called *like numbers*. We can always add or subtract like numbers according to the following rule.

> **ADDING DENOMINATE NUMBERS**
>
> STEP 1 Arrange the numbers so that the like units are in the same vertical column.
>
> STEP 2 Add in each column.
>
> STEP 3 Simplify if necessary.

The next example illustrates.

Example 4

Add: 5 ft 4 in, 6 ft 7 in, and 7 ft 9 in.

The columns here represent inches and feet.

	5 ft	4 in	Arrange in a vertical column.
	6 ft	7 in	
+	7 ft	9 in	
	18 ft	20 in	Add in each column.
=	19 ft	8 in	Simplify as before.

CHECK YOURSELF 3

Add: 3 h 15 min, 5 h 50 min, and 2 h 40 min.

To subtract denominate numbers, we have a similar rule.

> **SUBTRACTING DENOMINATE NUMBERS**
>
> STEP 1 Arrange the numbers so that the like units are in the same vertical column.
>
> STEP 2 Subtract in each column. You may have to borrow from the larger unit at this point.
>
> STEP 3 Simplify if necessary.

Consider the following example of subtracting denominate numbers.

Example 5

Subtract 3 lb 6 oz from 8 lb 13 oz.

	8 lb	13 oz	Arrange vertically.
−	3 lb	6 oz	
	5 lb	7 oz	Subtract in each column.

Section 12.2: Denominate Numbers

CHECK YOURSELF 4

Subtract 5 ft 9 in from 10 ft 11 in.

As step 2 points out, subtracting denominate numbers may involve borrowing.

Example 6

Subtract 5 ft 8 in from 9 ft 3 in.

9 ft 3 in
−5 ft 8 in

Do you see the problem? We cannot subtract in the inches column.

Borrowing with denominate numbers is not the same as in the place-value system where we always borrowed ten.

Here the "borrowed" number will depend on the units involved.

To complete the subtraction, we borrow 1 ft and rename.

~~9 ft   3 in~~   9 ft becomes 8 ft 12 in. Combine the 12 in with the original 3 in.

8 ft 15 in
−5 ft 8 in
3 ft 7 in   We can now subtract.

CHECK YOURSELF 5

Subtract 3 lb 9 oz from 8 lb 5 oz.

Certain types of problems will involve multiplying or dividing denominate numbers by abstract numbers, that is a number without a unit of measure attached. The following rule is used.

MULTIPLYING OR DIVIDING BY ABSTRACT NUMBERS

STEP 1 Multiply or divide each part of the denominate number by the abstract number.

STEP 2 Simplify if necessary.

Our final examples illustrate.

Example 7

4×5 in = 20 in or 1 ft 8 in.

Example 8

To multiply 3 × (2 ft 7 in), write

Multiply each part of the denominate number by 3.

```
  2 ft   7 in
×          3
─────────────
  6 ft  21 in
```

Simplify. The product is 7 ft 9 in.

CHECK YOURSELF 6

Multiply 5 lb 8 oz by 4.

Example 9

To divide 8 lb 12 oz by 4, write

Divide each part of the denominate number by 4.

$$\frac{8 \text{ lb } 12 \text{ oz}}{4}$$

= 2 lb 3 oz

CHECK YOURSELF 7

Divide 9 ft 6 in by 3.

CHECK YOURSELF ANSWERS

1. 8 ft 8 in.   **2.** 6 lb 8 oz.   **3.** 11 h 45 min.   **4.** 5 ft 2 in.
5. Rename 8 lb 5 oz as 7 lb 21 oz. Then subtract for the result, 4 lb 12 oz.
6. 20 lb 32 oz, or 22 lb.   **7.** 3 ft 2 in.

12.2 Exercises

Simplify:

1. 5 ft 20 in   **2.** 6 lb 20 oz   **3.** 5 qt 3 pt   **4.** 4 yd 40 in

5. 5 gal 9 qt   **6.** 5 min 90 s   **7.** 7 min 100 s   **8.** 9 h 80 min

Add:

9. 7 lb 9 oz
    <u>5 lb 11 oz</u>

10. 9 ft 7 in
     <u>3 ft 10 in</u>

11. 5 h 30 min
     3 h 50 min
     <u>2 h 45 min</u>

12. 5 yd 2 ft
     4 yd
     <u>6 yd 1 ft</u>

13. 5 lb 4 oz, 4 lb 12 oz, and 7 lb 9 oz.

14. 7 ft 8 in, 8 ft 5 in, and 9 ft 7 in.

Section 12.2: Denominate Numbers 513

Subtract:

15. 8 lb 13 oz
5 lb 8 oz

16. 7 ft 11 in
4 ft 3 in

17. 5 h 20 min
2 h 40 min

18. 5 gal 1 qt
1 gal 3 qt

19. Subtract 2 yd 2 ft from 5 yd 1 ft.

20. Subtract 3 h 45 min from 6 h 15 min.

Multiply:

21. 5 × 14 oz

22. 4 × 10 in

23. 4 × (5 ft 7 in)

24. 7 × (2 min 30 s)

Divide:

25. $\dfrac{8 \text{ ft } 10 \text{ in}}{2}$

26. $\dfrac{12 \text{ lb } 15 \text{ oz}}{3}$

27. $\dfrac{12 \text{ min } 40 \text{ s}}{4}$

28. $\dfrac{10 \text{ h } 50 \text{ min}}{5}$

Solve each of the following applications.

29. A railing for a deck requires pieces of cedar 5 ft 7 in, 12 ft 3 in, and 8 ft 4 in long. What is the total length of material that is needed?

30. Ted worked 3 hours 45 min on Monday, 5 hours 30 min on Wednesday, and 4 hours 15 min on Friday. How many hours did he work during the week?

31. A pattern requires a 2-ft 10-in length of fabric. If a 2-yd length is used, what length remains?

32. You use 1 lb 10 oz of hamburger from a package that weighs 3 lb 2 oz. How much is left over?

33. A picture frame is to be 2 ft 8 in long and 1 ft 9 in wide. A 9-ft piece of molding is available for the frame. Will this be enough for the frame?

34. A plumber needs three pieces of plastic pipe that are 4 ft 5 in long and 1 piece that is 2 ft 10 in long. He has a 16-ft piece of pipe. Is this enough for the job?

35. Mark uses 1 pt 9 fl oz and then 2 pt 10 fl oz from a container of film developer that holds 3 qt. How much of the developer remains?

36. Some flights limit passengers to 44 lb of checked-in luggage. Susan checks three pieces, weighing 20 lb 5 oz, 7 lb 8 oz, and 15 lb 7 oz. By how much was she under or over the limit?

37. Five packages weighing 3 lb 6 oz each are to be mailed. What is the total weight of the packages?

38. A bookshelf requires four boards 3 ft 8 in long and two boards 2 ft 10 in long. How much lumber will be needed for the bookshelf?

39. You can buy three 12-oz cans of peanuts for $3 or one large can containing 2 lb 8 oz for the same price. Which is the better buy?

40. Rich, Susan, and Marc agree to share the driving on a 12-hour trip. Rich has driven for 4 hours 45 min and Susan has driven for 3 hours 30 min. How long must Marc drive to complete the trip?

Answers

1. 6 ft 8 in   **3.** 6 qt 1 pt   **5.** 7 gal 1 qt   **7.** 8 min 40 s   **9.** 13 lb 4 oz   **11.** 12 h 5 min
13. 17 lb 9 oz   **15.** 3 lb 5 oz   **17.** 2 h 40 min   **19.** 2 yd 2 ft   **21.** 4 lb 6 oz   **23.** 22 ft 4 in
25. 4 ft 5 in   **27.** 3 min 10 s   **29.** 26 ft 2 in   **31.** 3 ft 2 in   **33.** Yes, 2 in remain   **35.** 1 pt 13 fl oz
37. 16 lb 14 oz   **39.** The three small cans contain 36 oz, or 2 lb 4 oz. The larger can is the better buy.

12.2 Supplementary Exercises

Simplify.

1. 4 ft 22 in   **2.** 4 lb 20 oz   **3.** 6 min 100 s   **4.** 3 h 80 min

Add.

5. 3 lb 9 oz
   5 lb 10 oz

6. 5 h 20 min
   3 h 40 min
   2 h 20 min

7. 4 yd, 5 yd, 2 ft, 2 ft, and 6 yd 1 ft.

Subtract.

8. 8 ft 10 in
   3 ft 5 in

9. 3 h 30 min
   1 h 50 min

10. 6 lb 12 oz from 9 lb 6 oz.

Multiply.

11. 4 × 10 oz

12. 3 × (1 h 25 min)

Divide:

13. $\dfrac{10 \text{ lb } 12 \text{ oz}}{2}$

14. $\dfrac{12 \text{ ft } 9 \text{ in}}{3}$

15. Mike wants to fill four candle molds with hot wax. The molds require 10 oz, 1 lb 8 oz, 1 lb 14 oz, and 2 lb of wax. How much wax will he need for the candles?

16. John worked 6 hours 15 min, 8 hours, 5 hours 50 min, 7 hours 30 min, and 6 hours during 1 week. What were the total hours worked?

17. A bookcase requires two boards of length 3 ft 10 in, two of length 2 ft 6 in, and one of length 3 ft 8 in. Will a 16-ft board give enough material?

18. A room requires two pieces of floor molding 12 ft 8 in long, one 6 ft 5 in long, and one 10 ft long. Will 42 ft of molding be enough for the job?

19. Jackie finished a long-distance race in 1 hour 55 min 30 seconds. George's time for the race was 2 hours 15 min 40 seconds. How much better was Jackie's time?

20. You use 1 pt 10 fl oz of liquid fertilizer from a full container that holds 3 pt 8 fl oz. How much of the liquid remains?

12.3 Finding Perimeter and Circumference

OBJECTIVES
1. To find the perimeter of a figure
2. To find the circumference of a circle
3. To apply perimeter and circumference

One application of our work with denominate numbers is in finding the *perimeter* of a figure.

> The perimeter of a figure is the distance around that figure.

If the figure has straight sides, the perimeter is the sum of the lengths of its sides.

Example 1

We wish to fence in the field shown in Figure 1. How much fencing will be needed?

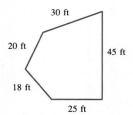

Figure 1

Solution The fencing needed is the perimeter (or the distance around) of the field. We must add the lengths of the five sides.

$20 + 30 + 45 + 25 + 18 = 138$ ft

So the perimeter is 138 ft.

CHECK YOURSELF 1

What is the perimeter of the region shown?

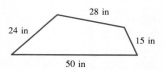

The perimeter of certain figures can be found more easily by using formulas. For example, in a rectangle, let L be the length and W the width, as shown in Figure 2.

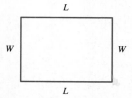

Figure 2

Section 12.3: Finding Perimeter and Circumference   517

Its perimeter is given by

FORMULA (1)

$P = 2L + 2W$ (rectangle)

The perimeter of a rectangle is two times its length plus two times its width.

The following example shows the use of this formula.

Example 2

A rectangle has length 11 in and width 8 in. What is its perimeter?

Solution By Formula (1),

$$P = 2L + 2W$$
$$= 2 \cdot 11 \text{ in} + 2 \cdot 8 \text{ in} \qquad \text{Replace } L \text{ with 11 in and } W \text{ with 8 in.}$$
$$= 22 \text{ in} + 16 \text{ in}$$
$$= 38 \text{ in}$$

CHECK YOURSELF 2

A living room is 13 by 17 ft. What is its perimeter?

Another common figure is a *square*. A square, shown in Figure 3, is a rectangle whose sides are all of equal length. All four sides have the same length S.

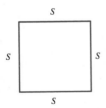

Figure 3

The perimeter of a square is given by

FORMULA (2)

$P = 4S$ (square)

Our next example illustrates how this formula is applied.

Example 3

Find the perimeter of a square if its sides are 5 ft long.

By Formula (2),

$P = 4 \cdot \underbrace{5 \text{ ft}}_{\text{Replace } S \text{ with 5 ft.}} = 20 \text{ ft}$

CHECK YOURSELF 3

You are putting fringe around a tablecloth that is 50 in square. How much fringe will you need?

A concept related to perimeter is *circumference*.

> The circumference of a circle is the distance around that circle.

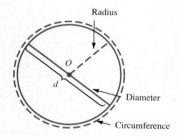

Figure 4

Let's define some terms. In the circle of Figure 4, d represents the *diameter*. This is the distance across the circle through its center (labeled with the letter O). The *radius, r,* is the distance from the center to a point on the circle. The diameter is always twice the radius.

It was discovered long ago that the ratio of the circumference of a circle to its diameter always stays the same. The ratio has a special name. It is named by the Greek letter π (pi). Pi is approximately $\frac{22}{7}$, or 3.14 to two decimal places. We can write the following formula:

The formula comes from the ratio

$\frac{C}{d} = \pi$

> **FORMULA (3)**
>
> $C = \pi d$ (circumference of a circle)

Example 4

A circle has a diameter of 4.5 ft, as shown in Figure 5. Find its circumference. (Use 3.14 for π.)

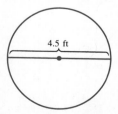

Figure 5

Section 12.3: Finding Perimeter and Circumference

Solution By Formula (3),

$$C = \pi d$$
$$\approx 3.14 \times 4.5 \text{ ft}$$
$$\approx 14.1 \text{ ft} \quad \text{(rounded to one decimal place)}$$

Since 3.14 is an approximation for pi, we can only say that the circumference is approximately 14.1 ft. The symbol $\approx$ means approximately.

CHECK YOURSELF 4

A circle has a diameter of $3\frac{1}{2}$ in. Find its circumference. (Use $\frac{22}{7}$ for π.)

Note In finding the circumference of a circle, you can use whichever approximation for pi you choose. If you are using a calculator and want more accuracy, you might use 3.1416. This is an approximation to four decimal places.

There is another useful formula for the circumference of a circle.

You needn't worry about running out of decimal places. The value for pi has been calculated to over 100,000,000 decimal places on a computer. The printout was some 20,000 pages long.

> **FORMULA (4)**
>
> $C = 2\pi r$ (circumference of a circle)

Since $d = 2r$ (the diameter is twice the radius) and $C = \pi d$, we have $C = \pi(2r)$ or $C = 2\pi r$

Example 5

What is the circumference of the circle shown in Figure 6? Use 3.14 for π.

Solution From Formula (4),

$$C = 2\pi r$$
$$\approx 2 \times 3.14 \times 8 \text{ in}$$
$$\approx 50.2 \text{ in} \quad \text{(rounded to one decimal place)}$$

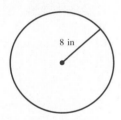

Figure 6

CHECK YOURSELF 5

Find the circumference of a circle with a radius of 2.5 in. Use 3.14 for π.

Sometimes we will want to combine the ideas of perimeter and circumference to solve a problem.

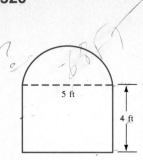

Figure 7

The distance around the semicircle is $\frac{1}{2}\pi d$.

Example 6

We wish to build a wrought iron frame gate according to the diagram in Figure 7. How many linear feet of material will be needed?

Solution The problem can be broken into two parts. The upper part of the frame is a semicircle (half a circle). The remaining part of the frame is just three sides of a rectangle.

Circumference (upper part) $\approx \frac{1}{2} \times 3.14 \times 5 \text{ ft} \approx 7.9 \text{ ft}$

Perimeter (lower part) $= 4 + 5 + 4 = 13 \text{ ft}$

Adding, we have

$7.9 + 13 = 20.9 \text{ ft}$

We will need approximately 20.9 linear feet of material.

CHECK YOURSELF ANSWERS

1. 117 in or 9 ft 9 in.
2. $P = 2 \cdot 17 \text{ ft} + 2 \cdot 13 \text{ ft}$
   $= 34 \text{ ft} + 26 \text{ ft}$
   $= 60 \text{ ft}$
3. P(the amount of fringe) $= 4 \cdot 50 \text{ in} = 200 \text{ in}$ or 16 ft 8 in.
4. $C \approx \frac{22}{7} \times 3\frac{1}{2} \text{ in} = \frac{\cancel{22}^{11}}{\cancel{7}} \times \frac{\cancel{7}}{\cancel{2}} \text{ in} = 11 \text{ in}$.
5. $C \approx 15.7 \text{ in}$.

12.3 Exercises

Find the perimeter or circumference of each figure. Use 3.14 for π and round your answer to one decimal place.

1.
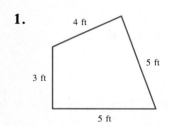

2.
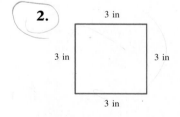

3.
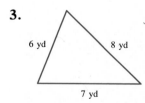

4.
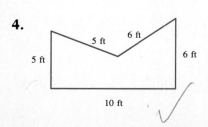

5.
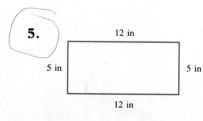

6.
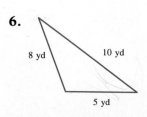

Section 12.3: Finding Perimeter and Circumference   521

7.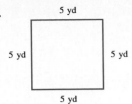

8.

9.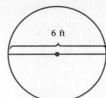

10.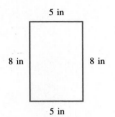

11.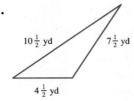

12.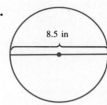

13.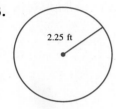

14.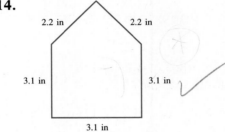

In Problems 15 and 16, use $\frac{22}{7}$ for π.

15.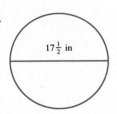

16.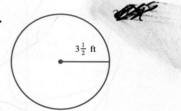

17. A rectangular picture window is 4 ft by 5 ft. A trim molding around the window costs 45¢ per foot. What will it cost to trim around the window?

18. You are fencing in a square backyard which is 40 ft by 40 ft. If the fencing costs $1.75 per foot, find the cost of the fencing that will be needed.

19. A door is 81 in high and 42 in wide. What will it cost to put weatherstripping along the sides and top of the door if the weatherstripping costs 40¢ a foot?

20. A poster measures 30 in by 36 in. Brenda chooses material that costs $1.25 per linear foot for framing. What will be her cost for framing material?

21. Marion is putting fringe around a rectangular tablecloth that measures 54 in by 48 in. The fringe is sold by the linear foot. How many feet will she need?

22. A house has the dimensions shown. If a rain gutter costs $2.25 per foot, what will it cost to put the rain gutter around the house? $405

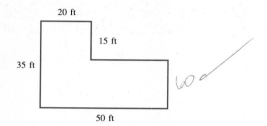

23. A path runs around a circular lake with a diameter of 1000 yd. Robert jogs around the lake three times for his morning run. How far has he run?

24. A circular rug is 6 ft in diameter. Binding for the edge costs $1.50 per linear yard. What will it cost to bind around the rug?

Find the perimeter of each figure. (The curves are semicircles.)

25.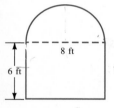

26.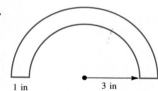

27.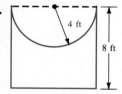

28.

Skillscan (Section 2.8)
Evaluate.

a. 5^2
b. 7^2
c. 3×4^2
d. 3×6^2

e. $(2.4)^2$
f. $3 \times (5.2)^2$
g. $\left(\dfrac{3}{4}\right)^2$
h. $3 \times \left(\dfrac{5}{2}\right)^2$

Section 12.3: Finding Perimeter and Circumference   523

Answers

1. 17 ft   3. 21 yd   5. 34 in   7. 20 yd   9. 18.8 ft   11. $22\frac{1}{2}$ yd   13. 14.1 ft   15. 55 in
17. $8.10   19. $6.80   21. 17 ft   23. 9420 yd   25. 32.6 ft   27. 36.6 ft   a. 25;   b. 49;   c. 48;
d. 108;   e. 5.76;   f. 81.12;   g. $\frac{9}{16}$;   h. $\frac{75}{4}$

12.3 Supplementary Exercises

Find the perimeter or circumference of each figure. Use 3.14 for π and round your answer to one decimal place.

1.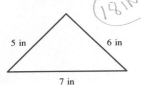

2.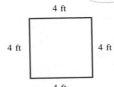

3.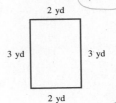

4.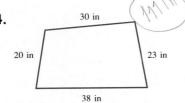

5.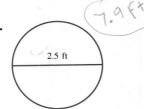

6.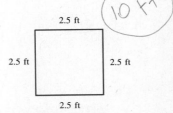

7.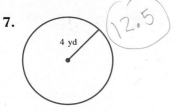

In Problem 8, use $\frac{22}{7}$ for π.

8.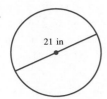

9. A room is 12 ft square. If a molding costs 35¢ per foot, what will it cost to put molding around the ceiling of the room?

10. You want to frame six photographs, each of which is 11 in by 14 in. How many linear feet of material will be needed for the frames?

11. If a wheel has a diameter of 20 in, how far will it travel in 3 revolutions?

12. A wire fence is to be placed around each of 12 new shrubs for protection. If each fence forms a circle with diameter 3 ft, how much fencing will be needed?

13. How far is one lap around the track shown?

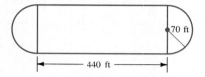

14. What is the perimeter of the field shown?

12.4 Finding Area

OBJECTIVES
1. To find the area of a figure
2. To apply area formulas

Figure 1

One square inch

The unit ft can be treated as though it were a number. So ft × ft can be written as ft². In this case, it is read "square feet."

Let's look now at the idea of *area*. Area is a measure that we give to a surface. It is measured in terms of *square units*. The area is the number of square units that are needed to cover the surface.

One standard unit of area measure is the *square inch*, written in². This is the measure of the surface contained in a square with sides of 1 in. See Figure 1.

Other units of area measure are the square foot (ft²) and the square yard (yd²).

Finding the area of a figure means finding the number of square units it contains. The simplest case is a rectangle. We found out how to calculate the area of a rectangle back in Chapter 2, but let's review briefly with an example.

Example 1

A rectangle has dimensions 5 in by 3 in. What is its area?

Section 12.4: Finding Area

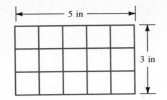

Figure 2

The length and width must be in terms of the same unit.

Solution Looking at Figure 2, we see that each row contains 5 in^2 and that there are 3 rows. The area is then 3×5 in^2 or 15 in^2.

In general, the area of a rectangle must be the product of its length and its width.

> **FORMULA (5)**
>
> $A = L \cdot W$ (the area of a rectangle)

Example 2

A room has dimensions 12 ft by 15 ft. Find its area.

Solution By Formula (5), with $L = 15$ ft and $W = 12$ ft, we have

$A = 15 \cdot 12 = 180$ ft^2

The area of the room is 180 ft^2.

CHECK YOURSELF 1

A desk top has dimensions 50 in by 25 in. What is the area of its surface?   1,250

We can also write a convenient formula for the area of a square. If the sides of the square have length S, we can write:

> **FORMULA (6)**
>
> $A = S \cdot S = S^2$ (the area of a square)

S^2 is read "S squared."

Example 3

You wish to cover a square table with a plastic laminate which costs 60¢ a square foot. If each side of the table measures 3 ft, what will it cost to cover the table?

Solution We first must find the area of the table. By Formula (6), with $S = 3$ ft, we have

Area = $(3 \text{ ft})^2 = 3 \text{ ft} \times 3 \text{ ft} = 9$ ft^2

Now multiply by the cost per square foot.

Cost = 9×60¢ = $5.40

CHECK YOURSELF 2

You wish to carpet a room that is a square, 4 yd by 4 yd, with carpet that costs $12 per square yard. What will be the total cost of the carpeting?

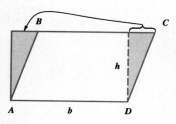

Figure 3

In Figure 3, ABCD is called a *parallelogram*. Its opposite sides are parallel and equal. Let's draw a line from D which forms a right angle with side BC. This cuts off one corner of the parallelogram. Now imagine that we move that corner over to the left side of the figure, as shown. This gives us a rectangle instead of a parallelogram. Since we haven't changed the area of the figure by moving the corner, the parallelogram has the same area as the rectangle, the product of the base and the height.

FORMULA (7)

$A = b \cdot h$ (area of a parallelogram)

Example 4

A parallelogram has the dimensions shown in Figure 4. What is its area?

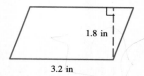

Figure 4

Solution By Formula (7), with $b = 3.2$ in and $h = 1.8$ in:

$A = 3.2 \text{ in} \times 1.8 \text{ in} = 5.76 \text{ in}^2$

CHECK YOURSELF 3

If the base of a parallelogram is $3\frac{1}{2}$ in and its height is $1\frac{1}{2}$ in, what is its area?

Another common geometric figure is the *triangle*. It has three sides. An example is triangle ABC in Figure 5.

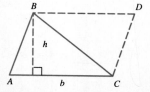

b is the base of the triangle.
h is the height or the *altitude* of the triangle.

Figure 5

Section 12.4: Finding Area 527

Once we have a formula for the area of a parallelogram, it is not hard to find the area of a triangle. If we draw the dotted lines from B to D and from C to D parallel to the sides of the triangle, we form a parallelogram. The area of the triangle is then one-half the area of the parallelogram [which is $b \cdot h$ by Formula (7)].

FORMULA (8)

$A = \dfrac{1}{2} \cdot b \cdot h$ (area of a triangle)

Example 5

A triangle has an altitude of 2.3 in, and its base is 3.4 in. What is its area?

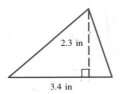

Figure 6

Solution By Formula (8),

$A = \dfrac{1}{2} \times 3.4 \text{ in} \times 2.3 \text{ in} = 3.91 \text{ in}^2$

CHECK YOURSELF 4

A triangle has a base 10 ft long and an altitude of 6 ft. Find its area.

The number pi (π), which we used in Section 12.3 to find circumference, is also used in finding the area of a circle. If r is the radius of a circle,

FORMULA (9)

$A = \pi r^2$ (area of a circle)

This is read, "Area equals pi r squared." You can multiply the radius by itself and then by pi.

Example 6

A circle has a radius of 7 in. What is its area?

153.9 IN

Figure 7

Solution We use $\frac{22}{7}$ for π. By Formula (9),

$$A \approx \frac{22}{7} \times (7 \text{ in})^2$$

$$\approx \frac{22}{\overset{1}{\cancel{7}}} \times \overset{1}{\cancel{7}} \text{ in} \times 7 \text{ in}$$

Again the area is an approximation because we use $\frac{22}{7}$, an approximation for π.

$$\approx 154 \text{ in}^2$$

CHECK YOURSELF 5

Find the area of a circle whose diameter is 4.8 in. Remember that the formula refers to the radius. Use 3.14 for π, and round your result to the nearest tenth of a square inch.   *72.4*

Sometimes we will want to convert from one square unit to another. For instance, look at one square yard, in Figure 8.

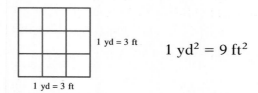

Figure 8

The table below gives some useful relationships.

SQUARE UNITS AND EQUIVALENTS
1 square foot (ft²) = 144 square inches (in²)
1 square yard (yd²) = 9 ft²
1 acre   = 4840 yd² = 43,560 ft²

Originally the acre was the area that could be plowed by a team of oxen in a day!

Example 7

A room has the dimensions 12 ft by 15 ft. How many square yards of linoleum will be needed to cover the floor?

Section 12.4: Finding Area

Solution

We first find the area in square feet; then convert to square yards.

$$A = 12 \text{ ft} \times 15 \text{ ft} = 180 \text{ ft}^2$$
$$= 180 \ (1 \text{ ft}^2)$$
$$= \overset{20}{\cancel{180}} \left(\frac{1}{\cancel{9}_1} \text{ yd}^2 \right) \quad \text{Replace 1 ft}^2 \text{ with } \frac{1}{9} \text{ yd}^2.$$
$$= 20 \text{ yd}^2$$

You can also make the conversion by multiplying by the unit ratio $\frac{1 \text{ yd}^2}{9 \text{ ft}^2}$, which is just 1.

CHECK YOURSELF 6

A hallway is 27 ft long by 4 ft wide. How many square yards of carpeting will be needed to carpet the hallway?

Example 8

A rectangular field is 220 yd long and 110 yd wide. Find its area in acres.

Solution

$$A = 220 \text{ yd} \times 110 \text{ yd} = 24{,}200 \text{ yd}^2$$
$$= 24{,}200 \ (1 \text{ yd}^2) \quad \text{Replace 1 yd}^2 \text{ with } \frac{1}{4840} \text{ acre.}$$
$$= \overset{5}{\cancel{24{,}200}} \left(\frac{1}{\cancel{4840}_1} \text{ acre} \right)$$
$$= 5 \text{ acres}$$

CHECK YOURSELF 7

A proposed site for an elementary school is 220 yd long by 198 yd wide. Find its area in acres.

CHECK YOURSELF ANSWERS

1. 1250 in^2.
2. $A = (4 \text{ yd})^2$
   $= 16 \text{ yd}^2$
   Cost $= \$12 \times 16 = \$192$
3. $A = 3\frac{1}{2} \text{ in} \times 1\frac{1}{2} \text{ in}$
   $= \frac{7}{2} \text{ in} \times \frac{3}{2} \text{ in}$
   $= 5\frac{1}{4} \text{ in}^2$
4. $A = \frac{1}{2} \times 10 \text{ ft} \times 6 \text{ ft}$
   $= 30 \text{ ft}^2$
5. The radius is half the diameter, and so r is 2.4 in.
   $A \approx 3.14 \times 2.4 \text{ in} \times 2.4 \text{ in}$
   $\approx 18.1 \text{ in}^2$
6. 12 yd^2.
7. 9 acres.

12.4 Exercises

Find the area of each figure. Use 3.14 for π, and round your answer to one decimal place.

1.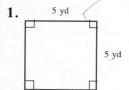

2.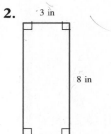

3.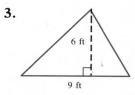

4.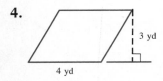

5.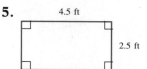

6.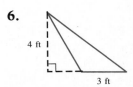

7.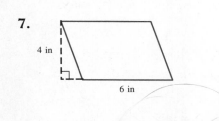

8.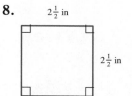

9.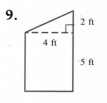

10.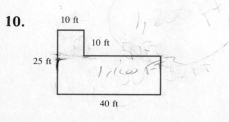

11.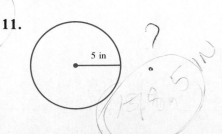

12.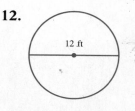

13. 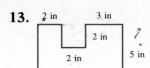   14. 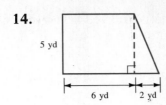   15.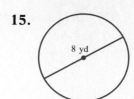

16.    17. 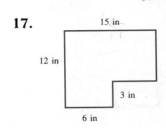   18.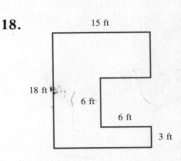

In Problems 19 and 20, use $\dfrac{22}{7}$ for π.

19. 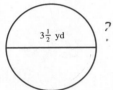   20.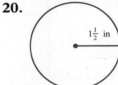

21. You wish to cover a bathroom floor with tiles 1 ft square which cost 60¢ each. If the bathroom is rectangular, $4\dfrac{1}{2}$ ft by 6 ft, how much will the tile cost?

22. A rectangular shed roof is 30 ft long and 20 ft wide. Roofing is sold in "squares" of 100 ft². How many squares will be needed to roof the shed?

23. An A-frame cabin has a triangular front with a base of 30 ft and a height of 20 ft. If the front is to be glass that costs $3 a square foot, what will be the cost of the glass?

24. Bill is painting two walls in a hallway that are 8 ft high by 24 ft long. The instructions on the paint can say that it will cover 400 ft²/gal. Will one gallon be enough for the job?

25. A roll of insulation will cover 80 ft². To insulate the ceiling of a room that is 12 ft by 20 ft, how many rolls will be needed?

26. A rectangular lawn has dimensions 60 ft by 40 ft. Instructions on a bottle of weed killer call for the use of 1 oz for every 100 ft² of grass. How many ounces will be needed for the lawn?

27. A room is 15 ft long by 12 feet wide with an 8-ft ceiling. There are two windows, one 6 ft by 4 ft and one 4 ft by 3 ft; and two doors, both 7 ft by 3 ft. What is the surface area of the walls that needs to be painted?

28. A circular piece of lawn has a radius of 25 ft. You have a bag of fertilizer which will cover 2000 ft² of lawn. Do you have enough?

29. A circular coffee table has a diameter of 3 ft. What will it cost to have the top refinished if the company charges $2 a square foot for the refinishing?

30. A circular terrace has a radius of 6 ft. If it will cost $12 per square yard to pave the terrace with brick, what will the total cost be?

31. A house addition is in the shape of a semicircle (a half circle) with radius 8 ft. What is its area?

32. A Tetra-Kite uses 12 pieces of plastic for its surface. Those pieces are triangular with base 12 in and height 12 in. How much material is needed for the kite?

33. You buy a lot which is 110 yd square. What is its size in acres?

34. You are making posters 24 in by 30 in. How many square feet of material will you need for six posters?

35. Andy is carpeting a recreation room which is 21 ft long by 15 ft wide. If the carpeting costs $12 per square yard, what will be the total cost of the carpet?

36. A shopping center is rectangular with dimensions 550 yd by 440 yd. What is its size in acres?

Find the area of the shaded part in each figure.

37.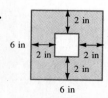

38.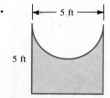

39.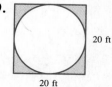

Section 12.4: Finding Area   533

40.

41.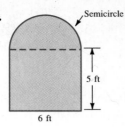
Semicircle

42.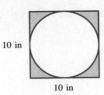

Answers
1. 25 yd²   3. 27 ft²   5. 11.3 ft²   7. 24 in²   9. 24 ft²   11. 78.5 in²   13. 31 in²   15. 50.2 yd²
17. 153 in²   19. $9\frac{5}{8}$ yd²   21. $16.20   23. $900   25. 3 rolls   27. 354 ft²   29. $14.13
31. 100.5 ft²   33. 2.5 acres   35. $420   37. 32 in²   39. 86 ft²   41. 44.1 ft²

12.4 Supplementary Exercises

Find the area of each figure. Use 3.14 for π, and round your answer to one decimal place.

1.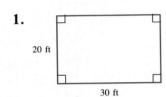

2.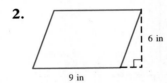

3.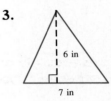

4.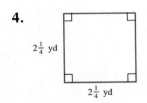

5.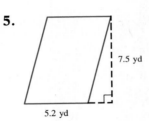

6.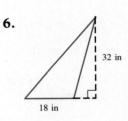

7.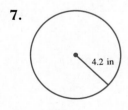

8.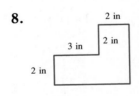

9.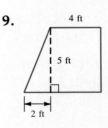

In Problem 10, use $\frac{22}{7}$ for π.

10.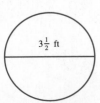

11. You are carpeting a room which measures 15 ft by 12 ft. If the carpeting costs $9 per square yard, how much will it cost to carpet the room?

12. A shopping center designs a triangular sign, 10 ft across its base and 8 ft high. If the material for the sign costs $3.50 a square foot, what will the material cost?

13. Which of these two cake pans is larger: a 9-in-diameter round pan or an 8-in square pan? What is the difference in their areas?

14. A sidewalk along two sides of a house has the dimensions shown below. What is the surface area of the sidewalk?

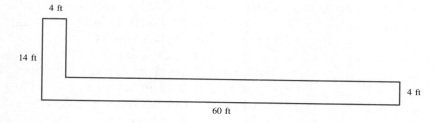

15. Diane wants to refinish a circular table that is 4 ft in diameter. If the refinishing costs $5 per square foot, what will the total cost be?

16. A corner fireplace hearth is in the shape of a quarter-circle with an 8-ft radius. If three bricks are needed for 1 ft² of area, how many bricks should be ordered?

17. Margaret has purchased a piece of land that is triangular in shape, with a base of 440 yd and a height 220 yd. What is the area of the land in acres?

18. A path around a garden has the dimensions shown below. What is the area of the path?

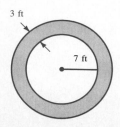

12.5 Finding Volume

OBJECTIVES
1. To find the volume of a solid
2. To apply volume formulas

Our last measurement section deals with finding *volumes*. The volume of a *solid* is the measure of the space contained in the solid.

A *solid* is a three-dimensional figure. It has length, width, and height.

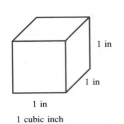

Figure 1

Volume is measured in *cubic units*. Examples are cubic inches (in^3), cubic feet (ft^3), and cubic yards (yd^3). A cubic inch, for instance, is the measure of the space contained in a cube that is one inch on each edge. See Figure 1.

In finding the volume of a figure, we want to know how many cubic units are contained in that figure. Let's start with a simple example, a *rectangular solid*. A rectangular solid is a very familiar figure. A box, a crate, and most rooms are all examples. Say that the dimensions of the solid are 5 in by 3 in by 2 in, as pictured in Figure 2. If we divide the solid into units of 1 in^3, we have two layers, each containing 3 by 5 units, or 15 in^3. Since there are two layers, the volume is 30 in^3.

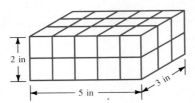

Figure 2

In general, we can see that the volume of a rectangular solid is the product of its length, width, and height.

FORMULA (10)

$V = L \cdot W \cdot H$ (volume of a rectangular solid)

Example 1

A crate has dimensions 3 ft by 4 ft by 2 ft. Find its volume.

Solution By Formula (10), with $L = 4$ ft, $W = 3$ ft, and $H = 2$ ft:

$V = 4 \text{ ft} \times 3 \text{ ft} \times 2 \text{ ft}$
$= 24 \text{ ft}^3$

We are not particularly worried about which is the length, which is the width, and which is the height, since the order in which we multiply won't change the result.

CHECK YOURSELF 1

A room is 15 ft long, 10 ft wide, and 8 ft high. What is its volume?

Another common geometric solid is the *cylinder*. One is shown in Figure 3. It is called a right circular cylinder because the top and the bottom are circles and the sides form a right angle with the base. Examples of such cylinders are cans, circular tanks, and any kind of tube or pipe. The volume of a right circular cylinder is the product of the area of its base πr^2 and its height h.

The base is just a circle with radius r, and so its area is πr^2.

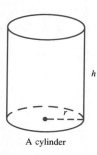

A cylinder

Figure 3

FORMULA (11)

$V = \pi r^2 h$ (volume of a cylinder)

Example 2

A tin can has the dimensions shown in Figure 4. What is the volume of the can?

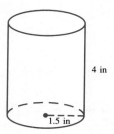

4 in

1.5 in

Figure 4

Solution By Formula (11),

$V \approx 3.14 \times (1.5 \text{ in})^2 \times 4 \text{ in}$
$\approx 28.3 \text{ in}^3$

Again this is an approximation for the volume because 3.14 is an approximation for π.

CHECK YOURSELF 2

A cylindrical water tank is 20 ft high and has a radius of 5 ft. What is its volume? Use 3.14 for π.

In the English system, volume or capacity is sometimes measured in pints, quarts, and gallons. Let's look at an example that relates these to cubic units.

Example 3

A paint can is 6 in tall and has a radius of $3\frac{1}{2}$ in. Will it hold a gallon of paint? **Note** 1 gal = 231 in^3.

Solution

First we find the volume, using $\frac{22}{7}$ for π.

$$V \approx \frac{22}{7} \times \left(\frac{7}{2} \text{ in}\right)^2 \times 6 \text{ in} \qquad \text{Write } 3\frac{1}{2} \text{ in as } \frac{7}{2} \text{ in.}$$

$$= 231 \text{ in}^3 \qquad \text{It will hold about 1 gal.}$$

CHECK YOURSELF 3

A cylindrical can has a radius of 7 in and a height of 12 in. What is its volume in gallons? Use $\frac{22}{7}$ for π.

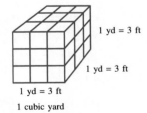

1 yd = 3 ft
1 yd = 3 ft
1 yd = 3 ft
1 cubic yard

For some applications, we will need to convert from one cubic unit to another. For instance, look at one cubic yard

1 yd^3 = 27 ft^3

The table below gives some useful relationships.

CUBIC UNITS AND EQUIVALENTS
1 cubic foot (ft^3) = 1728 cubic inches (in^3)
1 cubic yard (yd^3) = 27 ft^3

Example 4

A 6-in-thick concrete wall is to be 20 ft long by 3 ft high. How many cubic yards of concrete will be needed?

Solution We will first find the volume in terms of cubic feet.

$V = L \cdot W \cdot H$

$= 20 \text{ ft} \times \dfrac{1}{2} \text{ ft} \times 3 \text{ ft}$ Note that we convert the thickness, or width, from inches to feet.

$= 30 \text{ ft}^3$

$= \overset{10}{\cancel{30}} \text{ ft}^3 \left(\dfrac{1 \text{ yd}^3}{\underset{9}{\cancel{27} \text{ ft}^3}} \right)$ We have chosen to multiply by a unit ratio to convert from ft^3 to yd^3.

$= \dfrac{10}{9} \text{ yd}^3 = 1\dfrac{1}{9} \text{ yd}^3$

CHECK YOURSELF 4

A concrete walk, 4 in thick, 27 feet long, and 6 ft wide, will require how many cubic yards of concrete?

CHECK YOURSELF ANSWERS

1. $V = 15 \text{ ft} \times 10 \text{ ft} \times 8 \text{ ft}$
   $= 1200 \text{ ft}^3$
2. $V \approx 3.14 \times (5 \text{ ft})^2 \times 20 \text{ ft}$
   $\approx 1570 \text{ ft}^3$
3. 8 gal.
4. 2 yd^3.

12.5 Exercises

Find the volume of each solid shown. Use 3.14 for π, and round your answer to one decimal place.

1.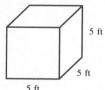

2.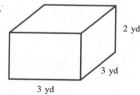

3.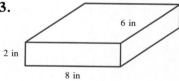

4.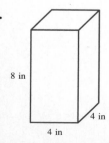

5.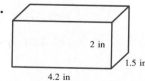

6.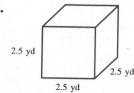

7.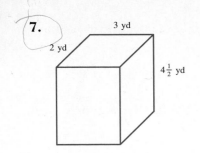

8.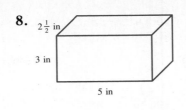

9.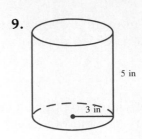

10.

11.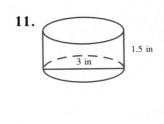

12.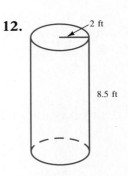

Solve each of the following applications.

13. A shipping container is 5 ft by 3 ft by 2 ft. What is its volume?

14. A cord of wood is 4 ft by 4 ft by 8 ft. What is its volume?

15. The inside dimensions of a grocery store's cooler are 6 ft by 6 ft by 3 ft. What is the capacity of the cooler in cubic yards?

16. How many gallons are there in 1 ft^3? (1 gal = 231 in^3.) Give your answer to the nearest tenth of a gallon.

17. A coffee can has a 4-in diameter and is 5 in tall. What is its volume?

18. A storage bin is 18 ft long, 6 ft wide, and 3 ft high. What is its volume in cubic feet? What is its volume in cubic yards?

19. A concrete wall is to be 4 in thick, 6 ft high, and 54 ft long. How many cubic yards of concrete will it require?

20. A water tank has a height of 20 ft and a diameter of 18 ft. What is its volume in cubic feet? How many gallons will it hold? *Hint:* See Problem 16.

21. A rectangular swimming pool is 30 ft by 20 ft with an average depth of 5 ft. How many gallons of water will it take to fill the pool? *Hint:* See Problem 16.

22. Nick wants to pour a concrete walk 54 ft long by 3 ft wide by 6 in thick. How many cubic yards of concrete should he order?

23. Julie wants to cover a rectangular garden, which measures 15 ft by 36 ft, with a 3-in layer of topsoil. How many cubic yards of topsoil should she order?

24. An oil drum has a radius of 20 in and a height of 5 ft. What is its volume in cubic feet?

25. A *cone* has one-third the volume of a cylinder with the same height and base. What is the volume of a cone with a height of 10 in and a base of radius 3 in?

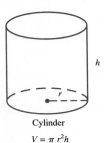
Cylinder
$V = \pi r^2 h$

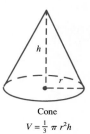
Cone
$V = \frac{1}{3} \pi r^2 h$

26. A *pyramid* has one-third the volume of a rectangular solid with the same height and base. Find the volume of a pyramid with a base 4 in by 4 in and a height of 6 in.

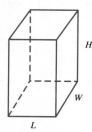
Rectangular solid
$V = L \cdot W \cdot H$

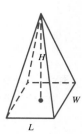
Pyramid
$V = \frac{1}{3} L \cdot W \cdot H$

Answers
1. 125 ft³   3. 96 in³   5. 12.6 in³   7. 27 yd³   9. 141.3 in³   11. 10.6 in³   13. 30 ft³   15. 4 yd³
17. 62.8 in³   19. 4 yd³   21. 22,500 gal   23. 5 yd³   25. 94.2 in³

12.5 Supplementary Exercises

Find the volume of each solid shown. Use 3.14 for π, and round your answer to one decimal place.

1.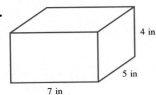

2.

3.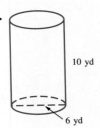

4.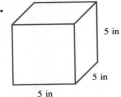

5.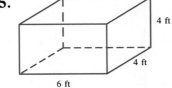

6.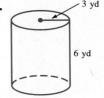

Solve each of the following applications.

7. A shipping crate is 6 ft long, 3 ft wide, and 3 ft tall. What is its volume in cubic yards?

8. A rectangular holding tank is 10 ft 6 in long, 8 ft wide, and 3 ft 6 in deep. What is its volume?

9. What is the volume of a cylindrical wastebasket that is 15 in high with a diameter of 12 in?

10. A container is 20 in tall and has a radius of 8 in. How many gallons will it hold? (1 gal = 231 in^3.) Give your answer to the nearest tenth of a gallon.

11. A pipe has an inside diameter of 12 in and a length of 10 ft. What is its volume?

12. A yard 30 ft by 36 ft is to be covered with 3 in of topsoil. How many cubic yards of topsoil will be required?

SELF-TEST for Chapter Twelve

The purpose of the Self-Test is to help you check your progress and review for a chapter test in class. Allow yourself about an hour to take the test. When you are done, check your answers in the back of the book. If you missed any problems, be sure to go back and review the appropriate sections in the chapter and do the supplementary exercises provided there.

In Problems 1 to 4, complete the statements.

1. 7 ft =   in

2. 240 min =   h

3. 3 pt =   fl oz

4. 64 oz =   lb

In Problems 5 and 6, simplify.

5. 4 ft 18 in

6. 6 min 90 s

In Problems 7 to 10, do the indicated operations.

7.   7 ft 9 in
   +3 ft 8 in

8.   7 lb 3 oz
   −4 lb 10 oz

9. 4 × (2 h 40 min)

10. $\dfrac{9 \text{ lb } 15 \text{ oz}}{3}$

In Problems 11 to 13, find the perimeter or circumference of the figures. Use 3.14 for π, and round your answer to one decimal place.

11.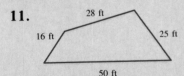

12.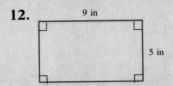

13.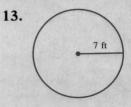

In Problems 14 and 15, solve the applications.

14. The Martins are fencing in a rectangular yard that is 120 ft long by 60 ft wide. If the fencing costs $2.50 per linear foot, what will be the total cost of the fencing?

15. You want to put a lace border around a circular tablecloth that is 60 inches in diameter. If you must buy the lace by the foot, and it costs 20¢ per foot, what will you pay for the lace?

In Problems 16 to 21, find the area of the figures. For the circles, use 3.14 for π, and round your answer to one decimal place.

16.

25 in
40 in

17.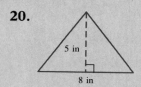

18.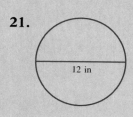

19.

3.5 ft
3.5 ft

20.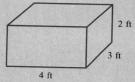

21.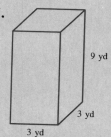

In Problems 22 to 24, solve the applications.

22. You wish to carpet a living room with carpeting that costs $15 per square yard. If the room measures 15 ft by 12 ft, how much will the carpeting cost?

23. A store designs a rectangular sign 20 ft long by 12 ft high. If the material for the sign costs $4.50 per square foot, what will the material cost?

24. A radio station can be heard in a circular area with a radius of 100 mi. What is the size of the listening area?

In Problems 25 to 27, find the volume of the solids. Use 3.14 for π, and round your answer to one decimal place.

25.

2 ft
3 ft
4 ft

26.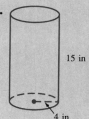

27.

9 yd
3 yd
3 yd

In Problems 28 to 30, solve the applications.

28. A room measures 12 ft by 18 ft by 8 ft. What is the volume of the room in cubic yards?

29. A pipe is 20 ft long and has an inside radius of 6 in. What is its volume in cubic feet?

30. You wish to cover an area 54 ft by 27 ft with a layer of topsoil 4 in thick. How many cubic yards of topsoil will you need?

PRETEST for Chapter Thirteen

The Metric System of Measurement

This pretest will point out any difficulties you may be having with the metric system. Do all the problems. Then check your answers on the following page.

1. 3 km = _____ m

2. 5 m = _____ cm

3. 8000 mm = _____ m

4. 5 in = _____ cm

5. 3 kg = _____ g

6. 6000 g = _____ kg

7. 5 kg = _____ lb

8. 2000 mL = _____ L

9. 5 L = _____ mL

10. The temperature in your classroom is (choose the most reasonable answer):
    30°C   20°C   10°C

ANSWERS TO PRETEST

For help with similar problems, turn to the section indicated.

1. 3000 (Section 13.1);
2. 500 (Section 13.1);
3. 8 (Section 13.1);
4. 12.7 (Section 13.1);
5. 3000 (Section 13.2);
6. 6 (Section 13.2);
7. 11 (Section 13.2);
8. 2 (Section 13.3);
9. 5000 (Section 13.3);
10. 20°C (Section 13.4)

Chapter Thirteen

The Metric System of Measurement

13.1 Metric Units of Length

OBJECTIVES
1. To know the meaning of the metric prefixes
2. To use metric units of length

Even in the United States, the metric system is used in science, medicine, and many other areas.

In the last chapter we studied the English system of measurement, which is used in the United States and a few other countries. Our work in this chapter will concentrate on the *metric system*, used throughout the rest of the world.

The basic unit of length in the metric system is also spelled metre; *either way is correct.*

The entire system is based on one unit of length, the *meter*. In the eighteenth century the meter was defined to be one ten-millionth of the distance from the north pole to the equator. Today the meter is scientifically defined in terms of a wavelength in the spectrum of krypton-86 gas.

In the metric system you don't have to worry about things like 12 in to a foot, 5280 ft to a mile, and all that.

One big advantage of the metric system is that you can convert from one unit to another by simply multiplying or dividing by powers of 10. This advantage and the need for uniformity throughout the world has led to legislation which will promote the use of the metric system in the United States. It is coming, so let's see how it works.

Let's start with measures of length and compare a basic English unit, the yard, with the meter (m).

The meter is one of the basic units of the International System of Units (abbreviated SI). This is a standardization of the metric system agreed to by scientists in 1960.

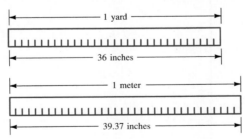

As you can see, the meter is just slightly longer than the yard. It is used for measuring the same things you might measure in feet or yards. Just to get a feel for the size of the meter:

A room might be 6 meters (6 m) long.

There is a standard pattern of abbreviation in the metric system. We will introduce the abbreviation for each term as we go along. The abbreviation for meter is m (no period!).

A building lot could be 30 m wide.

A fence is 2 m tall.

547

CHECK YOURSELF 1

Try to estimate the following lengths in meters.

(1) A traffic lane is _____ m wide.
(2) A small car is _____ m long.
(3) You are _____ m tall.

For other units of length, the meter is multiplied or divided by powers of 10. One commonly used unit is the *centimeter* (cm).

> The prefix *centi* means one-hundredth. This should be no surprise. What is our cent? It is one-hundredth of a dollar.

$$1 \text{ centimeter (cm)} = \frac{1}{100} \text{ meter (m)}$$

The drawing below relates the centimeter and the meter:

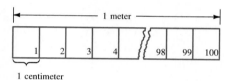

There are 100 cm to 1 m.

Just to give you an idea of the size of the centimeter, it is about the width of your little finger. There are about $2\frac{1}{2}$ cm to an inch, and the unit is used to measure small objects. To get used to the centimeter:

A small paperback book is 10 cm wide.

A playing card is 8 cm long.

A ballpoint pen is 16 cm long.

CHECK YOURSELF 2

Try to estimate each of the following. Then use a metric ruler to check your guess.

(1) This page is _____ cm long.
(2) A dollar bill is _____ cm long.
(3) The seat of the chair you are on is _____ cm from the floor.

To measure *very* small things, the millimeter is used. To give you an idea of its size, the *millimeter* is about the thickness of a new dime.

> The prefix *milli* means one-thousandth.

$$1 \text{ millimeter (mm)} = \frac{1}{1000} \text{ meter (m)}$$

Section 13.1: Metric Units of Length 549

The diagram below will help you see the relationships of the three units we have looked at.

Note that there are 10 millimeters to one centimeter.

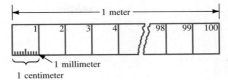

To get used to the millimeter:

Standard camera film is 35 mm wide.

A paper clip is 5 mm wide.

Some cigarettes are 100 mm long.

CHECK YOURSELF 3

Try to estimate each of the following. Then use a metric ruler to check your guess.

(1) Your pencil is ____ mm wide.
(2) The tabletop you are working on is ____ mm thick.

The *kilometer* is used to measure long distances. The kilometer is about six-tenths of a mile.

The prefix *kilo* means one thousand. You are already familiar with this. For instance, 1 kilowatt = 1000 watts

> 1 kilometer (km) = 1000 meters (m)

Using the kilometer:

The distance from New York to Boston is 325 km.

A distance run is 15 km.

Now that you have seen the four commonly used units of length in the metric system, let's review with the following exercise.

CHECK YOURSELF 4

Choose the most reasonable measure in each of the following statements.

(1) The width of a doorway: 50 mm, 1 m, or 50 cm.
(2) The length of your pencil: 20 m, 20 mm, or 20 cm.
(3) The distance from your house to school: 500 km, 5 km, or 50 m.
(4) The height of a basketball center: 2.2 m, 22 m, or 22 cm.
(5) The width of a matchbook: 30 cm, 30 mm, or 3 mm.

Of course, this is easy. All we need to do is move the decimal point to the right or left the required number of places. Again, that's the big advantage of the metric system.

Remember this rule: The *smaller* the unit, the *more* of them it takes, so *multiply*.

As we said earlier, to convert units of measure within the metric system, all we have to do is multiply or divide by the appropriate power of 10.

> To convert to a *smaller* unit of measure we will *multiply* by a power of 10, moving the decimal point *to the right*.

Example 1

5.2 m = 520 cm	Multiply by 100 to convert from meters to centimeters.
8 km = 8000 m	Multiply by 1000.
6.5 m = 6500 mm	Multiply by 1000.
2.5 cm = 25 mm	Multiply by 10.

CHECK YOURSELF 5

Complete the following. Remember, you don't need to do any calculation. Just move the decimal point the appropriate number of places, and write the answer.

(1) 3 km =   m
(2) 4.5 m =   cm
(3) 1.2 m =   mm
(4) 6.5 cm =   mm

Remember this rule: The *larger* the unit, the *fewer* of them it takes, so *divide*.

> To convert to a *larger* unit of measure, we will *divide* by a power of 10, moving the decimal point *to the left*.

Example 2

43 mm = 4.3 cm	Divide by 10.
3000 m = 3 km	Divide by 1000.
450 cm = 4.5 m	Divide by 100.

CHECK YOURSELF 6

Complete the following statements.

(1) 750 cm =   m
(2) 5000 m =   km
(3) 78 mm =   cm
(4) 3500 mm =   m

Section 13.1: Metric Units of Length 551

We have introduced all the commonly used units of linear measure in the metric system. There are other prefixes that can be used to form other linear measures. The prefix *deci* means one-tenth, *deka* means times 10, and *hecto* means times 100. Their use is illustrated in the following table:

USING METRIC PREFIXES

1 *milli*meter (mm) = $\frac{1}{1000}$ m

1 *centi*meter (cm) = $\frac{1}{100}$ m

1 *deci*meter (dm) = $\frac{1}{10}$ m

1 meter (m)

1 *deka*meter (dam) = 10 m

1 *hecto*meter (hm) = 100 m

1 *kilo*meter (km) = 1000 m

In converting between metric units, some students like to use the following chart.

To convert to smaller units ⟶

km	hm	dam	m	dm	cm	mm
1000 m	100 m	10 m	1 m	0.1 m	0.01 m	0.001 m

⟵ To convert to larger units

To convert between metric units, just move the decimal point to the left or right the number of places indicated by the chart.

Example 3

800 cm = ? m

Solution To convert from centimeters to meters, you can see from the chart that you must move the decimal point *two places to the left*.

800 cm = 8̬00 m = 8 m

Example 4

5000 m = ? km

Solution To convert from meters to kilometers, move the decimal point *three places to the left*.

5000 m = 5̬000 km = 5 km

Example 5

6 m = ? mm

Solution To convert from meters to millimeters, move the decimal point *three places to the right*.

6 m = 6000 mm = 6000 mm

While the important thing is to become familiar with the units of the metric system, you may also want to be able to convert from one system to the other.

The following table will be helpful in converting from the metric to the English system of measure.

> 1 m = 39.37 in
> 1 cm = 0.394 in
> 1 km = 0.62 mi

Example 6

Complete the following statement.

3 m = ? in

$$3 \text{ m} = 3 \text{ m} \left(\frac{39.37 \text{ in}}{1 \text{ m}} \right) = 118.11 \text{ in}$$

This is a unit ratio equal to 1.

There are all kinds of ways to convert between the systems. You can just multiply or divide by the appropriate factor. We have chosen to use the unit ratio idea introduced in the last chapter.

Example 7

Complete the following statement.

25 km = ? mi

$$25 \text{ km} = 25 \text{ km} \left(\frac{0.62 \text{ mi}}{1 \text{ km}} \right) = 15.5 \text{ mi}$$

CHECK YOURSELF 7

Complete the following statement.

5 cm = ___ in

Guidebooks for American travelers give this tip for a quick conversion from kilometers to miles: Multiply the number of kilome-

Section 13.1: Metric Units of Length   553

ters by 6 and drop the last digit. The result will be the approximate number of miles.

Example 8

A speed limit sign reads 90 km/h.

What this does is multiply the number of kilometers by 0.6.

90 × 6 = 540   Multiply by 6 and drop the last digit, in this case 0.

Drop this digit.

90 km/h is approximately 54 mi/h.

You may also want to be able to convert units of the English system to those of the metric system. You can use the following conversion factors.

---
1 in = 2.54 cm
1 yd = 0.914 m
1 mi = 1.6 km

Example 9

Complete the following statements.

(a) 5 in = ? cm

$$5 \text{ in} = 5 \text{ in} \left(\frac{2.54 \text{ cm}}{1 \text{ in}} \right)$$ $\frac{2.54 \text{ cm}}{1 \text{ in}}$ is a unit ratio equal to 1.

$$= 12.7 \text{ cm}$$

(b) 12 mi = ? km

$$12 \text{ mi} = 12 \text{ mi} \left(\frac{1.6 \text{ km}}{1 \text{ mi}} \right) = 19.2 \text{ km}$$

(c) 5 yd = ? m

$$5 \text{ yd} = 5 \text{ yd} \left(\frac{0.914 \text{ m}}{1 \text{ yd}} \right) = 4.57 \text{ m}$$

CHECK YOURSELF 8

Complete the following statements.

(1) 8 in =   cm
(2) 20 mi =   km

CHECK YOURSELF ANSWERS

1. (1) About 3 m; (2) perhaps 5 m; (3) you are probably between 1.5 and 2 m tall.
2. (1) About 28 cm; (2) almost 16 cm; (3) about 45 cm.
3. (1) About 8 mm; (2) probably between 25 and 30 mm.
4. (1) 1 m; (2) 20 cm; (3) 5 km; (4) 2.2 m; (5) 30 mm.
5. (1) 3000 m; (2) 450 cm; (3) 1200 mm; (4) 65 mm.
6. (1) 7.5 m; (2) 5 km; (3) 7.8 cm; (4) 3.5 m.
7. 1.97 in.
8. (1) 20.32 cm; (2) 32 km.

13.1 Exercises

Choose the most reasonable measure.

1. The height of a ceiling.
   (*a*) 25 m
   (*b*) 2.5 m
   (*c*) 25 cm

2. The diameter of a quarter.
   (*a*) 24 mm
   (*b*) 2.4 mm
   (*c*) 24 cm

3. The height of a kitchen counter.
   (*a*) 9 m
   (*b*) 9 cm
   (*c*) 90 cm

4. The diagonal measure of a television screen.
   (*a*) 50 mm
   (*b*) 50 cm
   (*c*) 5 m

5. The height of a two-story building.
   (*a*) 7 m
   (*b*) 70 m
   (*c*) 70 cm

6. An hour's drive on a freeway.
   (*a*) 9 km
   (*b*) 90 m
   (*c*) 90 km

7. The width of a roll of Scotch tape.
   (*a*) 1.27 mm
   (*b*) 12.7 mm
   (*c*) 12.7 cm

8. The width of a sheet of typing paper.
   (*a*) 21.6 cm
   (*b*) 21.6 mm
   (*c*) 2.16 cm

9. The thickness of window glass.
   (*a*) 5 mm
   (*b*) 5 cm
   (*c*) 50 mm

10. The height of a refrigerator.
    (*a*) 16 m
    (*b*) 16 cm
    (*c*) 160 cm

11. The length of a ballpoint pen.
    (*a*) 16 mm
    (*b*) 16 m
    (*c*) 16 cm

12. The width of a hand-calculator key.
    (*a*) 1.2 mm
    (*b*) 12 mm
    (*c*) 12 cm

Section 13.1: Metric Units of Length

Complete each statement using a metric unit of length.

13. A playing card is 6 _____ wide.

14. The diameter of a penny is 19 _____

15. A doorway is 2 _____ high.

16. A table knife is 22 _____ long.

17. A basketball court is 28 _____ long.

18. A commercial jet flies 800 _____ per hour.

19. The width of a nail file is 12 _____

20. The distance from New York to Washington, D.C. is 360 _____

21. A recreation room is 6 _____ long.

22. A ruler is 22 _____ wide.

23. A long-distance run is 35 _____

24. A paperback book is 11 _____ wide.

Complete each statement. (You can use the tables on pages 552 and 553.)

25. 4000 mm = _____ m   26. 5 in = _____ cm   27. 8 m = _____ cm

28. 55 mm = _____ cm   29. 150 km = _____ mi   30. 500 cm = _____ m

31. 9 cm = _____ in   32. 150 mi = _____ km   33. 3000 m = _____ km

34. 5 yd = _____ m   35. 8 cm = _____ mm   36. 45 cm = _____ mm

37. 5 km =   m   **38.** 4000 m =   km   **39.** 5 m =   mm

40. 7 km =   m

Use a metric ruler to measure the necessary dimensions and complete the statements.

41. The perimeter of the parallelogram is   cm.

42. The perimeter of the triangle is   mm.

43. The perimeter of the rectangle below is   cm.

44. Its area is   cm².

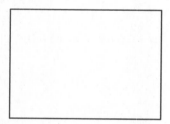

45. The perimeter of the square below is   mm.

46. Its area is   mm².

Section 13.1: Metric Units of Length   557

47. The circumference of the circle below is _____ mm.

48. Its area is _____ mm².

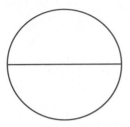

Answers
1. b **3.** c **5.** a **7.** b **9.** a **11.** c **13.** cm **15.** m **17.** m **19.** mm **21.** m **23.** km
25. 4 **27.** 800 **29.** 93 **31.** $9 \text{ cm} = 9 \text{ cm} \left(\dfrac{0.394 \text{ in}}{1 \text{ cm}} \right) = 3.546 \text{ in}$ **33.** 3 **35.** 80 **37.** 5000
39. 5000 **41.** 12 **43.** 14 **45.** 100 **47.** 94.2

13.1 Supplementary Exercises

Choose the most reasonable measure.

1. A marathon race.
(a) 40 km
(b) 400 km
(c) 400 m

2. The distance around your wrist.
(a) 15 mm
(b) 15 cm
(c) 1.5 m

3. The diameter of a penny.
(a) 19 cm
(b) 1.9 mm
(c) 19 mm

4. The width of a portable television screen.
(a) 28 mm
(b) 28 cm
(c) 2.8 m

5. The height of a doorway.
(a) 200 mm
(b) 20 m
(c) 2 m

6. The length of your car key.
(a) 60 mm
(b) 60 cm
(c) 6 m

Complete each statement using a metric unit of length.

7. A matchbook is 39 _____ wide.

8. The distance from San Francisco to Los Angeles is 618

9. A 1-lb coffee can has a diameter of 10 _____.

10. A fence is 2 _____ high.

11. A living room is 5 _____ long.

12. A pencil is 19 _____ long.

Complete each statement. (You can use the tables on pages 552 and 553.)

13. 2 km = _____ m
14. 3 in = _____ cm
15. 25 mm = _____ cm

16. 8 m = _____ mm
17. 6 cm = _____ in
18. 3 mi = _____ km

19. 3000 mm = _____ m
20. 2 m = _____ cm

Use a metric ruler to measure the necessary dimensions and complete the statements.

21. The perimeter of this parallelogram is _____ cm.

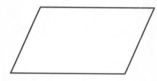

22. The perimeter of this rectangle is _____ mm.

23. The area of this rectangle is _____ mm².

24. The circumference of this circle is _____ cm.

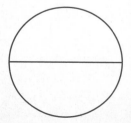

13.2 Metric Units of Weight

OBJECTIVE
To use metric units of weight

The gram is defined in terms of the meter. It is the weight of 1 cm³ of water at 4°C.

The basic unit of weight in the metric system is a very small unit called the *gram*. Think of a paper clip. It weighs roughly 1 gram. About 28 of these grams will make 1 ounce in the English system. The gram is most often used to measure items that are fairly light.

For heavier items, a more convenient unit of weight is the *kilogram*. From the prefix "kilo" you should know:

Technically the gram is a unit of mass rather than weight. Weight is the force of gravity on an object. Thus astronauts weigh less *on the moon than on earth even though their masses are unchanged. For common use on earth, the terms, mass and weight, are still used interchangeably.*

> 1 kilogram (kg) = 1000 grams (g)

A kilogram is just a bit more than 2 lb.

To help you get used to these units:

The gram was considered too small a unit, and so the kilogram *is the basic unit of mass in SI.*

The weight of a box of breakfast cereal is 320 g.

A collie dog weighs 35 kg.

A dinner fork weighs 50 g.

A woman might weigh 50 kg.

The weight of a nickel is 5 g.

Now try the following exercises.

CHECK YOURSELF 1

Choose the most reasonable measure.

(1) A penny: 30 g, 3 g, or 3 kg.
(2) A bar of soap: 120 g, 12 g, or 1.2 kg.
(3) A car: 5000 kg, 1000 kg, or 5000 g.
(4) A newborn baby: 35 g, 35 kg, or 3.5 kg.
(5) A bag of potato chips: 170 g, 1.7 kg, or 17 g.
(6) A tape cassette: 20 g, 2 kg, or 2 g.

Another metric unit of weight that you will encounter is the *milligram*. From the prefix you should see that

The prefix milli *means* $\frac{1}{1000}$.

> 1 milligram (mg) = $\frac{1}{1000}$ gram (g)

This is an extremely small unit. However, you will see it used. One case is in medicine for measuring drug amounts. For example, an aspirin weighs 300 mg.

Just as with units of length, converting metric units of weight is simply a matter of moving the decimal point. The following chart will help.

To convert to smaller units ⟶

kg*	hg	dag	g*	dg	cg	mg*
1000 g	100 g	10 g	1 g	0.1 g	0.01 g	0.001 g

⟵ To convert to larger units

*These are the units in common use.

Example 1

Complete the following statements.

(a) 7 kg = _____ g

We are converting to a *smaller* unit.

Move the decimal point *three places to the right* (to multiply by 1000).

7 kg = 7000 g = 7000 g

(b) 5000 mg = _____ g

We are converting to a *larger* unit.

Move the decimal point *three places to the left* (to divide by 1000).

5000 mg = 5 000. g = 5 g

CHECK YOURSELF 2

(1) 3000 g = _____ kg
(2) 500 cg = _____ g

Once again, you may want to convert between units of weight in the two systems. The following gives the necessary information.

Converting from metric to English units:

1 kilogram (kg) = 2.2 pounds (lb)

Converting from English to metric units:

1 lb = 0.45 kg

1 ounce (oz) = 28 grams (g)

Section 13.2: Metric Units of Weight

Example 2

A beef roast weighs 3 kg. What is its weight in pounds?

$$3 \text{ kg} = 3 \text{ kg} \left(\frac{2.2 \text{ lb}}{1 \text{ kg}} \right) = 6.6 \text{ lb}$$

CHECK YOURSELF 3

A radio weighs 4 kg. What is its weight in pounds?

Example 3

A jar of spices weighs $1\frac{1}{2}$ oz. What is its weight in grams?

$$1\frac{1}{2} \text{ oz} = 1.5 \text{ oz} = 1.5 \text{ oz} \left(\frac{28 \text{ g}}{1 \text{ oz}} \right) = 42 \text{ g}$$

CHECK YOURSELF 4

A bag of peanuts weighs $3\frac{1}{2}$ oz. What is its weight in grams?

Example 4

A package weighs 5 lb. What is its weight in kilograms?

$$5 \text{ lb} = 5 \text{ lb} \left(\frac{0.45 \text{ kg}}{1 \text{ lb}} \right) = 2.25 \text{ kg}$$

CHECK YOURSELF 5

If your cat weighs 8 lb, what is her weight in kilograms?

CHECK YOURSELF ANSWERS

1. (1) 3 g; (2) 120 g; (3) 1000 kg; (4) 3.5 kg; (5) 170 g; (6) 20 g.
2. (1) 3; (2) 5.   **3.** 8.8 lb.   **4.** 98 g.   **5.** 3.6 kg.

13.2 Exercises

Choose the most reasonable measure of weight.

1. A nickel.
   (a) 5 kg
   (b) 5 g
   (c) 50 g

2. A portable television set.
   (a) 8 g
   (b) 8 kg
   (c) 80 kg

3. A flashlight battery.
   (a) 8 g
   (b) 8 kg
   (c) 80 g

4. A 10-year-old boy.
   (a) 30 kg
   (b) 3 kg
   (c) 300 g

5. A Volkswagen Rabbit.
   (a) 100 kg
   (b) 1000 kg
   (c) 1000 g

6. A 10-lb bag of flour.
   (a) 45 kg
   (b) 4.5 kg
   (c) 45 g

7. A dinner fork.
   (a) 50 g
   (b) 5 g
   (c) 5 kg

8. A can of spices.
   (a) 3 g
   (b) 300 g
   (c) 30 g

9. A slice of bread.
   (a) 2 g
   (b) 20 g
   (c) 2 kg

10. A house paintbrush.
    (a) 120 g
    (b) 12 kg
    (c) 12 g

11. A sugar cube.
    (a) 2 mg
    (b) 20 g
    (c) 2 g

12. A salt shaker.
    (a) 10 g
    (b) 100 g
    (c) 1 g

Complete each statement using a metric unit of weight.

13. A marshmallow weighs 5 _____ .

14. A toaster weighs 2 _____ .

15. A _____ is $\frac{1}{1000}$ of a gram.

16. A bag of peanuts weighs 100 _____ .

17. An electric razor weighs 250 _____ .

18. A soup spoon weighs 50 _____ .

Section 13.2: Metric Units of Weight

19. A heavyweight boxer weighs 98   .

20. A vitamin C tablet weighs 500   .

21. A cigarette lighter weighs 30   .

22. A clock radio weighs 1.5   .

23. A household broom weighs 300   .

24. A 60-W light globe weighs 25   .

Complete each statement. (You can use the table on page 560.)

25. 8 kg =   g   **26.** 5000 mg =   g   **27.** 6 lb =   kg

28. 3 kg =   g   **29.** 8 oz =   g   **30.** 6 kg =   lb

31. 3 g =   mg   **32.** 2000 g =   kg

Answers

1. 5 g   **3.** 80 g   **5.** 1000 kg   **7.** 50 g   **9.** 20 g   **11.** 2 g   **13.** g   **15.** mg   **17.** g   **19.** kg
21. g   **23.** g   **25.** 8000   **27.** 6 lb = 6 lb $\left(\frac{0.45 \text{ kg}}{1 \text{ lb}}\right)$ = 2.7 kg   **29.** 224   **31.** 3000

13.2 Supplementary Exercises

Choose the most reasonable measure of weight.

1. A quarter.
 (a) 6 g
 (b) 6 kg
 (c) 60 g

2. A tube of toothpaste.
 (a) 20 kg
 (b) 200 g
 (c) 20 g

3. A refrigerator.
 (a) 120 kg
 (b) 1200 kg
 (c) 12 kg

4. A paperback book.
 (a) 1.2 kg
 (b) 120 g
 (c) 12 g

5. A package of cigarettes.
 (a) 2.6 g
 (b) 26 g
 (c) 2.6 kg

6. A bicycle.
 (a) 0.6 kg
 (b) 60 kg
 (c) 6 kg

Complete each statement using a metric unit of weight.

7. A loaf of bread weighs 500 _____ .

8. A compact car weighs 900 _____ .

9. A television set weighs 25 _____ .

10. A centigram would be $\frac{1}{100}$ of a _____ .

11. A table knife weighs 70 _____ .

12. A football player's weight might be listed as 115 _____ .

Complete each statement. (You can use the table on page 560.)

13. 5 kg = _____ g

14. 2000 g = _____ kg

15. 5 lb = _____ kg

16. 2000 mg = _____ g

13.3 Metric Units of Volume

OBJECTIVE
To use metric units of volume

The liter is related to the meter. It is defined as the volume of a cube 10 cm on each edge, so 1 L = 1000 cm³.

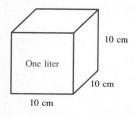

This unit of volume is also spelled *litre*. Either spelling is correct.

In measuring volume or capacity in the English system, we use the liquid measures ounces, pints, quarts, and gallons.

In the metric system, the basic unit of volume is the *liter* (L). The liter is just slightly more than the quart and would be used for soft drinks (many are already sold here by the liter), milk, oil, gasoline, and so on.

The metric unit that is used to measure smaller volumes is the *milliliter* (mL). From the prefix we know that it is one-thousandth of a liter.

1 liter (L) = 1000 milliliters (mL)

Section 13.3: Metric Units of Volume 565

The milliliter is the volume of a cube 1 cm on each edge. So 1 mL is equal to 1 cm^3. These units can be used interchangeably.

Note Scientists give measurements of volume in terms of cubic centimeters (cm^3). The liter is *not* an SI unit.

To get used to the metric units of volume:

A teaspoon is about 5 mL or 5 cm^3.

A 6-oz cup of coffee is about 180 mL.

A quart of milk is 946 mL (just less than a liter).

A gallon is just less than 4 L.

Now try these exercises.

CHECK YOURSELF 1

Choose the most reasonable measure.

(1) A can of soup: 3 L, 30 mL, or 300 mL.
(2) A pint of cream: 4.73 L, 473 mL, or 47.3 mL.
(3) A home-heating oil tank: 100 L, 1000 L, or 1000 mL.
(4) A tablespoon: 150 mL, 1.5 L, or 15 mL.

Converting metric units of volume is again just a matter of moving the decimal point. A chart similar to the ones you saw earlier may be helpful.

To convert to smaller units ⟶

kL	hL	daL	L*	dL	cL*	mL*
1000	100	10	1	0.1	0.01	0.001

⟵ To convert to larger units

*The most commonly used units. We have shown the other units simply to indicate that the prefixes and abbreviations are used in a consistent fashion.

Example 1

Complete the following statement.

4 L =   mL

We are converting to a *smaller* unit.

From the chart above, we see that we should move the decimal point three places to the right (to multiply by 1000).

4 L = 4000̬ mL = 4000 mL

Example 2

Complete the following statement.

3500 mL =   L

We are converting to a larger unit.

Move the decimal point three places to the left (to divide by 1000).

3500 mL = 3,500 L = 3.5 L

Example 3

Complete the following statement.

30 cL =   mL

We are converting to a smaller unit.

Move the decimal point one place to the right (to multiply by 10).

30 cL = 30.0 mL = 300 mL

CHECK YOURSELF 2

Complete the following statements.

(1) 5 L =   mL
(2) 7500 mL =   L
(3) 550 mL =   cL

If you need to convert between units of volume in the English and metric systems, you can use the following table.

> Converting from metric to English units:
>
> 1 liter (L) = 1.06 quarts (qt)
>
> Converting from English to metric units:
>
> 1 quart (qt) = 0.95 L
>
> 1 fluid ounce (fl oz) = 30 mL
>   = 30 cubic centimeters (cm^3)

The following examples show the use of these conversion factors.

Example 4

A bottle's volume is 3 L. What is its capacity in quarts?

$$3 \text{ L} = 3 \text{ L} \left(\frac{1.06 \text{ qt}}{1 \text{ L}} \right) = 3.18 \text{ qt}$$

Note that $\frac{1.06 \text{ qt}}{1 \text{ L}}$ is a unit ratio equal to 1.

CHECK YOURSELF 3

A soft drink is bottled in a 2-L container. What does it contain in quarts?

Section 13.3: Metric Units of Volume 567

Example 5

A gasoline tank holds 16 gal. How many liters will it hold?

$$16 \text{ gal} = 16 \text{ gal} \left(\frac{4 \text{ qt}}{1 \text{ gal}}\right) = 64 \text{ qt} \qquad \text{First we convert to quarts.}$$

$$= 64 \text{ qt} \left(\frac{0.95 \text{ L}}{1 \text{ qt}}\right) = 60.8 \text{ L} \qquad \text{Then multiply by the unit ratio.}$$

CHECK YOURSELF 4

A hot water tank will hold 40 gal. What is its volume in liters?

Example 6

A chemistry experiment calls for 8 fl oz of acid. We are using metric measures. How many milliliters of acid should we have?

$$8 \text{ fl oz} = 8 \text{ fl oz} \left(\frac{30 \text{ mL}}{1 \text{ fl oz}}\right) = 240 \text{ mL}$$

CHECK YOURSELF 5

A laboratory technician prepares a solution requiring 12 fl oz of alcohol. In a metric measure, how many cubic centimeters of alcohol are needed?

CHECK YOURSELF ANSWERS

1. (1) 300 mL; (2) 473 mL; (3) 1000 L; (4) 15 mL.
2. (1) 5000; (2) 7.5; (3) 55.
3. 2.12 qt.   4. 152 L.   5. 360 cm^3.

13.3 Exercises

Choose the most reasonable measure of volume.

1. A bottle of wine.
   (a) 75 mL
   (b) 7.5 L
   (c) 750 mL

2. A gallon of gasoline.
   (a) 400 mL
   (b) 4 L
   (c) 40 L

3. A bottle of perfume.
   (a) 15 mL
   (b) 150 mL
   (c) 1.5 L

4. A can of frozen orange juice.
   (a) 1.5 L
   (b) 150 mL
   (c) 15 mL

5. A hot-water heater.
   (a) 200 mL
   (b) 50 L
   (c) 200 L

6. An oil drum.
   (a) 220 L
   (b) 220 mL
   (c) 22 L

7. A bottle of ink.
   (a) 60 cm³
   (b) 6 cm³
   (c) 600 cm³

8. A cup of tea.
   (a) 18 mL
   (b) 180 mL
   (c) 18 L

9. A jar of mustard.
   (a) 150 mL
   (b) 15 L
   (c) 15 mL

10. A bottle of aftershave lotion.
    (a) 50 mL
    (b) 5 L
    (c) 5 mL

11. A cream pitcher.
    (a) 12 mL
    (b) 120 mL
    (c) 1.2 L

12. One tablespoon.
    (a) 1.5 mL
    (b) 1.5 L
    (c) 15 mL

Complete each statement using a metric unit of volume.

13. A can of tomato soup is 300 _____ .

14. A _____ would be $\frac{1}{100}$ of a liter.

15. A saucepan holds 1.5 _____ .

16. A thermos bottle contains 500 _____ of liquid.

17. A coffee pot holds 720 _____ .

18. A garbage can will hold 120 _____ .

19. A car's engine capacity is 2000 cm³. It is advertised as a 2.0 _____ model.

20. A bottle of vanilla extract contains 60 _____ .

21. A _____ would be $\frac{1}{10}$ of a centiliter.

Section 13.3: Metric Units of Volume 569

22. A can of softdrink is 35 _____ .

23. A garden sprinkler delivers 8 _____ of water per minute.

24. A kiloliter would be 1000 _____ .

Complete each statement. (You can use the table on page 566.)

25. 6 L = _____ mL   **26.** 3000 cm³ = _____ L   **27.** 4 qt = _____ L

28. 8 L = _____ qt   **29.** 5000 mL = _____ L   **30.** 8 fl oz = _____ mL

31. 5 L = _____ cm³   **32.** 2 L = _____ cL   **33.** 75 cL = _____ mL

34. 5 kL = _____ L   **35.** 5 L = _____ cL   **36.** 400 mL = _____ cL

Answers
1. *c*   3. *a*   5. *c*   7. *a*   9. *a*   11. *b*   13. mL (or cm³)   15. L   17. mL   19. L   21. mL
23. L   25. 6000   27. 4 qt = 4 qt $\left(\dfrac{0.95 \text{ L}}{1 \text{ qt}}\right)$ = 3.8 L   29. 5   31. 5000   33. 750   35. 500

13.3 Supplementary Exercises

Choose the most reasonable measure of volume.

1. A bottle of cough syrup.
 (a) 100 cm³
 (b) 10 cm³
 (c) 1 L

2. A watering can.
 (a) 10 L
 (b) 100 L
 (c) 100 mL

3. The gas tank of your car.
 (a) 500 mL
 (b) 5 L
 (c) 50 L

4. A bottle of eye drops.
 (a) 18 cm³
 (b) 180 cm³
 (c) 1.8 L

5. A can of soft drink.
 (a) 3.5 L
 (b) 350 mL
 (c) 35 mL

6. A punch bowl.
 (a) 200 L
 (b) 20 L
 (c) 200 mL

Complete each statement using a metric unit of volume.

7. A glass of beer holds 250 _____ .

8. The crankcase of an automobile takes 5.5 _____ of oil.

9. A pressure cooker holds 20 _____ .

10. The correct dosage for a cough medicine is 40 _____ .

11. A bottle of iodine holds 20 _____ .

12. A large mixing bowl holds 6 _____ .

Complete each statement. (You can use the table on page 566.)

13. 5 L = _____ mL   **14.** 6000 cm³ = _____ L   **15.** 9 L = _____ qt

16. 10 fl oz = _____ mL   **17.** 3000 mL = _____ L   **18.** 2 qt = _____ L

13.4 Measuring Temperature in the Metric System

OBJECTIVE
To use the Celsius temperature scale

If you listen to the weather report today, you may hear the temperature given in two ways. You are used to temperature given in degrees Fahrenheit (°F). Let's look at the temperature measure from the metric system.

The basic unit for temperature in the metric system is the *degree Celsius* (°C). The Celsius scale uses two basic points of reference:

1. The freezing point of water is 0°C.
2. The boiling point of water is 100°C.

Figure 1 compares the two temperature scales.

The Celsius scale used to be called the centigrade scale. But the name Celsius (which comes from the Swedish astronomer Anders Celsius) is more common today.

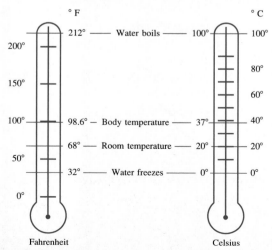

Figure 1

Section 13.4: Measuring Temperature in the Metric System

CHECK YOURSELF 1

Choose the most reasonable temperature reading.

(1) Inside your refrigerator
    4°C   40°C   20°C
(2) A nice day at the beach
    25°C   65°C   45°C
(3) A snowy day
    10°C   20°C   −5°C   *Note:* This means 5 below 0°C.
(4) The oven temperature for baking bread
    100°C   180°C   50°C

It is really more important for you to get used to temperatures in Celsius than to worry about converting back and forth between the two scales. If you do want to know how it is done, look at the following formulas.

TO CHANGE CELSIUS TO FAHRENHEIT

$$°F = \frac{9}{5}(°C) + 32$$

Example 1

Complete the statement 20°C = _____ °F.

Since $°F = \frac{9}{5}(°C) + 32$,

$$°F = \frac{9}{5} \cdot 20 + 32 \quad \text{Replace °C with 20.}$$

$$= \frac{9}{\cancel{5}} \cdot \overset{4}{\cancel{20}} + 32$$

$$= 36 + 32 \quad \text{Multiply first, then add.}$$

$$= 68°F$$

TO CHANGE FAHRENHEIT TO CELSIUS

$$°C = \frac{5}{9}(°F - 32)$$

Example 2

Complete the statement 77°F = ___ °C.

Since °C = $\frac{5}{9}$(°F − 32),

°C = $\frac{5}{9}$(77 − 32)   Replace °F with 77.

= $\frac{5}{\underset{1}{\cancel{9}}} \cdot \underset{5}{\cancel{45}}$ Subtract inside the parentheses, then multiply.

= 25°C

CHECK YOURSELF 2

Complete each statement.

(1) 40°C = ___ °F
(2) 41°F = ___ °C

CHECK YOURSELF ANSWERS

1. (1) 4°C; (2) 25°C; (3) −5°C; (4) 180°C.
2. (1) 104   (2) 5.

13.4 Exercises

Choose the most reasonable temperature measure.

1. A glass of iced tea.
   (a) 4°C
   (b) 14°C
   (c) 24°C

2. The setting for your hot water heater.
   (a) 75°C
   (b) 55°C
   (c) 35°C

3. A winter day in Fairbanks, Alaska.
   (a) 15°C
   (b) 5°C
   (c) −15°C

4. Your body temperature.
   (a) 37°C
   (b) 27°C
   (c) 47°C

5. The temperature in your classroom.
   (a) 20°C
   (b) 30°C
   (c) 10°C

6. The temperature of a swimming pool.
   (a) 32°C
   (b) 12°C
   (c) 22°C

Section 13.4: Measuring Temperature in the Metric System

7. A hot summer day in Las Vegas.
   (a) 25°C
   (b) 45°C
   (c) 35°C

8. A warm spring day at the beach.
   (a) 15°C
   (b) 35°C
   (c) 25°C

9. Bathwater.
   (a) 20°C
   (b) 40°C
   (c) 30°C

10. A cup of hot cocoa.
    (a) 50°C
    (b) 70°C
    (c) 30°C

11. The temperature outside of an airliner at 30,000 ft.
    (a) 5°C
    (b) −5°C
    (c) −25°C

12. The oven temperature setting for cooking a T.V. dinner.
    (a) 85°C
    (b) 185°C
    (c) 135°C

Complete each statement.

13. 212°F = _____ °C

14. 0°C = _____ °F

15. 50°C = _____ °F

16. 41°F = _____ °C

17. 50°F = _____ °C

18. 75°C = _____ °F

19. 30°C = _____ °F

20. 15°C = _____ °F

21. 20°C = _____ °F

22. 77°F = _____ °C

23. 104°F = _____ °C

24. 35°C = _____ °F

25. The newspaper gives the high temperatures in Paris as 15°C. What is the equivalent Fahrenheit reading?

26. The manual for your new Italian sports car says that the operating temperature should be 85°C. What is the Fahrenheit temperature?

27. The high temperature in New York was 77°F. How would this be reported in a London newspaper?

28. A pastry recipe in a German cookbook calls for an oven temperature of 180°C. How should you set your oven on the Fahrenheit scale?

Answers

1. *a* 3. *c* 5. *a* 7. *b* 9. *b* 11. *c* 13. *b* 15. 122 17. 10 19. 86 21. 68 23. 40
25. 59°F 27. 25°C

13.4 Supplementary Exercises

Choose the most reasonable temperature measure.

1. A child's fever.
   (*a*) 48°C
   (*b*) 28°C
   (*c*) 38°C

2. A cold winter day in Minneapolis.
   (*a*) −10°C
   (*b*) 10°C
   (*c*) 20°C

3. A cold Coca Cola.
   (*a*) 10°C
   (*b*) 0°C
   (*c*) 4°C

4. A summer noon in Death Valley.
   (*a*) 40°C
   (*b*) 50°C
   (*c*) 30°C

5. A comfortable fall afternoon.
   (*a*) 30°C
   (*b*) 20°C
   (*c*) 10°C

6. The operating water temperature in your car radiator.
   (*a*) 85°C
   (*b*) 65°C
   (*c*) 105°C

Complete each statement.

7. 32°F = _____ °C

8. 100°C = _____ °F

9. 25°C = _____ °F

10. 50°F = _____ °C

11. 68°F = _____ °C

12. 5°C = _____ °F

13. A bread recipe in a French cookbook says to bake the bread at 190°C. At what temperature should you set the Fahrenheit scale on your oven?

14. The newspaper lists the high temperature for a given day in Athens as 30°C. What is the equivalent Fahrenheit reading?

SELF-TEST for Chapter Thirteen

The purpose of the Self-Test is to help you check your progress and review for a chapter test in class. Allow yourself about an hour to take the test. When you are done, check your answers in the back of the book. If you missed any problems, be sure to go back and review the appropriate sections in the chapter and do the supplementary exercises provided there.

In Problems 1 and 2, choose the reasonable measure.

1. The width of your hand.
   (a) 50 cm
   (b) 10 cm
   (c) 1 m

2. The speed limit on a freeway.
   (a) 9 km/h
   (b) 90 km/h
   (c) 90 m/h

In Problems 3 and 4, use a metric unit of length to complete the statements.

3. A tennis court is 24 _____ long.

4. The width of your pencil is about 6 _____ .

In Problems 5 to 10, complete the statements.

5. 3 km = _____ m
6. 5 m = _____ mm
7. 800 cm = _____ m

8. 5 km = _____ mi
9. 5 in = _____ cm
10. 50 mi = _____ km

In Problems 11 and 12, choose the most reasonable measure.

11. A package of gum.
    (a) 3 g
    (b) 3 kg
    (c) 30 g

12. A football player.
    (a) 12 kg
    (b) 120 kg
    (c) 120 g

In Problems 13 and 14, use a metric unit of weight to complete the statements.

13. A living room sofa weighs 80 _____ .

14. A box of breakfast food weighs 750 _____ .

In Problems 15 to 18, complete the statements.

15. 3 kg =   g

16. 3 lb =   kg

17. 5 oz =   g

18. 7 g =   mg

In Problems 19 and 20, choose the most reasonable measure.

19. The gas tank of your car.
    (a) 600 L
    (b) 6 L
    (c) 60 L

20. A small can of tomato juice.
    (a) 4 L
    (b) 400 mL
    (c) 40 mL

In Problems 21 and 22, use a metric unit of volume to complete the statements.

21. A drinking glass holds 250   .

22. An oil drum holds 200   of oil.

In Problems 23 to 26, complete the statements.

23. 2 L =   mL

24. 300 cL =   L

25. 5 qt =   L

26. 8 fl oz =   mL

In Problems 27 and 28, choose the most reasonable measure.

27. A hot cup of coffee.
    (a) 20°C
    (b) 40°C
    (c) 60°C

28. The temperature in your freezer.
    (a) 0°C
    (b) −10°C
    (c) 10°C

In Problems 29 and 30, complete the statements.

29. 25°C =   °F

30. 95°F =   °C

SUMMARY for Part Five

Measurement

The English System of Measurement

The English System
The system of measurement in common use in the United States.

English Units of Measure and Equivalents

Length

1 foot (ft) = 12 inches (in)
1 yard (yd) = 3 ft

Volume

1 pint (pt) = 16 fluid ounces (fl oz)
1 quart (qt) = 2 pt
1 gallon (gal) = 4 qt

Weight

1 pound (lb) = 16 ounces (oz)
1 ton = 2000 lb

Time

1 minute (min) = 60 seconds (s)
1 hour (h) = 60 min
1 day = 24 h
1 week = 7 days

Denominate Numbers

Denominate Number
A number with a unit of measure attached. Numbers without units attached are called *abstract numbers*.

> 5 ft and 10 m are denominate numbers.
>
> 7 is an abstract number

To Add Like Denominate Numbers

1. Arrange the numbers so that the like units are in the same vertical column.
2. Add in each column.
3. Simplify if necessary.

> To add 4 ft 7 in and 5 ft 10 in:
>
>   4 ft 7 in
> + 5 ft 10 in
>   9 ft 17 in
> or 10 ft 5 in

To Subtract Like Denominate Numbers

1. Arrange the numbers so that the like units are in the same vertical column.
2. Subtract in each column. You may have to borrow from the larger unit at this point.
3. Simplify if necessary.

> To subtract:
>   5 h 20 min
> − 3 h 40 min
>
> Borrow and rename:
>   4 h 80 min
> − 3 h 40 min
>   1 h 40 min

577

To Multiply or Divide Denominate Numbers by Abstract Numbers

1. Multiply or divide each part of the denominate number by the abstract number.
2. Simplify if necessary.

$2 \times (3 \text{ yd } 2 \text{ ft}) = 6 \text{ yd } 4 \text{ ft},$
or $7 \text{ yd } 1 \text{ ft}$

Finding a Perimeter, Circumference, Area, or Volume

Finding Perimeter or Circumference

The *perimeter* or *circumference* of a figure is the distance around the figure. The following formulas are used:

Formula (1)   Perimeter of a Rectangle

$P = 2L + 2W$

Formula (2)   Perimeter of a Square

$P = 4S$

Formula (3)   Circumference of a Circle

$C = \pi d$

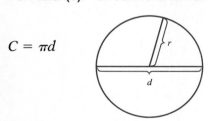

Formula (4)   Circumference of a Circle

$C = 2\pi r$

Finding Area

The *area* of a figure is the number of square units enclosed by the figure. The following formulas are used.

Formula (5)   Area of a Rectangle

$A = L \cdot W$

Formula (6)   Area of a Square

$A = S^2$

Formula (7)   Area of a Parallelogram

$A = b \cdot h$

Formula (8)   Area of a Triangle

$A = \dfrac{1}{2} \cdot b \cdot h$

Formula (9)   Area of a Circle

$A = \pi r^2$

Finding Volume

The *volume of* a solid is the number of cubic units contained in the solid. The following formulas are used.

Formula (10)   Volume of a Rectangular Solid

$V = L \cdot W \cdot H$

Formula (11)   Volume of a Cylinder

$V = \pi r^2 h$

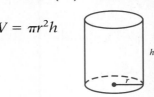

The Metric System of Measurement

The Metric System
The system of measurement used throughout the rest of the world.

Basic Metric Units

Length
meter (m)

Mass or Weight
gram (g)

Volume
liter (L)

Temperature
degree Celsius (°C)

Basic Metric Prefixes

*These are the most commonly used and should be memorized.

*milli means $\dfrac{1}{1000}$ *kilo means 1000

*centi means $\dfrac{1}{100}$ hecto means 100

deci means $\dfrac{1}{10}$ deka means 10

Converting Metric Units

You can use the following chart.

To convert to smaller units ⟶

km	hm	dam	m	dm	cm	mm
1000 m	100 m	10 m	1 m	0.1 m	0.01 m	0.001 m

⟵ To convert to larger units

To convert between metric units, just move the decimal point the same number of places to the left or right as indicated by the chart.

500 cm = ___?___ m

To convert from centimeters to meters, move the decimal point two places to the *left*.

5⌣00. cm = 5 m

Conversions between units of volume (liters) or units of weight (grams) work in exactly the same fashion.

3 L = ___?___ mL

To convert from liters to milliliters, move the decimal point three places to the *right*.

3 L = 3000⌣ mL = 3000 mL

Conversions Between the Systems

Converting from Metric to English Units

1 meter (m) = 39.37 inches (in)
1 centimeter (cm) = 0.394 in
1 kilometer (km) = 0.62 mile (mi)
1 kilogram (kg) = 2.2 pounds (lb)
1 liter (L) = 1.06 quarts (qt)

$°F = \frac{9}{5}(°C) + 32$

Converting from English to Metric Units

1 inch (in) = 2.54 centimeters (cm)
1 yard (yd) = 0.914 meter (m)
1 mile (mi) = 1.6 kilometers (km)
1 quart (qt) = 0.95 liter (L)
1 fluid ounce (fl oz) = 30 milliliters (mL)
= 30 cubic centimeters (cm^3)

1 pound (lb) = 0.45 kilogram (kg)
1 ounce (oz) = 28 grams (g)

$°C = \frac{5}{9}(°F - 32)$

SUMMARY EXERCISES for Part Five

You should now be reviewing the material in Part Five of the text. The following exercises will help in that process. Work all the exercises carefully. Then check your answers in the back of the book. References are provided there to the chapter and section for each problem. If you made an error, go back and review the related material and do the supplementary exercises for that section.

[12.1] In Problems 1 to 4, complete each statement.

1. 6 ft = _____ in

2. 5 lb = _____ oz

3. 4 min = _____ s

4. 5 gal = _____ qt

[12.2] In Problems 5 and 6, simplify.

5. 7 ft 22 in

6. 7 lb 18 oz

[12.2] In Problems 7 to 10, do the indicated operations.

7. 5 ft 8 in
 + 6 ft 10 in

8. 5 lb 3 oz
 − 2 lb 10 oz

9. 3 × (2 h 40 min)

10. $\dfrac{10 \text{ min } 45 \text{ s}}{5}$

[12.2] In Problems 11 and 12, solve the applications.

11. A plan for a bookcase requires three pieces of lumber 2 ft 8 in long and two pieces 3 ft 4 in long. What is the total length of material that is needed?

12. You can buy three bottles of dishwashing liquid, each containing 1 pt 6 fl oz, on sale for $2.40. For the same price you can buy a large container holding 2 qt. Which is the better buy?

[12.3] In Problems 13 to 16, find the perimeter or circumference of each figure.

13.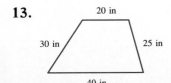

14.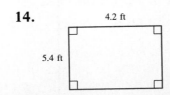

581

15.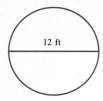

Use 3.14 for π, and round your answer to one decimal place.

16.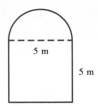

Use 3.14 for π, and round your answer to one decimal place.

[12.3] In Problems 17 and 18, solve the applications.

17. Binding for the edge of a rug costs 40¢ a foot. If a rug is 10 ft by 12 ft, how much will it cost to bind all the sides of the rug?

18. An above-ground swimming pool is circular with an 8-ft radius. If fencing costs $3.50 per yard, what will it cost to fence in the pool?

[12.4] In Problems 19 to 22, find the area of each figure.

19.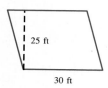

20.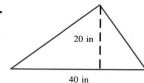

21.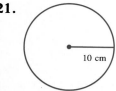

22.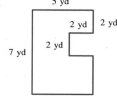

[12.4] In Problems 23 to 25, solve the application.

23. How many square feet of vinyl floor covering will be needed to cover the floor of a room which is 10 ft by 18 ft? How many square yards will be needed?

24. A rectangular roof for a house addition measures 15 ft by 30 ft. A roofer will charge $175 per "square" (100 ft²). Find the cost of the roofing for the addition.

25. A semicircular patio has a radius of 6 ft. If it will cost $18 per square yard to pave the patio with brick, find the total cost of paving the patio.

[2.5] In Problems 26 and 27, find the volume of each solid.

26.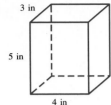

27.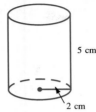

Use 3.14 for π, and round your result to one decimal place.

[2.5] In Problems 28 to 31, solve the application.

28. A shipping carton measures 5 ft by 3 ft by $3\frac{1}{2}$ ft. What is its volume?

29. How many cubic yards of concrete will be required for a concrete walk that is 6 in thick, 3 ft wide, and 72 ft long?

30. A cylindrical water tank is 10 ft deep and has an inside diameter of 8 ft. What is its volume?

31. A container has a 7-in radius and is 20 in high. How many gallons will it hold? (1 gal = 231 in³.) Use $\frac{22}{7}$ for π, and round to the nearest tenth of a gallon.

[13.1]–[13.4] In Problems 32 to 40, choose the most reasonable metric measure.

32. The width of a fence gate.
    (a) 1.5 m   (b) 15 cm   (c) 15 m

33. The volume of a home aquarium.
    (a) 40 L   (b) 400 L   (c) 400 mL

34. The temperature of a glass of iced tea.
    (a) 4°C   (b) 0°C   (c) 10°C

35. The width of a regular postage stamp.
    (a) 22 cm   (b) 22 mm   (c) 2.2 m

36. The weight of a sledgehammer.
    (a) 80 g   (b) 80 kg   (c) 8 kg

37. The length of a dinner knife.
    (a) 20 cm   (b) 2 cm   (c) 2 m

38. The weight of one playing card.
    (a) 2 mg   (b) 2 g   (c) 20 g

39. The volume of a bottle of "white-out" fluid.
    (a) 200 mL   (b) 2 mL   (c) 20 mL

40. The weight of a subcompact car.
    (a) 100 kg   (b) 10 kg   (c) 1000 kg

[13.1]–[13.4] In Problems 41 to 52, complete each statement.

41. 5 km = _____ m

42. 3 L = _____ mL

43. 30 cL = _____ mL

44. 400 cm = _____ m

45. 3000 cm³ = _____ L

46. 2000 mg = _____ g

47. 8 m = _____ cm

48. 3 g = _____ mg

49. 2000 mm = _____ m

50. 750 mL = _____ cL

51. 2 L = _____ cL

52. 30 cm = _____ mm

In Problems 53 to 60, complete each statement. (You can use the table on pages 578 and 579.)

53. 8 cm = _____ in

54. 4 lb = _____ kg

55. 50°C = _____ °F

56. 6 L = _____ qt

57. 12 fl oz = _____ mL

58. 50 km = _____ mi

59. 68°F = _____ °C

60. 5 oz = _____ g

CUMULATIVE TEST for Part Five

This test is provided to help you in the process of review over Chapters 12 and 13. Answers are provided in the back. If you missed any problems, be sure to go back and review the appropriate chapter sections.

In Problems 1 to 4, complete each statement.

1. 8 ft = _____ in

2. 6 lb = _____ oz

3. 5 min = _____ s

4. 7 gal = _____ qt

In Problem 5, simplify.

5. 8 ft 20 in

In Problems 6 and 7, do the indicated operations.

6.   7 lb   9 oz
  + 3 lb 12 oz

7.   4 min 10 s
  − 2 min 35 s

In Problem 8, solve the application.

8. A plan for a kitchen cabinet calls for four pieces of plywood 2 ft 8 in long and two pieces that are 3 ft 10 in long. What is the total length that will be needed?

In Problems 9 to 12, find the perimeter or the circumference of each figure.

9.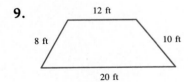

10.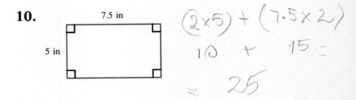

11.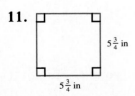

12.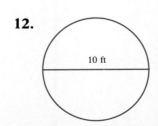

585

In Problem 13, solve the application.

13. A room measures 12 ft by 15 ft and there are two 3-ft door openings. If floor molding costs 75¢ per foot, what will it cost to put molding around the room?

In Problems 14 to 17, find the area of each figure.

14.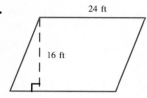

15.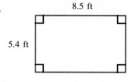

16.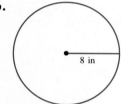

17.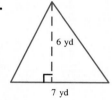

Use 3.14 for π, and round your answer to one decimal place.

In Problem 18, solve the application.

18. Matilda has chosen a vinyl floor covering costing $16 per square yard. If her kitchen is 12 ft long by 12 ft wide, how much will the floor covering cost?

In Problems 19 and 20, find the volume of each solid.

19.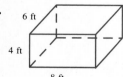

20.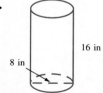

Use 3.14 for π, and round your result to one decimal place.

In Problems 21 and 22, solve the applications.

21. A concrete walk which is 4 in thick, 3 ft wide, and 54 ft long will require how many cubic yards of concrete?

22. A garden sprayer has a 6-in diameter and is 36 in tall. How many gallons will it hold? (1 gal = 231 in^3.) Give your answer to the nearest tenth of a gallon.

In Problems 23 to 30, choose the most reasonable metric measure.

23. The height of a flagpole.
  (*a*) 160 m
  (*b*) 160 cm
  (*c*) 16 m

24. The volume of a large mixing bowl.
  (*a*) 6 L
  (*b*) 60 cL
  (*c*) 60 L

25. The temperature on a snowy winter day.
  (*a*) 5°C
  (*b*) −5°C
  (*c*) 15°C

26. The weight of your camera.
  (*a*) 500 g
  (*b*) 50 g
  (*c*) 5 kg

27. The width of your desk.
  (*a*) 12 cm
  (*b*) 120 cm
  (*c*) 12 m

28. The weight of a pair of sunglasses.
  (*a*) 50 mg
  (*b*) 50 g
  (*c*) 500 g

29. The volume of a glass of orange juice.
  (*a*) 120 mL
  (*b*) 12 mL
  (*c*) 1.2 L

30. The weight of an empty styrofoam cup.
  (*a*) 400 g
  (*b*) 40 g
  (*c*) 400 mg

In Problems 31 to 36, complete each statement.

31. 8 km =   m   **32.** 3000 mg =   g   **33.** 500 cm =   m

34. 25 cL =   mL   **35.** 3 L =   mL   **36.** 2 kg =   g

In Problems 37 to 40, complete each statement. (You can use the table on pages 578 and 579).

37. 8 kg =   lb   **38.** 12 in =   cm

39. 8 oz =   g   **40.** 77°F =   °C

Part 6

An Introduction to Algebra

PRETEST for Chapter Fourteen

The Integers

This pretest will point out any difficulties you may be having with signed numbers. Do all the problems. Then check your answers on the following page.

1. The absolute value of −10 is   .

2. The opposite of −7 is   .

3. $10 + (-8) =$

4. $-5 + (-7) =$

5. $-10 - 8 =$

6. $8 - (-2) =$

7. $(-9)(6) =$

8. $(-8)(-9) =$

9. $\dfrac{-24}{-6} =$

10. $40 \div (-8) =$

ANSWERS TO PRETEST

For help with similar problems, turn to the section indicated.

1. 10 (Section 14.1);
2. 7 (Section 14.1);
3. 2 (Section 14.2);
4. −12 (Section 14.2);
5. −18 (Section 14.3);
6. 10 (Section 14.3);
7. −54 (Section 14.4);
8. 72 (Section 14.4);
9. 4 (Section 14.5);
10. −5 (Section 14.5)

Chapter Fourteen
The Integers

14.1 Signed Numbers

OBJECTIVES
1. To use signed numbers
2. To find the opposite of a number
3. To find the absolute value of a number

Up until now you have been working with two groups of numbers, whole numbers and fractions or decimals. There are many applications in arithmetic that cannot be solved with these numbers.

Suppose you want to represent a temperature of 10° below zero, a debt of $50, or an altitude 100 ft below sea level. These situations involve a new set of numbers called *negative numbers*.

Back in the first chapter we represented the whole numbers by points on a number line.

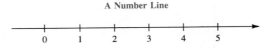

A Number Line

To form a number line, choose a convenient unit of measure and mark off equally spaced points from the origin (the point corresponding to 0).

We now want to extend the number line by marking off units to the *left* of 0.

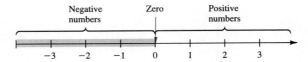

Zero is neither positive nor negative.

Since −3 is to the left of 0, it is a negative number. Read −3 as "negative three."

Numbers to the right of 0 on the number line are called *positive numbers*. Numbers to the left of 0 are called *negative numbers*.

To indicate a negative number, we use a negative sign (−) in front of the number. Positive numbers may be written with a plus sign (+) or with no sign at all, in which case the number is understood to be positive.

Example 1

If no sign appears, a number (other than 0) is positive.

+6 is a positive number.

−9 is a negative number.

5 is a positive number.

0 is neither positive nor negative.

593

Together, the sets of positive and negative numbers are called *signed numbers*.

An important idea in our work with signed numbers is the *opposite* of a number. Every number has an opposite.

> The *opposite* of a number corresponds to a point the same distance from 0 as the given number, but in the opposite direction.

Example 2

The opposite of a *positive* number is *negative*.

The opposite of 5 is −5.

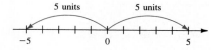

We can see that both numbers are located 5 units from 0.

CHECK YOURSELF 1

What is the opposite of 8?

Now let's look at the opposite of a negative number.

Example 3

The opposite of a *negative* number is *positive*.

The opposite of −3 is 3.

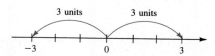

Both numbers correspond to points that are 3 units from 0.

CHECK YOURSELF 2

What is the opposite of −9?

We can indicate the opposite of a number by placing a negative sign in front of the number.

Section 14.1: Signed Numbers

Example 4

Place a negative sign in front of the number.

We write the opposite of 5 as −5.

You can now think of −5 in two ways: as negative 5, and as the opposite of 5.

Using the same idea, we can write the opposite of a negative number.

Example 5

Again place a negative sign in front of the number.

The opposite of −3 is −(−3).

Since we know from looking at the number line that the opposite of −3 is 3, this means that

$$-(-3) = 3$$

So the opposite of a negative number must be positive.

Let's summarize our results:

> 1. The opposite of a positive number is negative.
> 2. The opposite of a negative number is positive.
> 3. The opposite of 0 is 0.

CHECK YOURSELF 3

Complete the following statements.

(1) The opposite of 10 is   .
(2) The opposite of −7 is   .
(3) −(−8) =   .

We also need to define the *absolute value* or magnitude of a signed number.

> The *absolute value* of a signed number is the distance (on the number line) between the number and 0.

Example 6

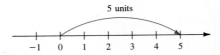

The absolute value of 5 is 5.   5 is five units from 0.

Example 7

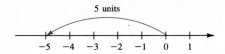

The absolute value of −5 is 5.   −5 is also five units from 0.

We usually write the absolute value of a number by placing vertical bars before and after the number. From Examples 6 and 7, we can write

|5| = 5   and |−5| = 5

CHECK YOURSELF 4

Complete the following statements.

(1) The absolute value of 9 is
(2) The absolute value of −12 is
(3) |−6| =
(4) |15| =

So far, all the numbers we have looked at in this section have been either natural numbers, zero, or the negatives of natural numbers. These numbers make up the set of integers.

> The set of integers consists of the natural numbers, their negatives, and zero.

Example 8

3 is an integer.

−5 is an integer.   −5 is the negative of the natural number 5.

0 is an integer.

$\frac{3}{4}$ and 0.5 are *not* integers.   Common or decimal fractions do *not* belong to the set of integers.

CHECK YOURSELF 5

Which of the following are integers?

7, 0, $\frac{4}{7}$, −5, 0.2

Of course, there are also many negative numbers that are not integers. For instance, negative fractions or decimals are not integers.

Example 9

$-\frac{2}{3}$, -0.3, and -3.5 are negative numbers. Look at their location on the number line.

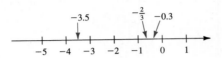

$-\frac{2}{3}$, -0.3, and -3.5 are *not* integers. The set of integers includes the natural numbers, 0, and the negatives of the natural numbers, *not* common or decimal fractions.

CHECK YOURSELF ANSWERS

1. -8.   2. 9.   3. (1) -10; (2) 7; (3) 8.
4. (1) 9; (2) 12; (3) 6; (4) 15.
5. 7, 0, -5.

14.1 Exercises

Indicate whether the following statements are true or false.

1. The opposite of 7 is -7.

2. The opposite of -10 is -10.

3. -9 is an integer.

4. 5 is an integer.

5. The opposite of -11 is 11.

6. The absolute value of 8 is 8.

7. $|-6| = -6$

8. $-(-30) = -30$

9. -12 is not an integer.

10. The opposite of -18 is 18.

11. $|7| = -7$

12. The absolute value of -10 is -10.

13. $-(-8) = 8$

14. $\frac{2}{3}$ is not an integer.

15. $|-20| = 20$

16. The absolute value of -3 is 3.

17. $\frac{3}{5}$ is an integer.

18. 0.8 is an integer.

19. 0.15 is not an integer.

20. $|-9| = -9$

21. $\frac{5}{7}$ is not an integer.

22. 0.23 is not an integer.

23. $-(-7) = -7$

24. The opposite of 15 is -15.

Complete the following statements.

25. The absolute value of -10 is _____.

26. $-(-16) =$ _____.

27. $|-20| =$ _____.

28. The absolute value of -12 is _____.

29. The absolute value of 7 is _____.

30. The opposite of -9 is _____.

31. The opposite of 30 is _____.

32. $|-15| =$ _____.

33. $-(-6) =$ _____.

34. The absolute value of 0 is _____.

35. $|50| =$ _____.

36. The opposite of 18 is _____.

Complete each of the following statements using the symbols <, =, or >.

37. -2 ___ -3

38. -10 ___ -5

39. -20 ___ -10

40. -15 ___ -14

41. $|3|$ ___ 3

42. $|-5|$ ___ -5

43. -4 ___ $|-4|$

44. 7 ___ $|7|$

Skillscan (Section 7.5)
Perform the indicated operations.

a. $2\dfrac{1}{4} + 5\dfrac{1}{4}$

b. $3\dfrac{7}{8} - 2\dfrac{5}{8}$

c. $2\dfrac{7}{10} + 3\dfrac{5}{10}$

d. $4\dfrac{1}{6} - 2\dfrac{5}{6}$

e. $5\dfrac{5}{12} - 2\dfrac{7}{12}$

f. $7\dfrac{5}{8} + 3\dfrac{7}{8}$

Answers
Solutions for the even-numbered exercises are provided in the back of the book.
1. T   3. T   5. T   7. F   9. F   11. F   13. T   15. T   17. F   19. T   21. T   23. F
25. 10   27. 20   29. 7   31. −30   33. 6   35. 50   37. −2 > −3   39. −20 < −10   41. |3| = 3
43. −4 < |−4| a. $7\dfrac{1}{2}$ b. $1\dfrac{1}{4}$ c. $6\dfrac{1}{5}$ d. $1\dfrac{1}{3}$ e. $2\dfrac{5}{6}$ f. $11\dfrac{1}{2}$

14.1 Supplementary Exercises

Indicate whether the following statements are true or false.

1. The opposite of 9 is −9.

2. The absolute value of −5 is 5.

3. The opposite of −10 is 10.

4. The opposite of −7 is −7.

5. The absolute value of −18 is −18.

6. −3 is an integer.

7. −(−7) = 7

8. |−8| = 8

9. |50| = −50

10. 0.8 is not an integer.

11. 0.5 is an integer.

12. $\dfrac{3}{4}$ is an integer.

Complete the following statements.

13. The absolute value of −12 is

14. |−15| =

15. The opposite of −8 is

16. The absolute value of 15 is

17. The opposite of 20 is

18. −(−18) =

14.2 Adding Signed Numbers

OBJECTIVE
To add signed numbers

Now that we have introduced signed numbers, let's see how the basic operations are performed when signed numbers are involved. We'll start with addition.

Remember that in the first chapter we used the number line to picture addition. Using that approach, we add positive numbers by moving to the *right* on the number line.

Example 1

To find the sum 2 + 3:

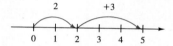

Start at 0 and move two units *to the right*. Then move three more units *to the right* to find the sum. So

2 + 3 = 5

To use the number line in adding signed numbers, we will move *to the right* for positive numbers, but *to the left* for negative numbers. Let's work through an example.

Example 2

To avoid confusion we always use parentheses when two signs are together. So write 5 + (−2), not 5 + −2.

To find the sum 5 + (−2), first move five units *to the right* of 0:

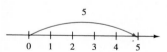

Then count two units *to the left* to add negative 2.

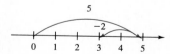

We see that

5 + (−2) = 3

CHECK YOURSELF 1

Use the number line to find the sum

$8 + (-3)$

Example 3

To find the sum $-2 + 5$:

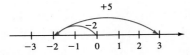

Since the first addend is negative, we start by moving left.

First move two units *to the left* of 0. Then, to add 5, move five units back *to the right*. We see that

$-2 + 5 = 3$

CHECK YOURSELF 2

Find the sum

$-7 + 9$

Example 4

To find the sum $4 + (-5)$:

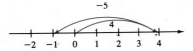

Move four units *to the right* of 0. Then move five units back *to the left* to add negative 5. So

$4 + (-5) = -1$

CHECK YOURSELF 3

Find the sum

$3 + (-7)$.

We can also use the number line to picture addition when two negative numbers are involved. The following example illustrates.

Example 5

To find the sum $-2 + (-3)$:

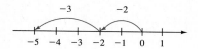

Move two units *to the left* of 0. Then move three more units *to the left* to add negative 3. We see that

$$-2 + (-3) = -5$$

CHECK YOURSELF 4

Find the sum

$$-7 + (-5)$$

You may have noticed some patterns in the addition of the previous examples. These patterns will let you do much of the addition mentally. Look at the following rule:

> **TO ADD SIGNED NUMBERS**
>
> 1. If two numbers have the same sign, add their absolute values. Give the sum the sign of the original numbers.

This means that the sum of two positive numbers is positive and the sum of two negative numbers is negative.

Example 6

$5 + 2 = 7$ The sum of two positive numbers is positive.

$-2 + (-6) = -8$ Add the absolute values ($2 + 6 = 8$). Give the sum the sign of the original numbers.

CHECK YOURSELF 5

Find the sums.

(1) $6 + 7$
(2) $-8 + (-7)$

You have no doubt also noticed that when you add a positive and a negative number, sometimes the answer is positive and sometimes it is negative. This depends on which number has the largest absolute value (the distance from 0). This leads us to the second part of our addition rule.

Section 14.2: Adding Signed Numbers

> **TO ADD SIGNED NUMBERS**
>
> **2.** If two numbers have different signs, subtract the smaller absolute value from the larger. Give the sum the sign of the number with the larger absolute value.

Example 7

$6 + (-2) = 4$ The numbers have different signs. Subtract the absolute values ($6 - 2 = 4$). The result is positive, since 6 has the larger absolute value.

Example 8

$4 + (-7) = -3$ Subtract the absolute values ($7 - 4 = 3$). The result is negative, since -7 has the larger absolute value.

In general, the sum of a positive and a negative integer with the same absolute value is 0.

$8 + (-8) = 0$ Subtract the absolute values ($8 - 8 = 0$). Don't worry about the sign, since 0 is neither positive nor negative.

CHECK YOURSELF 6

Find the sums.

(1) $3 + (-8)$ (2) $-9 + 15$ (3) $9 + (-9)$

Earlier we listed some properties of addition. To extend these, what about the order in which we add two integers? Let's look at this question in the following example.

Example 9

$-2 + 7 = 7 + (-2) = 5$
$-3 + (-4) = -4 + (-3) = -7$

Note that in both cases the order in which we add the integers does not affect the sum. This leads us to:

> **THE COMMUTATIVE PROPERTY**
>
> The *order* in which we add integers does not change the sum. Addition of integers is *commutative*. In symbols, for any integers *a* and *b*:
>
> $a + b = b + a$

CHECK YOURSELF 7

Show that $-8 + 2 = 2 + (-8)$

What if we want to add more than two integers? Another property of addition will be helpful. Look at the following example.

Example 10

To find the sum $2 + (-3) + (-4)$, first:

$\underline{2 + (-3)} + (-4)$ Add the first two integers.
$= \quad -1 \quad + (-4)$ Then add the third to that sum.
$= \quad -5$

Now let's try a second approach.

$2 + \underline{[(-3) + (-4)]}$ This time, add the second and third integers.
$= 2 + \quad (-7)$ Then add the first integer to that sum.
$= -5$

Do you see that it makes no difference which way we group integers in addition? The final sum is not changed. We can state:

THE ASSOCIATIVE PROPERTY

The way we *group* the integers does not change the sum. Addition of integers is *associative*. In symbols, for any integers *a*, *b*, and *c*:

$(a + b) + c = a + (b + c)$

CHECK YOURSELF 8

Show that $-2 + (-3 + 5) = [-2 + (-3)] + 5$

Let's look at two more properties of addition on the integers.

THE ADDITIVE IDENTITY

The sum of any integer and 0 is just that integer. Because of this special property, 0 is called the *additive identity*. In symbols, for any integer *a*:

$a + 0 = a$

Example 11

$-8 + 0 = -8$
$0 + (-20) = -20$

CHECK YOURSELF 9

Find the sum.

$-6 + 0$

We saw in the last section that every integer has an opposite. The opposite of an integer is also called the *additive inverse* of that integer. The additive inverse of any integer a is $-a$.

Example 12

We can also call -6 the opposite of 6.

The additive inverse of 6 is -6.

The additive inverse of -8 is $-(-8)$ or 8.

THE ADDITIVE INVERSE

The sum of any integer and its additive inverse is 0. In symbols, for any integer a,

$a + (-a) = 0$

Example 13

$6 + (-6) = 0$
$-8 + 8 = 0$

CHECK YOURSELF 10

Find the sum.

$9 + (-9)$

So far we have limited ourselves to addition on the integers. The process is the same if we want to add other negative numbers.

Example 14

$$3\frac{3}{4} + \left(-2\frac{1}{4}\right) = 1\frac{1}{2}$$

Subtract the absolute values $\left(3\frac{3}{4} - 2\frac{1}{4} = 1\frac{2}{4} = 1\frac{1}{2}\right)$. The sum is positive since $3\frac{3}{4}$ has the larger absolute value.

$$-0.5 + (-0.2) = -0.7$$

Add the absolute values $(0.5 + 0.2 = 0.7)$. The sum is negative.

CHECK YOURSELF 11

Find the sums.

(1) $-2\frac{1}{2} + \left(-3\frac{1}{2}\right)$ (2) $5.3 + (-4.3)$

We can also add more than two signed numbers by repeated use of the addition rule. Our final example illustrates.

Example 15

$$\underbrace{2.5 + (-1.5)} + (-0.5) \quad \text{Add the first two numbers.}$$
$$= \quad 1 \quad + (-0.5) \quad \text{Now add the third.}$$
$$= \quad 0.5$$

Note that by the associative property, the sum would be the same if the numbers were grouped differently for the addition. Try adding the second two numbers in the above example as the first step.

CHECK YOURSELF 12

Find the sum.

$-3.5 + (-1.5) + (-2.5)$

CHECK YOURSELF ANSWERS

1. 5. 2. 2. 3. -4. 4. -12.
5. (1) 13; (2) -15. 6. (1) -5; (2) 6; (3) 0.
7. $-6 = -6$. 8. $0 = 0$. 9. -6. 10. 0.
11. (1) -6; (2) 1. 12. -7.5.

14.2 Exercises

Add.

1. $-6 + (-5)$
2. $3 + 9$
3. $8 + (-4)$
4. $-6 + (-7)$

5. $4 + (-6)$
6. $9 + (-2)$
7. $7 + 9$
8. $-5 + 9$

9. $(-11) + 5$
10. $5 + (-8)$
11. $-8 + (-7)$
12. $8 + (-7)$

13. $(-16) + 15$
14. $(-12) + 10$
15. $-8 + 0$
16. $7 + (-7)$

17. $-9 + 10$
18. $-6 + 8$
19. $-4 + 4$
20. $5 + (-20)$

21. $7 + (-13)$
22. $0 + (-10)$
23. $-8 + 5$
24. $-7 + 3$

25. $6 + (-6)$
26. $-9 + 9$
27. $(-10) + (-6)$
28. $-18 + (-7)$

29. $5\frac{3}{8} + \left(-3\frac{1}{8}\right)$
30. $(0.7) + (-1.2)$
31. $-3.8 + 7.2$
32. $-4\frac{1}{10} + \left(-5\frac{3}{10}\right)$

33. $-4\frac{9}{16} + 7\frac{7}{16}$
34. $-3.8 + 2.9$
35. $-1.5 + (-0.3)$
36. $2\frac{1}{2} + (-3)$

37. $4 + (-7) + (-5)$
38. $-7 + 8 + (-6)$
39. $-2 + (-6) + (-4)$

40. $12 + (-6) + (-4)$
41. $-3 + (-7) + 5 + (-2)$
42. $7 + (-8) + (-9) + 10$

43. $-7 + (-3) + (-4) + 8$
44. $-8 + (-5) + (-4) + 7$

Skillscan (Sections 8.2 and 8.3)
Perform the indicated operations.

a. $3.25 + 1.75$
b. $8.65 - 5.35$
c. $5.25 - 3.75$

d. $2.85 + 3.45$
e. $8.65 - 6.95$
f. $1.25 + 5.85$

Answers

1. −11 3. 4 5. −2 7. 16 9. −6 11. −15 13. −1 15. −8 17. 1 19. 0 21. −6
23. −3 25. 0 27. −16 29. $2\frac{1}{4}$ 31. 3.4 33. $2\frac{7}{8}$ 35. −1.8 37. −8 39. −12 41. −7
43. −6 a. 5 b. 3.3 c. 1.5 d. 6.3 e. 1.7 f. 7.1

14.2 Supplementary Exercises

Add.

1. $6 + (-3)$
2. $11 + (-7)$
3. $7 + (-7)$
4. $7 + (-2)$

5. $8 + (-9)$
6. $5 + (-9)$
7. $-6 + (-3)$
8. $-12 + 12$

9. $-8 + 8$
10. $-5 + (-9)$
11. $-10 + 7$
12. $0 + (-10)$

13. $-7 + 0$
14. $9 + (-5)$
15. $3\frac{5}{8} + \left(-2\frac{1}{8}\right)$
16. $-2.7 + 0.7$

17. $-3.5 + (-1.5)$
18. $-2\frac{3}{16} + 1\frac{1}{16}$
19. $5 + (-8) + (-3)$
20. $-8 + 7 + (-5)$

21. $-8 + (-7) + 10 + 5$
22. $(-6) + (-3) + (-5) + 10$

14.3 Subtracting Signed Numbers

OBJECTIVE
To subtract signed numbers using the definition of subtraction

In the first chapter, subtraction was called the *inverse* operation to addition. This is extremely useful now because it means that any subtraction problem can be written as a problem in addition. Let's see how it works with the following rule.

TO SUBTRACT SIGNED NUMBERS

To subtract signed numbers, add the first number and the *opposite* of the number being subtracted. In symbols by definition:

$a - b = a + (-b)$

To find the difference $a - b$, we add a and the opposite of b.

Section 14.3: Subtracting Signed Numbers   609

Our first example illustrates.

Example 1

(a) To subtract $5 - 3$:

$$5 - 3 = 5 + (-3) = 2$$

To subtract 3, we can add the opposite of 3.

The opposite of 3

(b) To subtract $2 - 5$:

$$2 - 5 = 2 + (-5) = -3$$

The opposite of 5

CHECK YOURSELF 1

Find each difference using the definition of subtraction.

(1) $8 - 3$ (2) $7 - 9$

Let's look at two further examples.

Example 2

By the definition, add the opposite of 4, -4, to the value -3.

(a) To subtract $-3 - 4$:

$$-3 - 4 = -3 + (-4) = -7$$

The opposite of 4

(b) To subtract $-10 - 15$:

$$-10 - 15 = -10 + (-15) = -25$$

The opposite of 15

CHECK YOURSELF 2

Find each difference.

(1) $-5 - 9$ (2) $-12 - 6$

Now let's see how the definition is applied in subtracting a negative number.

Example 3

Find each difference.

By the definition of subtraction, we add the opposite of −3. Remember, the opposite of −3 is 3.

(a) $5 - (-3) = 5 + 3 = 8$

  ↑ The opposite of −3

(b) $7 - (-8) = 7 + 8 = 15$

  ↑ The opposite of −8

(c) $-9 - (-5) = -9 + 5 = -4$

  ↑ The opposite of −5

CHECK YOURSELF 3

Find each difference.

(1) $7 - (-5)$ (2) $5 - (-9)$ (3) $-10 - (-8)$

Signed numbers other than integers are subtracted in exactly the same way. Our final examples illustrate.

Example 4

$3.5 - (-2.25) = 3.5 + 2.25 = 5.75$

CHECK YOURSELF 4

Find the difference.

$-4.3 - (-7.3)$

Example 5

$-2\frac{3}{4} - \left(-1\frac{1}{4}\right) = -2\frac{3}{4} + 1\frac{1}{4} = -1\frac{1}{2}$

Section 14.3: Subtracting Signed Numbers

CHECK YOURSELF 5
Find the difference.

$$5\frac{7}{8} - \left(-2\frac{3}{8}\right)$$

CHECK YOURSELF ANSWERS

1. (1) 5; (2) −2. 2. (1) −14; (2) −18. 3. (1) 12; (2) 14; (3) −2.
4. 3. 5. $8\frac{1}{4}$.

14.3 Exercises

Subtract.

1. $5 - 7$
2. $7 - 5$
3. $9 - 3$
4. $4 - 9$

5. $-8 - 3$
6. $-15 - 6$
7. $-12 - 8$
8. $9 - 15$

9. $-22 - 18$
10. $-25 - 15$
11. $3 - (-2)$
12. $4 - (-2)$

13. $-2 - (-3)$
14. $-9 - (-6)$
15. $-5 - (-5)$
16. $7 - (-9)$

17. $10 - (-5)$
18. $-8 - (-8)$
19. $38 - (-12)$
20. $50 - (-25)$

21. $-15 - (-25)$
22. $-20 - (-30)$
23. $-25 - (-15)$
24. $-30 - (-20)$

25. $-0.5 - 1.5$
26. $0.25 - 0.75$
27. $3.5 - (-2.5)$
28. $-3.2 - (-1.2)$

29. $5 - \left(-3\frac{1}{2}\right)$
30. $-2 - 1\frac{1}{2}$
31. $-2\frac{1}{4} - \left(-3\frac{3}{4}\right)$
32. $3\frac{1}{4} - \left(-1\frac{3}{4}\right)$

33. $-7 - (-5) - 6$
34. $-5 - (-8) - 10$
35. $-10 - 8 - (-7)$
36. $3 - (-9) - 10$

37. The temperature in Chicago dropped from 22°F at 4 P.M. to −11°F at midnight. What was the drop in temperature?

38. Charley's checking account had $225 deposited at the beginning of 1 month. After writing checks for the month, the account was $65 *overdrawn*. What amount of checks did he write during the month?

39. Micki entered the elevator on the 34th floor. From that point the elevator went up 12 floors, down 27 floors, down 6 floors, and up 15 floors before she got off. On what floor did she get off the elevator?

40. A submarine dove to a depth of 500 ft below the ocean's surface. It then dives another 217 ft before climbing 140 ft. What is the depth of the submarine?

Skillscan (Section 6.3)
Multiply.

a. $\dfrac{2}{3} \cdot \dfrac{6}{7}$ **b.** $\dfrac{7}{8} \cdot \dfrac{4}{21}$ **c.** $\dfrac{9}{10} \cdot \dfrac{5}{18}$

d. $\dfrac{5}{12} \cdot \dfrac{9}{20}$ **e.** $\dfrac{10}{21} \cdot \dfrac{14}{15}$ **f.** $\dfrac{21}{25} \cdot \dfrac{5}{28}$

Answers
1. −2   **3.** 6   **5.** −11   **7.** −20   **9.** −40   **11.** 3 − (−2) = 3 + 2 = 5   **13.** 1   **15.** 0   **17.** 15
19. 50   **21.** 10   **23.** −10   **25.** −2   **27.** 6   **29.** $8\dfrac{1}{2}$ **31.** $1\dfrac{1}{2}$ **33.** −8   **35.** −11   **37.** 33°F
39. 28th floor   **a.** $\dfrac{4}{7}$ **b.** $\dfrac{1}{6}$ **c.** $\dfrac{1}{4}$ **d.** $\dfrac{3}{16}$ **e.** $\dfrac{4}{9}$ **f.** $\dfrac{3}{20}$

14.3 Supplementary Exercises

Subtract.

1. 6 − 9   **2.** 10 − 15   **3.** −9 − 5

4. −11 − 7   **5.** 50 − 60   **6.** 32 − 48

7. 8 − (−3)   **8.** 5 − (−3)   **9.** −6 − (−9)

10. $-4 - (-7)$

11. $45 - (-15)$

12. $-3 - (-3)$

13. $-4.2 - 1.8$

14. $-2.8 - (-0.8)$

15. $3\frac{1}{2} - \left(-1\frac{1}{2}\right)$

16. $-2\frac{1}{4} - \left(-1\frac{1}{4}\right)$

17. $-2 - (-4) - 5$

18. $-3 - (-8) - (-2)$

14.4 Multiplying Signed Numbers

OBJECTIVE
To multiply signed numbers

In an earlier section of this chapter we looked at addition involving signed numbers. When we first considered multiplication, we thought of it as repeated addition. Let's see what our work with addition can tell us about the multiplication of signed numbers.

Example 1

We can interpret multiplication as repeated addition to find a product.

$3 \cdot 4 = 4 + 4 + 4 = 12$

We can use the same idea to find products involving negative numbers.

Example 2

Note that we use parentheses ( ) to indicate multiplication when negative numbers are involved.

$(3)(-4) = (-4) + (-4) + (-4) = -12$
$(4)(-5) = (-5) + (-5) + (-5) + (-5) = -20$

Looking at the products we found by repeated addition in Example 2 should suggest our first rule for multiplying signed numbers.

> **TO MULTIPLY SIGNED NUMBERS**
> 1. The product of two numbers with different signs is negative.

The rule is easy to use. To multiply two numbers with different signs, just multiply their absolute values and attach a negative sign to the product.

Example 3

Find each product.

$(5)(-6) = -30$
$(10)(-12) = -120$
$(-7)(9) = -63$
$(1.5)(-0.3) = -0.45$

The product must have two decimal places.

$\left(-\dfrac{5}{8}\right)\left(\dfrac{4}{15}\right) = -\left(\dfrac{\overset{1}{\cancel{5}}}{\underset{2}{\cancel{8}}} \times \dfrac{\overset{1}{\cancel{4}}}{\underset{3}{\cancel{15}}}\right)$

$= -\dfrac{1}{6}$

The product is negative. You can simplify as before in finding the product.

CHECK YOURSELF 1

Find each product.

(1) $(15)(-5)$ (2) $(-0.8)(0.2)$ (3) $\left(-\dfrac{2}{3}\right)\left(\dfrac{6}{7}\right)$

The product of two negative numbers is harder to visualize. The following pattern may help you see how we can determine the sign of the product.

Decreasing by 1:

$(3)(-2) = -6$
$(2)(-2) = -4$
$(1)(-2) = -2$
$(0)(-2) = 0$
$(-1)(-2) = 2$
$(-2)(-2) = 4$

Do you see that the product is *increasing* by 2 each time?

If you would like a more detailed explanation about why the product of two negative numbers must be positive, see page 617.

This suggests that the product of two negative numbers is positive, and this is in fact the case. To extend our multiplication rule:

We already know that the product of two positive numbers is positive.

TO MULTIPLY SIGNED NUMBERS

2. The product of two numbers with the same sign is positive.

Example 4

$8 \cdot 7 = 56$
$(-9)(-6) = 54$
$(-0.5)(-2) = 1$

Since the numbers have the same sign, the product is positive.

Section 14.4: Multiplying Signed Numbers 615

CHECK YOURSELF 2

Find each product.

(1) $(5)(7)$ (2) $(-8)(-6)$ (3) $(9)(-6)$ (4) $(-1.5)(-4)$

Be Careful! $(-8)(-6)$ tells you to multiply. The parentheses are *next to* each other. The expression $-8 - 6$ tells you to subtract. The numbers are *separated* by the operation sign.

To multiply more than two signed numbers, just apply the multiplication rule repeatedly.

Example 5

$$\begin{aligned}
&\underbrace{(5)(-7)}(-3)(-2) \\
=\ &\underbrace{(-35)(-3)}(-2) \\
=\ &\quad (105)(-2) \\
=\ &\quad\ -210
\end{aligned}$$

$(5)(-7) = -35$

$(-35)(-3) = 105$

CHECK YOURSELF 3

Find the product.

$(-4)(3)(-2)(5)$

We saw earlier that the commutative and associative properties for addition could be extended to signed numbers. The same is true for multiplication. What about the order in which we multiply? Look at the following examples.

Example 6

$(-5)(7) = (7)(-5) = -35$
$(-6)(-7) = (-7)(-6) = 42$

The order in which we multiply does not affect the product. We can write:

> **THE COMMUTATIVE PROPERTY**
>
> The order in which we multiply does not change the product. Multiplication of integers is *commutative*. In symbols, for any integers *a* and *b*:
>
> a × b = b × a

CHECK YOURSELF 4

Show that $(-8)(-5) = (-5)(-8)$

What about the way we group numbers in multiplication? Look at Example 7.

Example 7

The symbols [ ] are called *brackets* and are used to group numbers in the same way as parentheses.

$[(3)(-7)](-2)$ or $(3)[(-7)(-2)]$
$= (-21)(-2)$ $= (3)(14)$
$= 42$ $= 42$

We group the first two numbers on the left and the second two numbers on the right. Note that the product is the same in either case. We can state:

THE ASSOCIATIVE PROPERTY

The way we *group* the numbers does not change the product. Multiplication of integers is *associative*. In symbols, for any integers *a*, *b*, and *c*:

$(a \times b) \times c = a \times (b \times c)$

CHECK YOURSELF 5

Show that $[(2)(-6)](-3) = (2)[(-6)(-3)]$

Two numbers, 0 and 1, have special properties in multiplication.

THE MULTIPLICATIVE IDENTITY

The product of 1 and any integer is that integer. We call 1 the *multiplicative identity*. In symbols, for any integer *a*:

$a \times 1 = a$

Example 8

$(-8)(1) = -8$
$(1)(-15) = -15$

Section 14.4: Multiplying Signed Numbers   617

CHECK YOURSELF 6
Find the product.

$(-10)(1)$

What about multiplication by 0? As we saw earlier:

MULTIPLYING BY ZERO

The product of 0 and any integer is 0. In symbols, for any integer a:

$a \times 0 = 0$

Example 9

$(-9)(0) = 0$
$(0)(-23) = 0$

CHECK YOURSELF 7
Find the product.

$(0)(-12)$

THE PRODUCT OF TWO NEGATIVE NUMBERS

The following argument shows why the product of two negative numbers must be positive.

From our earlier work, we know that a number added to its opposite is 0.	$5 + (-5) = 0$
Multiply both sides of the statement by -3:	$(-3)[5 + (-5)] = (-3)(0)$
A number multiplied by 0 is 0, so on the right we have 0.	$(-3)[5 + (-5)] = 0$
We can now use the distributive property on the left.	$(-3)(5) + (-3)(-5) = 0$
Since we know that $(-3)(5) = -15$, the statement becomes	$-15 + (-3)(-5) = 0$

We now have a statement of the form $-15 + \square = 0$. This asks, "What number must we add to -15 to get 0, where $\square$ is the value of $(-3)(-5)$." The answer is, of course, 15. This means that

$(-3)(-5) = 15$ The product must be positive.

It doesn't matter what numbers we use in the argument. The product of two negative numbers will always be positive.

CHECK YOURSELF ANSWERS

1. (1) −75; (2) −0.16; (3) $-\frac{4}{7}$.   2. (1) 35; (2) 48; (3) −54; (4) 6.
3. 120.   4. 40 = 40.   5. 36 = 36.   6. −10.   7. 0.

14.4 Exercises

Multiply.

1. $7 \cdot 8$
2. $(6)(-12)$
3. $(4)(-3)$
4. $15 \cdot 5$

5. $(-8)(9)$
6. $(-8)(3)$
7. $(-7)(-6)$
8. $(-12)(-2)$

9. $(-10)(0)$
10. $(10)(-10)$
11. $(-8)(-8)$
12. $(0)(-50)$

13. $(20)(-4)$
14. $(-25)(-8)$
15. $(-9)(-12)$
16. $(-9)(-9)$

17. $(-20)(1)$
18. $(1)(-30)$
19. $(-40)(5)$
20. $(-25)(5)$

21. $(1.8)(-0.2)$
22. $(-2.4)(-0.5)$
23. $\left(-\frac{7}{10}\right)\left(-\frac{5}{14}\right)$
24. $\left(-\frac{3}{8}\right)\left(\frac{4}{9}\right)$

25. $(-0.5)(-1.2)$
26. $(2.5)(-0.3)$
27. $\left(\frac{5}{8}\right)\left(-\frac{4}{15}\right)$

28. $\left(-\frac{8}{21}\right)\left(-\frac{7}{4}\right)$
29. $(-3)^2$
30. $(-2)^3$

31. -3^2
32. -2^3
33. $(-4)^3$

34. $(-3)^4$
35. $(-2)^4$
36. -5^2

37. $(-5)(3)(-8)$
38. $(4)(-3)(-5)$
39. $(-2)(-8)(-5)$

40. $(-7)(-5)(-2)$
41. $(2)(-5)(-3)(-5)$
42. $(-2)(-5)(-5)(-6)$

43. $(-4)(-3)(-6)(-2)$
44. $(-8)(3)(-2)(5)$

Recall that the formula for converting a Fahrenheit temperature reading to a reading on the Celsius scale is

$$°C = \frac{5}{9}(°F - 32)$$

45. Find the Celsius reading which corresponds to 5°F.

46. Find the Celsius reading which corresponds to −4°F.

Skillscan (Section 9.2)
Find each quotient.

a. $\dfrac{7.5}{1.5}$

b. $\dfrac{14.4}{2.4}$

c. $\dfrac{7.2}{0.6}$

d. $\dfrac{10.5}{0.7}$

e. $\dfrac{45}{1.25}$

f. $\dfrac{32.5}{1.3}$

Answers
1. 56   3. −12   5. −72   7. 42   9. 0   11. 64   13. −80   15. 108   17. −20   19. −200
21. −0.36   23. $\frac{1}{4}$ 25. 0.6   27. $-\frac{1}{6}$ 29. $(-3)^2 = (-3)(-3) = 9$ 31. −9   33. −64   35. 16
37. 120   39. −80   41. −150   43. 144   45. −15°C   a. 5   b. 6   c. 12   d. 15   e. 36
f. 25

14.4 Supplementary Exercises

Multiply.

1. $8 \cdot 9$

2. $(7)(-9)$

3. $(-6)(7)$

4. $(-12)(-7)$

5. $(10)(-8)$

6. $8 \cdot 6$

7. $(-11)(1)$

8. $(-10)(7)$

9. $(-15)(-7)$

10. $(-5)(0)$

11. $(1.8)(-0.5)$

12. $\left(-\dfrac{5}{6}\right)\left(\dfrac{9}{20}\right)$

13. $\left(-\dfrac{5}{8}\right)\left(-\dfrac{4}{9}\right)$

14. $(-0.25)(12)$

15. $(-5)^2$

16. -7^2

17. $(-3)^3$

18. $(-2)^4$

19. $(-3)(-7)(-2)$ **20.** $(-4)(3)(-6)$ **21.** $(-5)(3)(-2)(-2)$

22. $(5)(-3)(4)(-2)$

14.5 Dividing Signed Numbers

OBJECTIVE
To divide signed numbers

In our earlier work we said that multiplication and division are related operations. So every division problem can be stated as an equivalent multiplication problem.

Example 1

We used this earlier to check our division!

$8 \div 4 = 2$ Since $8 = 4 \cdot 2$

$\dfrac{12}{3} = 4$ Since $12 = 3 \cdot 4$

Since the operations are related, the rule of signs for multiplication that we stated in the last section is also true for division.

TO DIVIDE SIGNED NUMBERS
1. If two numbers have the same sign, the quotient is positive.
2. If two numbers have different signs, the quotient is negative.

Example 2

The numbers 20 and −5 have different signs, and so the quotient is negative.

$20 \div (-5) = -4$ Since $20 = (-5)(-4)$

Example 3

The two numbers have the same sign, and so the quotient is positive.

$\dfrac{-20}{-5} = 4$ Since $-20 = (-5)(4)$

CHECK YOURSELF 1

Find each quotient.

(1) $\dfrac{48}{-6}$ (2) $(-50) \div (-5)$

Section 14.5: Dividing Signed Numbers

Example 4

$-30 \div 6 = -5$ Since $-30 = (6)(-5)$

As you would expect, division with fractions or decimals uses the same rule for signs. Our next examples illustrate.

Example 5

First note that the quotient is positive. Then invert the divisor and multiply.

$$\left(-\frac{3}{5}\right) \div \left(-\frac{9}{20}\right) = \frac{\overset{1}{\cancel{3}}}{\underset{1}{\cancel{5}}} \cdot \frac{\overset{4}{\cancel{20}}}{\underset{3}{\cancel{9}}}$$

$$= \frac{4}{3} = 1\frac{1}{3}$$

Example 6

The numbers have different signs so the quotient is negative.

$-3.5 \div 0.7 = -5$

CHECK YOURSELF 2

Find each quotient.

(1) $-\frac{5}{8} \div \frac{3}{4}$ (2) $-4.2 \div (-0.6)$

Be very careful when 0 is involved in a division problem. Remember that 0 divided by any nonzero number is just 0. However, division *by 0* is not allowed.

Example 7

$0 \div 7 = 0$ $\frac{0}{-4} = 0$

A statement like $-9 \div 0$ has no meaning. There is no answer to the problem. Just write "undefined."

$-9 \div 0$ is undefined. $\frac{-5}{0}$ is undefined.

CHECK YOURSELF 3

Find the quotient if possible.

(1) $\frac{0}{-7}$ (2) $\frac{-12}{0}$

CHECK YOURSELF ANSWERS

1. (1) −8; (2) 10. **2.** (1) −$\frac{5}{6}$; (2) 7. **3.** (1) 0; (2) undefined.

14.5 Exercises

Divide.

1. $15 \div (-3)$

2. $\dfrac{35}{7}$

3. $\dfrac{48}{8}$

4. $-20 \div (-2)$

5. $\dfrac{-50}{5}$

6. $-36 \div 6$

7. $\dfrac{-24}{-3}$

8. $\dfrac{42}{-6}$

9. $-20 \div 5$

10. $-45 \div 9$

11. $-72 \div 8$

12. $\dfrac{-60}{-12}$

13. $\dfrac{60}{-15}$

14. $70 \div (-10)$

15. $18 \div (-1)$

16. $\dfrac{-250}{-25}$

17. $\dfrac{0}{-9}$

18. $\dfrac{-12}{0}$

19. $-144 \div (-12)$

20. $\dfrac{0}{-10}$

21. $-7 \div 0$

22. $\dfrac{-25}{1}$

23. $\dfrac{-150}{6}$

24. $\dfrac{-80}{-16}$

25. $-4.5 \div (-0.9)$

26. $-\dfrac{2}{3} \div \dfrac{4}{9}$

27. $-\dfrac{7}{9} \div \left(-\dfrac{14}{3}\right)$

28. $(-0.8) \div (-0.4)$

29. $\dfrac{7}{10} \div \left(-\dfrac{14}{25}\right)$

30. $\dfrac{0.75}{-0.5}$

31. $\dfrac{-7.5}{1.5}$

32. $-\dfrac{5}{8} \div \left(-\dfrac{5}{16}\right)$

To evaluate an expression involving a fraction (indicating division), we evaluate the numerator and then the denominator. We then divide the numerator by the denominator as the last step. Using this approach, find the value of each of the following expressions.

33. $\dfrac{5 - 15}{2 + 3}$ **34.** $\dfrac{4 - (-8)}{2 - 5}$ **35.** $\dfrac{-6 + 18}{-2 - 4}$

36. $\dfrac{-4 - 21}{3 - 8}$ **37.** $\dfrac{(5)(-12)}{(-3)(5)}$ **38.** $\dfrac{(-8)(-3)}{(2)(-4)}$

Answers
1. −5 **3.** 6 **5.** −10 **7.** 8 **9.** −4 **11.** −9 **13.** −4 **15.** −18 **17.** 0 **19.** 12
21. Undefined **23.** −25 **25.** 5 **27.** $\dfrac{1}{6}$ **29.** $-\dfrac{5}{4}$ **31.** −5 **33.** −2 **35.** −2 **37.** 4

14.5 Supplementary Exercises

Divide.

1. $12 \div (-4)$ **2.** $\dfrac{-45}{9}$ **3.** $\dfrac{24}{3}$

4. $-60 \div (-5)$ **5.** $-72 \div 8$ **6.** $\dfrac{81}{9}$

7. $-42 \div (-7)$ **8.** $\dfrac{-12}{0}$ **9.** $\dfrac{0}{-3}$

10. $48 \div (-6)$ **11.** $\dfrac{-8}{-1}$ **12.** $-48 \div (-4)$

13. $-2.5 \div 0.5$ **14.** $-\dfrac{5}{8} \div \left(-\dfrac{3}{4}\right)$ **15.** $\dfrac{3}{10} \div \left(-\dfrac{4}{5}\right)$

16. $(-4.5) \div (-0.5)$

SELF-TEST for Chapter Fourteen

The purpose of the Self-Test is to help you check your progress and review for a chapter test in class. Allow yourself about an hour to take the test. When you are done, check your answers in the back of the book. If you missed any problems, be sure to go back and review the appropriate sections in the chapter and do the supplementary exercises provided there.

In Problems 1 to 4, complete the statements.

1. The absolute value of -9 is

2. $-(-8) =$

3. $|-7| =$

4. The opposite of -10 is

In Problems 5 to 12, add.

5. $9 + (-3) =$

6. $8 + (-10) =$

7. $-7 + (-5) =$

8. $-7 + 7 =$

9. $-9 + 0 =$

10. $1\frac{3}{4} + \left(-2\frac{1}{4}\right) =$

11. $-3 + (-7) + 12 =$

12. $-2.8 + (-1.2) =$

In Problems 13 to 18, subtract.

13. $8 - 11 =$

14. $-3 - 7 =$

15. $7 - (-3) =$

16. $-5 - (-9) =$

17. $4.2 - (-3.8) =$

18. $-2\frac{1}{2} - (-3) =$

625

In Problems 19 to 24, multiply.

19. $(-7)(4) =$

20. $(-9)(-6) =$

21. $(-2.5)(-0.3) =$

22. $(-12)(0) =$

23. $(2)(-3)(-8) =$

24. $(-6)^2 =$

In Problems 25 to 30, divide.

25. $20 \div (-5) =$

26. $\dfrac{-28}{-4} =$

27. $\dfrac{-50}{10} =$

28. $-5.5 \div 1.1 =$

29. $0 \div (-7) =$

30. $-12 \div 0 =$

PRETEST for Chapter Fifteen

Algebraic Expressions and Equations

This pretest will point out any difficulties you may be having with algebraic expressions and equations. Do all the problems. Then check your answers on the following page.

1. Evaluate $7xy$ if $x = 2$ and $y = 3$.

2. Evaluate $\dfrac{5a - 2b}{b - a}$ if $a = 4$ and $b = -2$.

3. Solve for x: $7x = 42$.

4. Solve for x: $2x + 3 = 9$.

5. Combine like terms: $-4a + 9a =$

6. Solve for x: $3x - 7 = 2x$.

7. Solve for x: $7x - 4 = 5x + 2$.

8. Solve for x: $5 - 4x = 9 - 2x$.

9. Write using symbols: The quotient when 3 times a is divided by 3 more than a.

10. One number is 8 more than another. If the sum of the numbers is 50, find the two numbers.

ANSWERS TO PRETEST

For help with similar problems, turn to the section indicated.

1. 42 (Section 15.1);
2. −4 (Section 15.1);
3. 6 (Section 15.2);
4. 3 (Section 15.2);
5. $5a$ (Section 15.3);
6. 7 (Section 15.3);
7. 3 (Section 15.3);
8. −2 (Section 15.3)
9. $\dfrac{3a}{a+3}$ (Section 15.4);
10. 21, 29 (Section 15.4)

Chapter Fifteen

Algebraic Expressions and Equations

15.1 Evaluating Algebraic Expressions

OBJECTIVE
To evaluate algebraic expressions given specified values for the variables

Remember that a letter used to represent a number is called a *variable*.

The next step in getting ready for algebra involves using *algebraic* or *literal expressions*. You have already seen examples of these expressions, which use letters to represent numbers.

Example 1

Some algebraic expressions that we have used are

The variables are b and h. $b \cdot h$ The area of a rectangle.

The variable is F. $\frac{5}{9}(°F - 32)$ Temperature conversion from degrees Fahrenheit to degrees Celsius

Often you will want to find the value of an expression. You can do this when you know the number values for each of the letters or variables. The process is called *evaluating the expression*.

To evaluate an expression, just replace each variable or letter with its number value. Then do the necessary operations.

Example 2

Note: In algebra, writing letters together, such as *xy*, means multiplication, or "x times y." You must supply the "times" sign (·) with numbers, because 2 · 3 is entirely different from 23. Of course 2 × 3 and (2)(3) also mean "2 times 3."

Suppose that $x = 2$ and $y = 3$. To evaluate xy, replace x with 2 and y with 3. Then multiply.

$xy = 2 \cdot 3 = 6$

Example 3

If $m = 3$ and $n = 7$, evaluate $4mn$.

$4mn = 4 \cdot 3 \cdot 7 = 84$ Replace m with 3 and n with 7. Then multiply.

CHECK YOURSELF 1

If $a = 5$ and $b = 8$, evaluate $5ab$.

You may want to return to Section 3.6 and review the order of operations before going on.

Often, evaluating an expression will involve applying the rules for the order of operations. The following examples illustrate.

Example 4

If $x = 3$ and $y = 8$, evaluate $3x + 2y$.

Remember: Do any multiplication (or division) *before* any addition (or subtraction).

$$\begin{aligned} 3x + 2y &= 3 \cdot 3 + 2 \cdot 8 \quad &&\text{Replace } x \text{ with 3 and } y \text{ with 8.} \\ &= 9 + 16 \quad &&\text{Multiply, } then \text{ add.} \\ &= 25 \end{aligned}$$

CHECK YOURSELF 2

If $m = 4$ and $n = 5$, evaluate $5m - 3n$

Example 5

Evaluate $5m^2$ when $m = 2$.

Remember: If the expression involves a square, a cube, or some other power, find that value first. Then proceed as before.

$5m^2 = 5 \cdot 2^2 = 5 \cdot 4 = 20$ Evaluate the power, *then* multiply.

CHECK YOURSELF 3

Evaluate $3a^2$ when $a = 4$.

Example 6

If $a = 5$ and $b = 4$, find the value of $4a^2 - 3b$.

$$\begin{aligned} 4a^2 - 3b &= 4 \cdot 5^2 - 3 \cdot 4 \quad &&\text{Replace } a \text{ with 5 and } b \text{ with 4.} \\ &= 4 \cdot 25 - 3 \cdot 4 \quad &&\text{Evaluate the power first.} \\ &= 100 - 12 \quad &&\text{Multiply.} \\ &= 88 \quad &&\text{Finally, do the subtraction.} \end{aligned}$$

CHECK YOURSELF 4

Evaluate $3x^2 + 5y^3$ if $x = 3$ and $y = 2$.

If the variables of an expression have negative values, you should be especially careful with the rule for signs. Consider the next example.

Example 7

Evaluate $5x - 3y$ if $x = 8$ and $y = -3$.

$5x - 3y = 5 \cdot 8 - (3)(-3)$ Multiply first. Be sure to use the rule of signs for multiplication.
$ = 40 - (-9)$ Now subtract.
$ = 40 + 9 = 49$

CHECK YOURSELF 5

If $p = 5$ and $r = -4$, evaluate $4p - 7r$.

Our next example shows a negative number raised to a power.

Example 8

Find the value of $3x^2$ if $x = -2$.

$3x^2 = (3)(-2)^2$ Evaluate the power first:
$ = 3 \cdot 4$ Here $(-2)^2 = (-2)(-2) = 4$
$ = 12$

CHECK YOURSELF 6

Find the value of $2a^2 - 5b$ when $a = -2$ and $b = 4$.

Expressions involving parentheses can be evaluated easily if you remember our earlier rule: *Evaluate the expression in the parentheses first.* Then continue to evaluate the expression as before.

Example 9

Evaluate $3(x + y)$ when $x = 7$ and $y = 4$.

$3(x + y) = 3(7 + 4)$ Replace x with 7 and y with 4.
$ = 3(11)$ Evaluate the expression inside the parentheses as the first step.
$ = 33$ Then multiply.

CHECK YOURSELF 7

Evaluate $2(m + n)$ when $m = 3$ and $n = 4$.

Example 10

Evaluate $a(3b - c)$ when $a = 5$, $b = 3$, and $c = 6$.

$$\begin{aligned} a(3b - c) &= 5(3 \cdot 3 - 6) \\ &= 5(9 - 6) \\ &= 5(3) \\ &= 15 \end{aligned}$$

Multiply first and then subtract inside the parentheses.

As these examples illustrate, it is especially important for you to work carefully, step by step. Don't try to combine steps at this point. This will help you avoid errors.

CHECK YOURSELF 8

Evaluate $x(2y - z)$ when $x = 4$, $y = 2$, and $z = -3$.

Example 11

Find the value of $3(x^2 - y^2)$ if $x = 5$ and $y = 4$.

$$\begin{aligned} 3(x^2 - y^2) &= 3(5^2 - 4^2) \\ &= 3(25 - 16) \\ &= 3 \cdot 9 \\ &= 27 \end{aligned}$$

Evaluate the squares in the parentheses as your first step. Now subtract.

Multiply as the last step.

CHECK YOURSELF 9

Evaluate $x(2y^2 - 3z^3)$ if $x = 4$, $y = 3$, and $z = 2$.

Many expressions in algebra involve fractions. To evaluate a fraction, start by evaluating the numerator. Then evaluate the denominator. As the *last step*, divide the numerator by the denominator.

Example 12

Evaluate $\dfrac{2ab}{c}$ if $a = 5$, $b = -3$, and $c = 10$.

$$\dfrac{2ab}{c} = \dfrac{(2)(5)(-3)}{10} = \dfrac{-30}{10}$$

Evaluate the numerator. Be careful to apply the rules for signs in multiplication.

$$= -3$$

Divide the numerator by the denominator.

CHECK YOURSELF 10

Evaluate $\dfrac{5xy}{z}$ if $x = -4$, $y = -3$, and $z = 12$.

Example 13

Evaluate $\dfrac{3x^2 + y}{2x - 3y}$ if $x = 2$ and $y = 3$.

$$\dfrac{3x^2 + y}{2x - 3y} = \dfrac{3 \cdot 2^2 + 3}{2 \cdot 2 - 3 \cdot 3}$$

$$= \dfrac{15}{-5}$$

$$= -3$$

CHECK YOURSELF 11

Evaluate

$$\dfrac{2m^2 + 3n}{2m - n}$$

if $m = -3$ and $n = 3$.

CHECK YOURSELF ANSWERS

1. $5ab = 5 \cdot 5 \cdot 8 = 200$.   2. $5m - 3n = 5 \cdot 4 - 3 \cdot 5 = 5$.   3. 48.
4. $3x^2 + 5y^3 = 3 \cdot 3^2 + 5 \cdot 2^3$ 5. $4p - 7r = 4 \cdot 5 - (7)(-4)$
   $ = 3 \cdot 9 + 5 \cdot 8$ $ = 20 + 28$
   $ = 27 + 40 = 67$ $ = 48$
6. $2a^2 - 5b = 2(-2)^2 - 5 \cdot 4$ 7. 14.   8. 28.
   $ = 2 \cdot 4 - 5 \cdot 4 = -12$
9. $x(2y^2 - 3z^3) = 4(2 \cdot 3^2 - 3 \cdot 2^3)$ 10. $\dfrac{5xy}{z} = \dfrac{(5)(-4)(-3)}{12} = 5$.
   $ = 4(2 \cdot 9 - 3 \cdot 8)$
   $ = 4(18 - 24)$
   $ = 4(-6) = -24$
11. -3

15.1 Exercises

Evaluate the following expressions if $a = 5$, $b = 4$, $c = -2$, and $d = -3$.

1. $6a$
2. $5b$
3. $-2b$
4. $-4c$

5. $5ab$
6. $3bc$
7. $6cd$
8. $-3ad$

9. $2a + 3b$
10. $5a + 2b$
11. $3c + 4d$
12. $2a + 5d$

13. $3a - 4c$
14. $4b - 2d$
15. $5c - 2d$
16. $3b - 4c$

17. a^2
18. d^2
19. c^3
20. b^3

21. $2b^2$
22. $3c^2$
23. $a^2 + 2d^2$
24. $b^2 - 3c^2$

Evaluate the following expressions if $m = -2$, $n = 4$, $s = -3$, and $t = 6$.

25. $2(n + t)$
26. $3(m + n)$
27. $3(s - n)$
28. $5(m + s)$

29. $4(s + t)$
30. $6(t - s)$
31. $5(n + 3s)$
32. $6(2m + 3s)$

33. $m(2n + t)$
34. $s(3n + 2t)$
35. $2(n^2 + t^2)$
36. $4(m^2 + n^2)$

37. $3(2n^2 + s^2)$
38. $5(t^2 - 2m^2)$
39. $\dfrac{3mn}{t}$
40. $\dfrac{2nt}{3m}$

41. $\dfrac{5s + t}{s}$
42. $\dfrac{3t - m}{2n + s}$
43. $\dfrac{3m^2 - 4s}{3n - t}$
44. $\dfrac{2s^2 + 3t}{s - t}$

There are many applications of our work in evaluating expressions. One such application is in the use of formulas. The following are just a few examples.

45. A formula from physics for distance is $d = rt$. Find d if $r = 55$ and $t = 4$.

46. A formula from business for interest is $I = Prt$. Find I if $P = 3000$, $r = 0.06$, and $t = 3$.

47. A formula from physics for the height of an object is $h = 64t - 16t^2$. Find h if $t = 2$.

48. A formula from geometry for the area of a trapezoid is $A = \dfrac{1}{2}h(a + b)$. Find A if $h = 8$, $a = 10$ and $b = 16$.

49. A formula from business for the amount in an account is $A = P(1 + rt)$. Find A if $P = 5000$, $r = 0.08$, and $t = 2$.

50. A formula from mathematics for a sum is $S = \dfrac{n}{2}[a + (n - 1)d]$. Find S if $a = 5$, $d = 3$, and $n = 20$.

Skillscan (Section 14.2)
Add.

a. $5 + (-5)$

b. $-2 + 5$

c. $-7 + 7$

d. $-8 + 4$

e. $9 + (-9)$

f. $10 + (-4)$

Answers
Solutions for the even-numbered exercises are provided in the back of the book.
1. 30 **3.** −8 **5.** 100 **7.** 36 **9.** $2a + 3b = 2 \cdot 5 + 3 \cdot 4 = 10 + 22 = 22$ **11.** −18 **13.** 23
15. −4 **17.** 25 **19.** −8 **21.** 32 **23.** $a^2 + 2d^2 = 5^2 + 2(-3)^2 = 25 + 2 \cdot 9 = 25 + 18 = 43$ **25.** 20
27. −21 **29.** 12 **31.** −25 **33.** $m(2n + t) = (-2)(2 \cdot 4 + 6) = (-2)(8 + 6) = (-2)(14) = -28$ **35.** 104
37. 123 **39.** $\dfrac{3mn}{t} = \dfrac{(3)(-2)(4)}{6} = \dfrac{-24}{6} = -4$ **41.** 3 **43.** 4 **45.** 220 **47.** 64 **49.** 5800
a. 0 **b.** 3 **c.** 0 **d.** −4 **e.** 0 **f.** 6

15.1 Supplementary Exercises

Evaluate the following expressions if $x = 4$, $y = 3$, $z = -3$, and $w = -2$.

1. $5x$

2. $3z$

3. $4xy$

4. $5yz$

5. $4x + 3w$

6. $6w - 2z$

7. $3z - 4w$

8. z^2

9. $3y^2$

10. $2z^3$

11. $x^2 + 2z^2$

12. $y^2 - 2w^2$

Evaluate the following expressions if $a = -3$, $b = 5$, $c = -2$, and $d = 4$.

13. $3(b + d)$

14. $2(b - c)$

15. $2(a + c)$

16. $5(c - d)$

17. $a(2b + c)$

18. $d(b^2 - c^2)$

19. $\dfrac{2ab}{c}$

20. $\dfrac{3bd}{5c}$

21. $\dfrac{2b - c}{a + b}$

22. $\dfrac{3b - a}{2d + c}$

15.2 Solving Equations

OBJECTIVE
To solve equations using the addition and multiplication rules

In the last section we learned how to evaluate algebraic expressions by replacing the letters or variables in the expression with numerical values. Sometimes algebraic expressions are used in equations. An *equation* is a statement that two expressions are equal.

Example 1

This equation says that "$3x - 1$ is equal to $2x + 5$." As you can see, an equals sign (=) separates the two sides of the equation.

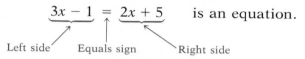 is an equation.

One of the most important ideas in algebra is finding the *solution* of an equation. An equation may be either true or false depending on the value given to the variable.

We sometimes say that the solution satisfies the equation.

> A *solution* for an equation is a numerical value for the variable or letter that makes the equation a true statement.

This means that the sides of the equation must be equal when the letter is replaced by the number. This is illustrated in the following example.

Example 2

Is 6 a solution for the following equation?

$$3x - 1 = 2x + 5$$

To find out, replace x with 6.

$$3 \cdot 6 - 1 \stackrel{?}{=} 2 \cdot 6 + 5$$
$$18 - 1 \stackrel{?}{=} 12 + 5$$
$$17 = 17$$

6 satisfies the equation.

Since the two sides are equal, 6 is a solution.

Example 3

Is 2 a solution for the following equation?

$$2x + 5 = x + 3$$

Section 15.2: Solving Equations

2 does not satisfy the equation.

Replace x with 2.

$$2 \cdot 2 + 5 \stackrel{?}{=} 2 + 3$$
$$4 + 5 \stackrel{?}{=} 2 + 3$$
$$9 \neq 5 \quad \text{This is read "nine is not equal to five."}$$

Since the sides are not equal, 2 is *not* a solution for the equation.

Is -2 a solution? This time replace x with -2.

$$(2)(-2) + 5 \stackrel{?}{=} -2 + 3$$
$$-4 + 5 \stackrel{?}{=} -2 + 3$$
$$1 = 1 \quad \text{(true)}$$

-2 satisfies the equation.

Since the sides now have the same value, -2 is a solution.

CHECK YOURSELF 1

For the equation $3x + 1 = 2x - 3$,

(1) Is 2 a solution?   (2) Is -4 a solution?

Note You may be wondering whether an equation can have more than one solution. It certainly can.

The equation $x^2 = 4$ has two solutions, 2 and -2, because

$$2^2 = 4 \quad \text{and} \quad (-2)^2 = 4$$

However, in this section we will be looking at *linear equations in one variable*. These are equations in one variable or letter (we have used x) which appears to the first power. No other powers (x^2, x^3, etc.) can appear. These linear equations in one variable will have at most *one solution*.

Example 4

$x + 5 = 8$ is a linear equation in one variable, x. The solution for the equation is 3, since $3 + 5 = 8$.

You can easily find the solution for an equation like the one in Example 5 by guessing the answer to the question, "What plus five is eight?"

> There is nothing wrong with guessing the answer when you can. It is just that in more difficult equations, guessing is a lot harder, and we need methods to handle these cases.

For more complicated equations, you will need something more than guesswork. So let's start to develop a set of rules that will allow you to solve any linear equation. First we will need a definition.

> Equations that have exactly the same solution are called *equivalent equations*.

Example 5

> Check for yourself that 3 satisfies each equation.

$3x + 2 = 11$, $3x = 9$, and $x = 3$ are all equivalent equations because they all have the same solution, 3.

We say that a linear equation is *solved* when the variable is alone on the left side of the equation and only a number appears on the right.

> An equation is solved when we write it as an equivalent equation of the form:
>
> $x = \square$ where $\square$ is some number

Here is the first rule we will need for solving equations:

> **SOLVING EQUATIONS**
>
> THE ADDITION RULE Adding or subtracting the same term on each side of an equation gives an equivalent equation.

> Remember, an equation is a statement that the two sides are equal. Adding or subtracting the same thing from the two sides maintains the equality or balance.

Let's work through our first example of the use of this rule.

Example 6

Solve $x - 3 = 9$.

> Remember: $-3 + 3 = 0$.

$$\begin{array}{rl} x - 3 = & 9 \\ + 3 & + 3 \\ \hline x + 0 = & 12 \\ x = & 12 \end{array}$$

Adding 3 "undoes" the subtraction and removes everything but the variable, x, from the left side.

Since $x + 0 = x$, you do not need to write the 0.

Since 12 is a solution for the equivalent equation, $x = 12$, it is a solution for our original equation. We have found the solution 12 for the equation $x - 3 = 9$. It is always a good idea to check your work. Replace x with 12 in the original equation.

$$12 - 3 \stackrel{?}{=} 9$$
$$9 = 9$$

This is a true statement, so 12 is a solution for the original equation.

Section 15.2: Solving Equations

CHECK YOURSELF 2

Solve and check.

$x - 5 = 8$

Example 7

Solve and check $x - 5 = -2$.

$$\begin{array}{r} x - 5 = -2 \\ +5 +5 \\ \hline x = 3 \end{array}$$

Add 5 to *both* sides. This undoes the subtraction on the left leaving x alone on that side of the equation.

Replace x with 3 in the original equation to check.

$3 - 5 \stackrel{?}{=} -2$
$ -2 = -2$ This is true, so 3 is the solution.

CHECK YOURSELF 3

Solve and check.

$x - 7 = -4$

Let's look at a slightly different example. This time we will need to subtract the same number from both sides of the equation.

Example 8

Solve and check $x + 5 = 9$.

$$\begin{array}{r} x + 5 = 9 \\ -5 -5 \\ \hline x = 4 \end{array}$$

Here 5 is *added* to the variable on the left side. *Subtract* 5 from both sides to undo the addition.

To check:

$4 + 5 \stackrel{?}{=} 9$
$ 9 = 9$

CHECK YOURSELF 4

Solve and check.

(1) $x + 8 = 12$ (2) $x + 6 = -3$

Now let's look at a different type of equation. What if we know that $6x = 18$? Adding or subtracting numbers won't help. We need a second rule for solving equations.

SOLVING EQUATIONS

THE MULTIPLICATION RULE Multiplying or dividing both sides of an equation by the same nonzero number gives an equivalent equation.

Again, as long as you do the *same* thing to *both* sides of the equation, the balance is maintained.

Let's see how this second rule is used to solve equations.

Example 9

Solve and check $5x = 15$.

Solution On the left, x is multiplied by 5. We can use division to undo that multiplication.

$$\frac{5x}{5} = \frac{15}{5}$$ Use the multiplication rule to divide both sides by 5.

$$\frac{\cancel{5}x}{\cancel{5}} = 3$$ We divide numerator and denominator by 5, leaving x alone on the left side. The right side simplifies to 3.

$$x = 3$$

Note: $\frac{\cancel{5}x}{\cancel{5}} = 1x = x$

To check, replace x with 3 in the original equation.

$5 \cdot 3 \stackrel{?}{=} 15$

$15 = 15$ True, so 3 is the solution.

CHECK YOURSELF 5

Solve and check.

$8x = 32$

Example 10

Solve and check $4x = -28$.

Solution Here x is being multiplied by 4. The 4 is removed by dividing by 4.

$$\frac{\cancel{4}x}{\cancel{4}} = \frac{-28}{4}$$ Divide *both* sides by 4.

$$x = -7$$ This leaves x alone on the left. The right side reduces to -7. Be careful of the rule for signs!

Section 15.2: Solving Equations

To check, replace x with -7 in the original equation.

$$4x = -28$$
$$(4)(-7) \stackrel{?}{=} -28$$
$$-28 = -28 \quad \text{True, so } -7 \text{ is the solution.}$$

CHECK YOURSELF 6

Solve and check.

$5x = -30$

Example 11

Solve and check $-5x = 30$.

$$\frac{-5x}{-5} = \frac{30}{-5} \quad \text{Here we must divide both sides by } -5 \text{ to leave } x \text{ alone on the left side.}$$
$$x = -6$$

To check, we replace x with -6.

$$-5x = 30$$
$$(-5)(-6) \stackrel{?}{=} 30$$
$$30 = 30 \quad \text{True, so } -6 \text{ is the solution.}$$

CHECK YOURSELF 7

Solve and check.

$-7x = 35$

Now let's look at an example that will require multiplying both sides of the equation by the same number.

Example 12

Solve and check $\dfrac{x}{3} = 5$.

Solution Here the variable is *divided* by 3. We can use multiplication to undo the division. Use the multiplication rule to multiply both sides by 3.

$$3\left(\frac{x}{3}\right) = 3 \cdot 5 \quad \text{Multiply both sides by 3.}$$

$$3\left(\frac{x}{3}\right) = 15 \quad \text{We divide numerator and denominator by 3 on the left leaving } x \text{ alone on that side.}$$

$$x = 15$$

To check, replace x with 15 in the original equation.

$$\frac{x}{3} = 5$$

$$\frac{15}{3} \stackrel{?}{=} 5$$

$5 = 5$ True, so 15 is the solution.

Example 13

Solve and check $\frac{x}{5} = -12$.

Solution In this case, multiply both sides by 5 to undo the division.

$$5\left(\frac{x}{5}\right) = (5)(-12)$$

$$x = -60$$

To check, replace x with -60.

$$\frac{x}{5} = -12$$

$$\frac{-60}{5} \stackrel{?}{=} -12$$

$-12 = -12$ True, so -60 is the solution.

CHECK YOURSELF 8

Solve and check.

(1) $\frac{x}{4} = 8$ (2) $\frac{x}{5} = -20$

In all the equations we have looked at so far, either the addition or multiplication rule was used in finding the solution. Sometimes we will need both rules in solving an equation.

Example 14

Solve and check $2x + 3 = 7$.

Solution In this equation, the variable x is multiplied by 2. Then 3 is added. The two operations mean that we will need both of our

Remember that "solving for x" means finding an equivalent equation with x by itself on one side of the equation and numbers on the other.

rules in order to solve for x. Start with the addition rule. We subtract 3 to undo the addition on the left.

$$\begin{array}{rr} 2x + 3 = & 7 \\ -3 & -3 \\ \hline 2x = & 4 \end{array}$$

Now use the multiplication rule. We divide both sides by 2 to undo the multiplication.

$$\frac{2x}{2} = \frac{4}{2}$$

$$x = 2$$

To check, replace x with 2 in the original equation.

$$2x + 3 = 7$$
$$2 \cdot 2 + 3 \stackrel{?}{=} 7$$
$$4 + 3 \stackrel{?}{=} 7$$
$$7 = 7 \quad \text{True, so 2 is the solution.}$$

CHECK YOURSELF 9

Solve and check.

$3x + 5 = 14$

As our examples have shown, we always start with the addition rule to undo any addition or subtraction. Then we apply the multiplication rule to undo any multiplication or division. Here is a summary of those steps.

SOLVING EQUATIONS

STEP 1 Add or subtract the same term on each side of the equation until the term involving the variable is on one side of the equation.

STEP 2 Multiply or divide both sides of the equation by the same nonzero number so that the variable is left by itself on one side of the equation.

STEP 3 Check your solution in the original equation.

Example 15

Solve and check $3x - 5 = 10$.

$$\begin{aligned} 3x - 5 &= 10 \\ +5 & +5 \\ \hline 3x &= 15 \end{aligned}$$

Add 5 to both sides to undo the subtraction.

$$\frac{3x}{3} = \frac{15}{3}$$

Now divide both sides by 3.

$$x = 5$$

To check, replace x with 5.

Note: We always return to the *original equation* for our check.

$$\begin{aligned} 3x - 5 &= 10 \\ 3 \cdot 5 - 5 &\stackrel{?}{=} 10 \\ 15 - 5 &\stackrel{?}{=} 10 \\ 10 &= 10 \end{aligned}$$

True, so 5 is the solution.

CHECK YOURSELF 10

Solve and check.

$4x - 7 = 5$

The variable may appear in any position in an equation. Just apply the rules carefully. The process of solving the equations remains much the same.

Example 16

Solve and check $3 - 2x = 9$.

$$\begin{aligned} 3 - 2x &= 9 \\ -3 & -3 \\ \hline -2x &= 6 \end{aligned}$$

Subtract 3 from both sides.

$$\frac{-2x}{-2} = \frac{6}{-2}$$

Now divide both sides by -2 to leave x alone on the left side of the equation.

$$x = -3$$

To check, replace x with -3.

$$\begin{aligned} 3 - 2x &= 9 \\ 3 - 2(-3) &\stackrel{?}{=} 9 \\ 3 + 6 &\stackrel{?}{=} 9 \\ 9 &= 9 \end{aligned}$$

Multiply first. Be careful with the rules for signs.

True, so -3 is the solution.

Section 15.2: Solving Equations 645

CHECK YOURSELF 11

Solve and check.

$10 - 3x = 1$

You may also have to combine multiplication with addition or subtraction to solve an equation. Our final example illustrates.

Example 17

Solve and check $\dfrac{x}{3} - 5 = 2$.

$$\dfrac{x}{3} - 5 = 2$$
$$\underline{+5 \quad +5} \qquad \text{Add 5 to both sides.}$$
$$\dfrac{x}{3} = 7$$

$$3\left(\dfrac{x}{3}\right) = 3 \cdot 7 \qquad \text{Now multiply both sides by 3.}$$
$$x = 21$$

To check, replace x with 21.

$$\dfrac{x}{3} - 5 = 2$$

$$\dfrac{21}{3} - 5 \stackrel{?}{=} 2$$

$$7 - 5 \stackrel{?}{=} 2$$

$$2 = 2 \qquad \text{True, so 21 is the solution.}$$

CHECK YOURSELF 12

Solve and check.

$\dfrac{x}{4} + 3 = 7$

CHECK YOURSELF ANSWERS

1. (1) 2 is not a solution; (2) −4 is a solution. 2. 13. 3. 3.
4. (1) 4; (2) −9. 5. 4. 6. −6. 7. −5. 8. (1) 32; (2) −100. 9. 3.
10. $4x - 7 = 5$
    $\underline{+7 \quad +7}$
    $4x = 12$
    $\dfrac{4x}{4} = \dfrac{12}{4}$ Check: $4 \cdot 3 - 7 \stackrel{?}{=} 5$
    $x = 3$ $12 - 7 \stackrel{?}{=} 5$
                        $5 = 5$

11. $10 - 3x = 1$
    $\underline{-10 \quad\quad -10}$
    $-3x = -9$
    $\dfrac{-3x}{-3} = \dfrac{-9}{-3}$
    $x = 3$

12. 16.

15.2 Exercises

Solve and check.

1. $x - 3 = 4$
2. $x + 7 = 12$
3. $x + 4 = 10$
4. $x - 5 = 3$
5. $x + 5 = -5$
6. $x - 7 = -3$
7. $x - 6 = -5$
8. $x + 4 = -2$
9. $x + 8 = -2$
10. $x - 5 = -7$
11. $7x = 28$
12. $9x = 45$
13. $-10x = -30$
14. $-8x = 72$
15. $6x = -42$
16. $3x = -36$
17. $-5x = 30$
18. $-7x = -35$
19. $\dfrac{x}{4} = 6$
20. $\dfrac{x}{8} = 4$
21. $\dfrac{x}{5} = -10$
22. $\dfrac{x}{9} = -4$
23. $2x + 5 = 9$
24. $3x - 4 = 5$
25. $4x - 5 = 7$
26. $5x + 7 = 12$
27. $3x - 10 = 17$
28. $4x - 1 = -21$
29. $4x - 3 = -11$
30. $6x + 5 = -7$
31. $5x + 6 = -14$
32. $3x - 2 = -11$
33. $5 - 3x = -16$
34. $3 - 4x = -17$
35. $6 - 5x = -9$
36. $7 - 2x = 11$
37. $\dfrac{x}{3} - 5 = 3$
38. $\dfrac{x}{4} + 4 = 7$
39. $\dfrac{x}{5} + 7 = 4$
40. $\dfrac{x}{6} - 5 = -12$

Skillscan (Section 2.6)
Evaluate the expressions.

a. $(4 + 5)3$

b. $(10 - 7)5$

c. $(8 - 2)4$

d. $(7 + 3)6$

e. $(10 + 2)7$

f. $(15 - 11)4$

Answers

1. 7   **3.** 6   **5.** −10   **7.** $x - 6 = -5$ Check: $1 - 6 = -5$ **9.** −10   **11.** 4   **13.** 3
$$\begin{aligned}+6 \quad +6 \\ \hline x \quad = \quad 1\end{aligned}$$
$\quad\quad -5 = -5$

15. $\dfrac{6x}{6} = \dfrac{-42}{6}$; $x = -7$; check: $(6)(-7) \stackrel{?}{=} -42$; $-42 = -42$ **17.** −6

19. $\dfrac{x}{4} = 6$; $4\left(\dfrac{x}{4}\right) = 4 \cdot 6$; $x = 24$; check: $\dfrac{24}{4} \stackrel{?}{=} 6$; $6 = 6$ **21.** −50   **23.** 2   **25.** 3   **27.** 9

29. $4x - 3 = -11$ $\dfrac{4x}{4} = \dfrac{-8}{4}$; $x = -2$;   check: $4(-2) - 3 \stackrel{?}{=} -11$ **31.** −4   **33.** 7   **35.** 3
$\quad\quad +3 = +3$
$\quad\quad \overline{4x \quad = -8}$
$\quad\quad\quad\quad\quad\quad\quad\quad\quad\quad\quad\quad\quad -11 = -11$

37. 24   **39.** −15   **a.** 27   **b.** 15   **c.** 24   **d.** 60   **e.** 84   **f.** 16

15.2 Supplementary Exercises

Solve and check.

1. $x - 5 = 6$

2. $x + 6 = 4$

3. $x + 3 = 9$

4. $x + 5 = -3$

5. $x - 4 = -8$

6. $9x = 72$

7. $-8x = 40$

8. $-5x = -50$

9. $-8x = 56$

10. $\dfrac{x}{4} = 5$

11. $\dfrac{x}{9} = -4$

12. $2x + 3 = 9$

13. $4x - 5 = 27$

14. $5x - 4 = 16$

15. $3x - 5 = -20$

16. $3x - 4 = -25$

17. $4 - 3x = -2$

18. $9 - 2x = 19$

19. $\dfrac{x}{4} + 8 = 5$

20. $\dfrac{x}{3} - 4 = 6$

15.3 More on Solving Equations

OBJECTIVE
To solve equations with more than one variable term

$3x - 5 = 2x + 4$ has two terms involving the variable x.

We are now going to look at some equations that are a bit more complicated because they have more than one term involving the unknown or variable. Let's start with a definition of what we mean by the word *term*.

> A *term* is a number, or the product of a number and one or more variables, raised to a power.

Each sign (+ or −) is a part of the term that follows the sign.

Example 1

(*a*) The expression $5x^2$ has one term.

(*b*) $3a + 2b$ has two terms, $3a$ and $2b$.

Each term "owns" the sign that precedes it.

(*c*) $4x^3 - 2y + 1$ has three terms: $4x^3$, $-2y$, and 1.

Note In the term $5x^2$, we call 5 the *numerical coefficient*, or just the coefficient.

CHECK YOURSELF 1

List the terms of each expression.

(1) $2b^4$ (2) $5m + 3n$ (3) $2s^2 - 3t - 6$

If the variable appears in more than one term of an equation, we must use the idea of *combining like terms* in finding the solution for the equation.

> *Like terms* have the same variable raised to the same power.

Section 15.3: More on Solving Equations 649

Example 2

$2x$ and $5x$ are like terms.	They have the same variable, x, raised to the same power, in this case the first power.
$3a$ and $-7a$ are like terms.	
$3x$ and $2y$ are *not* like terms.	The terms involve different variables or letters.
$5x$ and $2x^2$ are *not* like terms.	The terms have different powers of the variable.

Like terms can always be combined into a single term.

Example 3

$\underbrace{2x + 5x}_{\text{Combine}} = 7x$

To see why this works:
$2x = \underbrace{x + x}_{\text{Two } x\text{'s}}$

$5x = \underbrace{x + x + x + x + x}_{\text{Five } x\text{'s}}$

so $2x + 5x =$
$\underbrace{x + x + x + x + x + x + x}_{\text{Seven } x\text{'s}}$

or $2x + 5x = 7x$.

Technically this uses the distributive law, which we studied in Chapter 2:

$2x + 5x = (2 + 5)x = 7x$

and that law leads us to the following rule:

> **TO COMBINE LIKE TERMS**
> STEP 1 Add or subtract the coefficients.
> STEP 2 Attach the common variable.

Remember, coefficients are the numbers that multiply the letters or variables.

Example 4

$6x + 7x = ?$

Add the coefficients.

$6 + 7 = 13$

Then attach the common variable, x.

$6x + 7x = 13x$

$-5a + 2a = -3a$	Add: $-5 + 2 = -3$. Then attach the common variable, a.
$-3m + 4m = 1m = m$	Add: $-3 + 4 = 1$. then attach the common variable, m.
$-6t + 6t = 0t = 0$	Add: $-6 + 6 = 0$. Since 0 times something is 0, we write the result as 0.

Since $1m = m$, a coefficient of 1 is not written.

CHECK YOURSELF 2

Combine like terms.

(1) $5x + 9x$ (2) $-7x + 8x$

Combining like terms may also involve subtracting the coefficients.

Example 5

$8m - 5m = 3m$ Subtract: $8 - 5 = 3$. Then attach the common variable, m.

$-4y - 2y = -6y$ Subtract: $-4 - 2 = -6$. Then attach the common variable, y.

$7x - 8x = -1x = -x$ Subtract: $7 - 8 = -1$. Then attach the common variable, x.

We can indicate a coefficient of negative 1 by placing a negative sign in front of the variable.

CHECK YOURSELF 3

Combine like terms.

(1) $10y - 7y$ (2) $-5b - 3b$

We are now ready to use the idea of combining like terms in solving equations. Let's work through an example.

Example 6

Solve and check $4x = 3x + 5$.

Solution We start by subtracting $3x$ from both sides of the equation. Do you see why? Remember that an equation is solved when we have an equation of the form

x = some number

The letter or variable is by itself

$$\begin{array}{r} 4x = 3x + 5 \\ -3x \quad -3x \\ \hline x = \quad\quad 5 \end{array}$$

Subtracting $3x$ from both sides removes $3x$ from the *right side* of the equation, because $3x - 3x = 0$. On the left we have $4x - 3x$, or x. This leaves the letter on the left and a number on the right, which is what you want.

To check, replace x with 5 in the original equation.

$4x = 3x + 5$
$4 \cdot 5 \stackrel{?}{=} 3 \cdot 5 + 5$
$20 \stackrel{?}{=} 15 + 5$ Multiply first.
$20 = 20$ Now add. The result is true, so that 5 is the solution.

Section 15.3: More on Solving Equations   651

CHECK YOURSELF 4

Solve and check.

$8x = 7x - 9$

You may have to apply the rules for solving equations more than once in finding the solution of an equation. Look at the following example.

Example 7

Solve and check $8x - 7 = 7x$.

Solution Begin by subtracting $7x$ from both sides of the equation.

$$\begin{array}{rl} 8x - 7 = & 7x \\ -7x & -7x \\ \hline x - 7 = & 0 \end{array}$$

This removes $7x$ from the right side of the equation.

We could have solved this equation by first adding 7 to both sides of the equation and then subtracting 7x. The result would be the same.

Now add 7 to both sides.

$$\begin{array}{rl} x - 7 = & 0 \\ +7 & +7 \\ \hline x = & 7 \end{array}$$

Adding 7 undoes the subtraction on the left side.

To check, replace x with 7 in the original equation.

$$8x - 7 = 7x$$
$$8 \cdot 7 - 7 \stackrel{?}{=} 7 \cdot 7$$
$$56 - 7 \stackrel{?}{=} 49$$
$$49 = 49$$

This is true, and so 7 is the solution.

CHECK YOURSELF 5

Solve and check.

$5x + 3 = 4x$

An equation may involve more than one term in x *and* more than one number. Let's work through an example.

Example 8

Solve and check $5x - 7 = 4x + 3$.

Solution In this case we start by subtracting $4x$ from both sides of the equation.

Subtracting 4x removes 4x from the right side of the equation.

$$\begin{array}{rl} 5x - 7 = & 4x + 3 \\ -4x & -4x \\ \hline x - 7 = & 3 \end{array}$$

We now have the equation $x - 7 = 3$. This can be solved by the methods of the last section.

Adding 7 undoes the subtraction on the left side of the equation and leaves x alone on the left.

$$\begin{array}{rr} x - 7 = & 3 \\ +\ 7 & +\ 7 \\ \hline x\ \ \ \ = & 10 \end{array}$$

The solution for the equation should be 10. To check, replace x with 10 in the original equation.

$$5x - 7 = 4x + 3$$
$$5 \cdot 10 - 7 \stackrel{?}{=} 4 \cdot 10 + 3$$
$$50\ -7 \stackrel{?}{=}\ 40\ +3$$
$$43 = 43 \quad \text{This is true, and so 10 is the solution.}$$

CHECK YOURSELF 6

Solve and check.

$$4x - 5 = 3x + 2$$

Example 9

Solve and check $7x + 3 = 6x - 2$.

Solution We start the solution by subtracting $6x$ from both sides of the equation. This removes the term in x on the right.

$$\begin{array}{rr} 7x + 3 = & 6x - 2 \\ -\ 6x & -\ 6x \\ \hline x + 3 = & -2 \end{array}$$

Now subtract 3 from both sides. This will leave x alone on the left.

$$\begin{array}{rr} x + 3 = & -2 \\ -\ 3 & -\ 3 \\ \hline x\ \ \ \ = & -5 \end{array}$$

To check, replace x with -5 in the original equation.

$$7x + 3 = 6x - 2$$
$$(7)(-5) + 3 \stackrel{?}{=} (6)(-5) - 2$$
$$-35\ + 3 \stackrel{?}{=}\ -30\ -2$$
$$-32 = -32 \quad \text{This is true, and so } -5 \text{ is the solution.}$$

Section 15.3: More on Solving Equations

CHECK YOURSELF 7

Solve and check.

$6x + 5 = 5x + 2$

In solving some equations, we may also have to divide both sides of the equation by some number. Let's work through an example in which we will have to use division.

Example 10

Solve and check $7x - 3 = 4x + 9$.

Solution To start, subtract $4x$ from both sides.

$$\begin{array}{r} 7x - 3 = 4x + 9 \\ -4x -4x \\ \hline 3x - 3 = 9 \end{array}$$

This removes $4x$ from the right side of the equation.

Now add 3 to both sides.

$$\begin{array}{r} 3x - 3 = 9 \\ +3 +3 \\ \hline 3x = 12 \end{array}$$

Adding 3 undoes the subtraction on the left side.

Divide both sides by 3.

We divide by 3, the coefficient of x, to leave x alone on the left side.

$$\frac{\cancel{3}x}{\cancel{3}} = \frac{12}{3}$$

$x = 4$

To check, replace x with 4 in the original equation.

$$7x - 3 = 4x + 9$$
$$7 \cdot 4 - 3 \stackrel{?}{=} 4 \cdot 4 + 9$$
$$28 - 3 \stackrel{?}{=} 16 + 9$$
$$25 = 25 \quad \text{True, so 4 is the solution.}$$

CHECK YOURSELF 8

Solve and check.

$8x + 5 = 3x - 30$

To solve some equations, we may have to divide both sides by a negative number. The next example illustrates.

Example 11

Solve and check $3x - 7 = 5x + 9$.

Solution Start by subtracting $5x$ from both sides.

$$\begin{aligned} 3x - 7 &= 5x + 9 \\ -5x & -5x \\ \hline -2x - 7 &= 9 \end{aligned}$$

Subtracting $5x$ from both sides removes the term in x from the right side of the equation.

Now add 7 to both sides.

$$\begin{aligned} -2x - 7 &= 9 \\ +7 & +7 \\ \hline -2x &= 16 \end{aligned}$$

Adding 7 undoes the subtraction on the left side of the equation.

Divide both sides by -2.

$$\frac{-2x}{-2} = \frac{16}{-2}$$
$$x = -8$$

To check, replace x with -8 in the original equation.

$$\begin{aligned} 3x - 7 &= 5x + 9 \\ (3)(-8) - 7 &\stackrel{?}{=} (5)(-8) + 9 \\ -24 - 7 &\stackrel{?}{=} -40 + 9 \\ -31 &= -31 \end{aligned}$$

True, so -8 is the solution.

CHECK YOURSELF 9

Solve and check.

$4x - 6 = 7x + 9$

There is seldom only one "right way" to solve an equation. Let's look at an alternative method of solving the equation of Example 11. This time we will isolate x on the *right side* of the equation.

Example 12

$$3x - 7 = 5x + 9$$
$$-3x \qquad -3x$$
$$\overline{-7 = 2x + 9}$$
$$-9 \qquad -9$$
$$\overline{-16 = 2x}$$

$$\frac{-16}{2} = \frac{2x}{2}$$

$$-8 = x$$

This time, start by subtracting $3x$ from both sides. This removes the term in x from the left side.

Now subtract 9 to undo the addition on the right side.

Now divide by 2. This will leave x alone on the *right* side of the equation.

Note: If $-8 = x$, then $x = -8$.

Check the solution, -8, as before.

CHECK YOURSELF 10

Solve and check.

$4x - 6 = 7x + 9$

(Solve with the variable on the right this time.)

Some students prefer the approach of Example 12 because it avoids the negative coefficient of x. Others like the first method because it leaves the term in x on the left side of the equation. Compare both methods to see which you find easier.

Let's summarize the steps of solving an equation in which the variable appears on both sides of the equation.

SOLVING EQUATIONS

STEP 1 Add or subtract the same term on each side of the equation until all terms involving the variable are on one side of the equation and the numbers are on the other.

STEP 2 Multiply or divide both sides of the equation by the same nonzero number so that the variable is left by itself on one side of the equation.

STEP 3 Check your solution in the original equation.

Let's work through one more complete example of solving an equation.

Example 13

Solve and check.

$5 - 2x = 8 - 4x$

Start by adding $4x$ to both sides of the equation.

This will leave the terms in x on the left.

$$\begin{array}{r} 5 - 2x = 8 - 4x \\ + 4x + 4x \\ \hline 5 + 2x = 8 \end{array}$$

Now subtract 5 from both sides.

This will place the numbers on the right.

$$\begin{array}{r} 5 + 2x = 8 \\ -5 -5 \\ \hline 2x = 3 \end{array}$$

Finally, divide both sides by 2.

Dividing by 2 leaves x by itself on the left.

$$\frac{2x}{2} = \frac{3}{2}$$

$$x = \frac{3}{2}$$

To check, replace x with $\frac{3}{2}$ in the original equation.

$$5 - 2x = 8 - 4x$$

$$5 - 2\left(\frac{3}{2}\right) \stackrel{?}{=} 8 - 4\left(\frac{3}{2}\right)$$

$$5 - 3 \stackrel{?}{=} 8 - 6$$

$$2 = 2 \quad \text{This is true, and so } \frac{3}{2} \text{ is the solution.}$$

CHECK YOURSELF 11

Solve and check.

$6 - 3x = 7 - 6x$

CHECK YOURSELF ANSWERS

1. (1) $2b^4$; (2) $5m, 3n$; (3) $2s^2, -3t, -6$.
2. (1) $14x$; (2) x.   **3.** (1) $3y$; (2) $-8b$.   **4.** -9.
5. Subtract $4x$ from both sides.   **6.** 7.   **7.** -3.   **8.** -7.   **9.** -5.

$$\begin{array}{r} 5x + 3 = 4x \\ -4x -4x \\ \hline x + 3 = 0 \end{array}$$

10. -5.   **11.** $\frac{1}{3}$.

Subtract 3 from both sides.

$$\begin{array}{r} x + 3 = 0 \\ -3 -3 \\ \hline x = -3 \end{array}$$

15.3 Exercises

Combine the like terms.

1. $7a + 5a$
2. $9b - 4b$
3. $12b - 9b$
4. $7r + 9r$

5. $5x - 5x$
6. $-2s + 3s$
7. $-4y + 8y$
8. $-8w + 5w$

9. $-4m - 6m$
10. $-8x + 7x$
11. $7n - 9n$
12. $-4y + 4y$

13. $5a - 10a$
14. $12p - 15p$
15. $-6b + 7b$
16. $-9a - 5a$

Solve and check.

17. $5x = 4x + 3$
18. $8x = 7x - 5$
19. $7x = 6x - 4$

20. $4x = 3x + 8$
21. $7x - 12 = 5x$
22. $5x + 18 = 3x$

23. $7x + 4 = 6x - 3$
24. $8x - 3 = 7x + 7$
25. $3x - 5 = 2x + 7$

26. $5x + 3 = 4x - 2$
27. $9x - 5 = 8x - 7$
28. $9x - 9 = 8x - 13$

29. $6x + 3 = 4x + 17$
30. $5x - 6 = 2x + 3$
31. $7x - 5 = 4x + 10$

32. $8x + 5 = 4x - 1$
33. $9x - 3 = 6x - 1$
34. $12x + 9 = 5x - 26$

35. $5x - 3 = 6x - 10$
36. $4x - 8 = 5x + 2$
37. $5x + 3 = 8x - 9$

38. $9x + 8 = 14x - 22$
39. $8x + 5 = 12x + 2$
40. $6x - 25 = 9x - 30$

41. $4 - 3x = 10 - 5x$
42. $9 + 2x = 23 - 5x$
43. $5 + 4x = 17 + 7x$

44. $5 - 7x = 35 + 3x$
45. $6 + 3x = 14 - 9x$
46. $6 - 5x = 27 - 2x$

47. $9 - 3x = 49 + 5x$
48. $19 + 10x = 21 + 5x$

Answers

1. $12a$ 3. $3b$ 5. 0 7. $4y$ 9. $-10m$ 11. $-2n$ 13. $-5a$ 15. b 17. 3 19. -4 21. 6
23. -7 25. 12 27.
$$\begin{aligned} 9x - 5 &= 8x - 7 \\ -8x & \quad -8x \\ \hline x - 5 &= -7 \\ +5 & \quad +5 \\ \hline x &= -2 \end{aligned}$$
Check: $(9)(-2) - 5 = (8)(-2) - 7$
$-18 - 5 = -16 - 7$
$-23 = -23$
29. 7 31. 5 33. $\frac{2}{3}$

35. 7 37.
$$\begin{aligned} 5x + 3 &= 8x - 9 \\ -8x & \quad -8x \\ \hline -3x + 3 &= -9 \\ -3 & \quad -3 \\ \hline -3x &= -12 \end{aligned}$$
$\frac{-3x}{-3} = \frac{-12}{-3}$; $x = 4$ 39. $\frac{3}{4}$

41.
$$\begin{aligned} 4 - 3x &= 10 - 5x \\ +5x & \quad +5x \\ \hline 4 + 2x &= 10 \\ -4 & \quad -4 \\ \hline 2x &= 6 \end{aligned}$$
$\frac{2x}{2} = \frac{6}{2}$; $x = 3$ 43. -4 45. $\frac{2}{3}$ 47. -5

15.3 Supplementary Exercises

Combine the like terms.

1. $6x + 5x$
2. $9a - 5a$
3. $6y - 6y$

4. $-5b + 6b$
5. $-6a - 3a$
6. $5x - 7x$

7. $-9b + 8b$
8. $-3m + 3m$

Solve and check.

9. $6x = 5x + 4$
10. $7x = 6x - 3$
11. $7x + 8 = 5x$

12. $6x - 4 = 5x + 3$
13. $7x + 5 = 6x - 3$
14. $9x + 8 = 8x + 5$

15. $7x + 4 = 5x + 12$
16. $6x - 11 = 3x - 9$
17. $8x - 9 = 5x - 27$

18. $5x - 4 = 6x - 8$
19. $7x + 3 = 8x + 8$
20. $4x + 5 = 6x + 10$

21. $9 - 2x = 3 - 4x$
22. $7 + 3x = 22 - 2x$
23. $7 + 6x = 13 - 3x$

24. $8 - 5x = 20 - 2x$

15.4 Applying Algebra

OBJECTIVES
1. To translate a word phrase to an algebraic expression
2. To solve applications using algebraic equations

An essential skill for applying algebra to problem solving is being able to translate a phrase in words to an algebraic expression. Let's review the meaning of certain important phrases.

You are familiar with the symbols used to indicate the four fundamental operations of arithmetic. Now look at how these operations are indicated in algebra.

In arithmetic: + denotes addition, − denotes subtraction, × denotes multiplication, ÷ denotes division.

ADDITION

$x + y$ means the *sum* of x and y or x *plus* y.

Example 1

Some other words that tell you to add are "more than" and "increased by."

(a) The *sum* of a and 5 is written as $a + 5$.
(b) L *plus* W is written as $L + W$.
(c) 4 *more than* y is written as $y + 4$
(d) w *increased by* 8 is written as $w + 8$

CHECK YOURSELF 1

Write, using symbols.

(1) The sum of z and 9   (2) r plus s
(3) 5 more than w (4) m increased by 8

Let's look at how subtraction is indicated in algebra.

SUBTRACTION

$x - y$ means the *difference* of x and y or x *minus* y.

Example 2

Some other words that mean to subtract are "decreased by" and "less than."

(a) a *minus* b is written $a - b$.
(b) The *difference* of w and 5 is written $w - 5$.
(c) y *decreased by* 10 is written $y - 10$.
(d) 8 *less than* p is written $p - 8$.

CHECK YOURSELF 2

Write, using symbols.

(1) r minus s (2) The difference of m and 9
(3) z decreased by 7 (4) 5 less than a

You have seen that the operations of addition and subtraction are written exactly the same way in algebra as they were in arithmetic. Multiplication presents a slight difficulty because the sign × looks like the letter x. So in algebra we use other symbols to show multiplication to avoid any confusion. Here are some ways to write multiplication.

Note: x and y are called the factors of the product xy.

MULTIPLICATION

A raised dot $x \cdot y$
Parentheses $(x)(y)$ These all indicate the *product* of x and y.
Writing the letters next to each other xy

Example 3

(a) The *product* of 8 and a is written as $8 \cdot a$, $(8)(a)$, or $8a$. The last expression, $8a$, is the most common way of writing the product. *Note:* You can place two letters next to each other or a number and a letter next to each other to show multiplication. But you *cannot* place two numbers side by side to show multiplication: 37 means the number "thirty-seven," not 3 times 7.
(b) 3 *times* 7 can be written as $3 \cdot 7$ or $(3)(7)$.
(c) *Twice* b is written as $2b$
(d) The product of 3, r, and s is written as $3rs$.
(e) 5 more than the product of 4 and n is written as $4n + 5$.

CHECK YOURSELF 3

Write, using symbols.

(1) a times b (2) The product of m and n
(3) The product of s and 9   (4) The product of 6, c, and d.
(5) 4 more than the product of 5 and w

Section 15.4: Applying Algebra 661

Now let's look at the operation of division. Think of the variety of ways you showed division in arithmetic. Besides the division sign ÷, you used the long division symbol ⟌ and the fraction notation. For example, to indicate the quotient when 9 is divided by 3, you could write

$$9 \div 3 \quad \text{or} \quad 3\overline{)9} \quad \text{or} \quad \frac{9}{3}$$

In algebra the fraction form is usually used.

DIVISION

$\frac{x}{y}$ means *x divided by y* or the *quotient* when *x* is divided by *y*.

Example 4

(a) z divided by 4 is written as $\frac{z}{4}$.

(b) The quotient of m plus n divided by 5 is written as

$$\frac{m + n}{5}$$

(c) The sum r plus s divided by the difference r minus s is written as

$$\frac{r + s}{r - s}$$

CHECK YOURSELF 4

Write, using symbols.

(1) a divided by b
(2) The quotient when x minus y is divided by 8
(3) The difference p minus 3 divided by the sum p plus 3

Earlier in this chapter you learned how to solve a variety of equations. Now, combining that skill with our work in this section on translating word phrases to algebra will allow us to solve certain applications or word problems in algebra.

To help you feel more comfortable when solving word problems, we are going to present a step-by-step approach that will (*with practice*) allow you to organize your work, and organization is the key to the solution of these problems.

> You will no doubt notice that these steps are very similar to those used earlier to solve word problems in arithmetic.

> **TO SOLVE WORD PROBLEMS**
>
> **STEP 1** Read the problem carefully. Then reread it to decide what you are asked to find.
>
> **STEP 2** Choose a letter to represent one of the unknowns in the problem. Then represent all the unknowns of the problem, using the same letter.
>
> **STEP 3** Translate the problem to the language of algebra to form an equation.
>
> **STEP 4** Solve the equation.
>
> **STEP 5** Answer the question of the original problem. Then verify your solution by returning to the original problem.

Let's work through an example using this five-step process.

Example 5

The sum of twice a number and 8 is 20. What is the number?

Step 1 *Read carefully.* You must find the unknown number.

Step 2 *Choose letters or variables.* Let x represent the unknown number.

Step 3 *Translate* to the language of algebra.

$$\underbrace{2x}_{\text{Twice } x} \underbrace{+\ 8}_{\text{The sum of}} \underbrace{=}_{\text{Is}} 20$$

Step 4 *Solve the equation.*

Subtract 8.
$$2x + 8 = 20$$
$$-\ 8 = -8$$
$$2x\ \ \ = 12$$

Divide by 2.
$$\frac{2x}{2} = \frac{12}{2}$$
$$x = 6$$

The unknown number is 6.

> Always return to the *original problem* to check your result and *not* to the equation of step 3. This will prevent possible errors!

Step 5 *Check your solution.* Is the sum of twice 6 and 8 equal to 20? Yes! (12 + 8 = 20) We have checked our solution.

Section 15.4: Applying Algebra

CHECK YOURSELF 5

The sum of 3 times a number and 10 is 43. What is the number?

Example 6

One number is 7 more than another. If the sum of the numbers is 37, what are the two numbers?

Step 1 You must find the two unknown numbers.

Step 2 Let x represent the first number. Then

$\underbrace{x + 7}_{\text{7 more than } x.}$ represents the second number

Step 3
$\underbrace{x + x + 7}_{\text{The sum of the two numbers}} = 37$

Step 4

Combine the like terms on the left side, and solve as before.

$$x + x + 7 = 37$$
$$2x + 7 = 37$$
$$-7 -7$$
$$2x = 30$$
$$\frac{2x}{2} = \frac{30}{2}$$
$$x = 15$$

The numbers are $x = 15$ and $x + 7 = 22$.

Step 5 Since 22 is 7 more than 15, and the sum (15 + 22) is 37, we have satisfied the conditions of the original problem.

CHECK YOURSELF 6

One number is 12 more than another number. The sum of the numbers is 52. Find the two numbers.

A similar approach will allow you to solve a variety of applications. Consider the following example.

Example 7

There were 55 more yes votes than no votes on an election measure. If 735 votes were cast in all, how many yes votes were there? How many no votes?

Step 1 We want to find the number of yes votes and the number of no votes.

Step 2 Let x be the number of no votes. Then

$\underbrace{x + 55}_{\text{55 more than}}$ is the number of yes votes.

Step 3

$$\underbrace{x}_{\text{No votes}} + \underbrace{x + 55}_{\text{Yes votes}} = \underbrace{735}_{\text{The total number of votes}}$$

Step 4

Combine the like term on the left.

$$\begin{aligned} x + x + 55 &= 735 \\ 2x + 55 &= 735 \qquad \text{Combine like terms on the left.} \\ -55 \quad & -55 \qquad \text{Subtract 55.} \\ \hline 2x &= 680 \\ \frac{2x}{2} &= \frac{680}{2} \qquad \text{Divide by 2.} \\ x &= 340 \end{aligned}$$

No votes (x) = 340
Yes votes ($x + 55$) = 395

Step 5 340 no votes plus 395 yes votes equals 735 total votes. The solution is verified.

CHECK YOURSELF 7

Margaret earns $180 per month more than Steven. If they earn a total of $3500 per month, what are their monthly salaries?

Many word problems involve geometric figures and measurements. Let's look at one final example.

Example 8

The perimeter of a rectangle is 46 cm. If the length is 3 cm more than the width, what are the dimensions of the rectangle?

Step 1 We want to find the dimensions (length and width) of the rectangle.

Step 2 Let x be the width. Then

$x + 3$ is the length.
↑
3 more than

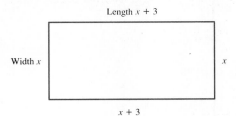

Step 3 The perimeter (the distance around) is 46 cm, so

$$x + x + 3 + x + x + 3 = 46$$

Width, Length, Width, Length

Step 4

$$x + x + 3 + x + x + 3 = 46$$
$$4x + 6 = 46$$
$$-6 \quad -6$$
$$4x = 40$$
$$\frac{4x}{4} = \frac{40}{4}$$
$$x = 10$$

Combine the like terms on the left. Then solve as before.

The width $(x) = 10$ cm. The length $(x + 3) = 13$ cm.

Step 5 To check: The perimeter is $10 + 13 + 10 + 13$, or 46 cm. The solution is verified.

CHECK YOURSELF 8

The perimeter of a rectangle is 54 in. If the length of the rectangle is twice the width, find the dimensions of the rectangle.

That is all we propose to do with word problems in algebra at this time. Remember to use the five-step approach as you work on the problems for this section. We hope that you find it helpful.

Always draw a sketch at this point where figures are involved. It will help you form an equation in step 3.

CHECK YOURSELF ANSWERS

1. (1) $z + 9$; (2) $r + s$; (3) $w + 5$; (4) $m + 8$.
2. (1) $r - s$; (2) $m - 9$; (3) $z - 7$; (4) $a - 5$.
3. (1) ab; (2) mn; (3) $9s$; (4) $6cd$; (5) $5w + 4$.
4. (1) $\frac{a}{b}$; (2) $\frac{x - y}{8}$; (3) $\frac{p - 3}{p + 3}$.
5. The equation is $3x + 10 = 43$. The number is 11.
6. 20 and 32.
7. Margaret: $1840; Steven: $1660.
8. Width: 9 in; length: 18 in.

15.4 Exercises

Write each of the following phrases using symbols.

1. The sum of r and s
2. a plus 7
3. w plus z
4. The sum of p and q
5. x increased by 2
6. 5 more than m
7. 10 more than y
8. b increased by 8
9. a minus b
10. 5 less than s
11. x decreased by 9
12. r minus 3
13. 6 less than r
14. y decreased by 5
15. w times z
16. The product of 9 and d
17. The product of 5 and t
18. 8 times a
19. The product of 9, p, and q
20. The product of 7, r, and s
21. The sum of twice x and y
22. The sum of 5 times c and d
23. x divided by 5
24. The quotient when m is divided by 8
25. The quotient when a plus b is divided by 7
26. The difference r minus s, divided by 7
27. The difference of p and q, divided by 4
28. The sum of b and 5, divided by 7
29. The sum of a and 3, divided by the difference of a and 3
30. The difference of m and n, divided by the sum of m and n

Section 15.4: Applying Algebra

Write each of the following phrases using symbols. Use the variable x to represent the number in each case.

31. 7 more than a number

32. A number increased by 9

33. 12 less than a number

34. A number decreased by 10

35. 8 times a number

36. Twice a number

37. 5 more than 3 times a number

38. 5 times a number decreased by 10

39. The quotient of a number and 9

40. A number divided by 3

41. The sum of a number and 7, divided by 6

42. The quotient when 5 less than a number is divided by 3

43. 8 more than a number divided by 8 less than that same number

44. The quotient when 3 less than a number is divided by 3 more than that same number

Solve the following word problems. Be sure to label the unknowns and to show the equation you use for the solution.

45. The sum of 2 times a number and 7 is 35. What is the number?

46. 3 times a number, increased by 8, is 62. Find the number.

47. 5 times a number, minus 12, is 78. Find the number.

48. 4 times a number, decreased by 20, is 32. What is the number?

49. One number is 5 more than another. If the sum of the two numbers is 55, what are the two numbers?

50. One number is 3 times another. The sum of the two numbers is 64. Find the two numbers.

51. One number is 7 less than another. If the sum of the two numbers is 33, what are the two numbers?

52. One number is 12 more than another and their sum is 52. Find the two numbers.

53. In an election, the winning candidate had 160 more votes than the loser. If the total number of votes cast was 3260, how many votes did each candidate receive?

54. Janet earns $150 more per month than Frank. If their monthly salaries total $3550, what amount does each earn?

55. A washer-dryer combination costs $650. If the washer costs $70 more than the dryer, what does each appliance cost?

56. William has a board that is 96 in long. He wishes to cut the board into two pieces so that one piece will be 8 in longer than the other. What should be the length of each piece?

57. The length of a rectangle is 5 cm more than its width. The perimeter is 98 cm. What are the dimensions? (*Hint:* Remember to draw a sketch whenever geometric figures are involved in a word problem.)

58. The length of a rectangle is 3 times its width. If the perimeter of the rectangle is 64 in, find the length and width of the rectangle.

59. The width of a rectangle is 5 in less than its length. If the perimeter of the rectangle is 78 in, find the width and length of the rectangle.

60. The length of a rectangle is 4 times its width and the perimeter is 80 ft. Find the dimensions of the rectangle.

Answers

1. $r + s$ **3.** $w + z$ **5.** $x + 2$ **7.** $y + 10$ **9.** $a - b$ **11.** $x - 9$ **13.** $r - 6$ **15.** wz **17.** $5t$ **19.** $9pq$ **21.** $2x + y$ **23.** $\frac{x}{5}$ **25.** $\frac{a+b}{7}$ **27.** $\frac{p-q}{4}$ **29.** $\frac{a+3}{a-3}$ **31.** $x + 7$ **33.** $x - 12$ **35.** $8x$ **37.** $3x + 5$ **39.** $\frac{x}{9}$ **41.** $\frac{x+7}{6}$ **43.** $\frac{x+8}{x-8}$ **45.** 14   **47.** 18   **49.** 25, 30   **51.** 20, 13   **53.** 1710, 1550   **55.** Washer: $360; dryer: $290   **57.** 22 cm, 27 cm   **59.** 17 in, 22 in

15.4 Supplementary Exercises

Write each of the following phrases using symbols.

1. The sum of m and n
2. b plus 9
3. y increased by 8

4. 6 more than s
5. r decreased by 5
6. 9 less than a

7. w minus z
8. The product of 5 and c
9. 7 times b

10. The product of 12, x, and y
11. The sum of twice a and b
12. a divided by 8

13. The quotient when r is divided by 6
14. The sum of m and n, divided by 3

15. Three more than p divided by three less than p

Write each of the following phrases using symbols. Use the variable x to represent the number in each case.

16. 9 more than a number.
17. A number decreased by 12

18. Three times a number
19. 8 more than twice a number

20. A number divided by 7

21. The quotient when 2 less than a number is divided by 3

22. 5 more than a number divided by 5 less than that same number

Solve the following word problems. Be sure to label the unknowns and to show the equation you use for the solution.

23. The sum of three times a number and 10 is 28. Find the number.

24. If 5 times a number, decreased by 7, is 58, what is the number?

25. One number is 8 more than another. If the sum of the two numbers is 92, find the two numbers.

26. One number is four times another. If the sum of the two numbers is 90, what are the two numbers?

27. In a recent election, there were 280 more yes votes than no votes. If 1580 total votes were cast, how many yes votes were there?

28. A desk and chair cost $470. If the desk cost $150 more than the chair, find the cost of each item.

29. The length of a rectangle is 4 ft more than its width. If the perimeter of the rectangle is 72 ft, find the dimensions of the rectangle.

30. The width of a rectangle is 3 in less than its length and the perimeter of the rectangle is 74 in. Find the width and length of the rectangle.

SELF-TEST for Chapter Fifteen

The purpose of the Self-Test is to help you check your progress and review for a chapter test in class. Allow yourself about an hour to take the test. When you are done, check your answers in the back of the book. If you missed any problems, be sure to go back and review the appropriate sections in the chapter and do the supplementary exercises provided there.

In Problems 1 to 6, evaluate the expressions if $x = 5$, $y = 2$, $z = -3$, and $w = -4$.

1. $2xy$
2. $2x + 7z$
3. $3z^2$
4. $4(x + 3w)$
5. $\dfrac{2w}{y}$
6. $\dfrac{2x - w}{2x + z}$

In Problems 7 to 12, solve each equation and check the solution.

7. $x - 5 = 3$
8. $x - 4 = -7$
9. $8x = 48$
10. $\dfrac{x}{7} = 8$
11. $3x - 5 = -32$
12. $15 - 2x = 3$

In Problems 13 to 15, combine like terms.

13. $5x + 3x =$
14. $10a - 9a =$
15. $-7b - 4b =$

In Problems 16 to 23, solve each equation and check the solution.

16. $6x = 5x + 6$
17. $7x + 5 = 6x - 2$
18. $5x + 7 = 2x + 31$
19. $8x - 3 = 4x - 15$
20. $4x + 3 = 5x + 11$
21. $5x + 2 = 8x - 7$
22. $3x + 16 = 5x - 2$
23. $5 - 2x = 49 - 6x$

In Problems 24 to 28, write each phrase using symbols.

24. 10 more than a

25. r decreased by 8

26. The product of 3, x, and y

27. The sum of twice m and 3.

28. The quotient when 5 more than w is divided by 8.

In Problems 29 and 30, solve the word problems.

29. One number is three times another. If the sum of the two numbers is 48, what are the two numbers?

30. The length of a rectangle is 5 cm more than its width. If the perimeter of the rectangle is 86 cm, find the dimensions of the rectangle.

SUMMARY for Part Six

An Introduction to Algebra

Signed Numbers—The Terms

Positive Numbers Numbers used to name points to the right of 0 on the number line.

Negative Numbers Numbers used to name points to the left of 0 on the number line.

Signed Numbers A set containing both positive and negative numbers.

Opposites Two numbers are opposites if the name points the same distance from 0 on the number line, but in opposite directions.

The opposite of a positive number is negative.

The opposite of a negative number is positive.

0 is its own opposite.

Integers The set consisting of the natural numbers, their opposites, and 0.

Absolute Value The distance on the number line between the point named by a number and 0.

The absolute value of a number is always positive or 0.

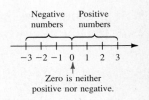

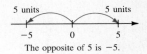

The opposite of 5 is −5.

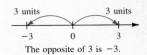

The opposite of 3 is −3.

The integers are
$\{\ldots, -3, -2, -1, 0, 1, 2, 3, \ldots\}$

The absolute value of a number a is written $|a|$.
$|7| = 7 \quad |-8| = 8$

Operations on Signed Numbers

To Add Signed Numbers

1. If two numbers have the same sign, add their absolute values. Give the sum the sign of the original numbers.
2. If two numbers have different signs, subtract the smaller absolute value from the larger. Give the sum the sign of the number with the larger absolute value.

$5 + 8 = 13$
$-3 + (-7) = -10$
$5 + (-3) = 2$
$7 + (-9) = -2$

673

To Subtract Signed Numbers

To subtract signed numbers, add the first number and the opposite of the number being subtracted.

$4 - (-2) = 4 + 2 = 6$

The opposite of -2.

To Multiply Signed Numbers

To multiply signed numbers, multiply the absolute values of the numbers. Then attach a sign to the product according to the following rules:

1. If the numbers have the same sign, the product is positive.
2. If the numbers have different signs, the product is negative.

$5 \cdot 7 = 35$
$(-4)(-6) = 24$
$(8)(-7) = -56$

To Divide Signed Numbers

To divide signed numbers, divide the absolute values of the numbers. Then attach a sign to the quotient according to the following rules:

1. If the numbers have the same sign, the quotient is positive.
2. If the numbers have different signs, the quotient is negative.

$\frac{-8}{-2} = 4$

$27 \div (-3) = -9$

$\frac{-16}{8} = -2$

Algebraic Expressions

Algebraic Expressions An expression which contains numbers and letters (called variables).

$2x + 3y$ is an algebraic expression. The variables are x and y.

Evaluating Algebraic Expressions

To evaluate an algebraic expression:

1. Replace each variable or letter with its number value.
2. Do the necessary arithmetic, following the rules for the order of operations.

Evaluate $2x + 3y$ if $x = 5$ and $y = -2$.

$ 2x + 3y$
$= 2 \cdot 5 + (3)(-2)$
$= 10 - 6 = 4$

Combining Like Terms

To combine like terms:

1. Add or subtract the coefficients (the numbers multiplying the variables).
2. Attach the common variable.

$5x + 2x = 7x$
$ 5 + 2$

$8a - 5a = 3a$
$ 8 - 5$

Algebraic Equations

Algebraic Equation A statement that two expressions are equal. An equation may be either true or false.

$5x - 1 = 3x + 7$ is an algebraic equation.

Solution A numerical value for the variable or letter that makes the equation a true statement.

4 is a solution for the equation above because

$$5 \cdot 4 - 1 \stackrel{?}{=} 3 \cdot 4 + 7$$
$$20 - 1 \stackrel{?}{=} 12 + 7$$
$$19 = 19$$

Equivalent Equations Equations that have exactly the same solutions.

$$2x + 1 = x + 5$$
$$x + 1 = 5$$
and $x = 4$

are equivalent equations. They all have the solution 4.

Rules for Writing Equivalent Equations

1. Adding or subtracting the same term on each side of an equation gives an equivalent equation.
2. Multiplying or dividing both sides of an equation by the same nonzero number gives an equivalent equation.

Solve and check:

$7x - 2 = 3x + 10$.

$$\begin{array}{r} 7x - 2 = 3x + 10 \\ -3x -3x \\ \hline 4x - 2 = 10 \\ + 2 + 2 \\ \hline 4x = 12 \end{array}$$

Solving Equations

1. Add or subtract the same term on each side of the equation until all terms involving the variable are on one side of the equation and the numbers are on the other.
2. Multiply or divide both sides of the equation by the same nonzero number so that the variable is left by itself on one side of the equation.
3. Check your solution in the original equation.

$$\frac{4x}{4} = \frac{12}{4}$$

$x = 3$

To check:

$$7 \cdot 3 - 2 \stackrel{?}{=} 3 \cdot 3 + 10$$
$$19 = 19$$

Applying Algebra

The Notation

Addition $x + y$ means the *sum* of x and y or x *plus* y. Some other words indicating addition are "more than" and "increased by."

The sum of x and 5 is $x + 5$.
7 more than a is $a + 7$.
b increased by 3 is $b + 3$.

Subtraction $x - y$ means the *difference* of x and y or x *minus* y. Some other words indicating subtraction are "less than" and "decreased by."

The difference of x and 3 is $x - 3$.

5 less than p is $p - 5$.
a decreased by 4 is $a - 4$.

Multiplication

$\left.\begin{array}{l} x \cdot y \\ (x)(y) \\ xy \end{array}\right\}$ These all mean the *product* of x and y or x *times* y.

The product of m and n is mn.
The product of 2 and the sum of a and b is $2(a + b)$.

Division $\dfrac{x}{y}$ means x *divided by* y or the *quotient* when x is divided by y.

n divided by 5 is $\dfrac{n}{5}$.

The sum of a and b, divided by 3, is $\dfrac{a + b}{3}$.

Solving Applications

Use the following five-step process to solve word problems:

1. Read the problem carefully. Then reread it to decide what you are asked to find.
2. Choose a letter to represent one of the unknowns in the problem. Then represent all the unknowns of the problem, using the same letter.
3. Translate the problem to the language of algebra to form an equation.
4. Solve the equation.
5. Answer the question of the original problem. Then verify your solution by returning to the original problem.

SUMMARY EXERCISES for Part Six

You should now be reviewing the material in Part Six of the text. The following exercises will help in that process. Work all the exercises carefully. Then check your answers in the back of the book. References are provided to the chapter and section for each problem. If you made an error, go back and review the related material and do the supplementary exercises for that section.

[14.1] In Problems 1 to 4, complete the statements.

1. The absolute value of 12 is

2. The opposite of −8 is

3. $|-3| =$

4. $-(-20) =$

[14.2] In Problems 7 to 10, add.

5. $15 + (-7) =$

6. $4 + (-9) =$

7. $-8 + (-3) =$

8. $2\frac{1}{2} + (-2) =$

9. $-3.75 + (-1.25) =$

10. $5 + (-6) + (-3)$

[14.3] In Problems 11 to 16, subtract.

11. $15 - 20 =$

12. $-10 - 5 =$

13. $2 - (-3) =$

14. $-7 - (-3) =$

15. $5\frac{3}{4} - \left(-\frac{3}{4}\right) =$

16. $-2.75 - 1.25 =$

[14.4] In Problems 17 to 22, multiply.

17. $(-12)(-3) =$

18. $(-10)(8) =$

19. $(-0.5)(3) =$

20. $\left(-\frac{3}{8}\right)\left(-\frac{4}{5}\right) =$

21. $(-4)^2 =$

22. $(-2)(7)(-3) =$

[14.5] In Problems 23 to 28, divide.

23. $48 \div (-12) =$

24. $\dfrac{-33}{-3} =$

25. $-9 \div 0 =$

26. $0.75 \div (-10) =$

27. $-\dfrac{7}{9} \div \left(-\dfrac{2}{3}\right) =$

28. $0 \div (-12) =$

[15.1] In Problems 29 to 34, evaluate the expressions if $w = 3$, $x = 2$, $y = -4$, and $z = -8$.

29. $3xy$

30. $3x - 4y$

31. $x^2 + z^2$

32. $3(x + 2y)$

33. $\dfrac{2x}{y}$

34. $\dfrac{y - 3z}{2y + w}$

[15.2] In Problems 35 to 40, solve each equation and check your solution.

35. $x - 3 = 5$

36. $x + 7 = -2$

37. $-9x = 81$

38. $3x - 7 = 8$

39. $5 - 8x = 11$

40. $\dfrac{x}{4} - 7 = 3$

[15.3] In Problem 41 to 44, combine like terms.

41. $7a + 9a$

42. $12b - 15b$

43. $-10y + 5y$

44. $-4z + 5z$

[15.3] In Problems 45 to 50, solve each equation and check your solution.

45. $5x + 12 = 4x$

46. $4x - 12 = 5x + 3$

47. $5x + 3 = 2x + 12$

48. $9x + 7 = 3x - 29$

49. $2x - 17 = 6x - 11$

50. $9 + 4x = 41 - 4x$

[5.4] In Problems 51 to 56, write each phrase using symbols.

51. y increased by 5

52. 9 more than s

53. The product of 5, c, and d.

54. The sum of twice w and 3

55. r divided by 7

56. The difference of s and t, divided by 4.

[5.4] In Problems 57 to 60, solve each word problem.

57. The sum of three times a number and 5 is 44. Find the number.

58. One number is 12 more than another. If the sum of the two numbers is 26, what are the two numbers?

59. Nick earns $80 more per week than Janice. If their combined salary is $660 per week, how much does each earn?

60. The length of a rectangle is 8 cm more than its width and its perimeter is 64 cm. Find the dimensions of the rectangle.

CUMULATIVE TEST for Part Six

This test is provided to help you in the process of review over Chapters 14 and 15. Answers are provided in the back of the book. If you missed any problems, be sure to go back and review the appropriate chapter sections.

In Problems 1 to 4, complete the statements.

1. The opposite of 20 is

2. The absolute value of -7 is

3. $-(-12) =$

4. $|-20| =$

In Problems 5 to 8, add.

5. $-12 + (-6)$

6. $-7 + 7$

7. $3\frac{2}{3} + \left(-2\frac{1}{3}\right)$

8. $9 + (-4) + (-7)$

In Problems 9 to 12, subtract.

9. $-5 - 8$

10. $8 - (-5)$

11. $-8 - (-8)$

12. $3.25 - (-1.75)$

In Problems 13 to 16, multiply.

13. $(-8)(-12)$

14. $(-6)(15)$

15. $(-1.25)(-4)$

16. $(-5)(-3)(-4)$

In Problems 17 to 20, divide.

17. $\dfrac{-56}{-8}$

18. $-\dfrac{5}{9} \div \dfrac{20}{3}$

19. $0.75 \div (-0.3)$

20. $-2.5 \div 0$

In Problems 21 to 24, evaluate the expressions if $a = 5$, $b = -3$, $c = 4$, and $d = -2$.

21. $6ad$

22. $3b^2$

23. $3(c - 2d)$

24. $\dfrac{2a - 7d}{a - b}$

In Problems 25 to 27, solve each equation and check your solution.

25. $x - 7 = -15$

26. $5x - 8 = 27$

27. $6 - 7x = -15$

In Problems 28 and 29, combine like terms.

28. $-12r + 7r$

29. $-8m + 9m$

In Problems 30 to 32, solve each equation and check your solution.

30. $5x - 7 = 2x + 14$

31. $8x - 2 = 3x - 32$

32. $4x + 13 = 10x + 28$

In Problems 33 to 38, write each phrase using symbols.

33. y increased by 9

34. 9 less than w

35. The product of 5, a, and b

36. The sum of twice s and 5

37. The quotient when z is divided by 8

38. The quotient when 6 more than m is divided by 6

In Problems 39 and 40, solve the word problems.

39. One number is 14 more than another. If the sum of the two numbers is 64, what are the two numbers?

40. The width of a rectangle is 7 ft less than its length. If the perimeter of the rectangle is 62 ft, find the dimensions of the rectangle.

Answers to Even-Numbered Section Exercises and All Chapter Self-Tests, Summary Exercises, and Cumulative Tests

Exercises 1.1
2. $(5 \times 100) + (4 \times 10) + 9$ **4.** $(2 \times 10,000) + (7 \times 100) + (2 \times 10) + 1$ **6.** Thousands **8.** Ten thousands
10. Hundred thousands **12.** Hundred millions **14.** twenty-three thousand, five hundred sixty-seven
16. Five hundred two thousand, two hundred **18.** 307,359 **20.** 1,230,000,000

Exercises 1.2
2. 7 (addend), 8 (addend), 15 (sum) **4.** 7 **6.** 14 **8.** 7 **10.** 16 **12.** 14 **14.** 8 **16.** 9 **18.** 13 **20.** 13
22. 12 **24.** 12 **26.** 12 **28.** 16 **30.** 11 **32.** 18

Exercises 1.3
2. 15 **4.** 15 **6.** 15 **8.** 15 **10.** 7 **12.** 5 **14.** 15 **16.** 23 **18.** Associative property
20. Commutative property **22.** Additive identity **24.** Associative property

Exercises 1.4
2. 18 **4.** 95 **6.** 687 **8.** 769 **10.** 5898 **12.** 7986 **14.** 9208 **16.** 47,728 **18.** 87 **20.** 2798 **22.** 589
24. 2889 **26.** 29 **28.** 168 **30.** 2475 **32.** 668 **34.** 569 pins **36.** 357 mi

Exercises 1.5
2. 72 **4.** 153 **6.** 154 **8.** 450 **10.** 1355 **12.** 779 **14.** 1224 **16.** 13,222 **18.** 5281 **20.** 6320
22. 112,330 **24.** 119,655 **26.** 414 **28.** 316 **30.** 3787 **32.** 42,022 **34.** 32, 38, 44, 50 **36.** 41, 49, 57, 65

Exercises Using Your Calculator
2. 846 **4.** 1299 **6.** 40,134 **8.** 1125 **10.** 45,694 **12.** 2,598,960

Exercises 1.6
2. 70 **4.** 580 **6.** 700 **8.** 6700 **10.** 4000 **12.** 40,000 **14.** 39,000 **16.** 600,000 **18.** 930,000
20. Estimate, 280; actual sum, 278 **22.** Estimate, 330; actual sum, 326 **24.** Estimate, 4600; actual sum, 4614
26. Estimate, 7500; actual sum, 7503 **28.** Estimate, 10,000; actual sum, 9925
30. Estimate, 35,000; actual sum, 35,255 **32.** $0 < 5$ **34.** $20 > 15$ **36.** $3000 > 2000$

Exercises 1.7
2. 7 (minuend), 5 (subtrahend), 2 (difference), $2 + 5 = 7$ **4.** 43 **6.** 54 **8.** 52 **10.** 273 **12.** 320 **14.** 2034
16. 2610 **18.** 56,572 **20.** 32,202 **22.** 35 **24.** 110 **26.** 140 **28.** $704 **30.** $210

Exercises 1.8
2. 37 **4.** 17 **6.** 435 **8.** 384 **10.** 275 **12.** 177 **14.** 337 **16.** 2874 **18.** 1957 **20.** 2775 **22.** 3458
24. 1651 **26.** 27,838 **28.** 24,800 **30.** 13,355 **32.**

Monthly income	$1620
House payment	343
Balance	1277
Car payment	183
Balance	1094
Food	312
Balance	782
Clothing	89
Amount remaining	693

34. Incorrect **36.** Correct

38. 90, 73, 56, 39 **40.** All the sums are 34. **42.**

4	3	8
9	5	1
2	7	6

44.

7	12	1	14
2	13	8	11
16	3	10	5
9	6	15	4

Exercises Using Your Calculator

2. 187   **4.** 2164   **6.** 7685   **8.** 595,766   **10.** 28,385   **12.** 178,750   **14.** 170   **16.** 29   **18.** 402
20. trip 1, 1032; trip 2, 988; trip 3, 718; total, 2738   **22.** 90,501 square miles   **24.** 6480 square miles

Exercises 1.9

2. $8725   **4.** 122 points   **6.** 1076 people   **8.** $3335   **10.** $171   **12.** 736 miles   **14.** 174 students   **16.** $28
18. 43 points   **20.** 7595 miles   **22.** 236,687 square miles

Chapter 1 Self-Test

1. Ten thousands (Sec. 1.1)   **2.** Twenty-three thousand, five hundred forty-three (Sec. 1.1)   **3.** 408,520,000 (Sec. 1.1)
4. Commutative property (Sec. 1.3)   **5.** Associative property (Sec. 1.3)   **6.** Additive identity (Sec. 1.3)
7. 565 (Sec. 1.4)   **8.** 4558 (Sec. 1.4)   **9.** 7173 (Sec. 1.5)   **10.** 1918 (Sec. 1.5)   **11.** 2731 (Sec. 1.5)
12. 13,103 (Sec. 1.5)   **13.** 25,979 (Sec. 1.5)   **14.** 21,696 (Sec. 1.5)   **15.** 8550, 8500, 9000 (Sec. 1.6)
16. 2970, 3000, 3000 (Sec. 1.6)   **17.** Incorrect (Sec. 1.6)   **18.** Correct (Sec. 1.6)   **19.** 12 < 14 (Sec. 1.6)
20. 500 > 400 (Sec. 1.6)   **21.** 235 (Sec. 1.7)   **22.** 12,220 (Sec. 1.7)   **23.** 429 (Sec. 1.8)   **24.** 3239 (Sec. 1.8)
25. 30,770 (Sec. 1.8)   **26.** 40,555 (Sec. 1.8)   **27.** 1572 (Sec. 1.8)   **28.** 68,565 people (Sec. 1.9)
29. 3528 sq mi (Sec. 1.9)   **30.** 72 lb (Sec. 1.9)

Exercises 2.1

2. 20   **4.** Factors, product   **6.** 1, 2, 3, 4, 6, 9, 12, 18, 36   **8.** Multiples   **10.** 28   **12.** 45   **14.** 36   **16.** 7
18. 0   **20.** 32   **22.** 56   **24.** 35   **26.** 63   **28.** 42   **30.** 48   **32.** 81   **34.** 49   **36.** 25

Exercises 2.2

2. Multiplicative identity   **4.** Commutative property   **6.** Multiplication property of zero
8. Distributive property of multiplication over addition   **10.** Associative property   **12.** Multiplicative identity
14. Commutative property   **16.** Multiplication property of zero

Exercises 2.3

2. 96   **4.** 265   **6.** 8127   **8.** 4445   **10.** 35,126   **12.** 62,037   **14.** 183,144   **16.** 155,075   **18.** 2754   **20.** 8806
22. 1636   **24.** 117,260

Exercises 2.4

2. 2842   **4.** 5100   **6.** 19,293   **8.** 342,300   **10.** 27,900   **12.** 228,762   **14.** 203,740   **16.** 666,333
18. 1,254,047   **20.** 6,451,794   **22.** 18,816   **24.** 163,812

Exercises 2.5

2. 670   **4.** 73,000   **6.** 345,600   **8.** 420,000   **10.** 2320   **12.** 816,300   **14.** 112,220   **16.** 1,340,900   **18.** 16,890
20. 428,400   **22.** 335,000   **24.** 848,000   **26.** 900   **28.** 4200   **30.** 80,000   **32.** 350,000

Exercises 2.6

2. 16   **4.** 1   **6.** 39   **8.** 23   **10.** 0   **12.** 58   **14.** 72   **16.** 28   **18.** 12   **20.** 40   **22.** 98   **24.** 98

Exercises Using Your Calculator

2. 2450   **4.** 233,700   **6.** 61,017,734   **8.** 2880   **10.** 12,956,736   **12.** 675,990   **14.** 29   **16.** 3   **18.** 3   **20.** 88
22. 88
24.
$ 8 $ 15
 16   31
 32   63
 64   127
128   255
256   511
512   1023

Exercises 2.7

2. 825 cal   **4.** 504 mi   **6.** 272 cars   **8.** 336 cars   **10.** 14,500 copies   **12.** $18,512   **14.** 588 sq cm   **16.** $252
18. $90   **20.** $5340   **22.** $69   **24.** 172 people   **26.** 863,040 ways   **28.** 31,536,000 seconds

Exercises 2.8
2. 8 **4.** 25 **6.** 243 **8.** 256 **10.** 1 **12.** 7 **14.** 100 **16.** 10,000,000 **18.** 512 **20.** 36 **22.** 49 **24.** 48 **26.** 144 **28.** 28 **30.** 21 **32.** 69 **34.** Not a PT **36.** PT **38.** PT

Exercises Using Your Calculator
2. 64 **4.** 343 **6.** 1024 **8.** 9 **10.** 1 **12.** 1,000,000 **14.** 729 **16.** 390,625 **18.** 48 **20.** 89

Chapter 2 Self-Test
1. 1, 2, 3, 4, 6, 8, 12, 16, 24, 48 (Sec. 2.1) **2.** 6, 12, 18, 24, 30 (Sec. 2.1) **3.** Commutative property (Sec. 2.2)
4. Multiplicative identity (Sec. 2.2) **5.** Multiplication property of zero (Sec. 2.2) **6.** Associative property (Sec. 2.2)
7. Distributive property of multiplication over addition (Sec. 2.2) **8.** 174 (Sec. 2.3) **9.** 1911 (Sec. 2.3)
10. 4984 (Sec. 2.4) **11.** 55,414 (Sec. 2.4) **12.** 2,453,016 (Sec. 2.4) **13.** 53,000 (Sec. 2.5) **14.** 226,800 (Sec. 2.5)
15. 321,840 (Sec. 2.5) **16.** 150,000 (Sec. 2.5) **17.** 920,000 (Sec. 2.5) **18.** 17 (Sec. 2.6) **19.** 22 (Sec. 2.6)
20. 1 (Sec. 2.6) **21.** 75 (Sec. 2.6) **22.** 75 (Sec. 2.6) **23.** 252 people (Sec. 2.7) **24.** $308,750 (Sec. 2.7)
25. 3240 square meters (Sec. 2.7) **26.** $2244 (Sec. 2.7) **27.** 625 (Sec. 2.8) **28.** 1 (Sec. 2.8) **29.** 108 (Sec. 2.8)
30. 9 (Sec. 2.8)

Exercises 3.1
2. quotient, divisor, dividend **4.** 5 **6.** $35 = 5 \times 7$ **8.** $46 = 7 \times 6 + 4$ **10.** 6 **12.** 6 **14.** 7 **16.** 8 **18.** 15
20. 7 **22.** 5 r3 **24.** 5 r5 **26.** 6 r2 **28.** 4 r4 **30.** 8 r3 **32.** 11 r2

Exercises 3.2
2. 9 **4.** 1 **6.** 10 **8.** undefined **10.** 8 **12.** undefined **14.** 0 **16.** undefined **18.** 1 **20.** 0

Exercises 3.3
2. 8 r6 **4.** 14 r4 **6.** 58 **8.** 49 r3 **10.** 238 r1 **12.** 126 r2 **14.** 108 r2 **16.** 987 r3 **18.** 391 r8
20. 1117 r1 **22.** 1622 r2 **24.** 408 r4 **26.** 8657 r2 **28.** 12,184 r4

Exercises 3.4
2. 21 r2 **4.** 7 r36 **6.** 6 r48 **8.** 14 r1 **10.** 65 r35 **12.** 345 r20 **14.** 164 r37 **16.** 207 r26 **18.** 4 r837
20. 36 r68 **22.** 20 r110 **24.** 52 r159 **26.** 301 r120 **28.** 706 r126

Exercises 3.5
2. 15 r3 **4.** 13 r1 **6.** 243 **8.** 318 **10.** 131 r1 **12.** 142 r1 **14.** 488 r5 **16.** 406 r3 **18.** 1274 r5
20. 1262 r4 **22.** 8573 r1 **24.** 3128 r3

Exercises 3.6
2. 17 **4.** 6 **6.** 4 **8.** 24 **10.** 12 **12.** 4 **14.** 128 **16.** 12 **18.** 96 **20.** 4 **22.** 4 **24.** 4

Exercises Using Your Calculator
2. 78 **4.** 647 **6.** 2456 **8.** 16 **10.** 16 **12.** 25 **14.** 10 **16.** 3

Exercises 3.7
2. 32 sections **4.** 43 mi/gal **6.** $16 **8.** $24 **10.** $229 **12.** 62 mi/hr **14.** 579 cars **16.** $91 **18.** $1316
20. The remainder is always zero. **22.** 360

Exercises 3.8
2. 21 **4.** 13 **6.** 11 **8.** 169 mi **10.** 211 students **12.** 68 points **14.** 1987 by $2 **16.** 176

Chapter 3 Self-Test

1. Divisor, dividend, quotient, remainder (Sec. 3.1) **2.** 8 (Sec. 3.2) **3.** 1 (Sec. 3.2) **4.** 0 (Sec. 3.2)
5. Undefined (Sec. 3.2) **6.** 123 (Sec. 3.3) **7.** 492 r6 (Sec. 3.3) **8.** 3041 r2 (Sec. 3.3) **9.** 6 r23 (Sec. 3.4)
10. 14 r41 (Sec. 3.4) **11.** 76 r7 (Sec. 3.4) **12.** 24 r191 (Sec. 3.4) **13.** 22 r21 (Sec. 3.4) **14.** 209 r145 (Sec. 3.4)
15. 131 (Sec. 3.5) **16.** 118 r7 (Sec. 3.5) **17.** 649 r6 (Sec. 3.5) **18.** 5 (Sec. 3.6) **19.** 7 (Sec. 3.6) **20.** 3 (Sec. 3.6)
21. 16 (Sec. 3.6) **22.** 20 (Sec. 3.6) **23.** 3 (Sec. 3.6) **24.** $223 (Sec. 3.7) **25.** 35 mi/gal (Sec. 3.7)
26. $248 (Sec. 3.7) **27.** $28 (Sec. 3.7) **28.** 15 (Sec. 3.8) **29.** 204 riders (Sec. 3.8) **30.** 93 points (Sec. 3.8)

Exercises 4.1

2. 1, 2, 3, 6 **4.** 1, 2, 3, 4, 6, 12 **6.** 1, 3, 7, 21 **8.** 1, 2, 4, 8, 16, 32 **10.** 1, 2, 3, 6, 11, 22, 33, 66 **12.** 1, 37
14. prime **16.** prime **18.** composite **20.** prime **22.** composite **24.** composite **26.** 59, 61, 67, 71, 73
28. 3 and 5 **30.** 2 and 5 **32.** 2 and 3 **34.** 3 and 5 **36.** none **38.** 3 and 5 **40.** none

Exercises 4.2

2. 2×11 **4.** 5×7 **6.** $2 \times 3 \times 7$ **8.** 2×47 **10.** $2 \times 3 \times 3 \times 5$ **12.** $2 \times 2 \times 5 \times 5$ **14.** $2 \times 2 \times 2 \times 11$
16. $2 \times 2 \times 2 \times 2 \times 5 \times 5$ **18.** $2 \times 2 \times 3 \times 11$ **20.** $2 \times 3 \times 5 \times 11$ **22.** $2 \times 2 \times 5 \times 5 \times 5$ **24.** $2 \times 3 \times 3 \times 5 \times 13$
26. 3 and 5 **28.** 4 and 7

Exercises 4.3

2. 3 **4.** 2 **6.** 11 **8.** 14 **10.** 1 **12.** 12 **14.** 18 **16.** 35 **18.** 18 **20.** 15 **22.** 16 **24.** 36

Exercises 4.4

2. 15 **4.** 18 **6.** 36 **8.** 60 **10.** 30 **12.** 84 **14.** 50 **16.** 70 **18.** 72 **20.** 420 **22.** 72 **24.** 60
26. 84 **28.** 40 **30.** 120 **32.** 420

Chapter 4 Self-Test

1. 1, 2, 3, 6, 9, 18 (Sec. 4.1) **2.** 1, 2, 3, 6, 7, 14, 21, 42 (Sec. 4.1) **3.** 1, 17 (Sec. 4.1) **4.** 13, 29, 37 (Sec. 4.1)
5. 21, 51, 91 (Sec. 4.1) **6.** 41, 43, 47, 53, 59 (Sec. 4.1) **7.** 2, 3, and 5 (Sec. 4.1) **8.** none (Sec. 4.1)
9. 2 and 3 (Sec. 4.1) **10.** 3 and 5 (Sec. 4.1) **11.** $2 \times 3 \times 7$ (Sec. 4.2) **12.** $2 \times 2 \times 2 \times 3 \times 3$ (Sec. 4.2)
13. $2 \times 3 \times 5 \times 7$ (Sec. 4.2) **14.** $2 \times 2 \times 2 \times 3 \times 3 \times 11$ (Sec. 4.2) **15.** 5 (Sec. 4.3) **16.** 6 (Sec. 4.3) **17.** 8 (Sec. 4.3)
18. 1 (Sec. 4.3) **19.** 10 (Sec. 4.3) **20.** 4 (Sec. 4.3) **21.** 14 (Sec. 4.3) **22.** 22 (Sec. 4.3) **23.** 35 (Sec. 4.4)
24. 18 (Sec. 4.4) **25.** 60 (Sec. 4.4) **26.** 90 (Sec. 4.4) **27.** 20 (Sec. 4.4) **28.** 24 (Sec. 4.4) **29.** 168 (Sec. 4.4)
30. 180 (Sec. 4.4)

Part One Summary Exercises

1. Hundreds **2.** Hundred thousands **3.** Twenty-seven thousand, four hundred twenty-eight
4. Two hundred thousand, three hundred five **5.** 37,583 **6.** 300,400 (Sec. 1.1) **7.** Commutative property
8. Associative property (Sec. 1.3) **9.** 1389 **10.** 24,552 **11.** 14,722 (Sec. 1.5) **12.** 7000 **13.** 16,000
14. 550,000 (Sec. 1.6) **15.** 5900 (Sec. 1.6) **16.** $60 < 70$ **17.** $38 > 35$ (Sec. 1.6) **18.** 4478 **19.** 18,800
20. 1763 (Sec. 1.8) **21.** 969 people **22.** $536 (Sec. 1.9) **23.** Factor **24.** Multiple (Sec. 2.1)
25. Associative property of multiplication (Sec. 2.2) **26.** Commutative property of multiplication
27. Distributive property of multiplication over addition (Sec. 2.2) **28.** 1075 **29.** 154,602 **30.** 42,657 (Sec. 2.4)
31. 30,960 **32.** 600,000 (Sec. 2.5) **33.** 28 **34.** 36 **35.** 36 (Sec. 2.6) **36.** $16,380 (Sec. 2.7)
37. $630 (Sec. 2.7) **38.** 40 **39.** 1000 (Sec. 2.8) **40.** 0 **41.** Undefined (Sec. 3.2) **42.** 308 r5 (Sec. 3.3)
43. 55 r12 **44.** 106 r171 (Sec. 3.4) **45.** 427 r4 (Sec. 3.5) **46.** 4 (Sec. 3.6) **47.** 28 mi/gal **48.** $260 (Sec. 3.7)
49. 88 (Sec. 3.8) **50.** 1, 2, 4, 13, 26, 52 **51.** 1, 41 (Sec. 4.1)
52. Prime: 2, 5, 7, 11, 17, 23, 43; composite: 14, 21, 27, 39 (Sec. 4.1) **53.** 2 and 5 **54.** none (Sec. 4.1)
55. $2 \times 2 \times 2 \times 2 \times 3$ **56.** $2 \times 2 \times 3 \times 5 \times 7$ (Sec. 4.2) **57.** 5 **58.** 1 (Sec. 4.3) **59.** 54 **60.** 120 (Sec. 4.4)

Part One Cumulative Test

1. Hundred thousands **2.** Three hundred two thousand, five hundred twenty-five **3.** 2,430,000
4. Commutative property **5.** Additive identity **6.** Associative property **7.** 966 **8.** 23,351 **9.** 5900
10. 950,000 **11.** 7700 **12.** $49 > 47$ **13.** $80 < 90$ **14.** 3861 **15.** 17,465 **16.** 905 **17.** $7579
18. Associative property of multiplication **19.** Multiplicative identity
20. Distributive property of multiplication over addition **21.** 378,214 **22.** 686,000 **23.** 7695 **24.** 600,000
25. $1008 **26.** 67 r43 **27.** 103 r176 **28.** 105 **29.** 56 **30.** 36 **31.** 8 **32.** $58 **33.** 85
34. Prime: 5, 13, 17, 31; composite: 9, 22, 27, 45 **35.** 2 and 3 **36.** $2 \times 2 \times 2 \times 3 \times 11$ **37.** 12 **38.** 8 **39.** 60
40. 90

Answers

Exercises 5.1
2. 5 is the numerator, 12 is the denominator.   4. 7 is the numerator, 15 is the denominator.   6. $\frac{1}{3}$ 8. $\frac{3}{7}$ 10. $\frac{3}{8}$ 12. $\frac{8}{8}$ or 1   14. $\frac{4}{4}$ or 1   16. $\frac{2}{7}$ 18. $\frac{5}{9}$ 20. $\frac{2}{5}$ 22. $\frac{5}{9}, \frac{4}{9}$ 24. $4 \div 5$

Exercises 5.2
2. improper   4. proper   6. mixed number   8. improper   10. mixed number   12. improper   14. Proper   16. Improper   18. $1\frac{2}{3}$ 20. $2\frac{1}{4}$

Exercises 5.3
2. $2\frac{2}{3}$ 4. $1\frac{1}{6}$ 6. $3\frac{3}{8}$ 8. $4\frac{1}{6}$ 10. $8\frac{2}{7}$ 12. $12\frac{7}{12}$ 14. 20   16. 6   18. $\frac{13}{4}$ 20. $\frac{37}{8}$ 22. $\frac{8}{1}$ 24. $\frac{20}{9}$ 26. $\frac{73}{10}$ 28. $\frac{49}{4}$ 30. $\frac{601}{4}$ 32. $\frac{1003}{4}$

Exercises 5.4
2. Equivalent   4. Not equivalent   6. Not equivalent   8. Equivalent   10. Equivalent   12. Not equivalent

Exercises 5.5
2. $\frac{4}{5}$ 4. $\frac{3}{10}$ 6. $\frac{4}{5}$ 8. $\frac{7}{8}$ 10. $\frac{2}{5}$ 12. $\frac{3}{8}$ 14. $\frac{3}{5}$ 16. $\frac{1}{3}$ 18. $\frac{2}{3}$ 20. $\frac{8}{11}$ 22. $\frac{3}{7}$ 24. $\frac{7}{8}$ 26. $\frac{21}{32}$ 28. $\frac{42}{55}$ 30. $\frac{1}{10}$ 32. $\frac{1}{8}$ 34. $\frac{3}{10}$

Exercises 5.6
2. 5   4. 8   6. 40   8. 60   10. 16   12. 15   14. 48   16. 45   18. $\frac{4}{9}, \frac{5}{11}$ 20. $\frac{8}{9}, \frac{9}{10}$ 22. $\frac{7}{18}, \frac{5}{12}, \frac{3}{6}$ 24. $\frac{11}{32}, \frac{3}{8}, \frac{7}{16}$ 26. $\frac{5}{6} < \frac{11}{12}$ 28. $\frac{7}{10} < \frac{11}{15}$ 30. $\frac{5}{12} > \frac{7}{18}$ 32. $\frac{7}{12} < \frac{9}{15}$ 34. $\frac{25}{30}, \frac{24}{30}$ 36. $\frac{21}{36}, \frac{10}{36}$ 38. $\frac{6}{30}, \frac{10}{30}, \frac{5}{30}$ 40. $\frac{75}{120}, \frac{36}{120}, \frac{70}{120}$ 42. $\frac{3}{16}$ in   44. $\frac{3}{8}$ in

Chapter 5 Self-Test
1. $\frac{5}{6}$ ← numerator / ← denominator (Sec. 5.1)   2. $\frac{5}{8}$ ← numerator / ← denominator (Sec. 5.1)   3. $\frac{3}{5}$ ← numerator / ← denominator (Sec. 5.1)   4. Proper: $\frac{10}{11}, \frac{1}{8}$ Improper: $\frac{9}{5}, \frac{7}{7}, \frac{8}{1}$ Mixed number: $2\frac{3}{5}$ (Sec. 5.2)   5. $4\frac{1}{4}$ (Sec. 5.2)   6. $2\frac{3}{5}$ (Sec. 5.3)   7. $7\frac{2}{9}$ (Sec. 5.3)   8. 3 (Sec. 5.3)   9. 5 (Sec. 5.3)   10. $\frac{13}{4}$ (Sec. 5.3)   11. $\frac{17}{6}$ (Sec. 5.3)   12. $\frac{37}{7}$ (Sec. 5.3)   13. $\frac{93}{10}$ (Sec. 5.3)   14. Equivalent (Sec. 5.4)   15. Not equivalent (Sec. 5.4)   16. $\frac{7}{9}$ (Sec. 5.5)   17. $\frac{3}{4}$ (Sec. 5.5)   18. $\frac{3}{7}$ (Sec. 5.5)   19. $\frac{15}{32}$ (Sec. 5.5)   20. 16 (Sec. 5.6)   21. 36 (Sec. 5.6)   22. 105 (Sec. 5.6)   23. $\frac{5}{7}, \frac{7}{9}$ (Sec. 5.6)   24. $\frac{4}{15}, \frac{1}{3}, \frac{2}{5}$ 25. $\frac{5}{6} > \frac{4}{5}$ (Sec. 5.6)   26. $\frac{7}{12} < \frac{11}{18}$ (Sec. 5.6)   27. $\frac{8}{20}, \frac{15}{20}$ (Sec. 5.6)   28. $\frac{21}{24}, \frac{10}{24}$ (Sec. 5.6)   29. $\frac{9}{24}, \frac{6}{24}, \frac{20}{24}$ (Sec. 5.6)   30. $\frac{27}{60}, \frac{44}{60}, \frac{35}{60}$ (Sec. 5.6)

Exercises 6.1
2. $\frac{21}{88}$ 4. $1\frac{1}{15}$ 6. $\frac{3}{8}$ 8. $\frac{10}{33}$ 10. $\frac{7}{15}$ 12. $\frac{1}{12}$ 14. $1\frac{1}{6}$ 16. $3\frac{8}{9}$ 18. $7\frac{1}{2}$ 20. 14   22. $\frac{7}{15}$ 24. $\frac{1}{2}$

Exercises 6.2

2. $\frac{9}{10}$ 4. $1\frac{2}{5}$ 6. $1\frac{3}{5}$ 8. $1\frac{4}{5}$ 10. $5\frac{5}{6}$ 12. $4\frac{4}{21}$ 14. $4\frac{1}{2}$ 16. 9 18. $6\frac{2}{3}$ 20. 33 22. 28 24. 42

Exercises 6.3

2. $\frac{5}{21}$ 4. $\frac{22}{75}$ 6. $\frac{4}{9}$ 8. $\frac{8}{15}$ 10. $1\frac{3}{5}$ 12. $1\frac{1}{3}$ 14. $4\frac{1}{5}$ 16. 30 18. $7\frac{1}{2}$ 20. $6\frac{1}{2}$ 22. 28 24. $\frac{1}{6}$
26. $1\frac{2}{3}$ 28. 14 30. $\frac{3}{4}$ 32. 15 34. $1\frac{1}{14}$ 36. $1\frac{3}{5}$ 38. $54 40. $700; $1050 42. $\frac{5}{12}$ sq yd 44. $\frac{7}{15}$
46. 132 mi 48. $8\frac{1}{3}$ acres 50. $2\frac{1}{4}$ sq in 52. 24 54. 96

Exercises 6.4

2. $1\frac{1}{5}$ 4. $\frac{5}{6}$ 6. $\frac{55}{72}$ 8. $1\frac{7}{33}$ 10. $1\frac{1}{3}$ 12. $\frac{5}{12}$ 14. 40 16. $\frac{1}{8}$ 18. $\frac{1}{15}$ 20. $6\frac{2}{3}$ 22. $1\frac{1}{2}$ 24. $\frac{4}{5}$
26. $\frac{3}{14}$ 28. $\frac{1}{4}$ 30. $3\frac{3}{10}$ 32. $\frac{1}{15}$ 34. $\frac{5}{6}$ 36. $2\frac{2}{9}$ 38. $2\frac{4}{5}$ 40. 24 bowls 42. $12,000 44. 36 books
46. 26 shirts 48. 64 lots

Chapter 6 Self-Test

1. $\frac{12}{35}$ (Sec. 6.1) 2. $\frac{9}{16}$ (Sec. 6.1) 3. $\frac{2}{9}$ (Sec. 6.1) 4. $2\frac{1}{7}$ (Sec. 6.1) 5. $2\frac{2}{5}$ (Secs. 6.2, 6.3) 6. $9\frac{1}{5}$ (Secs. 6.2, 6.3)
7. $\frac{3}{5}$ (Sec. 6.3) 8. $\frac{2}{5}$ (Sec. 6.3) 9. $\frac{4}{15}$ (Sec. 6.3) 10. 3 (Sec. 6.3) 11. $6\frac{1}{9}$ (Sec. 6.3) 12. $7\frac{1}{5}$ (Sec. 6.3)
13. $4\frac{2}{3}$ (Sec. 6.3) 14. $\frac{1}{2}$ (Sec. 6.3) 15. $\frac{7}{12}$ yd (Sec. 6.3) 16. 99¢ (Sec. 6.3) 17. 20 sq yd (Sec. 6.3)
18. 190 mi (Sec. 6.3) 19. $1\frac{3}{32}$ (Sec. 6.4) 20. $\frac{5}{8}$ (Sec. 6.4) 21. $\frac{3}{32}$ (Sec. 6.4) 22. 16 (Sec. 6.4)
23. $3\frac{3}{14}$ (Sec. 6.4) 24. $\frac{2}{3}$ (Sec. 6.4) 25. $1\frac{1}{2}$ (Sec. 6.4) 26. $2\frac{2}{3}$ (Sec. 6.4) 27. 47 homes (Sec. 6.4)
28. 54 mi/h (Sec. 6.4) 29. 64 sheets (Sec. 6.4) 30. 48 books (Sec. 6.4)

Exercises 7.1

2. $\frac{5}{7}$ 4. $\frac{9}{16}$ 6. $\frac{2}{3}$ 8. $\frac{1}{2}$ 10. 1 12. $\frac{4}{5}$ 14. $1\frac{2}{5}$ 16. $1\frac{1}{2}$ 18. $1\frac{1}{3}$ 20. $\frac{7}{10}$ 22. $1\frac{7}{12}$ 24. $1\frac{7}{20}$
26. $\frac{4}{5}$ of a dollar 28. $\frac{2}{3}$ of an hour

Exercises 7.2

2. 15 4. 10 6. 30 8. 120 10. 60 12. 72 14. 144 16. 420 18. 60 20. 66 22. 54 24. 360

Exercises 7.3

2. $\frac{14}{15}$ 4. $\frac{1}{3}$ 6. $\frac{9}{10}$ 8. $\frac{3}{10}$ 10. $\frac{8}{15}$ 12. $\frac{27}{40}$ 14. $\frac{43}{60}$ 16. $\frac{5}{12}$ 18. $\frac{47}{75}$ 20. $1\frac{19}{150}$ 22. $\frac{47}{60}$
24. $1\frac{7}{24}$ 26. $1\frac{11}{20}$ 28. $2\frac{7}{120}$ 30. $1\frac{1}{8}$ in 32. $\frac{17}{24}, \frac{7}{24}$ 34. $1\frac{7}{8}$ in 36. $\frac{1}{4}$

Exercises 7.4

2. $\frac{3}{7}$ 4. $\frac{2}{5}$ 6. $\frac{1}{5}$ 8. $\frac{1}{3}$ 10. $\frac{11}{18}$ 12. $\frac{23}{42}$ 14. $\frac{3}{10}$ 16. $\frac{19}{60}$ 18. $\frac{13}{60}$ 20. $\frac{1}{12}$ 22. $\frac{1}{90}$ 24. $\frac{11}{48}$
26. $\frac{1}{5}$ 28. $\frac{9}{14}$ 30. $\frac{5}{16}$ in 32. $\frac{1}{16}$ lb 34. $\frac{7}{16}$ 36. Yes, $\frac{1}{3}$ gal will be left over after both jobs.

Answers

Exercises 7.5

2. $11\frac{2}{3}$ **4.** 7 **6.** $8\frac{3}{5}$ **8.** $3\frac{5}{12}$ **10.** $11\frac{11}{36}$ **12.** $7\frac{1}{40}$ **14.** $10\frac{19}{20}$ **16.** $16\frac{1}{18}$ **18.** $12\frac{11}{24}$ **20.** $2\frac{2}{3}$
22. $1\frac{1}{3}$ **24.** $4\frac{19}{30}$ **26.** $3\frac{41}{60}$ **28.** $6\frac{17}{21}$ **30.** $2\frac{1}{3}$ **32.** $3\frac{5}{9}$ **34.** 3 **36.** $1\frac{17}{30}$

Exercises 7.6

2. $6\frac{5}{8}$ lb **4.** $11\frac{1}{12}$ mi **6.** $5\frac{3}{8}$ in **8.** $5\frac{1}{8}$ in **10.** $\frac{7}{8}$ lb **12.** $13\frac{3}{8}$ yd **14.** $\frac{3}{4}$ in **16.** 4 in **18.** $2\frac{5}{12}$ mi
20. $3\frac{3}{8}$ mi

Chapter 7 Self-Test

1. $\frac{6}{7}$ (Sec. 7.1) **2.** $\frac{2}{3}$ (Sec. 7.1) **3.** $1\frac{1}{5}$ (Sec. 7.1) **4.** 60 (Sec. 7.2) **5.** 36 (Sec. 7.2) **6.** $\frac{9}{10}$ (Sec. 7.3)
7. $\frac{25}{42}$ (Sec. 7.3) **8.** $\frac{19}{24}$ (Sec. 7.3) **9.** $1\frac{11}{60}$ (Sec. 7.3) **10.** $1\frac{23}{40}$ (Sec. 7.3) **11.** $1\frac{5}{12}$ cup (Sec. 7.3) **12.** $\frac{1}{3}$ (Sec. 7.4)
13. $\frac{5}{12}$ (Sec. 7.4) **14.** $\frac{7}{20}$ (Sec. 7.4) **15.** $\frac{23}{30}$ (Sec. 7.4) **16.** $\frac{1}{4}$ h (Sec. 7.4) **17.** $7\frac{7}{10}$ (Sec. 7.5) **18.** $10\frac{1}{3}$ (Sec. 7.5)
19. $7\frac{11}{12}$ (Sec. 7.5) **20.** $12\frac{3}{40}$ (Sec. 7.5) **21.** $11\frac{5}{6}$ (Sec. 7.5) **22.** $13\frac{11}{20}$ (Sec. 7.5) **23.** $1\frac{3}{4}$ (Sec. 7.5)
24. $1\frac{11}{18}$ (Sec. 7.5) **25.** $3\frac{23}{24}$ (Sec. 7.5) **26.** $1\frac{8}{15}$ (Sec. 7.5) **27.** $5\frac{3}{4}$ h (Sec. 7.6) **28.** $24\frac{3}{4}$ in (Sec. 7.6)
29. $4\frac{3}{4}$ in (Sec. 7.6) **30.** $30\frac{5}{6}$ yd (Sec. 7.6)

Part Two Summary Exercises

1. Fraction $\frac{3}{8}$, numerator 3, denominator 8 **2.** Fraction $\frac{5}{6}$, numerator 5, denominator 6 (Sec. 5.1)
3. Proper: $\frac{2}{3}, \frac{7}{10}$; improper: $\frac{5}{4}, \frac{45}{8}, \frac{7}{7}, \frac{9}{1}, \frac{12}{5}$; mixed numbers: $2\frac{3}{7}, 3\frac{4}{5}, 5\frac{2}{9}$ (Sec. 5.2) **4.** $6\frac{5}{6}$ **5.** 4 **6.** $\frac{61}{8}$
7. $\frac{43}{10}$ (Sec. 5.3) **8.** Not equivalent **9.** Equivalent (Sec. 5.4) **10.** $\frac{2}{3}$ **11.** $\frac{3}{5}$ **12.** $\frac{7}{9}$ **13.** $\frac{16}{21}$ (Sec. 5.5)
14. 15 **15.** 32 (Sec. 5.6) **16.** $\frac{7}{12}, \frac{5}{8}$ **17.** $\frac{7}{10}, \frac{4}{5}, \frac{5}{6}$ (Sec. 5.6) **18.** $\frac{5}{12} > \frac{3}{8}$ **19.** $\frac{3}{7} = \frac{9}{21}$
20. $\frac{9}{16} < \frac{7}{12}$ (Sec. 5.6) **21.** $\frac{4}{24}, \frac{21}{24}$ **22.** $\frac{36}{120}, \frac{75}{120}, \frac{70}{120}$ (Sec. 5.6) **23.** $\frac{1}{9}$ **24.** $\frac{1}{6}$ **25.** $1\frac{1}{2}$ **26.** $2\frac{1}{8}$
27. $9\frac{3}{5}$ **28.** $11\frac{1}{3}$ **29.** 8 (Secs. 6.1, 6.2, 6.3) **30.** $\frac{2}{3}$ **31.** $\frac{5}{6}$ **32.** $\frac{3}{16}$ **33.** $1\frac{1}{2}$ (Sec. 6.4) **34.** $\frac{3}{7}$ **35.** 220 mi
36. $204 **37.** 52 mi/hr **38.** 48 lots (Sec. 6.3, 6.4) **39.** $\frac{7}{9}$ **40.** $1\frac{3}{5}$ **41.** $1\frac{4}{9}$ **42.** $\frac{31}{36}$ **43.** $1\frac{41}{60}$ (Sec. 7.1–7.3)
44. $\frac{1}{4}$ **45.** $\frac{7}{18}$ **46.** $\frac{7}{60}$ **47.** $\frac{7}{54}$ **48.** $\frac{35}{72}$ (Sec. 7.4) **49.** $10\frac{2}{7}$ **50.** $9\frac{37}{60}$ **51.** $9\frac{17}{24}$ **52.** $4\frac{1}{3}$ **53.** $6\frac{1}{24}$
54. $2\frac{19}{24}$ **55.** $4\frac{7}{10}$ (Sec. 7.5) **56.** $69\frac{1}{16}$ in **57.** $19\frac{9}{16}$ in **58.** $1\frac{3}{8}$ in **59.** $53\frac{9}{16}$ in **60.** Yes; $\frac{5}{12}$ yd (Sec. 7.6)

Part Two Cumulative Test

1. Fraction: $\frac{5}{8}$; numerator: 5; denominator: 8 **2.** Proper: $\frac{7}{12}, \frac{3}{7}$; improper: $\frac{10}{8}, \frac{9}{9}, \frac{7}{1}$; mixed number: $3\frac{1}{5}, 2\frac{2}{3}$
3. $2\frac{4}{5}$ **4.** 4 **5.** $\frac{13}{3}$ **6.** $\frac{63}{8}$ **7.** Equivalent **8.** Not equivalent **9.** $\frac{2}{3}$ **10.** $\frac{3}{8}$ **11.** $\frac{6}{11}, \frac{5}{9}$
12. $\frac{8}{15}, \frac{3}{5}, \frac{7}{10}$ **13.** $\frac{15}{24}, \frac{14}{24}$ **14.** $\frac{24}{36}, \frac{20}{36}, \frac{27}{36}$ **15.** $\frac{8}{27}$ **16.** $\frac{4}{15}$ **17.** $5\frac{2}{5}$ **18.** $22\frac{2}{3}$ **19.** $\frac{3}{4}$ **20.** $1\frac{1}{3}$
21. $4\frac{1}{2}$ **22.** $\frac{5}{6}$ **23.** $1\frac{1}{2}$ **24.** $540 **25.** 88 sheets **26.** $\frac{4}{5}$ **27.** $\frac{61}{75}$ **28.** $1\frac{31}{40}$ **29.** $\frac{1}{2}$ **30.** $\frac{5}{36}$ **31.** $\frac{5}{9}$
32. $6\frac{2}{7}$ **33.** $8\frac{1}{24}$ **34.** $4\frac{5}{9}$ **35.** $4\frac{1}{24}$ **36.** $3\frac{5}{8}$ **37.** $3\frac{13}{24}$ **38.** $14\frac{19}{30}$ hours **39.** $\frac{5}{8}$ in **40.** $1\frac{11}{12}$ hours

Exercises 8.1

2. Tenths   **4.** hundred thousandths   **6.** 0.371   **8.** 3.5   **10.** 7.0431   **12.** Three hundred seventy-one thousandths
14. Two hundred fifty-one ten-thousandths   **16.** Twenty-three and fifty-six thousandths   **18.** 0.0253   **20.** 12.245
22. $\frac{765}{100000}$ or $\frac{153}{20000}$ **24.** $4\frac{17}{10000}$ **26.** $0.75 < 0.752$ **28.** $2.451 > 2.45$ **30.** $4.98 < 5$ **32.** $0.235 = 0.2350$

Exercises 8.2

2. 3.22   **4.** 3.735   **6.** 28.29   **8.** 31.135   **10.** 22.605   **12.** 27.8947   **14.** 2.209   **16.** 46.86   **18.** 41.26
20. 280.101   **22.** 1.718   **24.** 67.28   **26.** 8.2 mi   **28.** 5.725 in   **30.** $100.15   **32.** $241.24

Exercises 8.3

2. 3.03   **4.** 28.78   **6.** 18.497   **8.** 2.63   **10.** 2.99   **12.** 36.85   **14.** 6.15   **16.** 32.375   **18.** 26.125   **20.** 6.88
22. 6.18   **24.** 7.62   **26.** 0.0575 in   **28.** $15.42   **30.** $25.98   **32.** 0.89 in below normal

34.

2.4	8.4	7.2
10.8	6	1.2
4.8	3.6	9.6

Exercises Using Your Calculator

2. 21.876   **4.** 12,807.13   **6.** 5.175   **8.** 10.385   **10.** 2.1925   **12.** $77.64 overdrawn   **14.** $974.46

Exercises 8.4

2. 27.95   **4.** 42.32   **6.** 1984.5   **8.** 4.275   **10.** 46.268   **12.** 0.7638   **14.** 3.6666   **16.** 21.576   **18.** 0.665
20. 0.046368   **22.** 204.16   **24.** 0.01918   **26.** $905.40   **28.** $252.45   **30.** $18.85   **32.** $337.60   **34.** $261.30
36. 11.8 cm

Exercises Using Your Calculator

2. 0.1067   **4.** 755.811   **6.** 196.66995   **8.** 0.000512   **10.** 0.271441   **12.** $224.64   **14.** $182.25

Exercises 8.5

2. 89.5   **4.** 24.1   **6.** 580   **8.** 0.25   **10.** 9500   **12.** 23,420   **14.** 360   **16.** 5800   **18.** 530 cm   **20.** $178.00

Exercises 8.6

2. 6.79   **4.** 5.8   **6.** 2.358   **8.** 1.5   **10.** 0.8536   **12.** 52.873   **14.** 12.547   **16.** 503.82   **18.** 18.6 gal
20. 14.76 in   **22.** $50   **24.** $24

Chapter 8 Self-Test

1. 0.431 (Sec. 8.1)   **2.** 5.13 (Sec. 8.1)   **3.** Four hundred thirty-one thousandths (Sec. 8.1)
4. Five and thirteen hundredths (Sec. 8.1)   **5.** $5.93 > 5.928$ (Sec. 8.1)   **6.** $2.149 < 2.15$ (Sec. 8.1)   **7.** 2.208 (Sec. 8.2)
8. 3.521 (Sec. 8.2)   **9.** 40.764 (Sec. 8.2)   **10.** 10.805 (Sec. 8.2)   **11.** 50.8 gal (Sec. 8.2)   **12.** 10.15 in (Sec. 8.2)
13. 2.58 (Sec. 8.3)   **14.** 4.875 (Sec. 8.3)   **15.** 6.515 (Sec. 8.3)   **16.** 937.3 mi (Sec. 8.3)   **17.** $279.57 (Sec. 8.3)
18. 1.75 in (Sec. 8.3)   **19.** 21.46 (Sec. 8.4)   **20.** 1.5718 (Sec. 8.4)   **21.** 0.0094 (Sec. 8.4)   **22.** 37.41 sq cm (Sec. 8.4)
23. $202.10 (Sec. 8.4)   **24.** 5.4 (Sec. 8.5)   **25.** 84,320 (Sec. 8.5)   **26.** $470 (Sec. 8.5)   **27.** 2.6 (Sec. 8.6)
28. 23.34 (Sec. 8.6)   **29.** 40.2 km (Sec. 8.6)   **30.** $110.81 (Sec. 8.6)

Exercises 9.1

2. 5.49   **4.** 0.92   **6.** 0.345   **8.** 4.384   **10.** 4.6   **12.** 3.26   **14.** 0.345   **16.** 2.816   **18.** 0.66   **20.** 0.093
22. 0.047   **24.** 0.09   **26.** $24.58   **28.** $.80 or 80¢   **30.** $15.78   **32.** $36.72   **34.** 48.1   **36.** 17.7

Exercises 9.2

2. 13.55   **4.** 4.6   **6.** 0.565   **8.** 21.4   **10.** 6.215   **12.** 2.45   **14.** 0.215   **16.** 12.8   **18.** 2.76   **20.** 0.254
22. 2.36   **24.** 0.018   **26.** 2.37   **28.** 0.295   **30.** $6.28   **32.** 640 nails   **34.** 16.4 cu ft   **36.** 4.65 in

Answers 691

Exercises 9.3
2. 0.023 **4.** 0.0672 **6.** 0.0592 **8.** 0.048 **10.** 0.00243 **12.** 0.02371 **14.** $235 **16.** $289 **18.** 0.75 liters
20. 76¢

Exercises Using Your Calculator
2. 58.5 **4.** 62.5 **6.** 23.48 **8.** 4.9 **10.** 1.59 **12.** 2.835 **14.** 230 **16.** 31.4 mi/gal

Exercises 9.4
2. 0.8 **4.** 0.3 **6.** 0.875 **8.** 0.55 **10.** 0.4375 **12.** 0.53125 **14.** 0.5625 **16.** 0.44 **18.** 0.58 **20.** 0.208
22. $0.\overline{4}$ **24.** $0.4\overline{6}$ **26.** 7.75 **28.** 4.4375 **30.** $\frac{7}{8} > 0.87$ **32.** $\frac{9}{25} < 0.4$

Exercises Using Your Calculator
2. 0.6875 **4.** 0.29 **6.** 0.147 **8.** $0.\overline{63}$ **10.** 3.8 **12.** 8.1875

Exercises 9.5
2. $\frac{3}{10}$ **4.** $\frac{3}{5}$ **6.** $\frac{97}{100}$ **8.** $\frac{379}{1000}$ **10.** $\frac{3}{4}$ **12.** $\frac{13}{20}$ **14.** $\frac{29}{250}$ **16.** $\frac{9}{40}$ **18.** $\frac{23}{40}$ **20.** $\frac{67}{1000}$ **22.** $\frac{17}{400}$
24. $5\frac{7}{10}$ **16.** $3\frac{31}{100}$ **28.** $15\frac{7}{20}$

Chapter 9 Self-Test
1. 2.75 (Sec. 9.1) **2.** 2.385 (Sec. 9.1) **3.** 0.46 (Sec. 9.1) **4.** 0.145 (Sec. 9.1) **5.** $23.28 (Sec. 9.1)
6. 0.65 in. (Sec. 9.1) **7.** 6.7 (Sec. 9.2) **8.** 3.225 (Sec. 9.2) **9.** 2.84 (Sec. 9.2) **10.** 5.53 (Sec. 9.2)
11. 2.02 (Sec. 9.2) **12.** 0.541 (Sec. 9.2) **13.** 30.3 mi/gal (Sec. 9.2) **14.** 29 shirts (Sec. 9.2) **15.** 45 lots (Sec. 9.2)
16. 3.857 (Sec. 9.3) **17.** 0.02847 (Sec. 9.3) **18.** 0.003795 (Sec. 9.3) **19.** $5.37 (Sec. 9.3) **20.** 0.828 liters (Sec. 9.3)
21. 0.875 (Sec. 9.4) **22.** 0.21875 (Sec. 9.4) **23.** 0.857 (Sec. 9.4) **24.** $0.\overline{63}$ (Sec. 9.4) **25.** 2.5625 (Sec. 9.4)
26. $\frac{29}{100}$ (Sec. 9.5) **27.** $\frac{14}{25}$ (Sec. 9.5) **28.** $\frac{71}{200}$ (Sec. 9.5) **29.** $\frac{313}{400}$ (Sec. 9.5) **30.** $2\frac{19}{25}$ (Sec. 9.5)

Part Three Summary Exercises
1. Hundredths **2.** Ten-thousandths (Sec. 8.1) **3.** 0.37 **4.** 0.0307 (Sec. 8.1) **5.** Seventy-one thousandths
6. Twelve and thirty-nine hundredths (Sec. 8.1) **7.** 4.5 **8.** 400.037 (Sec. 8.1) **9.** 0.79 > 0.785 **10.** 1.25 = 1.250
11. 12.8 < 13 **12.** 0.832 > 0.83 (Sec. 8.1) **13.** 3.47 **14.** 18.852 **15.** 37.728 **16.** 20.533 (Sec. 8.2) **17.** 22.2 mi
18. 6.15 cm (Sec. 8.2) **19.** 23.46 **20.** 4.245 **21.** 1.075 **22.** 6.62 (Sec. 8.3) **23.** $61.75 (Sec. 8.3)
24. $18.93 (Sec. 8.3) **25.** 16.416 **26.** 0.000261 **27.** 69.44 **28.** 0.0012275 (Sec. 8.4) **29.** $271.15 **30.** $287.50
31. $152.10 (Sec. 8.4) **32.** 52 **33.** 450 (Sec. 8.5) **34.** $5742 (Sec. 8.5) **35.** 5.84 **36.** 9.572
37. 4.876 (Sec. 8.6) **38.** 45.94 sq cm (Sec. 8.6) **39.** 0.385 **40.** 4.65 **41.** 0.322 (Sec. 9.1) **42.** $23.45
43. 39.3 mi/gal (Sec. 9.1) **44.** 2.66 **45.** 8.45 **46.** 1.3 **47.** 0.089 (Sec. 9.2) **48.** 54 lots
49. 29.8 mi/gal (Sec. 9.2) **50.** 0.76 **51.** 0.0807 **52.** 0.0457 (Sec. 9.3) **53.** $7.09 (Sec. 9.3) **54.** 0.4375
55. 0.429 **56.** $0.2\overline{6}$ **57.** 3.75 (Sec. 9.4) **58.** $\frac{21}{100}$ **59.** $\frac{21}{250}$ **60.** $5\frac{7}{25}$ (Sec. 9.5)

Part Three Cumulative Test
1. Ten-thousandths **2.** 0.049 **3.** Two and fifty-three hundredths **4.** 12.017 **5.** 0.889 < 0.89 **6.** 0.531 > 0.53
7. 16.64 **8.** 47.253 **9.** 12.803 **10.** 50.2 gal **11.** 10.54 **12.** 24.375 **13.** 3.888 **14.** $3.06 **15.** 17.437
16. 0.02793 **17.** 1.4575 **18.** 7.525 sq in **19.** 735 **20.** 12,570 **21.** $543 **22.** 0.598 **23.** 23.57
24. 10.05 ft **25.** 0.465 **26.** 2.35 **27.** 0.051 **28.** $76.96 **29.** 2.385 **30.** 7.35 **31.** 0.067 **32.** 32 lots
33. 0.004983 **34.** 0.00523 **35.** $573.40 **36.** 0.4375 **37.** 0.429 **38.** $0.\overline{63}$ **39.** $\frac{9}{125}$ **40.** $4\frac{11}{25}$

Exercises 10.1
2. $\frac{5}{4}$ **4.** $\frac{5}{12}$ **6.** $\frac{3}{2}$ **8.** $\frac{5}{8}$ **10.** $\frac{23}{36}$ **12.** $\frac{5}{4}$ **14.** $\frac{2}{3}$ **16.** $\frac{5}{6}$ **18.** $\frac{9}{32}$ **20.** $\frac{2}{9}$ **22.** $\frac{1}{3}$ **24.** $\frac{2}{5}$
26. $\frac{7 \text{ cups}}{4 \text{ loaves}}$ **28.** $\frac{\$53}{8 \text{ hr}}$ **30.** $\frac{3}{5}$ **32.** $\frac{48 \text{ mi}}{1 \text{ gal}}$

Exercises 10.2

2. 5 and 4 are the means; 2 and 10 are the extremes.   **4.** x and 20 are the means; 5 and 24 are the extremes.
6. 8 and 15 are the means; 3 and 40 are the extremes.   **8.** 9 and n are the means; 7 and 45 are the extremes.   **10.** T
12. T   **14.** F   **16.** T   **18.** F   **20.** T   **22.** F   **24.** T   **26.** T   **28.** T   **30.** F   **32.** T   **34.** T   **36.** F

Exercises 10.3

2. 2   **4.** 6   **6.** 5   **8.** 25   **10.** 30   **12.** 3   **14.** 5   **16.** 14   **18.** 28   **20.** 75   **22.** 27   **24.** 15   **26.** 3
28. 22   **30.** 30   **32.** 3   **34.** 100   **36.** 36   **38.** 100   **40.** 2   **42.** 15   **44.** 9

Exercises 10.4

2. $1.44   **4.** $9.48   **6.** $2.00   **8.** 2500 women   **10.** 180 chairs   **12.** $1200   **14.** 240 mi   **16.** $12.25
18. $5000   **20.** 9 lb   **22.** 10 ft   **24.** 20 ft   **26.** 1 in   **28.** 24 ft   **30.** $48   **32.** 20 km   **34.** 900 sq ft
36. 90 rolls

Exercises Using Your Calculator

2. 27.5   **4.** 11.72   **6.** 23.24   **8.** $263.77   **10.** 8 parts   **12.** 6320 items   **14.** 1.24 in   **16.** 10 oz   **18.** 7 oz
20. 1 qt 14 oz   **22.** 5 lb

Chapter 10 Self-Test

1. $\frac{5}{14}$ (Sec. 10.1)   **2.** $\frac{5}{2}$ (Sec. 10.1)   **3.** $\frac{2}{9}$ (Sec. 10.1)   **4.** $\frac{1}{16}$ (Sec. 10.1)   **5.** $\frac{23}{27}, \frac{23}{4}$ (Sec. 10.1)
6. 12 and 15 are the means; 5 and 36 are the extremes. (Sec. 10.2)
7. a and 15 are the means; 5 and 21 are the extremes (Sec. 10.2)   **8.** T (Sec. 10.2)   **9.** F (Sec. 10.2)   **10.** T (Sec. 10.2)
11. F (Sec. 10.2)   **12.** 2 (Sec. 10.3)   **13.** 12 (Sec. 10.3)   **14.** 25 (Sec. 10.3)   **15.** 45 (Sec. 10.3)   **16.** 20 (Sec. 10.3)
17. 18 (Sec. 10.3)   **18.** 3 (Sec. 10.3)   **19.** 48 (Sec. 10.3)   **20.** 6 (Sec. 10.3)   **21.** 16 (Sec. 10.3)
22. $2.28 (Sec. 10.4)   **23.** 644 points (Sec. 10.4)   **24.** 576 mi (Sec. 10.4)   **25.** 2880 no votes (Sec. 10.4)
26. 420 mi (Sec. 10.4)   **27.** 10 in (Sec. 10.4)   **28.** 12 ft (Sec. 10.4)   **29.** 576 mufflers (Sec. 10.4)
30. 24 teaspoons (Sec. 10.4)

Exercises 11.1

2. 60%   **4.** 40%   **6.** You receive 6% interest for one year.   **8.** 29% watched the event.
10. 35% of the cars were sold.   **12.** 36% of the employees are part-time.   **14.** 85% work on part-time jobs.
16. 76% have children attending public schools.

Exercises 11.2

2. $\frac{13}{100}$ **4.** $\frac{1}{5}$ **6.** $\frac{6}{25}$ **8.** $\frac{13}{25}$ **10.** $\frac{7}{20}$ **12.** $\frac{12}{25}$ **14.** $1\frac{2}{5}$ **16.** $1\frac{2}{3}$ **18.** 0.7   **20.** 0.75   **22.** 0.23
24. 0.07   **26.** 2.5   **28.** 1.1   **30.** 0.105   **32.** 0.035   **34.** 0.005   **36.** 0.0825

Exercises 11.3

2. 9%   **4.** 5%   **6.** 76%   **8.** 45%   **10.** 30%   **12.** 80%   **14.** 150%   **16.** 375%   **18.** $9\frac{1}{2}$% or 9.5%
20. $7\frac{1}{2}$% or 7.5%   **22.** 0.6%   **24.** 0.1%   **26.** 75%   **28.** 50%   **30.** 80%   **32.** $87\frac{1}{2}$% or 87.5%   **34.** 120%
36. $66\frac{2}{3}$%   **38.** $18\frac{3}{4}$% or 18.75%   **40.** 42.9%

Exercises 11.4

2. (A) 150, (R) 20%, (B) 750   **4.** (A) 90, (R) 30%, (B) 300   **6.** (A) 80, (R) what percent—the unknown, (B) 400
8. (A) 30, (R) what percent—the unknown, (B) 150   **10.** (A) What is—the unknown, (R) 60%, (B) 250
12. (A) 150, (R) 75%, (B) what number—the unknown   **14.** (A) $209, (R) 22%, (B) the salary—the unknown
16. (A) oz. of peanuts—the unknown, (R) 80%, (B) 16 oz   **18.** (A) 20, (R) what percent—the unknown, (B) 500
20. (A) what interest—the unknown, (R) 8.5%, (B) $5000

Exercises 11.5
2. 80 **4.** 480 **6.** 90 **8.** 6% **10.** 7% **12.** 14% **14.** 600 **16.** 800 **18.** 350 **20.** 500 **22.** 3200
24. 130% **26.** 106% **28.** 250 **30.** 900 **32.** 45 **34.** 17.5 **36.** 495 **38.** $33\frac{1}{3}$% **40.** 9.75% **42.** 8.5%
44. 400 **46.** 2000 **48.** 350 **50.** 750 **52.** 120 **54.** 6000

Exercises 11.6
2. 54 mL **4.** $5100 **6.** 9% **8.** 11% **10.** $200 **12.** $35,000 **14.** $190 **16.** 56 questions **18.** 16.5%
20. 16% **22.** 1500 students **24.** $1800 **26.** 963 **28.** $6840 **30.** 11.5% **32.** 9% **34.** $600 **36.** $1500
38. 9440 **40.** $262.50 **42.** $3434.70 **44.** $5955.08

Exercises Using Your Calculator
2. 575 **4.** 25.4% **6.** 86.24 **8.** 14,480 **10.** $53.27 **12.** 8.8% **14.** $7500 **16.** $62.99 **18.** $34.77
20. $20,442.50 **22.** $5875.20

Chapter 11 Self-Test
1. 80% (Sec. 11.1) **2.** $\frac{9}{100}$ (Sec. 11.2) **3.** $\frac{12}{25}$ (Sec. 11.2) **4.** 0.35 (Sec. 11.2) **5.** 0.02 (Sec. 11.2)
6. 1.3 (Sec. 11.2) **7.** 5% (Sec. 11.3) **8.** $7\frac{1}{2}$% or 7.5% (Sec. 11.3) **9.** 60% (Sec. 11.3)
10. $62\frac{1}{2}$% or 62.5% (Sec. 11.3) **11.** (A) 50, (R) 25%, (B) 200 (Sec. 11.4)
12. (A) What is—the unknown, (R) 8%, (B) 500 (Sec. 11.4) **13.** (A) $20, (R) 4%, (B) the purchase amount (Sec. 11.4)
14. 18.75 (Sec. 11.5) **15.** 500 (Sec. 11.5) **16.** 540 (Sec. 11.5) **17.** 20% (Sec. 11.5) **18.** $7\frac{1}{2}$% or 7.5% (Sec. 11.5)
19. 125% (Sec. 11.5) **20.** 400 (Sec. 11.5) **21.** 300 (Sec. 11.5) **22.** $2.64 (Sec. 11.6) **23.** 60 questions (Sec. 11.6)
24. $62.40 (Sec. 11.6) **25.** 12% (Sec. 11.6) **26.** 24% (Sec. 11.6) **27.** 15% (Sec. 11.6) **28.** $18,000 (Sec. 11.6)
29. 6400 students (Sec. 11.6) **30.** $8500 (Sec. 11.6)

Part Four Summary Exercises
1. $\frac{5}{13}$ **2.** $\frac{2}{3}$ **3.** $\frac{2}{5}$ **4.** $\frac{5}{24}$ **5.** $\frac{2}{3}$ (Sec. 10.1) **6.** F **7.** T **8.** T **9.** F **10.** T **11.** F (Sec. 10.2)
12. 2 **13.** 20 **14.** 4 **15.** 6 **16.** 16 **17.** 120 **18.** 30 **19.** 60 **20.** 0.5 (Sec. 10.3) **21.** $44
22. 256 freshmen **23.** 15 in **24.** 180 drives **25.** 45 parts **26.** 120 mi **27.** 70 grams **28.** 960 sq ft
29. 24 oz (Sec. 10.4) **30.** 75% (Sec. 11.1) **31.** $\frac{3}{100}$ **32.** $\frac{3}{20}$ **33.** $1\frac{3}{10}$ (Sec. 11.2) **34.** 0.45 **35.** 0.125
36. 2.25 (Sec. 11.2) **37.** 8% **38.** 12.5% **39.** 240% **40.** 40% **41.** $133\frac{1}{3}$% (Sec. 11.3) **42.** 2000 **43.** 140%
44. 240 **45.** 12.5% **46.** 5000 **47.** 75 **48.** 57.5 **49.** 45 **50.** 600 (Sec. 11.5) **51.** $1750 **52.** 22%
53. 7.5% **54.** $8800 **55.** $93 **56.** $1300 **57.** 680 students **58.** 6.5% **59.** $5362.50
60. $1100, $1199 (Sec. 11.6)

Part Four Cumulative Test
1. $\frac{2}{3}$ **2.** $\frac{3}{5}$ **3.** $\frac{3}{8}$ **4.** F **5.** T **6.** T **7.** 15 **8.** 3 **9.** 60 **10.** 100 **11.** $77 **12.** 555 mi
13. 32 employees **14.** 12 in **15.** 120 mi **16.** $168 **17.** 18 oz **18.** 70% **19.** $\frac{7}{20}$ **20.** $1\frac{3}{4}$ **21.** 0.55
22. 0.185 **23.** 12.5% **24.** 0.1% or $\frac{1}{10}$% **25.** 60% **26.** 62.5% **27.** 600 **28.** 30 **29.** 11% **30.** 255
31. 150 **32.** 175% **33.** 9.5% **34.** $17,500 **35.** She had 70%, yes **36.** 1650 students **37.** $1600
38. $85,500 **39.** 7.5% **40.** $323

Exercises 12.1
2. 32 **4.** 5 **6.** 96 **8.** 48 **10.** 360 **12.** 420 **14.** 15,840 **16.** 4 **18.** 6 **20.** 6 **22.** 144 **24.** 18
26. 5 **28.** 6 **30.** 5 **32.** 36 **34.** 2 **36.** 168

Exercises 12.2

2. 7 lb 4 oz **4.** 5 yd 4 in **6.** 6 min 30 sec **8.** 10 hr 20 min **10.** 13 ft 5 in **12.** 16 yd **14.** 25 ft 8 in **16.** 3 ft 8 in **18.** 3 gal 2 qt **20.** 2 hr 30 min **22.** 1 yd 4 in or 3 ft 4 in **24.** 17 min 30 sec **26.** 4 lb 5 oz **28.** 2 hr 10 min **30.** 13 hr 30 min **32.** 1 lb 8 oz **34.** No, 16 ft 1 in is needed **36.** Under 12 oz **38.** 20 ft 4 in **40.** 3 hr 45 min

Exercises 12.3

2. 12 in **4.** 32 ft **6.** 23 yd **8.** 50.2 ft **10.** 26 in **12.** 26.7 in **14.** 13.7 in **16.** 22 ft **18.** $280 **20.** $13.75 **22.** $382.50 **24.** $9.42 **26.** 24 in **28.** 35.7 in

Exercises 12.4

2. 24 in^2 **4.** 12 yd^2 **6.** 6 ft^2 **8.** $6\frac{1}{4}$ in^2 **10.** 700 ft^2 **12.** 113 ft^2 **14.** 35 yd^2 **16.** 201 ft^2 **18.** 234 ft^2 **20.** $7\frac{1}{14}$ in^2 **22.** 6 squares **24.** Yes, the area is 384 ft^2 **26.** 24 oz **28.** Yes, the area is 1962.5 ft^2 **30.** $150.72 **32.** 864 in^2 **34.** 30 ft^2 **36.** 50 acres **38.** 15.2 ft^2 **40.** 50.2 ft^2 **42.** 21.5 in^2

Exercises 12.5

2. 18 yd^3 **4.** 128 in^3 **6.** 15.6 yd^3 **8.** $37\frac{1}{2}$ in^3 **10.** 384.7 in^3 **12.** 106.8 ft^3 **14.** 128 ft^3 **16.** 7.5 gal **18.** 324 ft^3, 12 yd^3 **20.** 5086.8 ft^3, 38,151 gal **22.** 3 yd^3 **24.** 43.6 ft^3 **26.** 32 in^3

Chapter 12 Self-Test

1. 84 (Sec. 12.1) **2.** 4 (Sec. 12.1) **3.** 48 (Sec. 12.1) **4.** 4 (Sec. 12.1) **5.** 5 ft 6 in (Sec. 12.2) **6.** 7 min 30 sec (Sec. 12.2) **7.** 11 ft 5 in (Sec. 12.2) **8.** 2 lb 9 oz (Sec 12.2) **9.** 10 h 40 min (Sec. 12.2) **10.** 3 lb 5 oz (Sec. 12.2) **11.** 119 ft (Sec. 12.3) **12.** 28 in (Sec. 12.3) **13.** 44 ft (Sec. 12.3) **14.** $900 (Sec. 12.3) **15.** $3.20 (Sec. 12.3) **16.** 1000 in^2 (Sec. 12.4) **17.** $11\frac{1}{4}$ in^2 (Sec. 12.4) **18.** 28.3 yd^2 (Sec. 12.4) **19.** 12.3 ft^2 (Sec. 12.4) **20.** 20 in^2 (Sec. 12.4) **21.** 113 in^2 (Sec. 12.4) **22.** $300 (Sec. 12.4) **23.** $1080 (Sec. 12.4) **24.** 31,400 mi^2 (Sec. 12.4) **25.** 24 ft^3 (Sec. 12.5) **26.** 753.6 in^3 (Sec. 12.5) **27.** 81 yd^3 (Sec. 12.5) **28.** 64 yd^3 (Sec. 12.5) **29.** 15.7 ft^3 (Sec. 12.5) **30.** 18 yd^3 (Sec. 12.5)

Exercises 13.1

2. a **4.** b **6.** c **8.** a **10.** c **12.** b **14.** mm **16.** cm **18.** km **20.** km **22.** mm **24.** cm **26.** 12.7 **28.** 5.5 **30.** 5 **32.** 240 **34.** 4.57 **36.** 450 **38.** 4 **40.** 7000 **42.** 110 **44.** 12 **46.** 625 **48.** 706.5

Exercises 13.2

2. b **4.** a **6.** b **8.** c **10.** a **12.** b **14.** kg **16.** g **18.** g **20.** mg **22.** kg **24.** g **26.** 5 **28.** 3000 **30.** 13.2 **32.** 2

Exercises 13.3

2. b **4.** b **6.** a **8.** b **10.** a **12.** c **14.** cL **16.** mL **18.** L **20.** mL **22.** cL **24.** L **26.** 3 **28.** 8.48 **30.** 240 **32.** 200 **34.** 5000 **36.** 40

Exercises 13.4

2. b **4.** a **6.** c **8.** c **10.** a **12.** b **14.** 32 **16.** 5 **18.** 167 **20.** 59 **22.** 25 **24.** 95 **26.** 185°F **28.** 356°F

Chapter 13 Self-Test

1. b (Sec. 13.1) **2.** b (Sec. 13.1) **3.** m (Sec. 13.1) **4.** mm (Sec. 13.1) **5.** 3000 (Sec. 13.1) **6.** 5000 (Sec. 13.1) **7.** 8 (Sec. 13.1) **8.** 3.1 (Sec. 13.1) **9.** 12.7 (Sec. 13.1) **10.** 80 (Sec. 13.1) **11.** c (Sec. 13.2) **12.** b (Sec. 13.2) **13.** kg (Sec. 13.2) **14.** g (Sec. 13.2) **15.** 3000 (Sec. 13.2) **16.** 1.35 (Sec. 13.2) **17.** 140 (Sec. 13.2) **18.** 7000 (Sec. 13.2) **19.** c (Sec. 13.3) **20.** b (Sec. 13.3) **21.** mL (Sec. 13.3) **22.** L (Sec. 13.3) **23.** 2000 (Sec. 13.3) **24.** 3 (Sec. 13.3) **25.** 4.75 (Sec. 13.3) **26.** 240 (Sec. 13.3) **27.** c (Sec. 13.4) **28.** b (Sec. 13.4) **29.** 77 (Sec. 13.4) **30.** 35 (Sec. 13.4)

Part Five Summary Exercises

1. 72 **2.** 80 **3.** 240 **4.** 20 (Sec. 12.1) **5.** 8 ft 10 in **6.** 8 lb 2 oz (Sec. 12.2) **7.** 12 ft 6 in. **8.** 2 lb 9 oz
9. 8 hours **10.** 2 minutes 9 seconds (Sec. 12.2) **11.** 14 ft 8 in **12.** The three smaller bottles (Sec. 12.2) **13.** 115 in
14. 19.2 ft **15.** 37.7 ft **16.** 32.9 m (Sec. 12.3) **17.** $17.60 (Sec. 12.3) **18.** $58.61 (Sec. 12.3) **19.** 750 ft^2
20. 400 in^2 **21.** 314 cm^2 **22.** 31 yd^2 (Sec. 12.4) **23.** 180 ft^2 or 20 yd^2 (Sec. 12.4) **24.** $787.50
25. $113.04 (Sec. 12.4) **26.** 60 in^3 **27.** 62.8 cm^3 (Sec. 12.5) **28.** $52\frac{1}{2}$ ft^3 **29.** 4 yd^3 **30.** 502.4 ft^3
31. 13.3 gal (Sec. 12.5) **32.** 1.5 m (Sec. 13.1) **33.** 40 L (Sec. 13.3) **34.** 4°C (Sec. 13.4) **35.** 22 mm (Sec. 13.1)
36. 8 kg (Sec. 13.2) **37.** 20 cm (Sec. 13.1) **38.** 2 g (Sec. 13.2) **39.** 20 mL (Sec. 13.3) **40.** 1000 kg (Sec. 13.2)
41. 5000 (Sec. 13.1) **42.** 3000 (Sec. 13.3) **43.** 300 (Sec. 13.3) **44.** 4 (Sec. 13.1) **45.** 3 (Sec. 13.3)
46. 2 (Sec. 13.2) **47.** 800 (Sec. 13.1) **48.** 3000 (Sec. 13.2) **49.** 2 (Sec. 13.1) **50.** 75 (Sec. 13.3)
51. 200 (Sec. 13.3) **52.** 300 (Sec. 13.1) **53.** 3.152 (Sec. 13.1) **54.** 1.8 (Sec. 13.2) **55.** 122 (Sec. 13.4)
56. 6.36 (Sec. 13.3) **57.** 360 (Sec. 13.3) **58.** 31 (Sec. 13.1) **59.** 20 (Sec. 13.4) **60.** 140 (Sec. 13.2)

Part Five Cumulative Test

1. 96 **2.** 96 **3.** 300 **4.** 28 **5.** 9 ft 8 in **6.** 11 lb 5 oz **7.** 1 minute 35 seconds **8.** 18 ft 4 in **9.** 50 ft
10. 25 in **11.** 23 in **12.** 31.4 ft **13.** $36 **14.** 384 ft^2 **15.** 45.9 ft^2 **16.** 201 in^2 **17.** 21 yd^2 **18.** $256
19. 192 ft^3 **20.** 803.8 in^3 **21.** 2 yd^3 **22.** 4.4 gal **23.** 16 m **24.** 6 L **25.** −5°C **26.** 500 g **27.** 120 cm
28. 50 g **29.** 120 mL **30.** 400 mg **31.** 8000 **32.** 3 **33.** 5 **34.** 250 **35.** 3000 **36.** 2000 **37.** 17.6
38. 30.48 **39.** 224 **40.** 25

Exercises 14.1

2. F **4.** T **6.** T **8.** F **10.** T **12.** F **14.** T **16.** T **18.** F **20.** F **22.** T **24.** T **26.** 16
28. 12 **30.** 9 **32.** 15 **34.** 0 **36.** −18 **38.** −10 < −5 **40.** −15 < −14 **42.** |−5| > −5 **44.** 7 = |7|

Exercises 14.2

2. 12 **4.** −13 **6.** 7 **8.** 4 **10.** −3 **12.** 1 **14.** −2 **16.** 0 **18.** 2 **20.** −15 **22.** −10 **24.** −4
26. 0 **28.** −25 **30.** −0.5 **32.** $-9\frac{2}{5}$ **34.** −0.9 **36.** $-\frac{1}{2}$ **38.** −5 **40.** 2 **42.** 0 **44.** −10

Exercises 14.3

2. 2 **4.** −5 **6.** −21 **8.** −6 **10.** −40 **12.** 6 **14.** −3 **16.** 16 **18.** 0 **20.** 75 **22.** 10 **24.** −10
26. −0.5 **28.** −2 **30.** $-3\frac{1}{2}$ **32.** 5 **34.** −7 **36.** 2 **38.** $290 **40.** 577 ft below sea level

Exercises 14.4

2. −72 **4.** 75 **6.** −24 **8.** 24 **10.** −100 **12.** 0 **14.** 200 **16.** 81 **18.** −30 **20.** −125 **22.** 1.2
24. $-\frac{1}{6}$ **26.** −0.75 **28.** $\frac{2}{3}$ **30.** −8 **32.** −8 **34.** 81 **36.** −25 **38.** 60 **40.** −70 **42.** 300 **44.** 240
46. −20°C

Exercises 14.5

2. 5 **4.** 10 **6.** −6 **8.** −7 **10.** −5 **12.** 5 **14.** −7 **16.** 10 **18.** Undefined **20.** 0 **22.** −25 **24.** 5
26. $-\frac{3}{2}$ **28.** 2 **30.** −1.5 **32.** 2 **34.** −4 **36.** 5 **38.** −3

Chapter 14 Self-Test

1. 9 (Sec. 14.1) **2.** 8 (Sec. 14.1) **3.** 7 (Sec. 14.1) **14.** 10 (Sec. 14.1) **5.** 6 (Sec. 14.2) **6.** −2 (Sec. 14.2)
7. −12 (Sec. 14.2) **8.** 0 (Sec. 14.2) **9.** −9 (Sec. 14.2) **10.** $-\frac{1}{2}$ (Sec. 14.2) **11.** 2 (Sec. 14.2) **12.** −4 (Sec. 14.2)
13. −3 (Sec. 14.3) **14.** −10 (Sec. 14.3) **15.** 10 (Sec. 14.3) **16.** 4 (Sec. 14.3) **17.** 8 (Sec. 14.3) **18.** $\frac{1}{2}$ (Sec. 14.3)
19. −28 (Sec. 14.4) **20.** 54 (Sec. 14.4) **21.** 0.75 (Sec. 14.4) **22.** 0 (Sec. 14.4) **23.** 48 (Sec. 14.4)
24. 36 (Sec. 14.4) **25.** −4 (Sec. 14.5) **26.** 7 (Sec. 14.5) **27.** −5 (Sec. 14.5) **28.** −5 (Sec. 14.5) **29.** 0 (Sec. 14.5)
30. Undefined (Sec. 14.5)

Exercises 15.1

2. 20 **4.** 8 **6.** −24 **8.** 45 **10.** 33 **12.** −5 **14.** 22 **16.** 20 **18.** 9 **20.** 64 **22.** 12 **24.** 4 **26.** 6
28. −25 **30.** 54 **32.** −78 **34.** −72 **36.** 80 **38.** 140 **40.** −8 **42.** 4 **44.** −4 **46.** 540 **48.** 104
50. 620

Exercises 15.2

2. 5 **4.** 8 **6.** 4 **8.** −6 **10.** −2 **12.** 5 **14.** −9 **16.** −12 **18.** 5 **20.** 32 **22.** −36 **24.** 3 **26.** 1
28. −5 **30.** −2 **32.** −3 **34.** 5 **36.** −2 **38.** 12 **40.** −42

Exercises 15.3

2. $5b$ **4.** $16r$ **6.** s **8.** $-3w$ **10.** $-x$ **12.** 0 **14.** $-3p$ **16.** $-14a$ **18.** −5 **20.** 8 **22.** −9 **24.** 10
26. −5 **28.** −4 **30.** 3 **32.** $-\frac{3}{2}$ **34.** −5 **36.** −10 **38.** 6 **40.** $\frac{5}{3}$ **42.** 2 **44.** −3 **46.** −7 **48.** $\frac{2}{5}$

Exercises 15.4

2. $a + 7$ **4.** $p + q$ **6.** $m + 5$ **8.** $b + 8$ **10.** $s - 5$ **12.** $r - 3$ **14.** $y - 5$ **16.** $9d$ **18.** $8a$ **20.** $7rs$
22. $5c + d$ **24.** $\frac{m}{8}$ **26.** $\frac{r - s}{7}$ **28.** $\frac{b + 5}{7}$ **30.** $\frac{m - n}{m + n}$ **32.** $x + 9$ **34.** $x - 10$ **36.** $2x$ **38.** $5x - 10$
40. $\frac{x}{3}$ **42.** $\frac{x - 5}{3}$ **44.** $\frac{x - 3}{x + 3}$ **46.** 18 **48.** 13 **50.** 16, 48 **52.** 20, 32 **54.** Janet: $1850; Frank: $1700
56. 52 in, 44 in **58.** 24 in, 8 in **60.** 8 ft 32 ft

Chapter 15 Self-Test

1. 20 (Sec. 15.1) **2.** −11 (Sec. 15.1) **3.** 27 (Sec. 15.1) **4.** −28 (Sec. 15.1) **5.** −4 (Sec. 15.1) **6.** 2 (Sec. 15.1)
7. 8 (Sec. 15.2) **8.** −3 (Sec. 15.2) **9.** 6 (Sec. 15.2) **10.** 56 (Sec. 15.2) **11.** −9 (Sec. 15.2) **12.** 6 (Sec. 15.2)
13. $8x$ (Sec. 15.3) **14.** a (Sec. 15.3) **15.** $-11b$ (Sec. 15.3) **16.** 6 (Sec. 15.3) **17.** −7 (Sec. 15.3) **18.** 8 (Sec. 15.3)
19. −3 (Sec. 15.3) **20.** −8 (Sec. 15.3) **21.** 3 (Sec. 15.3) **22.** 9 (Sec. 15.3) **23.** 11 (Sec. 15.3)
24. $a + 10$ (Sec. 15.4) **25.** $r - 8$ (Sec. 15.4) **26.** $3xy$ (Sec. 15.4) **27.** $2m + 3$ (Sec. 15.4) **28.** $\frac{w + 5}{8}$ (Sec. 15.4)
29. 12, 36 (Sec. 15.4) **30.** 19 cm, 24 cm (Sec. 15.4)

Part Six Summary Exercises

1. 12 **2.** 8 **3.** 3 **4.** 20 (Sec. 14.1) **5.** 8 **6.** −5 **7.** −11 **8.** $\frac{1}{2}$ **9.** −5 **10.** −4 (Sec. 14.2) **11.** −5
12. −15 **13.** 5 **14.** −4 **15.** $6\frac{1}{2}$ **16.** −4 (Sec. 14.3) **17.** 36 **18.** −80 **19.** −1.5 **20.** $\frac{3}{10}$ **21.** 16
22. 42 (Sec. 14.4) **23.** −4 **24.** 11 **25.** Undefined **26.** −0.075 **27.** $\frac{7}{6}$ **28.** 0 (Sec. 14.5) **29.** −24 **30.** 22
31. 68 **32.** −18 **33.** −1 **34.** −4 (Sec. 15.1) **35.** 8 **36.** −9 **37.** −9 **38.** 5 **39.** $-\frac{3}{4}$
40. 40 (Sec. 15.2) **41.** $16a$ **42.** $-3b$ **43.** $-5y$ **44.** z (Sec. 15.3) **45.** −12 **46.** −15 **47.** 3 **48.** −6
49. $-\frac{3}{2}$ **50.** 4 (Sec. 15.3) **51.** $y + 5$ **52.** $s + 9$ **53.** $5cd$ **54.** $2w + 3$ **55.** $\frac{r}{7}$ **56.** $\frac{s - t}{4}$ (Sec. 15.4) **57.** 13
58. 7, 19 **59.** Nick: $370, Janice: $290 **60.** 12 cm, 20 cm (Sec. 15.4)

Part Six Cumulative Test

1. −20 **2.** 7 **3.** 12 **4.** 20 **5.** −18 **6.** 0 **7.** $1\frac{1}{3}$ **8.** −2 **9.** −13 **10.** 13 **11.** 0 **12.** 5 **13.** 96
14. −90 **15.** 5 **16.** −60 **17.** 7 **18.** $-\frac{1}{12}$ **19.** −2.5 **20.** Undefined **21.** −60 **22.** 27 **23.** 24 **24.** 3
25. −8 **26.** 7 **27.** 3 **28.** $-5r$ **29.** m **30.** 7 **31.** −6 **32.** $-\frac{5}{2}$ **33.** $y + 9$ **34.** $w - 9$ **35.** $5ab$
36. $2s + 5$ **37.** $\frac{z}{8}$ **38.** $\frac{m + 6}{6}$ **39.** 25, 39 **40.** 12 ft, 19 ft

INDEX

Absolute value, 595, 673
Abstract numbers, 509, 511, 579
Acre, 528
Addends, 10, 173
Addition:
  carrying in, 22, 322
  of decimals, 321, 387
    carrying in, 322
  of denominate numbers, 510, 579
  of fractions, 259, 267, 302
  of mixed numbers, 281, 302
  in the order of operations, 82, 97, 131, 175
  properties of, 13, 14, 15, 173, 603, 604, 605
  of signed numbers, 600, 602, 603, 673
  symbol for, 10, 659, 675
  of whole numbers, 17, 22
  in word problems, 48, 91
  with zero, 15, 173, 604
Addition facts, 11
  table of, 11
Additive identity, 15, 173, 604
Additive inverse, 605
Algebraic equations, 636, 675
Algebraic expressions, 629, 674
Amount:
  finding the, 458, 460, 490
  identifying the, 452, 490
Applications (see Word problems)
Approximate numbers, 345
Area, 524
  of a circle, 527, 580
  of a parallelogram, 526, 580
  of a rectangle, 525, 580
  of a square, 525, 580
  of a triangle, 527, 580
Arithmetic mean, 140
Associative property:
  of addition, 14, 173, 604
  of multiplication, 65, 174, 616
Average, 140

Bar notation in decimals, 376
Base:
  in exponents, 95, 175

Base *(Cont.)*:
  in percent: finding the, 459, 461, 491
    identifying the, 452, 490
Basic addition facts, 11
  table of, 11
Basic multiplication facts, 60
  table of, 61
Borrowing:
  with denominate numbers, 511, 579
  in subtraction, 39
Building fractions, 212, 300

Carrying:
  in adding decimals, 322
  in addition, 22
  in multiplication, 69
Celsius, Anders, 570
Celsius compared to Fahrenheit, 571, 579
Celsius temperature scale, 570
Centigrade, 570
Centigram (cg), 560
Centiliter (cL), 565
Centimeter (cm), 548, 578
Checking:
  in division, 109
  in solving proportions, 414
  in subtraction, 35
Circle:
  area of, 527, 580
  circumference of, 518, 519, 580
Circumference, 518, 519, 580
Coefficient, 648
Combining like terms, 649, 674
Commission, 471
Common denominator, 264
  adding fractions with a, 259, 302
  finding the least (*see* Least common denominator)
Common factor, 160
Common fraction, 189
  (*See also* Fractions)
Common multiple, 165
Commutative property:
  of addition, 13, 173

Commutative property *(Cont.)*:
  of multiplication, 64, 174, 615
Complex fraction, 245
Composite numbers, 150, 176
Cone, 540
Continued division, 156
Counting numbers, 10
Cross products, 203, 300
Cubic centimeters (cm^3), 565
Cubic units, 535, 537
Cylinder, 536
  volume of, 536, 580

Day, 505, 577
Decigram (dg), 560
Deciliter (dL), 565
Decimal equivalents, 374
Decimal form, 316, 387
Decimal place, 6, 29, 316, 387
Decimal place value system, 5, 315, 387
Decimal point, 315
Decimals, 315, 387
  addition of, 321, 387
    carrying in, 322
  bar notation in, 376
  and common fractions, 318, 374, 382, 389, 390
  dividing by powers of ten, 368, 389
  dividing by whole numbers, 355, 389
  as divisors, 361, 389
  multiplying, 335, 388
  multiplying by powers of ten, 77, 341, 388
  and percent, 444, 447, 490
  place in, 316, 387
  reading and writing, 317, 387
  repeating, 375
  rounding, 345, 388
  subtracting, 327, 388
  terminating, 375
  in word problems, 323, 329, 336, 346, 357, 364
Decimeter (dm), 551

697

Dekagram (dag), 560
Dekaliter (daL), 565
Dekameter (dam), 551
Denominate numbers, 508, 579
   and abstract numbers, 511, 579
   adding, 510, 579
   subtracting, 510, 579
Denominator, 189, 299
   common, 264
   least common (LCD), 264, 301
Diameter of a circle, 518
Difference, 35, 174
Digits, 5, 173
Distributive property of multiplication over addition, 66, 175
Dividend, 107, 175
   trial, 120
Divisibility tests, 151, 176
Division:
   by abstract numbers, 511, 579
   checking in, 109
   continued, 156
   of decimals, 355, 361, 389
   of fractions, 244, 301
   long, 115, 120
   of mixed numbers, 248, 301
   in the order of operations, 131, 175
   by powers of ten, 368, 389
   remainder in, 109, 175
   short, 127
   of signed numbers, 620, 674
   symbol for, 107, 661, 675
   of whole numbers, 107, 115, 120
   in word problems, 136
   with zero, 112, 175, 617
Divisor(s), 107, 149, 175
   decimals as, 361, 389
   trial, 120

English system of measurement, 505, 577
   converting to the metric system, 553, 560, 566, 579
   units of, 505, 577
Equations, 413, 636
   equivalent, 638, 675
   linear, 637
   solving, 638, 640, 643, 655, 675
Equivalent equations, 638, 675
Equivalent fractions, 202, 300
Estimation:
   in addition, 31
   with decimals, 348
   in division, 121
   with fractions, 241
   in multiplication, 79
   with percents, 466
   in subtraction, 42
Evaluating algebraic expressions, 629, 674
Expanded form of numeral, 6
Exponents, 95, 175
Extremes of a proportion, 407, 489

Factor, 59, 149, 174
   common, 160
   greatest common (GCF), 160, 176
   prime, 154
Fahrenheit compared to Celsius, 571, 578
Fahrenheit temperature scale, 570
Fluid ounces, 505, 577
Foot, 505, 577
Fractions, 189, 299
   adding like, 259, 302
   adding unlike, 267, 302
   building, 212, 300
   common, 189
   complex, 245
   and decimals, 318, 374, 382, 389, 390
   dividing, 244, 301
   equivalent, 202, 300
   fundamental principle of, 205, 212, 300
   improper, 195, 299
      and mixed numbers, 198, 199, 299
   inverting, 244
   like, 214, 259
   multiplying, 225, 233, 301
   and percent, 443, 448, 490
   proper, 195, 299
   raising to higher terms, 212, 300
   as a ratio, 401
   reciprocal of, 244
   simplifying, 205, 300
   subtracting like, 273, 302
   subtracting unlike, 274, 302
   in word problems, 236, 249, 270, 276, 289
Fundamental principle of fractions, 205, 212, 300

Gallon, 505, 577
Gram (g), 559, 577
Greatest common factor (GCF), 160, 176
   finding the, 161, 176

Hectogram (hg), 560
Hectoliter (hL), 565
Hectometer (hm), 551
Hour, 505, 577

Identity:
   for addition, 15, 173, 604
   for multiplication, 65, 175, 616
Improper fractions, 195, 299
   and mixed numbers, 198, 199, 299
Inch, 505, 577
Integers, 596, 673
   adding, 600, 602, 603, 673
   dividing, 620, 674
   multiplying, 613, 674
   subtracting, 608, 674
Interest, 468
   compound, 479

International System of Units (SI), 547
Inverting fractions, 244

Kilogram (kg), 559, 578, 579
Kiloliter (kL), 565, 578, 579
Kilometer (km), 549

Least common denominator (LCD), 264, 301
Least common multiple (LCM), 165, 177
Length:
   English units of, 505, 577
   metric units of, 547, 577
Like fractions, 214, 259
   adding, 259, 302
   subtracting, 273, 302
Like terms, 648
   combining, 649, 674
Linear equations, 637
Liter, (L), 564, 577, 578, 579
Literal expressions, 629
Long division, 115, 120
Lowest terms:
   of fractions, 205, 300
   of ratios, 402

Magic squares, 44
Mean, 140
Means of a proportion, 407, 489
Measurement, 505
   English system (*see* English system of measurement)
   metric system (*see* Metric system)
   (*See also specific units of measure*)
Meter (m), 547, 577, 578, 579
Metric system, 547
   converting to the English system, 552, 560, 566, 578
   converting units within, 550, 551, 560, 565, 578
   measuring temperature, 570, 577
   prefixes, 551, 578
   units of length, 547, 577
   units of volume, 564, 577
   units of weight (mass), 559, 577
Mile, 505, 577
Milligram (mg), 559
Milliliter (mL), 564
Millimeter (mm), 548
Minuend, 35
Minus sign, 35
Minutes, 505, 577
Mixed numbers, 195, 299
   adding, 281, 302
   dividing, 248, 301
   and improper fractions, 198, 199, 299
   multiplying, 229, 235, 301
   subtracting, 283, 302
   vertical arrangement for, 286
Multiple, 60, 164
   common, 165

Multiple *(Cont.)*:
   least common (LCM), 165, 177,
      finding the, 166, 177
Multiplication:
   by abstract numbers, 511
   carrying in, 69
   of decimals, 335, 388
   of fractions, 225, 233, 301
   of mixed numbers, 229, 235, 301
   in the order of operations, 82, 97,
      131, 175
   by powers of ten, 77, 341, 388
   properties of, 64, 65, 66, 174, 175,
      615, 616
   of signed numbers, 613, 674
   symbols for, 59, 660, 675
   of whole numbers, 59, 68, 72
   in word problems, 88
   with zero, 65, 175, 617
Multiplication facts, 60
   table of, 61
Multiplicative identity, 65, 175, 616

Natural numbers, 10
Negative numbers, 593, 673
   product of, 617
Number line, 10, 593
Number systems, 5
Numerals, 5, 173
   expanded form of, 6
Numerator, 189, 299

Opposite of a number, 594, 673
Order of operations:
   with addition or subtraction, 82,
      97, 131, 175
   with division, 131, 175
   with multiplication, 82, 97, 131,
      175
   with parentheses, 83, 97, 131, 175
   with powers, 97, 131, 175
Order on the whole numbers, 31, 174
Origin, 10
Ounce, 505, 577

Parallelogram, 526
   area of, 526, 580
Parentheses, 14, 83, 175, 631
Partial product, 72
Percent, 439, 495
   applications of, 468
   and decimals, 444, 447, 490
   and fractions, 443, 448, 490
   meaning of, 439
   symbol for, 439
Percent proportion, 460, 491
Perimeter, 515, 579
   finding, 516
   of a rectangle, 517, 580
   of a square, 517, 580
Pi (π), 518, 527
Pint, 505, 577
Place:
   in decimals, 316

Place *(Cont.)*:
   value, 6, 173, 316
Place-value system, 5, 315, 387
Plus, 10
Positive numbers, 593, 673
Pound, 505, 577
Powers of ten, 77
   dividing by, 368, 389
   multiplying by, 77, 341, 388
Powers of whole numbers, 95, 175
   in the order of operations, 97, 131,
      175
Prime factors, 154
Prime factorization, 155, 176
Prime numbers, 149, 176
Principal, 468
Product(s), 59, 174
   cross, 203
   of factors, 59, 154
   of means and extremes, 408, 489
   partial, 72
   (*See also* Multiplication)
Proper fraction, 195, 299
Properties:
   of addition, 13, 14, 15, 173, 603,
      604, 605
   of multiplication, 64, 65, 66, 174,
      175, 615, 616
Proportions, 406, 489
   applying, 421, 489
   percent, 460, 491
   rule, 408, 489
   solving, 412, 415, 489
   terms of, 407, 489
Pyramid, 540
Pythagorean triples, 99

Quart, 505, 577
Quotient, 107, 175 (*See also* Division)

Radius of a circle, 518
Raising fractions to higher terms,
   212, 300
Rate:
   of commission, 471
   finding the, 459, 461, 491
   identifying the, 452, 490
   of interest, 469
   tax, 472
Ratio, 401, 489
   simplifying, 402
Reciprocal of a fraction, 244
Rectangle:
   area of, 525, 580
   perimeter of, 517, 580
Rectangular solid, 535
   volume of, 535, 580
Reducing fractions, 205, 300
Remainder, 109, 175
Renaming:
   in measurement, 505
   in mixed numbers, 284
Repeating decimals, 375
Rounding:
   decimals, 345, 388

Rounding *(Cont.)*:
   whole numbers, 28, 174

Second, 505, 577
Short division, 127
SI (International System of Units),
   547
Signed numbers, 594, 673
   adding, 600, 602, 603, 673
   dividing, 620, 674
   multiplying, 613, 674
   subtracting, 608, 674
Similar figures, 419
Simplifying:
   fractions, 205, 300
   ratios, 402
Solid:
   cylindrical, 536, 580
      volume of, 536, 580
   definition of, 535
   rectangular, 535
      volume of, 535, 580
Solving:
   equations, 638, 640, 643, 655, 675
   proportions, 412, 415, 489
Square:
   area of, 525, 580
   perimeter of, 517, 580
   units, 524, 528
Subtraction:
   of decimals, 327, 388
   of denominate numbers, 510, 579
   of fractions, 273, 274, 302
   of mixed numbers, 283, 302
   in the order of operations, 82, 97,
      131, 175
   of signed numbers, 608, 674
   symbol for, 35, 659, 675
   of whole numbers, 35, 39
   in word problems, 48, 91
Subtrahend, 35, 174
Sum, 11, 173, 659
Symbols:
   for addition, 10, 659, 675
   for division, 107, 661, 675
   for multiplication, 59, 660, 675
   for percent, 439
   for subtraction, 35, 659, 676

Temperature, 570
Terminating decimal, 375
Terms:, 648
   like, combining, 649, 674
   lowest: of fractions, 205, 300
   of a proportion, 407, 489
Ton, 505, 577
Trial dividend, 120
Triangle: 527
   area of, 527, 580

Unit pricing, 430
Unit ratio, 506
Unlike fractions:
   adding, 267, 302

Unlike fractions *(Cont.)*:
  subtracting, 274, 302

Variables, 15, 66, 412, 629
Vertical arrangement for mixed
  numbers, 286
Volume, 535
  of cylinder, 536, 580
  English units of, 535, 577
  metric units of, 564, 577
  of rectangular solid, 535, 580

Week, 505, 577
Weight, 505, 577
  metric units of (mass), 559, 577

Whole numbers, 10
  adding, 17, 22
  dividing, 107, 115, 120
  dividing decimals by, 355, 389
  multiplying, 59, 68, 72
  order on the, 31, 174
  powers of, 97, 175
  reading and writing, 7, 8
  rounding, 28, 174
  subtracting, 35, 39
Word problems:
  with addition, 48, 91
  in algebra, 662, 676
  with division, 136
  involving decimals, 323, 329, 336, 346, 357, 364

Word problems *(Cont.)*:
  involving fractions, 236, 249, 270, 276, 289
  with multiplication, 88
  with proportions, 421, 489
  with subtraction, 48, 91

Yard, 505, 577

Zero, 7, 14
  in addition, 15, 173, 604
  in division, 112, 175, 621
  in multipliction, 65, 175, 617
  as a power, 96